Annice Macdonald

School Science Laboratories

School Science Laboratories

A Handbook of Design, Management and Organisation

W. F. Archenhold
E. W. Jenkins
C. Wood-Robinson

Centre for Studies in Science Education
University of Leeds

John Murray 50 Albemarle Street London

Acknowledgements

Our thanks are due to Mr. O. M. Stepan and the Institute of Physics for granting permission to reproduce Table 1.6, to Mr. O. M. Stepan for permission to reproduce Figure 1.4 and to the Controller of HMSO for permission to reproduce Figures 1.1 to 1.3, 1.5 to 1.9, and Table 1.5 from DES Building Bulletin 39.

Thanks are due also to the following for permission to reproduce photographs: North Kent Plastic Cages Ltd. (Figure 5.4); Griffin and George Ltd. (Figure 5.5); W. R. Prior and Co. Ltd. (Figures 5.18, 5.19, 5.23 and 5.24). Figures 5.20, 5.21 and 5.22a were taken by Mr R. Hobson.

Cover design *Craig Dodd*

Note on equipment suppliers

Most laboratory equipment and reagents referred to in this book may be obtained from one of the major general laboratory suppliers. For a list of suppliers see:

Archenhold, W. F., Jenkins, E. W., Wood-Robinson, C. (1977) *Addresses for Science Teachers*, Centre for Studies in Science Education, University of Leeds.
Wilson, R. W. (1974) *Useful Addresses for Science Teachers*, Edward Arnold.

Printed in Great Britain by William Clowes and Sons Ltd.
London, Beccles and Colchester

ISBN 0 7195 3436 4

Contents

Information about materials and equipment widely used in school science teaching in alphabetical order.

Air-tracks 74, ammeters 74, amplifiers 74, anatomical models 75, aquaria 75, asbestos 79, atmometers 79, autoclaves 79;

balances 80, barometers 81, batteries 82, beakers 83, blood-grouping apparatus 84, blood lancets 84, bottles 84, buffer solutions 84, burettes 85;

cages 86, capacitors 88, capillary tubing 90, cells 90, cellulose tubing 90, centrifuges 90, chart recorders 90, choice chambers 91, chromatography 91, colorimeters and spectrophotometers 92, conical flasks 92, corks 93, crucibles 93, crucible tongs 93;

deionisers 94, desiccators 94, disposable laboratory ware 95, dissecting apparatus 96, dynamics trolleys and runways 98;

ecology apparatus 98, electromagnetism 99, electrophoresis 99, electrostatics 99, environmental comparators 99, ergometers 100;

filter flasks 100, filter papers 100, filter pumps 101, force meters 101, freezing mixtures 101, fume cupboards 101, funnels 103, fuses 104;

galvanometers 104, gas analysis apparatus 104, gas cylinders 105, Geiger-Müller tubes 106, generators 107, glass rod 108, glass syringes 108, glass tubing 108, gloves 109, goggles 110;

haemacytometers 110;

incubators 112, indicators 113, interchangeable laboratory glassware 113;

kymographs 113;

lamps 114, lasers 115, leads 115, lenses 115;

magnetic stirrers 116, magnets 116, magnifiers 116, masses 117, measuring cylinders 117, mechanical stirrers 117, meters 118, microbiological apparatus 119, microscopes—monocular 120, microscope components 123, microscopes—binocular 126, microscope accessories 127, microscopy apparatus 130, mortars and pestles 132;

nets, collecting 132;

oil baths 134, oscilloscopes 134, ovens and drying cabinets 134;

pH meters 135, photography 135, pipettes 135, plant growth apparatus 136, plugs 136, pneumatic troughs 136, polystyrene spheres 137, potometers 137, power supplies 137;

radioactive materials 139, ratemeters 139, refrigerators 139, resistance wire 140, resistors 140, respirometers 140, retort stand bases and rods 140, rheostats 140;

safety spectacles and eye-shields 140, scalers 141, signal generators 141, silica ware 141, sinks 141, skeletal material 142, sockets 142, soil apparatus 142, solid state detectors 143, spirometers 143, stills 144, stopclocks/stopwatches 144, stoppers 144, strobe photography 145;

Teltron tubes 145, test tubes 145, test tube stands and racks 145, thermometers 145, three centimetre wave apparatus 147, ticker tape vibrators 147;

vacuum pumps 147, vacuum tubes 147, van de Graaff generators 148, vibration generators 148, voltmeters 148;

wash bottles 148, waste disposal 148, water baths 148, weights 149, wire 149.

Introduction

The programme of science curriculum development which began in the nineteen sixties and continued into the present decade has made available to schools a wider range of resources, materials and techniques than at any time in the past. Many of the items of equipment developed for school science teaching are expensive and are manufactured to several different designs. Some require regular and skilled maintenance if their working life is to be maximised. Others may be improvised as required from materials which are commonly available. Similarly, the variety of chemical solutions and living organisms employed in school science teaching is much greater than a decade or so ago and science teachers are able to call upon an impressive range of audio-visual aids, from television to the overhead projector, to help them realise their objectives. Not surprisingly, therefore, the science teacher is sometimes regarded as a 'learning manager' for his pupils, a view which emphasises his skill in deploying the resources available to him to provide appropriate learning experiences for his pupils.

This book is intended as a comprehensive introduction to these resources and to some of the factors which govern their use. While it is recognised that a given resource or technique may be used in a number of ways to fulfil a variety of objectives, no attempt has been made to compare different methods of teaching science or to evaluate multifarious science curricula.

Most science teaching, at least in secondary schools, is conducted within specialist accommodation and teachers are involved increasingly in the planning and fitting of school laboratories. For these reasons and because the work that can be done in a laboratory is heavily dependent upon the use that may be made of the accommodation available, the design of school laboratories and science blocks is considered in the first chapter of this book. The tasks of management and laboratory organisation are discussed in Chapter 2 and particular attention is given to the role of the Head of Department in planning, communication, administration and the delegation of responsibility.

Recent legislation in the field of health and safety has given added significance to the role of the science teacher in ensuring that school science laboratories remain safe places in which to teach and to learn. Chapter 3 is, therefore, devoted to a survey of the hazards likely to be encountered in the teaching of science and of the legal responsibilities of the teacher. It is also concerned with emphasising that safety is a positive activity which assumes that almost all accidents can be prevented by suitable knowledge and adequate planning. Chapter 4 provides an introduction to the range of audio-visual resources available to school science teachers and to some of the uses which may be made of them. Chapter 5 catalogues information about materials and equipment widely used in school science teaching and provides details of maintenance procedures where these are appropriate. Chapter 6 reviews some materials commonly used in the improvisation of apparatus, an undertaking which often has educational as well as economic merits. This chapter also includes an index to commonly available sources which describe the construction of individual items of equipment.

The alphabetical list of laboratory reagents, stains and culture media in Chapter 7 is likely to be adequate for almost all school purposes and Chapter 8 provides practical details of the culture and maintenance of a wide variety of living organisms. Chapter 9 offers practical instructions for a number of techniques of particular importance in the teaching of biology.

In the matters of units and nomenclature, we have been guided by the advice offered by the Association for Science Education in its publications *SI Units, Signs, Symbols and Abbreviations*, 1974 and *Chemical Nomenclature, Symbols and Terminology*, 1972. However, Imperial and other non-SI units are widely used outside the scientific community and even within that community there is considerable divergence of practice, particularly in the naming of substances. In writing this book, therefore, we have aimed at maximum clarity and in most instances where SI units or systematic names are not normally used, the SI and systematic equivalents have also been given.

We acknowledge with gratitude those who have helped in the writing and production of this book. We are particularly indebted to Colin Campbell, who read the entire typescript and offered much helpful comment; to Ken Everett, Allan Kavanagh, Hugh Perfect, Peter Scott and Olgierd Stepan for their advice on particular sections of the typescript, and to Mr. Roderick Concannon and Mrs. Allyson Rodway of John Murray for their patient encouragement. Mr. J. H. Baxter, Mr. K. G. Butler, Mr. R. Hobson and Mrs. K. Sharpe have also offered much helpful advice. Our thanks are also due to our typists, Mrs. D. M. Gilliard, Mrs. A. Terry and, in particular, Miss J. Dartnall who was responsible for preparing most of the typescript. Finally, we wish to acknowledge, in a general way, the very large number of sources upon which a work of this kind inevitably depends.

As always, responsibility for the final contents of the book must rest solely with the authors.

September 1977

W. F. Archenhold
E. W. Jenkins
C. Wood-Robinson

1 Design of school laboratories and science blocks

Introduction

The aim of this chapter is to provide some background information for science teachers who may be faced with the problem of advising on the design of new laboratories, and to alert others who may see in the suggestions offered (and reference material quoted) some means of improving the environment for the teaching of science in their school.

The design of science laboratories, or a science block, must be a co-operative exercise between the architects and those engaged in science education and administration. Many science teachers feel that there has been too little consultation in the past. Even if an attempt at consultation has been made, the head of the science department may not have had the necessary knowledge of design and management to comment constructively. Similarly, architects have usually had too little experience of the changes in school organisation and science curricula to help them decide whether a teacher's recommendation is far-seeing and acceptable.

Examples of fruitful co-operation between architects, science advisers and science teachers have been few in number, although the situation is improving. The publication of Building Bulletin 39 (DES, 1967) threw a welcome shaft of light onto the scene. The Bulletin deals with the design of a science block by the Development Group of the Architects and Building Branch of the Department of Education and Science in close collaboration with the science staff of Oxford School. Although the design is directed towards science provision for a boys' grammar school, the project gave the group an opportunity to investigate in some depth the requirements of modern science teaching in terms of the buildings, and most of the basic principles considered in the Bulletin, particularly the strong case made for flexibility in laboratory organisation, have stood the test of time.

Historical note

In the early part of this century the problem of providing laboratory accommodation at minimum cost was frequently solved by simply removing the partition wall between two standard size classrooms, each of which had an area of $7.3 \, m \times 6.1 \, m$ ($24 \, ft \times 20 \, ft$). This method produced the rectangular-shaped $7.3 \, m \times 12.2 \, m$ ($24 \, ft \times 40 \, ft$) laboratory still seen in schools today. The total floor area of $89 \, m^2$ ($960 \, ft^2$) for about thirty pupils led to the figure of approximately $3 \, m^2$ of floor area per pupil, although an area of about $3.5 \, m^2$ was generally allowed per student at sixth-form level.

One of the disadvantages of the rectangular-shaped laboratory is the large teacher/pupil distance for some pupils, a problem which was overcome in the early 1950s by designing laboratories of a nearly square shape, typically $9.2 \, m \times 9.7 \, m$ ($30 \, ft \times 32 \, ft$), which again provided a total floor area of $89 \, m^2$ ($960 \, ft^2$).

In the late 1950s and early 1960s, the Industrial Fund for the Advancement of Scientific Education in Schools (Savage, 1964) was set up to help some two hundred direct grant and independent schools in the United Kingdom to modernise and extend their laboratory provision. The size of laboratory which the Fund accepted as standard for 'elementary' laboratories (up to age sixteen) was again $89 \, m^2$ ($960 \, ft^2$). Sixth-form laboratories, to take a group of up to 16 students, were based on an area of $56 \, m^2$ ($600 \, ft^2$), while a floor area of $50 \, m^2$ ($540 \, ft^2$) was taken as the norm for demonstration rooms, to seat between thirty and sixty pupils for special lecture/demonstrations or gatherings such as meetings of the school's Scientific Society.

Savage considered six factors to be of particular importance when planning the layout of the above laboratories. These factors clearly reflect the main activities in a fairly 'traditional' approach to the teaching of science, i.e. pupils doing individual practical work, writing, or watching a lecture/demonstration. The six factors were:

a Linear bench space. The provision of 1 linear metre (3 ft) of bench space (invariably fixed) per pupil in 'elementary' laboratories and 1.3 linear metres (4 ft) per student in sixth-form laboratories.

b Circulation space. The provision of 1.7 m (5 ft) of space between the ends of benches and walls to provide circulation space and room for wall furniture.

c Storage. Space for cupboards and shelves in the laboratory for the storage of apparatus and other materials in common use.

d Permanent equipment. Space for the installation of 'non-movable' equipment, e.g. a still in chemistry laboratories or aquaria in biology laboratories.

e Wall space. Space for apparatus which needs to be fixed in a vertical plane, e.g. a parallelogram of forces board in physics laboratories.

f Demonstration bench. The provision of a demonstration bench in laboratories used for 'teaching' as well as individual practical work.

While some of these factors are just as important now as then, the provision of *fixed work benches* in all types of laboratories built before the middle 1960s has led to a somewhat inflexible laboratory environment which is less suitable for the more varied requirements of modern science curricula and changes in class organisation. Inevitably, many science teachers are now attempting to meet these changed demands in laboratories designed for a more traditional approach. While an imaginative staff will be able to overcome some of these 'environmental' constraints, it seems clear that in designing for the future *flexibility* is an important priority. The increased adoption of *movable work benches*, particularly in biology, physics and general science laboratories, is a significant development, as is the increasing use of trolleys for the movement of apparatus within a science department. A science block is an expensive facility and present-day designers must ensure, as far as is possible, that their design will enable the laboratories to be used to maximum efficiency even if the requirements of science teaching should change in the near future.

Time allocation for science

The number of periods allocated to science for the different year groups in a school will depend on several considerations, including the ages of the pupils, the need to achieve a degree of balance in the total curriculum, the requirements of the 'option scheme' operated by the school, and the need to prepare the 14–18-year-old students for a variety of external examinations.

The Association for Science Education (ASE, 1973) has recommended that *all* pupils in the 13–16 age range should continue with their study of science, and that their commitment to science should be *at least* 20 per cent of the available teaching time. A 20 per cent time allocation for pupils in the 13–16 age range is shown in Table 1.1 as 6, 9, and 9 periods in years three, four and five respectively. This represents an average of eight periods out of a 40-period week over three school years, which is consistent with the ASE recommendation of a minimum 20 per cent time allocation of science for pupils in this particular age range. It is common practice to timetable year groups in the 13–16 age range with at least one double period making up any additional time with single

or double periods. The ASE further recommends that twenty pupils should generally be the maximum number in this age range under the supervision of one member of staff—a situation which already prevails in Scotland.

Age	9+	10+	11+	12+	13+	14+	15+	16+	17+
Year			1	2	3	4	5	6	7
	Science periods per week per form								
Low	2	2	4	4	6	9	9	21*	21*
High	3	3	6	6	9	12	12	27*	27*

* Assumes 3 sciences being taught—7–9 periods per week per subject in each of the two years.

Table 1.1 Science allocation out of a 40-period week for different year groups

The time allocation for science for the 11–13 age range is generally taken as an *average* of about 12.5 per cent, which represents five periods in a 40-period week. While it is accepted that pupils in the 9–11 age range often study science as part of an integrated activity, it is helpful to quantify this work in terms of an equivalent *average* science provision of about 6 per cent, or 2.5 periods in a 40-period week. The time allocation for each Advanced Level subject such as biology, chemistry or physics, is generally taken as about 20 per cent of the available teaching time, representing eight periods in a 40-period week.

The ASE recommended *minimum* 20 per cent time allocation for the 13–16 age range and the generally accepted *average* time allocation for the 9–13 and 16–18 age ranges are expressed in Table 1.1 in terms of both a low and high provision of actual periods per week, based on a *40-period week*. The number of science periods would have to be correspondingly reduced in a school operating on a *35-period week*.

Number of laboratories

An analysis of the science courses offered by the school, the option schemes operated in years four and five, the number of periods of science likely to be taught in laboratories, lecture rooms or outdoors (rural studies) and the size and range of age groups in the school, are the basic parameters to be used in calculating the size of a science block.

In this section, the number of periods allocated to science out of a 40-period week for different year groups, as shown in Table 1.1, will be used to estimate the number of laboratories required for three types of school: **(a)** 9–13 middle school; **(b)** 11–16 school; **(c)** 13–18 school. Similar calculations may be done for schools containing pupils of any other age range.

The emphasis in modern science curricula, devised for all the age ranges considered here, is on pupil involvement in practical activities. While it is accepted that practical work will not always take up the whole of a lesson, it is assumed that all science lessons will be taught in laboratories. The teacher must judge whether the occasional lesson based almost wholly on teacher demonstrations, films or television would be more suitably taught in an associated lecture room, which should preferably be adjacent to the laboratories.

To maintain a laboratory at maximum efficiency, regular servicing is essential. The actual time required for servicing depends on the subject matter being taught, the storage system and the general management methods being used in the department (Chapter 2). In calculating laboratory provision, it is assumed that technical staff will have access to a laboratory for a minimum of one period per day as well as before and after school hours. It follows that if a laboratory is available for teaching for 35 out of the 40 periods per week, then the number of laboratory periods required to accommodate, for example 70 periods of science is, in fact, 80 periods, or in terms of the number of laboratories, 80/40 = 2 laboratories. Hence the laboratory provision based on the actual number of periods taught has to be scaled up by a factor of 8/7 to take into account the servicing of laboratories.

a 9–13 middle schools

Assuming a four-form entry school, science periods per form per week (see Table 1.1, page 2):

Low: 2+2+4+4 = 12
High: 3+3+6+6 = 18

Science periods per four forms per week:

Low: 12 × 4 = 48
High: 18 × 4 = 72

Number of laboratories required on the basis of a 40-period week:

Low: 48/40 = 1.2
High: 72/40 = 1.8

Number of laboratories required if a laboratory is left free for one period per day for servicing:

Low: 1.2 × 8/7 = 1.4
High: 1.8 × 8/7 = 2.1

Mean number of laboratories:

$$\text{Mean} = \frac{1.4 + 2.1}{2} = 1.7$$

The requirement is therefore *at least* one purpose-built science laboratory plus science areas for use by the 9+ and 10+ age groups. The provision of a preparation/store room, a science trolley and facilities for a chemical store must also be considered, as must provision for rural science in schools appropriately placed for such activities.

Table 1.2 summarises the results of similar calculations for 9–13 middle schools having entries of 2, 4, 6 and 8 forms respectively

Form entry	Science periods 9–13		Number of laboratories		Number of laboratories 1 free period/lab/day		
	Low	High	Low	High	Low	High	Mean
2	24	36	0.6	0.9	0.7	1.0	0.9
4	48	72	1.2	1.8	1.4	2.1	1.7
6	72	108	1.8	2.7	2.1	3.1	2.6
8	96	144	2.4	3.6	2.7	4.1	3.4

Table 1.2 Laboratory requirements for various size schools catering for the 9–13 age range

b 11–16 schools

The laboratory requirements for various sizes of 11–16 schools, based on the science period provision in a *40-period week* given in Table 1.1 (page 2), is shown in Table 1.3.

Form entry	Science periods 11–16		Number of laboratories		Number of laboratories 1 free period/lab/day		
	Low	High	Low	High	Low	High	Mean
4	128	180	3.2	4.5	3.7	5.1	4.4
6	192	270	4.8	6.7	5.5	7.7	6.6
8	256	360	6.4	9.0	7.3	10.3	8.8
10	320	450	8.0	11.2	9.1	12.8	11.0

Table 1.3 Laboratory requirements for various size schools catering for the 11–16 age range

It follows from Table 1.3 that, as a rough guide, the number of laboratories in 11–16 schools should at least equal the number of forms per year group in the school. The Association for Science Education (ASE, 1973) further recommends one additional associated room, which may be an adjacent lecture room, for every two-form entry in an 11–16 school.

Although the school does not have a sixth form, the specialist requirements of biology, chemistry, physics and possibly rural science or technology will need to be taken into account in planning the science block provision.

c 13–18 schools

The laboratory requirements for various size 13–18 schools are best calculated in two parts—provision for the 13–16 age range and for the sixth form.

Table 1.4 shows the results of calculations for various sizes of 13–16 schools, based on the science period provision in a 40-period week given in Table 1.1 (page 2).

Form entry	Science periods 13–16		Number of laboratories		Number of laboratories 1 free period/lab/day		
	Low	High	Low	High	Low	High	Mean
4	96	132	2.4	3.3	2.7	3.8	3.2
6	144	192	3.6	5.0	4.1	5.7	4.9
8	192	264	4.8	6.6	5.5	7.5	6.5
10	240	330	6.0	8.2	6.9	9.4	8.1

Table 1.4   Laboratory requirements for various size schools catering for the 13–16 age range

In calculating the laboratory space required by the sixth form, the important factors to consider are the types of courses offered, the time commitment to each course and the number of students following each course. If biology, chemistry and physics are taught separately to years six and seven, and the numbers of students justify only one set for each subject in each of these years, then using Table 1.1 (page 2), the provision per subject falls between 14 periods (low) and 18 periods (high) with a mean of 16 periods per subject in a specialist laboratory. Making the usual allowance for servicing, this is equivalent to 0.5 laboratories per subject, or '1.5' laboratories for the three science subjects in years six and seven. If pupil numbers in years six and seven are such that two sets are required per subject in each year, it follows that 3.0 laboratories are required for sixth-form work alone.

A typical eight-form entry 13–18 comprehensive school, requiring '1.5' laboratories for years six and seven, would therefore need a total of 6.5 + '1.5' = 8 laboratories, of which at least 3 laboratories must be designed to cater for the specialist demands of biology, chemistry and physics respectively. To be used efficiently, such specialist laboratories should be suitable for both the sixth form and pupils following specialist courses in years four and five.

The siting of laboratories

There is probably no widely applicable solution to the problem of deciding where to site laboratories within a school. In part, this is because no two schools will be identical in their needs and environmental surroundings. However, the following points should be borne in mind before coming to a decision.

a Science block. There are several advantages in having laboratories close together: the movements of staff and apparatus are reduced; mutual facilities such as a workshop and resources centre can be shared; the availability of technical assistance is increased; work of an interdisciplinary nature can be developed more easily; and, from a financial point of view, the installation of services such as gas, electricity, water and drainage, can

be kept to a minimum. A science block with two storeys or more should have a service lift or hoist to move equipment and apparatus. Steps along corridors connecting laboratories on the 'same level' must be avoided so that trolleys can be used without hindrance.

b Practical wing. The siting of general-purpose science laboratories and specialist biology, chemistry and physics laboratories in relation to accommodation occupied by housecraft, handicrafts, technology and mathematics, should be considered carefully. Links with other subject areas, with possible sharing of lecture rooms and support areas, can provide growth points for co-operation, and may be particularly beneficial to students engaged in project activities.

c Environmental considerations. The north–south orientation of laboratories has become more important as the ratio of window area to wall area has increased. While at least one sunny bench, with blinds to control the sunshine, is desirable for biology laboratories, the sun shining on non-luminous bunsen flames in chemistry laboratories can be particularly dangerous. The siting of windows adjacent to chalkboards or overhead-projector screens should also be avoided.

In countries near the equator there are advantages (Lewis, 1972) if windows face either north or south, as this will reduce the extent to which strong sunlight enters through the windows directly. Increased overhangs may also have to be used for sun protection. The direction of the prevailing wind, particularly during the hottest part of the day, must also be considered in hot countries, and it may be necessary to arrive at a compromise orientation of the laboratories.

Provision of special facilities

The planning of special indoor and outdoor science facilities should be considered at an early stage in the design process.

a Indoors

Science spaces are often some of the least attractive teaching areas in a school. The provision and use of *exhibition and display areas* in laboratories and in a central area or foyer, to stimulate interest and set high standards of display, can do much to counteract the impression gained by at least some pupils that science is 'restricted to the laboratory'.

These spaces may be used for the exhibition of materials related to work currently being done in the science department, and consideration should also be given to the provision of a small *museum area* or display case, in which older apparatus, manuscripts or specimens can be shown.

In addition to an adequate science section in the general school library, a *science library* containing background readers, periodicals and a variety of text books for reference, is an invaluable resource for students. Books for teachers are better kept in a *science resources area*, where duplicating facilities, audio-visual equipment and other materials are stored.

The siting of a *photographic dark room*, which is generally close to a specialist physics laboratory so that it can also be used for optical experiments, needs to take account of any special ventilation requirements. Another important consideration is the provision of *secure stores* for radioactive sources, poisons and flammable chemicals (see Chapter 3). This is particularly vital in open-plan designs where unsupervised pupils may gain access to the science area during the lunch break, or the public may be in the school for evening classes or other community activities.

The provision of a workshop and preparation rooms/ stores, is discussed on page 8.

b Outdoors

The provision and siting of outdoor facilities is of particular importance to the biology and rural science sections of a science department. Easy access to specialist facilities such as greenhouses, animal houses, ponds and experimental plots (Nuffield Advanced Biology, 1971) suggests that the biology and rural science laboratories should be at ground-floor level (Wyatt, 1965).

A three-quarters span *greenhouse* sharing a south-facing wall with the biology laboratory has the advantage of easy access. Climbing plants can be grown on the wall, which also acts as a 'heat-sink' so that sudden fluctuations of temperature are avoided. To aid maintenance, automatic ventilating and watering devices are useful, in addition to thermostatically controlled electrical heating.

An *animal room/house* should be kept at a temperature between 18 °C and 26 °C and should be shielded from direct sunlight as much as possible. Light can be provided by north-facing windows or roof lights, supplemented by artificial lighting. The house should be well ventilated but draught-proof, any ventilation apertures being covered with insect-proof mesh.

To aid drainage, the floor should slope slightly to a floor drainage grille. A hot and cold water supply is essential in an animal house, and shielded electric mains points should be provided so that animals requiring temperatures above the ambient room temperature can have heating devices in their cages (Wray, 1974). The cages may be stored on adjustable shelving.

The doors of the animal area should be fitted with a lock different from the school pass key as additional security, and the entrance must be wide enough for a trolley to be pushed in and out. Any wooden surfaces which may be gnawed should be protected with sheet metal to a height of one metre above floor level.

A secure *chemical store* must be provided for such volatile liquids as carbon disulphide, ether, petroleum ethers, methylbenzene (toluene), alcohol and propanone (acetone), which may cause fires by ignition of their vapours. The store can be outdoors or indoors, but the latter type must have at least one hour's fire resistance (DES, *Safety in Science Laboratories*, 1976). No more than $500 \, cm^3$ of any one volatile flammable liquid should be kept in bottles on laboratory benches or shelves.

Laboratory design criteria

What are the basic requirements for which science laboratories should be designed to meet the needs of present and future generations of pupils? The most important considerations would seem to be the *role* of science in the school, the *organisation* of science in the school and the *management* of resources (see Chapter 2) and storage facilities (see page 9). It seems evident that a non-specialist laboratory must be designed for a variety of possible activities, which are summarised below. Each section of activities contains a check list of design criteria. The particular requirements of specialist subjects are discussed on page 13.

a Practical work by individuals or small groups

i all pupils working on the same experiment;
ii all pupils moving from station to station in a 'circus' arrangement doing a different experiment at each station;
iii all pupils working on different experiments and reporting to the class during a discussion;
iv individual work based on an individualised approach;
v long-term practical work: extended 'standard' experiments; one- or two-week investigations of an open ended nature; projects which may extend over a term or even a year.

Design criteria: for each of these activities to be carried out efficiently requires the usual range of services (see page 11) near fixed or movable work benches, and the application of appropriate management principles (see Chapter 2) for the transfer of apparatus and availability of resources.

b Study activities

i writing up experiments, making observations and recording, drawing diagrams;
ii working from individualised resources;
iii writing notes from the blackboard or overhead projector.

Design criteria: recommended heights of working surfaces are 750 mm and 850 mm for seated and standing work respectively. Chairs and stools should be about 250 mm lower, i.e. 500 mm and 600 mm respectively. A

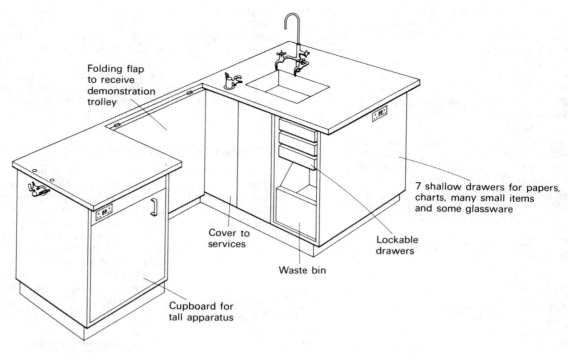

Folding flap
to receive
demonstration
trolley

7 shallow drawers for papers,
charts, many small items
and some glassware

Cover to
services

Lockable
drawers

Waste bin

Cupboard for
tall apparatus

Figure 1.1   A demonstration bench which can incorporate a trolley of the same height

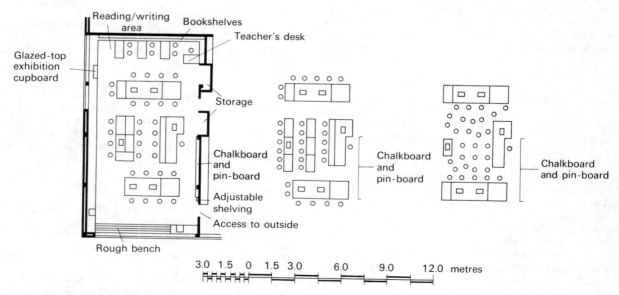

Reading/writing
area

Bookshelves

Teacher's desk

Glazed-top
exhibition
cupboard

Storage

Chalkboard
and
pin-board

Chalkboard
and
pin-board

Chalkboard
and pin-board

Adjustable
shelving

Access to outside

Rough bench

3.0 1.5 0   1.5 3.0   6.0   9.0   12.0 metres

Figure 1.2   The plan on the left shows an arrangement for practical work; the two plans on the right show alternative arrangements for watching a demonstration

recent study (DES, 1976) makes recommendations for the heights of tables and chairs for various age ranges, including 8–11-year-olds, 11–15-year-olds and 15-year-olds upwards. 'Clean' study activities need to be separated, whenever possible, from 'dirty' laboratory work.

c Lecture/demonstration

i from work benches at suitable points in the laboratory;
ii from a trolley placed at a suitable point.

Design criteria: the traditional long demonstration bench takes up considerable space and the height adds to the difficulty of enabling all pupils to get a clear view in a laboratory. Such a bench is perfectly acceptable in a lecture room with writing benches mounted in tiers, each bench being approximately 225 mm higher than the one in front of it. Demonstrations done at an ordinary workbench, at the special demonstration bench shown in Figure 1.1, or at a trolley placed in the demonstration bench or elsewhere, with the pupils seated at a suitable distance from the demonstration, again leads to the criterion of flexibility of furniture layout in parts of the laboratory. Figure 1.2 shows an arrangement of furniture and pupils for individual practical work/written assignments and two plans for watching a demonstration, one from close quarters and one when pupils are seated at their work benches.

d Discussion

i pupils in small groups;
ii pupils in a compact group helping with the design of an experiment, or discussing the social consequences of a scientific discovery;
iii pupils at their work benches discussing on a class basis the implications of some observations.

Design criteria: pupils at their work benches should be able to see the teacher and vice versa. This is also essential from the point of view of safety in laboratories. Movable tables, which can create space for a class discussion, are an advantage.

Figure 1.3 shows a 1.2 m × 0.6 m service station containing interchangeable plastic trays for kits of apparatus, a shelf for students' books and bags and a cupboard for a few standard items of equipment. The basic services of gas, water and mains electricity are provided and the unit is shown extended by the addition of movable tables, each of which is also 1.2 m long and 0.6 m wide.

In Figure 1.4, the service stations are located around the perimeter of a laboratory, with tables combined in different ways to show arrangements for pupils doing practical work (diagram a), pupils involved in study activities or watching a demonstration (diagram b), or engaged in a class discussion (diagram c).

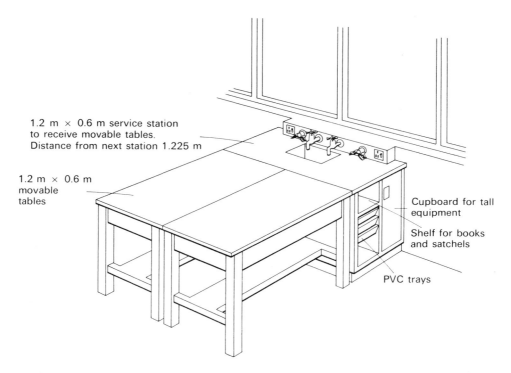

1.2 m × 0.6 m service station to receive movable tables. Distance from next station 1.225 m

1.2 m × 0.6 m movable tables

Cupboard for tall equipment

Shelf for books and satchels

PVC trays

Figure 1.3   A fixed wall service station extended by two movable tables

a

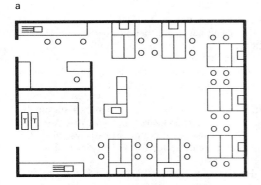

1.2m × 0.6m service station

b

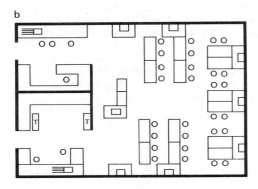

c

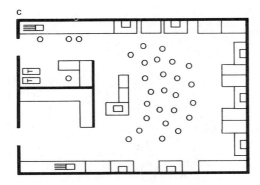

Approximate scale 1:215   T = trolley

Figure 1.4 Arrangements for a) *pupils doing practical work;* b) *pupils watching a demonstration;* c) *class discussion*
(NOTE: Area of laboratory = 69.7 m² (8.5 m × 8.2 m); area per pupil = 2.5 m² for a class of 28 pupils)

e Audio-visual activities

   i teacher using an overhead projector;
  ii teacher showing a film, usually with sound, using a 16 mm projector;
 iii teacher or pupils showing film loops on an 8 mm projector;
 iv teacher or pupils showing 35 mm slides, either by forward or back projection;
  v teacher and pupils watching a television programme or listening to a sound broadcast;
 vi teacher using closed-circuit television to aid a demonstration;
 vii teacher using an episcope or microprojector;
viii teacher showing a wallchart.

Design criteria: with improved optical design, most of these visual aids can be used under normal daylight conditions, although some shading may be desirable. The position of the screen, the number of people viewing, the relative positions of screen and windows, the seating arrangements in the laboratory, the availability of a projector stand or trolley and the nearness of suitable electric sockets and television aerial sockets, are all factors which need to be taken into account at the design stage.

For some purposes, the lighting in a laboratory has to be dimmed, and fire-proof blackout curtains, which fall behind a retaining rail 150 mm above the window sill along the window bench, are more suitable than Venetian blinds which tend to rattle when windows are opened for ventilation.

f Display/exhibition areas

  i charts, diagrams, reports;
 ii apparatus, models, projects;
iii light objects hung from the ceiling.

Design criteria: provision of pin-board in all display areas is essential. The use of movable screens covered with coloured cloth is convenient and adds colour to a display. Glazed-top exhibition cupboards and low tables are suitable for exhibitions of books, apparatus and specimens. Light objects may be suspended from ceiling hooks or from horizontal rods attached to a wall bracket.

Central workshop

The siting and design of a suitably equipped laboratory workshop for the science department should be considered at an early stage in the design of a science block.

There are three main functions of a workshop.

a Maintenance, repair and construction of apparatus

This would generally require basic workshop equipment, such as a rigid work bench with vice, lathe, a drill, a sharpening wheel and tools. It must also be equipped with the usual services, such as gas, hot and cold water, a large sink (50 cm × 30 cm × 20 cm) and electrical mains outlets.

The provision of basic glass working equipment is desirable, as is a fume cupboard, especially if it can be shared with another laboratory or a lecture room (Figure 1.5). For fume cupboard design, see page 101.

Figure 1.5 also illustrates how the siting of the workshop enables materials to be moved by hoist to and from the physics preparation room, which in this case is sited directly above the workshop.

b Store for bulk materials

The workshop must have suitable storage facilities, such as plastic guttering to store long glass tubing and adjustable shelves for pieces of timber, perspex and metal. A wall store for sheet materials is also desirable.

c Base for chief technician

The third area of the workshop can be arranged as an office for the chief technician, to contain a desk, visitors'

chairs for consultations with trade representatives and junior technicians, and a filing cabinet for order books, stock books, trade catalogues and requisition forms. A pin-board for notices is also desirable.

The consideration which is most likely to influence the siting of the workshop within the science area is the need for direct access to the workshop from a suitable service road. The delivery of goods and materials can be carried out most efficiently in this way. In Figure 1.5, the workshop has three exits: a wide door for deliveries from the service road, a door to the neighbouring lecture room, and a door to the corridor. All doorways must be wide enough for a trolley to pass through.

An outdoor chemical store for quantities of flammable liquids (see page 49) should be suitably placed near the service road leading to the workshop.

Preparation and storage facilities

The need for increased storage facilities in science departments has grown as teachers have encouraged both

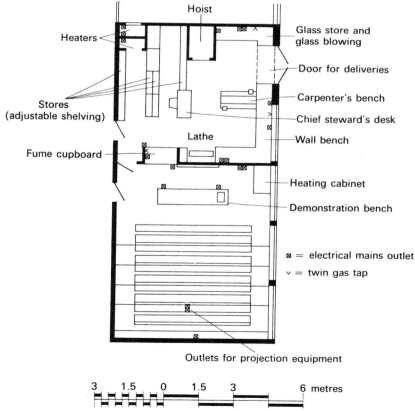

Figure 1.5 A workshop with access from a service road

younger pupils and sixth-form students to pursue individual projects and open-ended investigations. At the same time, an increasing number of schools have become comprehensive in their intake of pupils. A consequence of this is that laboratory facilities are being shared by a variety of classes for whom the aims of science teaching may differ widely and who may require quite different apparatus. The storage of apparatus in drawers and cupboards of fixed benches may still have its uses for the housing of simple sets of chemical apparatus, but, with the increasing use of strong but light movable tables, such general storage is gradually disappearing from many laboratories.

Most recent designs of laboratory blocks have included storerooms with preparation areas, from which materials can be distributed to the laboratories. Figure 1.6 shows preparation and storage facilities for a chemistry department, and Figure 1.7 gives details of a preparation room between two laboratories intended primarily for physics teaching.

A well organised storage system will enable easy collection of apparatus from the store and its transfer, generally by trolley, to the laboratory in readiness for the lesson. As science laboratories are often in use for almost all of the teaching day, it is evident that an efficient method of transferring apparatus within the department is central to the management function of the head of department.

Most storage systems involve the use of standardised sizes of plastic or wooden trays or plastic-coated wire baskets, whose dimensions differ only in depth. This means that different capacity trays can be slotted into the same storage frame or trolley frame. Trolleys can be purchased from the suppliers or constructed of slotted steel angle iron of a suitable gauge. A range of trolleys for apparatus and audio-visual aids, such as overhead projectors and film projectors, may be needed to fulfil a variety of requirements.

To minimise the problem of dust, sides and backs can be added to open storage frames, and it may be possible to adapt existing cupboards in older laboratories by fitting runners which will accept trays.

Some lockable, clearly labelled cupboards will always be required, particularly for expensive or dangerous items, the latter including poisons and radioactive sources.

Adjustable shelves also provide flexibility in a storeroom. It may be convenient to store apparatus in uniform and clearly labelled cardboard boxes which make maximum use of available shelf space. Kits of apparatus can be split so that boxes or trays contain only the items required for a particular experiment. A means of checking the apparatus at the beginning and the end of practical lessons is essential and this process is made easier if items are provided in standard numbers, or if components have special slots or spaces for the right number in the tray or box so that only a visual check is required.

The preparation area should be provided with the usual services such as gas and mains electricity, and a large and small sink, each with hot and cold water. Parts of the bench top should be covered with a heat-resistant material, and a small workshop bench with a supply of tools will enable simple repairs to be done as required.

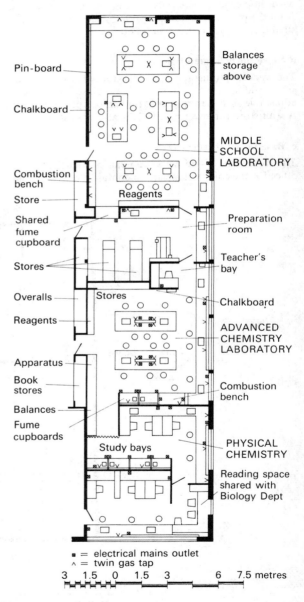

Figure 1.6 Preparation and storage facilities for a chemistry department

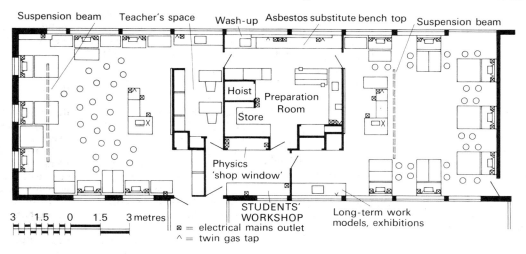

Suspension beam   Teacher's space   Wash-up   Asbestos substitute bench top   Suspension beam

Hoist
Preparation Room
Store
Physics 'shop window'

3   1.5   0   1.5   3 metres

STUDENTS' WORKSHOP

Long-term work models, exhibitions

⊠ = electrical mains outlet
∧ = twin gas tap

Figure 1.7   The first floor physics preparation room, 6.1 m × 5.5 m, contains a hand hoist linked to a ground floor central workshop

Services

There are two important aspects to the provision of electricity, gas and water supplies and of waste disposal facilities.

a Provision of the above services at suitable points and in the required quantities throughout the science department.
b Technical details concerned mainly with the design of distribution systems and the choice of materials.

For the science teacher, adequate and appropriate provision of services is of the utmost importance and basic to the organisation of practical activities. Requirements vary and are dependent on the age range of the pupils and the specialist demands of the subject. For instance, younger pupils using microscopes near water supplies could be in some danger if the only electricity mains outlets provided are at about 240V. For such pupils, a low voltage supply system can be installed, although mains outlets should also be provided for other purposes. Chemistry laboratories make special demands on water supplies and waste disposal, while physics laboratories need little water but a generous supply of electrical mains outlets.

Teachers should have easy access to *mains* gas, electricity and water controls in the laboratory.

The provision of services at different locations and for different functions needs to be considered at the design stage. Particular attention should be given to the provision of services:

  i at demonstration positions;
 ii at students' working positions, taking specialist requirements into account;
iii for special facilities, such as photographic dark rooms, greenhouses and animal houses;
 iv for workshops;
  v for preparation areas;
 vi for fittings such as fume cupboards, ovens, stills, aquaria and animal cages;
vii for resource centres;
viii for services such as television, radio, audio-visual aids (teacher and student use), telephones and clocks.

To increase flexibility in the arrangement of furniture in school laboratories, some designers have proposed the adoption of the *umbilical cord* system of service distribution, which is found in many industrial and university laboratories. Services are brought into the laboratory in an overhead service boom some two metres above floor level, and the various services are distributed to the working areas via flexible leads (electricity) and flexible nylon tubes (gas and water). Liquid waste passes from sinks along corrugated tubing via dilution traps to floor gullies.

For school use, the advantages of considerable flexibility in furniture layout need to be weighed against the disadvantages of leads and tubing obscuring the view across the laboratory and constituting a possible safety hazard. A compromise solution as regards cost and flexibility is to install permanently-fixed services in fixed benches or service stations sited along the sides of the laboratory (see Figure 1.3 on page 7), and to provide two or three additional service stations for electricity and gas in the middle of the laboratory. Alternatively, two or three additional electrical points may be provided near the middle of the laboratory by plugging extension leads into ceiling sockets, and small bottles of gas can be used to add to the number of gas points.

Technical details relating to distribution systems and the choice of materials are available in BS 3202 (BSI, 1959), at present being revised, and *Designing for Science* (DES, 1967). Judging by the large number of fires that have been reported as a result of the use of polythene, polypropylene and polystyrene for sinks, pipes and traps (see Chapter 5), it is essential that architects and designers should be extremely cautious about using new materials in school science departments. Building Bulletin 7 (DES, 1975) deals with *Fire and the Design of Schools* and should be consulted on all design aspects of fire precautions.

Environmental services and controls

The provision of a pleasant and comfortable working environment for science should be considered most carefully at the planning stage.

Heating and ventilation

The siting of fixtures such as radiators, heating ducts, heater cabinets and 'fresh air' heater cabinets, the latter having the facility to draw in fresh air from outside, should be such as to cause minimum interference with the efficient use of available space.

The type of heating system adopted will, in general, not be determined by the needs of the science department alone. However, there would appear to be advantages in adopting a hot air system for a science department because of the flexibility it allows in meeting particular needs. Fresh air heater cabinets, for example, are particularly recommended for use in chemistry laboratories and lecture rooms as they can be designed to produce a given number of air changes in a room per hour, the stale air and fumes being extracted along suitably lined ducts terminating in a fan-powered roof extraction unit operated by an electric switch.

Eight changes of air per hour are recommended for photographic dark rooms, which therefore also require a fan-powered extraction unit, as do fume cupboards (see Chapter 5). Rooms with a high proportion of fixed side benching can be heated by perimeter heating ducts, the outlet grilles being placed under windows at sill height, while heater cabinets can be used in rooms with little side benching.

The main disadvantage of the hot air system, in which air is blown over an array of hot copper tubes, is the noise of the fans. This noise varies from one design to another and architects should ensure that the fan specified meets acceptable acoustic criteria. The noise is often increased by inadequate servicing of the heater.

For the special heating needs of a greenhouse and an animal house see page 5.

Room temperatures are generally controlled by thermostats placed near doors. However, the temperature

and humidity in a laboratory in which certain experiments are being done can rise rapidly, and a science teacher needs to anticipate such fluctuations by increasing the ventilation to maintain comfortable working conditions.

Lighting

Much of the work done in a science department is visually more exacting than work done in ordinary classrooms. An illuminance of 500 lux is recommended for school laboratories, which may be achieved in practice by a combination of general lighting supplemented by directed localised lighting, either permanently installed or provided by the use of lamps on the work bench. For particularly intricate work, such as dissection, higher levels of illuminance are obtained by using either tungsten filament lamps or lamps fitted with miniature fluorescent tubes, the latter having a lower heat output than the tungsten filament lamps.

Many types of fluorescent lamps are now available, each producing its own characteristic type of light, and it is important to ensure that the particular lamps chosen blend with available daylight and enable colour changes, such as those in titrations, to be perceived without difficulty. Lamps mounted in luminaires (lamp fittings) which are fitted with directional control devices such as prismatic panels or louvres can prevent excessive glare.

The control of excessive natural light entering laboratories by suitable siting of the laboratories was discussed on page 4. However, it may be necessary to plan for light-grey fireproof curtains or roller blinds, housed in 'curtain/blind boxes', to obtain a fine control of lighting during the day. These can also be effective in controlling the amount of solar heat gain and in reducing the dangers associated with non-luminous bunsen flames.

Figure 1.8 shows the siting of artificial lighting and permanent supplementary artificial lighting in relation to the windows in an elementary chemistry laboratory.

Noise

Noise is a much under-estimated safety hazard; Building Bulletin 51 (DES, 1975) deals with *Acoustics in Educational Buildings* from a technical point of view. There are three main types of noise defect in science laboratories:

a Insufficient absorption of noise. This leads to excessive reverberation, making speech seem indistinct. It is a very common defect and can be eliminated to a large extent by an acoustically absorbent ceiling. In open-plan type laboratories, the backs of high 'dividers' should preferably be fitted with sound-absorbing materials, such as cloth curtains or acoustic 'ceiling' tiles.

b Excessive generation of noise. Floor tiling materials should be chosen to reduce the noise of stools scraping

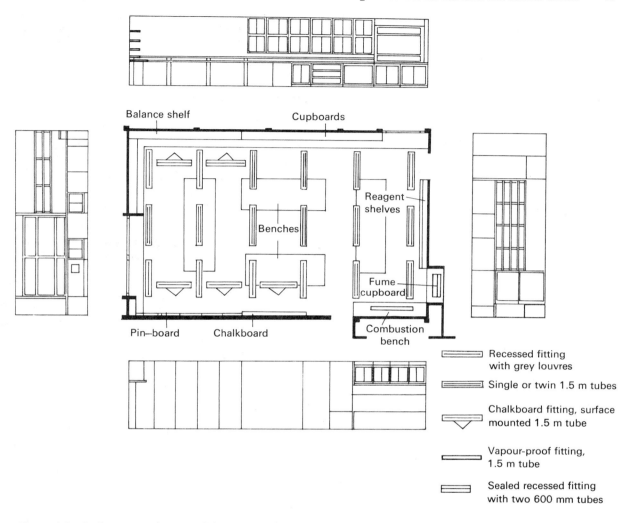

Figure 1.8   Lighting in a chemistry laboratory with permanent supplementary artificial lighting as required

over the floor, and the stools should be fitted with rubber feet.

c Noise by transmission. The attenuation of vibrations transmitted through the structure of the building is basically a problem for the architect, who should give it special attention.

Special subject requirements

Although some of the specialist requirements of each subject will also apply to the design of general purpose science laboratories, the items listed in this section will be of particular importance when designing specialist laboratories and preparation/store rooms for biology, chemistry and physics.

A BIOLOGY

Laboratories

   i easy access to the outside and to a greenhouse;
  ii at least one sunny bench with curtains/blinds to control the amount of sunshine;
 iii good artificial lighting to aid dissection;
 iv good ventilation for extraction of smells;
  v blackout for micro-projection;

vi low voltage supplies for microscopes and separate microscope lamps;
vii areas reserved for wet and dirty work;
viii bench area and services for permanent apparatus/specimens, e.g. aquaria, locust cages, plants.

Preparation rooms/stores

i space for a refrigerator and deep freeze;
ii space for potting tables;
iii siting of bins for food and refuse;
iv storage for chemicals;
v storage for glass cases with specimens.

B CHEMISTRY

Laboratories

i provision of fume cupboard;
ii balance bench, with adjacent electrical mains outlets;
iii good ventilation;

iv adequate control of direct sunlight;
v ample supplies of gas and water. This suggests some fixed benches in the room in addition to fixed side benches. Figure 1.9 shows a chemical bench with the electrical outlets separated from wet surfaces.

Preparation rooms/stores

i still for production of distilled water in hard water areas and an ion-exchange apparatus in soft water areas;
ii fume cupboard, usually shared with laboratory;
iii combustion bench;
iv glass working bench and appropriate tools;
v storage for chemicals;
vi separate storage for electrical equipment;
vii poison cupboard;
viii gas cylinder store;
ix fire-proof cupboard for volatile, flammable liquids if no outside chemical store is planned.

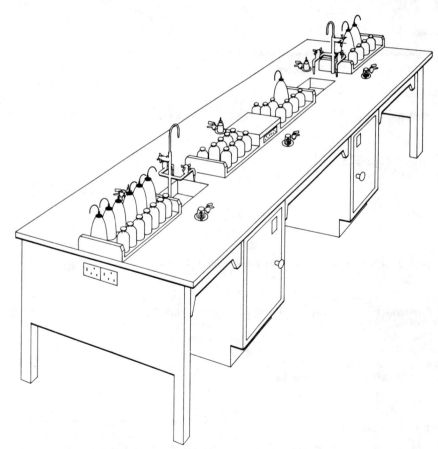

Figure 1.9   A chemistry bench showing the separation of electrical outlets from wet services. If the reagents need to be moved regularly, they should be placed on trays which fit into the spaces provided on the bench

C PHYSICS

Laboratories

i adequate supply of electrical mains outlets—a minimum of 24 pairs of outlets are recommended for the Nuffield O level Physics course;
ii a few sinks with both hot and cold water, and some twin gas taps;
iii blackout facilities;
iv provision of a suspension beam above clear floor space so that the full height of the laboratory can be used.

Preparation rooms/stores

i radioactive sources store;
ii adequate storage facilities for all types of apparatus, from vacuum pumps to fragile electron diffraction tubes and from kits containing over 500 items to small specialist items;
iii store for chemicals.

Open plan laboratories

Open plan laboratories are designed to provide a flexible organisational system which facilitates a number of teachers being involved with an appropriate number of classes (preferably fewer classes than teachers) in a team-teaching situation. By 'removing' walls, it is hoped to provide pupils—often of widely differing motivation and ability—with opportunities to work in appropriate groups on suitable experiments. As such, open plan laboratories should not be used for conventional class teaching.

Experience suggests that for an open plan system to succeed, the general ethos of the school must be supportive, the pupil:teacher ratio must be smaller than average, the teaching team members must have similar aims, and the provision of resources must be on an adequate scale. Even under optimum conditions, problems exist. For example, the transmission of noise throughout the open area can be extremely distracting and hazardous, despite the provision of sound baffles. Also, the opportunities for children to get into danger or cause mischief would seem to be increased, and precautions must, therefore, be taken to restrict access to the science area. Schools used as community centres in the evening must ensure that members of the public—who may be unaware of the possible dangers—are prevented from entering laboratories and stores.

Check lists

As a summary of many of the issues raised in this chapter, two check lists are provided.

Table 1.5 is a room layout check list, and deals with items which should be considered, under three headings: General, Services, and Fixed Furniture and Fittings. A layout for a laboratory can be tested using cut-out furniture shapes on squared paper, using a scale of 1:100 or 1:50. By this means, different layouts can be tried, not only of movable tables but also of fixed furniture and fittings. The consequences for the arrangement of pupils in the laboratory can also be assessed.

Table 1.6 summarises the issues which must be considered when planning a science department. Its purpose is to help the head of the science department, the science advisers and the architect to formulate appropriate design criteria.

Table 1.5 Check list for each room layout

Item	Check list
General	
Room names	Nomenclature and numbering
Wall space	Check that there is sufficient wall space in each room to which apparatus may be fixed, if this is required
Students' storage space	What provision, if any, is needed for storage of pupils' belongings?
Doors	(a) Clear width. Check against trolley or other movement between rooms, minimum door opening width (b) Locking of doors
Projection	(a) Method of projection (front, rear or overhead) (b) Type of projection (c) Size of screen and type (d) Minimum distance from projector to screen
Blackout	(a) Complete (b) If partial, describe acceptable standard
Fire-fighting and first-aid equipment	
Balance	Traditional or electric
Dangerous substances	Provision for storage
Workshop areas	Any special requirements for these areas
Display areas	What services are required? (see services)
Greenhouse	What services are required? (see services)

Item	Check list
Services	
Gas	(a) Number and placing of outlets
	(b) Type of outlets, e.g. 1-way, 2-way, etc
Electricity, mains	(a) Number and placing of socket outlets
	(b) Do any require to be switched sockets?
	(c) Do any sockets require indicating lights?
	(d) Do any sockets require to be greater capacity than 13 amp?
	(e) Is three-phase current required?
	(f) Is teacher control required in each room?
	(g) Individual light at work station
	i for teachers
	ii for pupils
Electricity, low voltage	(a) Number and placing of socket outlets
	(b) Consider alternative use of power packs taken off HV
	(c) Battery or accumulator supply
Water	(a) Hot
	(b) Cold
	(c) Type of outlet and location
	(d) Sinks—length, width, depth; are covers and draining boards needed?
	(e) Drip sockets
	(f) Distilled/demineralised water
Drainage	(a) Any particular comments on dilution pots and the problem of corrosive waste material
	(b) Drainage points in floor
Ventilation	
Fume cupboards	(a) Dimensions: clear width and depth
	(b) Gas: number and placing of outlets
	(c) Electricity: number and placing of outlets

Item	Check list
	(d) Cold water: type of outlet and position; drip socket (if not, sink with sizes)
	(e) Method of control of services
	(f) Worktop surface
	(g) Lighting—type of fitting
Other services	
Telephone	(a) Internal
	(b) Link with school switchboard
Television	(a) Closed system: if required, which rooms need input and outlets and what links with rest of school?
	(b) Broadcast reception: if required, will there be a central aerial for whole school? Which rooms require outlets?
Others	(a) Vacuum
	(b) Compressed air
Waste disposal	See pages 32 and 48
Fixed furniture and fittings	
Chalkboards	(a) Plain
	(b) Ruled or squared; give dimensions and colour of rulings
	(c) Fixed or roller
	(d) Local lighting—type of fitting
Pin-boards and peg boards	
Bench working tops	Finish: any special requirements (e.g. are chemicals to rest on worktop?)
Under-bench units	
Storage cupboards	(a) What is to be stored?
	(b) Type of doors: e.g. glazed, solid, sliding, hinged
Drawers/plastic trays	(a) Number
	(b) Length, width, depth
Book storage	Number and placing of text books and reference books
High-level storage	
Suspension beam	Size, maximum load

Size of a science department

Number and size of science groups
Number of periods of sciences
in laboratories
in lecture rooms
outdoors (biology, rural studies)

Links between branches of science and connections with other activities, indicating juxtaposition of laboratories, lecture rooms and supporting areas

Range and scale of special facilities

Indoors
exhibition spaces
science library
museum
students' workshop
photographic dark room

Outdoors
greenhouses
animal houses
ponds
experimental plots
large pieces of equipment
use of roof areas

Lecture–demonstration room

maximum size of audience
range of experimental facilities required
audio-visual aids, blackout, ventilation, noise levels

Activities in workshop/preparation/store and storage routine

Function and number of laboratory technicians, preferable location for preparation rooms, workshops (access from

road, access to preparation rooms, disposal of waste material), store rooms, stores for dangerous substances

Patterns of work in each teaching space

Lecture–demonstration
size and nature of demonstration
positions for demonstration
audio-visual aids
demonstrations on trolleys
amount of storage within demonstration bench
writing facilities

Practical work by students
size and nature of experiments
space for work in progress
selection of kits of equipment
methods of control of completeness of kits

Long-term work
amount of
areas required
dirty or clean
need for sunshine/shade
static or movable
how long is 'long term'?
services for

Special fittings
fume cupboards
combustion benches, ovens etc.
benches for balances and other fragile apparatus

Reference–writing areas: number of books, periodicals; working positions, if any
Storage needs and preferable/possible locations for
students' equipment
Long-term work
displays, pin-boards etc.

special apparatus such as balances, centrifuges, galvanometers, projectors, cameras etc. and any special requirements such as stability, height or illumination
Facilities for teachers (teacher's corner, teacher's experimental work)

Services

at lecture-demonstration positions
at students' individual working positions
at positions for students' long-term work
for special fittings
special provisions such as wash-up units
other services: telephones, clocks, TV, radio

Environmental services and controls

Heating
levels of temperature and humidity
preferable emission systems
possible positions of heat sources
special heating in greenhouses/ animal houses/culture rooms

Ventilation
rates
special needs
controls

Lighting
position
direction
intensity
quality
Noise
levels
sources
possible methods of attenuation

Table 1.6   Check list for a science department

Bibliography

AMA, ASE, AAM (1970) *The Teaching of Science in Secondary Schools*, John Murray.
ASE (1973) *Science for the 13–16 Age Group*, ASE, College Lane, Hatfield, Herts.
—(1975) *Provision of Temporary Laboratory Accommodation*, Education in Science No. 62, pp. 14–16.
—(1976) *Non-streamed Science, A Teachers' Guide*, Study Series, No. 7.
British Standards Institution (1959) *Laboratory Furniture and Fittings*, BS 3202, BSI.
Department of Education and Science (1966) *Middle Schools*, Building Bulletin 35, HMSO.
—(1967) *Lighting in Schools*, Building Bulletin 33, HMSO.
—(1967) *Designing for Science*, Building Bulletin 39, HMSO.
—(1973) *Furniture and Equipment, Working Heights and Zones for Practical Activities*, Building Bulletin 50, HMSO.
—(1976) *Safety in Science Laboratories*, Safety Series, No. 2, HMSO.
—(1976) *School Furniture, Standing and Sitting Position*, Building Bulletin 52, HMSO.

—(1975) *Fire and the Design of Schools*, Building Bulletin 7 (Fifth Edition), HMSO.

—(1975) *Acoustics in Educational Buildings*, Building Bulletin 51, HMSO.

Everett, K., Hughes, D. (1975) *A Guide to Laboratory Design*, Butterworths.

Laboratories Investigation Unit (1972) *Adaptable Furniture and Services for Education and Science*, Paper No. 6, DES, Elizabeth House, York Road, London SE1 7PH.

Lewis, J. L. (ed.) (1972) *Teaching School Physics*, a UNESCO Source Book, Longman/Penguin Books.

Lloyd, W. H. (1972) Laboratory Design, Part 1 *Education in Chemistry*, **9**, 4, pp. 142–145.

—(1972) Laboratory Design, Part II *Education in Chemistry*, **9**, 5, pp. 190–193.

National Science Teachers Association (1961) *School Facilities for Science Instruction*, NSTA, 1201 Sixteenth St. NW, Washington, DC, USA.

Nuffield Advanced Biological Science (1971) *Laboratory Book*, Longman/Penguin Books.

Nuffield O level Physics Project (1968) *Guide to Apparatus*, Longman/Penguin Books.

OECD Programme on Educational Building (1976) *Providing for Future Change: Adaptability and Flexibility in School Building*, Paris.

Richardson, E. J. (1967) *The Environment of Learning*, Nelson.

Savage, G. (1964) *The Planning and Equipment of School Science Blocks*, John Murray.

Schools Council (1969) *The Middle Years of Schooling from 8 to 13*, Working Paper No. 22, HMSO.

Science Teacher Education Project (1974a) *The Art of the Science Teacher*, McGraw-Hill, pp. 137–147.

—(1974b) *Theory into Practice*, McGraw-Hill, pp. 83–84.

Stepan, O. M. (1966) *Storage of Apparatus*, ASE and Nuffield Science Teaching Project.

—(1969) School Science Accommodation, a check list, *Physics Education*, **4**, 4, pp. 234–235.

Wray, J. D. (1974) *Animal Accommodation for Schools*, English Universities Press.

Wyatt, H. V. (ed.) (1965) *The Design of Biological Laboratories*, Institute of Biology, 41 Queens Gate, London SW7.

2 Management and laboratory organisation

Introduction

The aims of this chapter are to draw attention to basic management principles and organisational methods and to discuss their application and implementation in school science departments.

The terms 'science department' and 'head of the science department' can mean different things in different schools. The nature of the job of the head of department is affected by the size of the school, the age range and ability range of its pupils, the amount of teaching of specialised sciences and the degree of their integration. However, the fundamental responsibility of the head of department for the implementation of general school policy decisions regarding a subject area or subject areas is identical in all situations. The term 'head of science', as used in this chapter, may, therefore, be interpreted as describing the head of a single subject department (principal teacher in a subject) or a co-ordinating head or chairman of the science departments with responsibility delegated to the head of each subject department, e.g. biology, chemistry, physics and rural science.

The head of science as middle-manager

In assessing the management tasks facing the head of a school, comparisons are often drawn with the tasks facing a senior manager in industry. Neither can act in complete isolation if the institution is to realise its objectives. Before deciding on the general aims and more specific objectives for the school (or industrial concern), the head (or manager) will have consulted widely and held appropriate discussions with the heads of individual subject departments. In pursuing the objectives, the head has to make the best use of available human and material resources, and to delegate authority for the various subject areas to the appropriate heads of departments while retaining final responsibility. It is also necessary for him or her to gain the maximum support from assistant teachers, ancillary staff and pupils, as well as from parents, administrators and others involved in the education service. This requires not only adequate skills in personal relationships, but also the establishment and operation of lines of communication and suitable administrative structures for consultation and discussion.

The head of a subject department has similar management functions to the head of school, but on a different scale. The major responsibility is clearly the implementation of policy decisions affecting science in the school curriculum. Such a job can be compared with that of middle-management in industry, whose responsibility is the implementation of policy decisions and on whose success the future viability of the business or industrial concern will largely depend. It follows that the head of department must be allowed some time during the school day to devote to management duties, and be provided with such facilities—including an office and access to secretarial assistance—as will enable these duties to be carried out efficiently. The overall aim of the head of a science department should be the development of a stimulating and interesting environment in which pupils, science staff and ancillary staff can work with initiative towards appropriate goals.

Communication and discussion

The creation of an atmosphere which encourages discussion within a department, and the establishment and proper use of lines of communication between the head of science and members of the department, are two important factors which determine the vigour of a department's thinking.

A head of department is likely to discover as much of colleagues' attitudes through informal conversations as through the more formal contributions in a departmental meeting. The important point is that the head of department should be *available* for spontaneous discussions, not only in the laboratories or science block, but also in the school staff room. A science team which regularly spends its non-teaching time in its own corner of the school can easily become isolated from the rest of the school staff because informal contacts are not established. The need for informal contacts between science staff and their colleagues in other departments is thus an important consideration in the design of school science accommodation.

Informal meetings should be seen as a means of providing the basis for more formal departmental meetings. These formal gatherings require careful planning so that the topics are of relevance to the work of the department. The agenda, accompanied by prepared papers, must be circulated several days in advance of the meeting. Care should be taken that time is not wasted on routine administrative matters, such as examination marking arrangements, for which a circular to each member of the department may be a more appropriate means of communication. The use of a departmental noticeboard as an aid to communication of general notices and administrative matters is discussed in more detail on page 21.

The head of science department's responsibilities in keeping parents informed about the work of the science department are not always appreciated. This is particularly important at a time of major curriculum innovation, as parental interest and co-operation can do much to support such innovation.

Opportunities should be sought to invite parents to the school during an evening, when they can talk with the science staff, have a chance to put more formal questions to the head of the department and head of the school, and possibly observe pupils working with the 'new resources'. A short paper outlining the aims and policy of the science department might well be prepared prior to such a meeting, and should in any case be made available to all parents and science advisory staff as well as to local industrialists and employers with a particular interest in science.

Delegation and leadership

The head of science is likely to be a more effective leader if all members of the department feel themselves to be *involved* as a team in promoting the objectives of the department. To obtain such involvement, as many as possible should share in 'creative' work such as the production of teaching schemes for particular units of work, and in the 'organisational' and 'administrative' tasks.

In delegating specific responsibilities to the teaching staff within the fields of organisation and administration, the head of department needs both to diagnose and to deploy the diversity of special interests and strengths in the department, at the same time ensuring that the younger and less experienced members of the team are given opportunities to develop their expertise. (For a discussion of the role of ancillary staff in organisation and administration see page 23).

A list of areas for which members of the science staff have agreed to be responsible in an academic year should be known to all members of the department, and is likely to include the following (the list is not intended to cover all possible areas).

a Broadcasts
i radio;
ii television, including the organisation of video-

recording. (Legal requirements are discussed in Chapter 4, page 67).

b Displays and exhibitions

c Examinations
i internal: different members of staff might be given responsibility for the setting and marking of examinations for particular groups;
ii continuous school assessment for externally awarded certificates;
iii external: planning and organisation of practical examinations.

d Films
i maintenance of up-to-date information, including reviews and catalogues;
ii advance ordering and returning of films;
iii maintenance of a file on the educational merit of films for future use.

e Library
i book suggestions for the main school library;
ii departmental library—senior students can help;
iii staff reference library.

f Minuting secretary
i taking minutes at departmental meetings;
ii circulation of minutes after checking by the head of department, for approval by the next departmental meeting.

g Safety

h Science schemes
i combined science or integrated science for a group/set;
ii updating of single and combined subject schemes, particularly with regard to available resources and time spent on each section.

j Scientific society
i junior section;
ii senior section.

k Stationery materials
i for pupils;
ii for staff, including chalk.

l Visits

i specialised residential field trips;
ii day visits;
iii liaison with Local Education Authority or other appropriate body regarding insurance.

m Wall charts

i maintenance of wall charts, suggestions for additions;
ii regular changes in laboratories, in phase with topics being taught.

Office management

The tasks of organisation and administration within a science department are likely to be more effectively carried out if appropriate facilities, such as a head of department office and staff resources centre, are available.

The following points should be considered in relation to office management and office organisation.

Timetables

a Compilation of school timetable. The task of compilation of the school timetable is often delegated by the head to a senior member of the staff. It is essential that the special requirements of science—with particular regard to laboratory timetabling and use of double or triple periods—are made known and are stated explicitly. Problems associated with the movement of apparatus can be reduced if parallel classes are timetabled for the same laboratory—possibly consecutively. In some schools the head of science is allocated blocks of time for particular year groups and with the heads of subjects is responsible for the staff, laboratory and pupil group timetables, which will then be fitted into the school timetable.

b Display of timetables. Copies of the complete school timetable, the laboratory timetables and timetables of all teachers contributing to the teaching of science, should be displayed on a large pin-board for easy reference. Similarly, if student teachers are attached to the science department, their timetables should be displayed and kept up to date as additional classes are taken over. In this way, the head of science has a complete picture of staff commitments, and can advise on possible substitutions for an absent science teacher and/or make any necessary alterations in laboratory timetabling to suit a particular situation, such as school examinations or a practical examination.

Noticeboard

A noticeboard, preferably a large pin-board, should provide up-to-date information and be an aid to communication within the department. The following

section headings, preferably printed in large letters on coloured card, might be appropriate:

i school timetable;
ii science laboratories timetables;
iii science teachers' timetables;
iv student teachers' timetables;
v departmental notices (not for detailed administrative notices, which should be sent by circular to each member of the department);
vi inservice courses, national and local;
vii local meetings e.g. ASE, subject centres;
viii teachers' centre;
ix today (notices must be cleared each evening);
x URGENT;
xi visitors;
xii notices from examination boards; public examination timetable.

Filing system

Two filing cabinets, one for papers and correspondence of a confidential nature, and one for items to which the science staff and chief technician have open access, form a convenient way of storing and retrieving information.

a 'Confidential' filing cabinet. The confidential papers handled by the head of a science department include items concerned with examinations, such as confidential instructions to supervisors of externally set practical examinations; assessment cards for pupils; copies of references for pupils, science staff and ancillary staff; and communications with the head, advisory staff and parents. This filing cabinet should be kept locked.

b 'Open-access' filing cabinet. The type of materials conveniently stored in an open-access filing cabinet or in box files stored on shelves includes the following: school science syllabuses for the various years and groups within the years; past examination papers; information on careers in science and applied science; opportunities in further and higher education; safety file (see Chapter 3); catalogues for books and audio-visual resources; teaching topic files; curriculum information in various subject areas; regulations for visits to places such as factories and field centres; notes and syllabuses from examining boards.

Staff resources centre

Whereas the establishment of a resource area for pupil use will depend on the general aims and organisation of the science teaching in the school, a resources centre within the head-of-department office, for use by the staff, is a convenient way of centralising resources not kept in preparation areas or stores. The following might be included in such a resources centre.

a **Reference library.** This should contain copies of science textbooks being used in the school, as well as a selection of science books for staff reference. Other clearly labelled sections might include curriculum project books; Schools Council (or similar body) publications relating to the teaching of science; periodicals from the professional institutes and science teaching associations; books about teaching; books about science; scientific dictionaries; mathematical tables; data books; bound volumes of examination syllabuses and past papers.

b **Audio-visual aids.** These include materials for making wallcharts, visual aids and notices; overhead transparencies, pens and files of transparencies for general staff use; film and film-strip catalogues; films, film-strips and slides; radio and television broadcast notices; audio and video tapes for recording programmes for subsequent use; and copying software for reprographic processes. An introduction to a wide variety of audio-visual aids is given in Chapter 4.

c **Worksheet store.** Worksheets, information sheets and related printed materials need to be categorised according to type of material, year group, module specification and item number. Copies of each item are best stored in envelope or box files which are clearly labelled and numbered. The storage of skins requires special care—they should preferably be hung vertically or kept in separate folders in a large drawer to avoid crumpling. The construction of worksheets is discussed in Chapter 4.

d **Stationery.** Files, filepaper, graph paper and exercise books should all be kept in the resources centre. It is usual for the teacher to sign, stamp or rip off the bottom right hand corner of the cover of the full exercise book in exchange for a new one.

e **Duplicating facilities.** The various methods of duplication are discussed in Chapter 4. The provision of different types of duplicators in a school will depend on the size of the school and the use made of such facilities. It is quite usual for a spirit duplicator and thermal copier to be kept in the science department. The provision of a stencil duplicator and electronic scanning stencil cutter should also be given serious consideration—particularly if worksheets are commonly used.

f **Projection facilities.** A system must be established for the borrowing of film projectors, film-loop projectors and slide projectors, television and radio recordings, and overhead projectors.

Money management

Capitation allowance

The method adopted by the head of the school for allocating money to the various departments varies from school to school, but, whatever the system, the head of the science department has a crucial role in presenting to the head realistic annual estimates of departmental needs.

In preparing the annual estimates, the following points should be among those considered:

i *the global sum* should show the requirements of the various sections in the science department, e.g. biology, chemistry, physics, rural science;

ii *equipment costs*, including items of a non-consumable nature;

iii *running costs*, including items required to maintain levels of stock of chemicals and other expendable materials;

iv *stationery* requirements have increased in recent years, especially in schools making considerable use of worksheets;

v *books and audio-visual aids* shown separately will underline the need for books, as well as equipment, in a science department;

vi *workshop needs* should be prepared in consultation with the senior technician, who may be able to advise on the comparative costs of making or buying certain items of apparatus;

vii a *local purchase fund* is considered essential by most science teachers for the efficient running of the department. Such a fund should be reserved for the purchase of small items and perishable goods;

viii *living organism fund.* Living organisms are often needed at short notice and may not keep. If the supplier requires 'cash with order', it may be possible to deposit a lump sum with the supplier, so that the cost of an order can be deducted from this lump sum to avoid delay;

ix *replacement fund.* The lifetime of apparatus, particularly of kits of apparatus in almost continuous use by pupils, is not infinite, and it is sensible to earmark a certain percentage of the annual capitation allowance for the systematic replacement of apparatus. The need for medium- and long-term forward planning is an aspect of management which is only beginning to be acknowledged in schools. A flow chart of likely replacement needs for a period of at least five years ahead will indicate the short-term needs in terms of a long-term policy.

x *practical examinations.* The specification of apparatus and materials required for externally set practical examinations may be such that certain items have to be purchased at relatively short notice.

Just as the head of the school has to judge the competing claims of the various departmental estimates, so the head of the science department will need to prepare the science estimates in a 'climate of discussion' (Marland, 1971) with the various interests in the department, and, after an analysis of priorities, base the final estimates on the stated objectives of science teaching in the school.

Special estimates

To purchase very costly items of equipment or to finance a curriculum innovation, a special request may need to be formulated. In general, this involves the preparation of a brief based on a careful analysis of the reasons for the financial request, with an explicit statement of the educational ends which it is hoped to achieve. Such a brief needs to be based on discussions within the science department, with the head of the school and, where appropriate, the adviser for science. Local Authority science advisers often have a small pool of money at their disposal to meet such requests; similarly, the board of school governors or the parent-teacher association may be willing to meet such requests from the school.

Ancillary staff—role, training and job specification

The need for adequate laboratory assistance in a science department has been recognised for a long time (*Science in Secondary Schools*, 1960). Unfortunately, the need is not always met in practice (ASE, 1968). The Association for Science Education recommends one technician to every two laboratories (ASE, 1973).

The problem of assistance in laboratories is not only a question of the number of ancillary staff—both technical and secretarial—but also of their qualifications, skills and experience. It is one of the management functions of the head of the science department to analyse the competence of the ancillary team, and to specify the roles which each member of the team should play in the department. This job specification should be drawn up in conjunction with the chief technician who, by definition, should be suitably qualified and experienced to undertake such a joint task.

Role of the chief technician

A qualified and experienced chief technician will play a major role in helping with the management and organisation of a science department. It is essential for a chief technician to have a base—possibly in a central workshop (page 8)—with an office area and facilities for the maintenance, repair and construction of apparatus and the storage of bulk materials.

The major duties and responsibilities of a *chief technician* should be specified by the head of the science department, and are likely to include the following.

a Personnel management and welfare. Supervision and deployment of technical staff; training of junior technical staff; organisation of holiday rotas; attendance, time-keeping.

b Administration. Stock control, including annual stock check; requisitioning and receipt of supplies; petty cash accounts; breakage record books; legal records of poisons, flammable liquids and alcohol; storage, including labelling; organisation of distribution and transfer of materials and audio-visual aids within the science department; maintenance of filing system, catalogues and up-to-date price lists.

c Safety and security. Implementation of safety regulations; knowledge of legal safety requirements, e.g. Health and Safety at Work Act, 1974; safe disposal of chemical and biological waste materials; first aid; access to laboratories, stores, and preparation rooms by students; checking of services, windows, doors—locking up; reporting deterioration and damage of equipment to the head of the science department.

d Laboratory servicing. Preparation of materials, solutions, specimens and apparatus; cleaning of glassware; care of animals and plants; sterilisation of apparatus; biological techniques; keeping the science department at optimum efficiency.

e Maintenance, repair and construction. Keeping all equipment, including balances and audio-visual apparatus, in good working order; photographic work; museum mounting and displays; duplication and reprography.

The majority of the above duties are of an administrative nature and dependent on particular skills and knowledge, and are therefore appropriate for a chief technician, but it is also important to involve the other technical staff in many of these duties. Delegation to the technical staff is as important as within the teaching staff and encourages maximum involvement within the science department. The duties of the technical staff should be under the general supervision of the chief technician.

Training and qualifications of technicians

With the increasing size and complexity of school science departments, the demands made on technicians have also increased. This has made the need for training all the more important; it has also made a career for laboratory technicians more attractive.

The professional body which aims to promote the status and efficiency of technicians through the medium of training and qualifications is The Institute of Science Technology, 345 Gray's Inn Road, London WC1. The City and Guilds of London Institute courses 735 part I and 735 part II are run in many technical colleges on a day-release basis, and have for some time provided technicians with the opportunity to learn about such topics as the care and maintenance of apparatus and aspects of laboratory servicing and safety. The status of laboratory technician is achieved by gaining part 1; promotion to chief technician is usually dependent on passing part II and occasionally part III.

A revised scheme of courses and qualifications for technicians is planned to come into operation in 1978. A course of standard roughly equivalent to City and Guilds course 735 part I and *part* of part II, will lead to the Technician Education Council *Ordinary Technician's Certificate*, suitable for a person who is capable of undertaking a variety of jobs in the laboratory with the minimum of supervision. A course of standard roughly equivalent to City and Guilds part III and the *remaining part* of part II, will lead to the *Higher Technician's Certificate*, suitable for an experienced technician who also has the management skills to become a chief technician. Science teachers should make themselves familiar with the syllabuses leading to these qualifications, so that the skills of qualified technicians are not underestimated.

While the possession of qualifications is important, this should not reduce the responsibility of the head of department and science teaching staff to give 'on-the-job inservice training' as necessary. Technical staff should, for example, attend local exhibitions of scientific apparatus (page 27). They should feel that their services and ideas are an essential contributory factor to the strength of science teaching in the school.

Technician task analysis

The tasks normally undertaken by technicians are outlined on page 23 and have been specified in a number of publications (*Science in Secondary Schools*, 1960; Kramer, 1969; ASE, 1976).

The time *actually* spent on various activities was shown in a task analysis carried out in 1969 within a particular Local Education Authority (STEP, 1974b), in which the average ratio of technicians to science staff was 1 : 3. The relevant percentages are shown in Table 2.1.

Activity	Percentage time
Dismantling, washing, cleaning	20
Setting up apparatus	16
Preparing special experiments	11
Checking laboratories	10
Preparing solutions	9
Making and repairing apparatus	8
Looking after plants and animals	6
Assisting in laboratories	5
Duplicating and office work	4
Assisting with visual aids	3
Stocktaking	3
Working outside the science department	3
Ordering	2
Total	100

Table 2.1   Average percentage of time spent on the stated activities by technicians in a local task analysis

The analysis suggests that about one-third of the time was spent on skilled activities, one-third on semi-skilled activities, and one-third on unskilled activities. A task analysis of this type is bound to have its imperfections, but it may well indicate the needs of a department in terms of skilled and unskilled help, and enable a case to be made for additional or alternative ancillary help on grounds of cost effectiveness. For example, secretarial assistance might be more cost effective than using technical help if many worksheets are being produced in the department or, a part-timer might be able to wash apparatus and clear things away to enable a more highly paid technician to spend time on skilled activities.

Ordering materials for lessons

There are two main contributory factors to the smooth running of this aspect of laboratory management.

a A system of ordering. The practice of ordering materials for lessons varies from school to school but the system used must give technicians time to prepare the various resources. It is important that a means of communication between teacher and technician is established, which enables checks on possible clashes of requirements to be made in good time, particularly when investigations or projects are in progress. If the major part of the preparation has been done in good time, a technician is more likely to be able to cope with emergencies, so that the lesson can go ahead as planned.

Table 2.2 shows a section of a weekly request form for an individual member of staff, which runs from Monday, period 3, to the following Monday, period 2. The technician is given each teacher's form prior to school on Monday morning, and the teacher can retain a carbon copy for personal reference during the week.

The details of the entries should be as concise as possible. If the school is following one of the published courses, such as Nuffield Physics in the example in Table 2.2, then the year of the course, the experiment number and whether it is intended to do a demonstration (D) or class experiment (C), will specify the requirements. For other courses it is helpful to prepare a stock of numbered cards, listing the resources required for certain lessons so that a card reference number, e.g. C.Sc./15 in Table 2.2, specifies the basic requirements. It may be useful for a list to be available in the technician's room which shows the number of pupils (and working groups) in each class or set.

An alternative to the weekly request form is for teachers to write their individual lesson requirements on a separate sheet for each lesson. Such sheets should preferably be of uniform size and must indicate the day, period(s), class and laboratory. An advantage of this system is its flexibility, as the chief technician can sort the sheets according to day and laboratory, and allocate jobs to the ancillary staff in order of priority.

DAY	PERIODS	CLASS	LAB	REQUIREMENTS for Nuffield: year, class or dem. expt. no.

SCIENCE DEPARTMENT: Physics   NAME:
WEEKLY REQUEST FORM   WEEK BEG.:

DAY	PERIODS	CLASS	LAB	REQUIREMENTS for Nuffield: year, class or dem. expt. no.
Mon	3–4	L6	S1	Unit 4, Cl5a, b, c, d
	5–6	4A	T1	IV D110, D111 & FILM-LOOP 8
	7–8	3L	S2	III C30, C31
Tues	1–2	5P	S1	V C142, C143
	3		Prep	I want to try out Unit 4, D21
	5–6	4A	S2	IV C112a, C112c, C113
	7–8	2X	S4	Ref. Sheet C.Sc./15 & DRY ICE

Table 2.2 A section of a weekly request form

Some teachers prefer to write their requirements in a diary or log book, possibly using a different page divided into lessons for each laboratory or even a different book for each laboratory. Such a system requires the technician to prepare and date pages for two or three weeks in advance.

b Moving apparatus to and from the laboratories. The movement of resources to and from the laboratory can be greatly simplified by the use of trolleys. It further helps if apparatus is in bins or on trays, so that it is easier to move apparatus quickly and to make a visual check that all is in order before the lesson begins (and at the end).

In discussing the organisation of the movement of apparatus it must not be forgotten that pupils and students should be taught to become 'laboratory-minded'. They should assist with the distribution and collection of apparatus and accept some responsibility for ensuring that all apparatus has been collected in.

Storage and retrieval

Factors to be considered in planning the siting and design of preparation and storage facilities, with a view to maximising the efficiency of the movement of apparatus and resources within a science department, are discussed on page 9. Special subject requirements are discussed further on page 13, and safety aspects in Chapter 3.

The detailed organisation of the storage and retrieval systems in a particular school requires a most careful analysis of the facilities available and the aims of science teaching in the school. Additional factors to be considered include the age range of the pupils, the emphasis on teaching the separate sciences or an integrated scheme to various groups, and the degree to which pupils or students are encouraged in independent learning activities and the selection of apparatus. The organisation of storage systems should be kept under regular review by the head of department (and science staff) in consultation with the chief technician (and ancillary staff).

Whatever the system adopted, experience suggests that certain basic principles should be applied.

a Labelling. Every item must have a specific storage place. The item itself, and its container, should bear a distinguishing mark or number to correspond with the description and mark or number in the indexing system. The storage place should be clearly and boldly labelled, to assist with visual checks on the availability of an item and its retrieval and return after use. Self-checking devices, such as having eight spaces for eight items, or small boxes or sections of trays with a stated number of items, assist efficiency. Overcrowding must be avoided.

b Accessibility. The most frequently used items should be the most accessible. Certain simple pieces of apparatus, such as bunsen burners, tripods, stands, clamps and bosses or basic glassware, can be stored close to working positions in the laboratory itself, or near the entrance of the store—avoiding high or low shelves. Items which are particularly attractive to children should not be stored in the laboratory. Heavy items should be kept on lower shelves.

c Safety. The highest standards of safety must be maintained throughout the department. Ancillary staff should be particularly vigilant in keeping store gangways clear, avoiding the build-up of 'bits and pieces' near sinks and on tables in the preparation area, and maintaining standards which will help to educate pupils and students in good laboratory practice. Non-slip steps or a ladder should be used to reach high shelves—the use of stools for such a purpose is dangerous.

d Indexing and retrieval. As the number and range of resources in the department increase, so a carefully designed and simple indexing system becomes increasingly important. Alphabetical card-index files have the advantage of flexibility and are more easily kept up to date than book files. The following categories are likely to need separate card-index files: apparatus, the index possibly sectioned into biology, chemistry, physics, rural science etc.; books and periodicals; audio-visual equipment, film-loops, film-strips, slides and video-cassettes; worksheets; and topic resources. A certain amount of cross-indexing is inevitable and positively desirable if it increases the usefulness of the system.

e Storage problems. What is the best way to store items such as scissors, wall charts, and reels of wire? The following is based upon a discussion unit (STEP, 1974a) on 'Storing materials' and is intended to draw attention to some methods of overcoming storage problems.

Item	Possible solutions
1 Items stored haphazardly	Use of trays, boxes, trolley
2 Items in drawer	Drawer racks, partitioned drawers, polystyrene moulds
3 Drying glassware	Drying rack
4 Retort stands, bosses, clamps	Fixed by Terry clips
5 Glass tubing/rod	Tubing stored horizontally in plastic guttering
6 Charts	Stored on hangers
7 Metre rules	Specially built rack
8 Flasks, beakers	Use of plastic trays or specially shaped shelves
9 Winchesters	Transfer contents to plastic containers with taps on shelves plus guttering to catch drips
10 Burettes and pipettes	Use of Terry clips or slotted battens in drawer or on the back of a glass-fronted cupboard
11 Wire leads	Plugged into pegboard
12 Lenses	Use of tray with partition holder
13 Scissors, scalpels, needles	Shaped polystyrene foam or plywood in drawers
14 Reels of wire	Hung on pegs
15 Measuring cylinders	Foam rubber collars fitted to reduce breakage in store and in use
16 Shunts for meters	Wood battens with 4 mm holes suitably spaced
17 Microscopes	Lockable cupboard away from chalkboard
18 Microscope slides	Special boxes, clearly labelled

The storage of film-loops, film-strips, stencils etc. is discussed in Chapter 4.

f Storage of radioactive sources. Closed sources are supplied in individual small lead castles inside a hardwood box with a hinged lid. Sources must always be returned to their lead castles after use, using the special tongues which grip the 4 mm stem. The hardwood boxes should be stored in a metal container, such as a closed biscuit tin, inside a locked and suitably marked cupboard. For details of safety aspects and codes of practice for using radioactive sources see Chapters 3 and 5 (radioactive sources).

g Storage of flammable solvents. (See Chapter 3, page 49.)

Requisitioning and receipt of supplies

The maintenance of stock in a serviceable condition, the requisitioning of new resources within the limits of the yearly capitation allowance and the establishment of an efficient system for the dispatch of repairs and receipt of supplies, are among the more important management functions of the head of department.

In general, money is allocated for one financial year and cannot be carried forward into the next. There may be special circumstances, for example when a sum of money has been allocated for curriculum innovation spread over a number of years, so that the funds can be carried forward from one year to the next. It follows that requisitions should be prepared at the beginning of the financial year, always keeping some money in a *contingency fund* for breakages and other requirements later in the financial year.

A useful way of collecting suggestions for new resources is for the science staff, including the head of department, to note ideas in a *Science resources suggestion book* during the year. These suggestions should be dated, contain a clear specification, name the supplier (if appropriate) and give an indication of price. It may be helpful to categorise each suggestion as essential, highly desirable or a desirable luxury, so that priorities can be drawn up by the head of department before discussion of requisitions at a departmental meeting.

The following sections raise points regarding the choice of supplies (and suppliers), the placement of the order, the receipt of goods, and authorisation of payment.

The choice of supplies (and suppliers)

The amount of freedom granted to heads of science departments in the choice of their suppliers of apparatus varies and, in the case of some Local Education Authorities, may be dependent on the existence of a Buying Agency or similar body, which negotiates special terms with particular manufacturers. While such an agency provides a discount, the choice of apparatus may be restricted.

Good money management does not necessarily imply buying at the cheapest price. The versatility of the equipment, its compatability with existing stock, the quality and robustness of the design, the possibility of bulk purchase, as well as the price, should be carefully considered. The following sources provide valuable information on which a final decision can be based.

a Research and development groups. The Scottish Schools Science Equipment Research Centre (SSSERC), 103 Broughton Street, Edinburgh, Scotland, produces comparative reports on many items of equipment. Teachers in Scotland may request these reports, which are normally loaned for a period of one month. The Director will supply information on literature available from the Research Centre.

The Consortium of Local Education Authorities for the Provision of Science Equipment (CLEAPSE), Brunel University, Uxbridge, England, produces evaluations of currently available equipment, lists of equipment-suppliers and apparatus notes, all of which are available to teachers in member LEAs (mainly in the Midlands and South of England). A list of publications will be supplied by the Director, who will also respond to enquiries by telephone.

Reports of a comparative nature produced by the Association for Science Education Apparatus Committee are published from time to time in *Education in Science*. Members of the Committee generally draw attention to new apparatus at the ASE Annual Meeting (see also section **e**).

b Trade catalogues. All suppliers of scientific apparatus for schools provide catalogues containing descriptions and dimensions of individual items, often accompanied by photographs. Similarly, publishers send out lists of books with a brief synopsis of the contents of each book. Price lists should be kept up to date, and quotations obtained for large orders.

c Trade representatives. Representatives visit most schools on a regular if infrequent basis; in addition, they are generally prepared to make a visit by request to show particular items of apparatus being considered by the school.

d Advertisements. Advertisements and reviews of apparatus and books appear in most of the scientific journals normally read by science teachers.

e ASE Annual Meeting. The Annual Meeting of the Association for Science Education in the United Kingdom is an occasion for extensive manufacturers' and publishers' exhibitions. It provides a splendid oppor-

tunity to meet trade representatives, to compare similar items of apparatus produced by different manufacturers, and to glance through books displayed by the publishers. In recent years, manufacturers' workshops have been organised at the Annual Meeting to enable teachers to gain experience of equipment new to the market.

f Local exhibitions. Exhibitions of resources for science teaching are often arranged at local level, for example at Teachers' Centres, Subject Centres, or in connection with meetings of science teachers.

g Book inspection service. Publishers provide an inspection service, and generally allow retention of the inspection copy for staff use if an order is placed for a stated minimum number of copies of the book.

Placement of the order

Much time and correspondence is saved if items are identified correctly on the requisition form. The following are points which should be borne in mind.

a Departmental order book. If orders are written in a departmental order book, using the same style as that required for the official school requisition form, then this provides a useful record which may be used to check on the various stages from placing the order to authorising payment for the receipt of the goods. The departmental books should include columns for the date of the order, the identification of the items (see section **b**), estimated price, actual price, date of receipt and date of the invoice.

b Identification of equipment. The description of each separate item must include the manufacturer's or buying agency's catalogue reference number, the quantity required and a precise description as given in the catalogue. Care needs to be taken in specifying the quantities. Chemicals, for example, are often sold in minimum quantities, and pupils' kits are often sold in full kits, half kits or quarter kits. Other items may be sold in packets of five or ten, with a discount for orders above a given number.

c Identification of books. This should include the International Standard Book Number (ISBN), the number of copies, title, author, publisher, type of binding (if choice), edition and year of publication.

d Practical examinations. Suppliers may provide special stickers to identify 'examination orders', which are then given priority clearance. Such orders should be sent off at the earliest opportunity to allow time for administration and the delivery of the goods.

e Biological materials. Certain biological specimens or materials may be required on a particular day. Such orders should be specially marked after obtaining an assurance from the supplier that the goods can be delivered on a specified day.

Receipt of goods

It is the job of the chief technician to accept goods and sign for their receipt, though it is important that the system for the receipt of goods is well known by all the science teachers and ancillary staff. Goods should be checked as soon as possible against the delivery note and the order in the departmental order book. Each piece of equipment should be checked to ensure that it is in full working order. Any discrepancies, broken or damaged items, incorrect quantities, or wrong goods supplied, must be noted immediately so that the head of department can inform the supplier—stating the order number, its date, the delivery note number, the date of delivery and giving full details of discrepancies or breakages. The method of returning empty crates needs to be arranged with the supplier.

The items on a requisition are not necessarily all delivered on the same day. By writing the date of delivery of an item in the departmental order book, it is possible to see which items are still 'to follow'.

Authorisation of payment

When the head of the science department is satisfied that an order has been completed satisfactorily, the invoice relating to that order can be passed to the head of the school, who will authorise payment. Difficulties over payment can arise if orders are placed too near the end of the financial year, especially if there is some delay in the receipt of the goods.

A careful account must also be kept, preferably in the departmental order book, of any 'petty cash' items bought through the *local purchase fund*, so that the record of departmental expenditure is complete. In many schools the petty cash fund for the school is administered by the school office, where a safe is usually located. Receipts for science petty cash items generally have to be signed by the head of the science department, before the money is refunded.

Security in a science department

Security may be distinguished from safety, although certain aspects of security have a direct bearing on safety.

Although security is a matter of general school policy, the responsibility for implementing this policy within the science department is invariably delegated to the head of the department, who in turn will need to ensure that the whole school staff is aware of the particular security requirements within the science department. These particular requirements should be clearly specified in written form after detailed discussion between the head of department, the science staff and ancillaries. In view of the valuable and potentially dangerous materials normally found in laboratories and stores, science teachers have a special responsibility to take all reasonable precautions to safeguard pupils and property, and yet cause a minimum of inconvenience or restriction of work.

The special requirements of science include the following.

Access to laboratories

A clear policy is needed on the unlocking and locking of laboratory doors or doors leading to open plan accommodation. As a general principle, doors should be kept locked until the teacher is ready to supervise the class entering the laboratory. Similarly, doors should be locked when a class has been dismissed at the end of a lesson, the teacher or technician being the last to leave the laboratory. A system for the receipt of piles of homework books needs to be established in the light of the 'access policy'.

Much good experimental work is often done in lunch periods or after normal school hours, but here again it is essential that adult supervision is provided and that the head of department is fully aware of such arrangements.

Access to stores/preparation rooms

These rooms normally contain potentially dangerous materials such as flammable chemicals, poisons and radioactive sources, as well as expensive and specialist items of equipment. Access by pupils should be prohibited—although particular circumstances in a school may make it possible for sixth-form students to have access to stores under the supervision of a technician.

Services

In addition to windows and lights, science teachers and technicians need to check that gas, water and electricity are switched off after lessons and at the end of the day. Balances plugged into the mains must be switched off, and specialist facilities, such as fume cupboards and workshop services, should also be checked regularly.

Animals and plants

The provision of a key for the animal area and greenhouse different from the school pass key provides additional security.

Safety

Publications stressing rules for safety and encouraging good conduct in laboratories are available from many sources. A pamphlet, *Safety in the Laboratory* (ASE,

1972) written by a working party of teachers, may be purchased in bulk from the Association for Science Education and distributed to individual pupils. Publications from the Department of Education and Science and the Scottish Education Department are listed in Chapter 3, in which the various aspects of safety in school science teaching are discussed in detail.

The implementation of thoughtful and positive attitudes to safe laboratory work by teachers, technicians and pupils is a vital aspect of laboratory management and organisation.

Bibliography

ASE (1976) The role of the science technician in schools and colleges, *Education in Science*, 66, pp. 12–13.

—(1973) *Science for the 13–16 Age Group*, ASE, Hatfield, Herts.

—(1968) Supply of laboratory technicians, *Education in Science*, 28, pp. 33–43.

—(1972) *Safety in the Lab*, ASE, Hatfield, Herts.

Guy, K. (second edition 1973) *Laboratory Organisation and Administration*, Butterworths & Co. (Publishers) Ltd.

Husbands, W. H. & Thomas, R. V. (1972) *Handbook for School Chemistry Laboratory Technicians*, Pergamon Press Ltd.

Kramer, L. M. J. (1969) Aspects of managing school laboratories, *School Science Review*, 172, **50**, pp. 492–583.

Lewis, J. L. (ed.) (1972) *Teaching School Physics*, a UNESCO source book, Longman/Penguin Books.

Marland, M. (1971) *Head of Department*, Heinemann Educational Books, London.

Science in Secondary Schools (1960) *Ministry of Education Pamphlet No. 38*, HMSO, pp. 160–161.

Science Teacher Education Project (1974a) *Activities and Experiences*, McGraw Hill Book Co. (UK) Ltd.

—(1974b) *The Art of the Science Teacher*, McGraw Hill, pp. 137–147.

Stepan, O. M. (1966) *Storage of Apparatus*, ASE and Nuffield Science Teaching Project.

3 Safety in school science teaching

Introduction

In the absence of appropriate national statistics, it is not possible to identify the most common causes and sites of the accidents which have occurred in schools. However, sample surveys indicate that most accidents have occurred in playgrounds, with playing fields as the next commonest site, and that domestic science and handicraft rooms, together with science laboratories and swimming baths have been among the safest places in which to teach and learn. It is also known that the incidence of reported accidents shows marked geographical variation and that, even within one Local Education Authority, schools appear to have widely differing accident rates.

Factors affecting level of risk in schools

It is clearly important that the level of risk to teachers and pupils working in school science laboratories be maintained at as low a level as possible. The following are among the factors which may have contributed to any change in the level of risk in recent years.

a The raising of the school leaving age. The raising of the school leaving age means that more pupils are being asked to conduct more experiments than at any other time in the past.

b Changes in teaching methods. Many experiments are now of an open-ended nature. This may be a source of danger when pupils do not have the appropriate experience to distinguish between proper experimentation and hazardous exploration.

c Developments in apparatus and techniques. The range of equipment and techniques used in schools has increased considerably in recent years. Radioactive and X-ray sources, lasers, high-voltage power packs, and microbiological materials are now relatively common in secondary schools. The hazards associated with some items of equipment, e.g. lasers and X-ray sources, are indicated by the fact that their use is governed by relevant Administrative Memoranda issued by the Department of Education and Science (see pages 40–42).

d Laboratory design. The safety of pupils, teachers and others who work in school laboratories (e.g. technicians,

cleaners), is of fundamental importance in laboratory design. It is inevitable that some laboratories are safer places to work in than others, but open-plan laboratories pose particular hazards which need to be contained by appropriate design features. Some aspects of safety and laboratory design are considered on page 31. (See also page 5.)

e The development of integrated science courses. At the present stage of development and implementation of integrated courses in science, it is inevitable that teachers whose training is in a specialist subject will be called upon to use techniques or equipment with which they would not normally be expected to be familiar; thus a physics graduate is unlikely to be as aware as a biology graduate of the hazards associated with working with microorganisms and he is less likely to be able to deal appropriately with such incidents as may arise.

f Increased awareness of hazards. Many of the hazards associated with teaching practical science in a school are, of course, well established, although the consequences of exposing children or teachers to given concentrations of materials under the conditions prevailing in school laboratories remain largely unknown. New hazards are constantly being suspected or identified, and knowledge of these dangers forms the basis of the advice given periodically to science teachers by the Department of Education and Science, the Scottish Education Department and the Association for Science Education (see the Bibliography at the end of this Chapter). It is essential that science teachers familiarise themselves with the hazards known to be associated with any given procedure, material, or items of equipment and that such knowledge be kept up to date.

Because many hazards remain unidentified, the case for proper laboratory technique and for adequate hygiene is overwhelming.

Safety as a positive activity

The maintenance of the highest possible standards of safety in school science teaching involves more than a consideration of safety precautions. Laboratory safety is best regarded as a positive undertaking which is an integral part of every activity in which the science teacher is engaged with his pupils. Lesson planning must, therefore, take account not only of the most likely accidents but also of every possible accident which could occur in a given situation and of the appropriate steps to deal with them should they arise. The acquisition by pupils of safety conscious attitudes is an important part of a science teacher's task, and teaching and learning activities should be designed to minimise all safety hazards. The teacher's good example is important but, in many instances, it is unlikely to be enough. The reasons underlying good and bad laboratory practice need to be stated explicitly and the safety component of experimental design stressed whenever it is appropriate to do so, e.g. the advantages of a dropping funnel compared with a thistle funnel in the preparation of hydrogen and the easier control of liquid/liquid reactions compared with liquid/solid mixtures. At all times, it is important to remember that the teacher is vastly more experienced than the pupil, so that what is common sense to the teacher may well be outside the range of experience of the pupil and therefore far from obvious to him.

An accident follows a sequence of events governed by a pupil's knowledge, experience and manipulative skills, and by the hazards present in the environment within which he is required to work. The crucial link to be broken is the commission of an unsafe act or the succumbing to a particular physical hazard. Pupils must, therefore, be made aware of the hazards involved in any operation they are asked to undertake and, in the interests of both safety and efficiency, be taught to work systematically. Routine has an important stabilising effect on laboratory discipline and pupils must understand clearly the laboratory rules and the thinking which lies behind them.

Some LEAs may suggest draft laboratory rules and/or insist that copies of the laboratory rules be given to pupils, who sign for their receipt and stick the copies in their laboratory work books. The rules should be committed to memory, as it is too late to consult them in an emergency. At the very least, the rules will make reference to what pupils *must not do*, but more positive advice and instruction should also be given. Use may be made of the wide range of techniques which now exist to develop pupils' awareness of hazards and to foster appropriate safety-conscious skills and attitudes, e.g. safety quizzes, simulation and games exercises, and 'spot the hazard' pictures. Attention is also drawn to the film produced for pupils by the DES, entitled *In the Movies, it Doesn't Hurt*.

Eye protection

Insufficient use is made in schools of eye shields and protective spectacles. Effective and inexpensive plastic spectacles are readily obtainable for use by both teachers and pupils. Where necessary, these spectacles may be sterilised by washing with diluted Dettol or by enclosing them in a plastic bag which is then filled with sulphur dioxide. Antiseptic cleaning cloths are also available. Teachers are referred to the following advice of the DES:

'Goggles *must* be worn whenever there is any risk to the eyes. Risks may arise from powder spills and splashes as well as from explosions. Even tiny explosions such as those caused by heating small quantities of potassium manganate (VII) regularly cause accidents to eyes. This means that whenever any operation with chemicals is performed, goggles must be worn, whether in chemistry lessons, biology lessons, physics lessons, geology lessons, or indeed in home economics or craft.'

Goggles or spectacles must be designed not only to provide adequate protection against chemical attack but also to resist the impact of an explosion.

For a discussion of the use and storage of safety spectacles in the school laboratory, see *SSR*, 1973, 190, **55**, pages 190–191, which also lists some suppliers. Attention is also drawn to the Protection of Eyes Regulations, 1974 (which came into force on 10th April 1975) and to the relevant British Standards specifications for eye protectors, BS 1542 (1960) and BS 2092 (1967). See also pages 47 and 140.

Safety and laboratory design

Nowadays it is usual for science teachers to be consulted about the design or modification of laboratories or science blocks in which they are to work. The following aspects of laboratory design are particularly important from the safety point of view.

a The design must be such that it is possible to supervise adequately *all* the pupils in the class, particularly when they are engaged in experimental work.

b If the laboratory includes a demonstration bench, this should be located and designed so that pupils can sit or stand around it without overcrowding. The ASE recommends that pupils be kept 'at a safe distance (6 ft away)' from demonstration experiments conducted by the teacher.

c The laboratory floor should be resistant to minor spillages and, if polished or varnished, must not be slippery. It must also be free from cracks.

d The wall and floor fittings must not protrude unreasonably into the laboratory walking space.

e Provision must be made for the storage of pupils' satchels and bags and for jackets and other clothing that may be removed when pupils engage in practical work.

f It must be possible to open the laboratory windows and to lower any blackout or other blinds without having to climb onto benches to do so. Large windows which swing open about a horizontal axis are unsatisfactory in laboratories. It is important that a steady temperature is maintained throughout the year; there is some evidence that a temperature of 20 °C (68 °F) represents an optimum working temperature from a safety point of view. Temperature range and control are also important where living organisms are being cultured (see Chapter 8).

g At least one fume cupboard and/or hood will be required in a laboratory where noxious or unpleasant gases and vapours are produced. Such fume cupboards must be quiet and efficient in operation and their design is of particular importance (see page 101). The extraction of air via a fume cupboard should be coupled with an appropriate laboratory air flow. In this way, mechanical ventilation of a laboratory becomes possible.

h The design of the laboratory or science block must be such that the teacher has ready access to the mains gas, water and electricity controls, which should be located near the exit door of the laboratory. It is also important that pupils know the position of these controls so that they may deal with an emergency which involves the teacher or which arises during his temporary absence. The mains controls must be clearly and unambiguously labelled. It is helpful if a system of colour coding is adopted (BS 1710 *Identification of Pipelines*) to distinguish between gas, electricity and water supply lines. It is particularly important that electricity supplies of different voltages and phases should not be confused, e.g. low voltage d.c. and mains a.c. supplies. Adequate labelling and the use of non-interchangeable fittings (plugs and sockets) are essential, minimum safeguards.

j The laboratory design must incorporate adequate provision for the collection of waste materials. Separate and clearly labelled waste containers will be needed for broken glass, biological material, wet material such as filter paper, and for solid chemicals. Care must be taken in mixing excess chemicals in this way and with suitable economies of scale, little of such waste need be produced.

k The laboratory must be fitted with adequate fire-fighting equipment which is suitably located. The various types of fire extinguisher are described on page 46. As well as having suitable fire extinguishers, every laboratory should be equipped with a sand bucket and scoop and with a fire blanket.

l First aid kits must be provided in adequate numbers and kept in places accessible to teachers, technical staff and pupils. Pupils must be made aware of the location and contents of the kits and should be taught how to use them. Responsibility for maintaining the stock of first aid kits must be clearly defined.

The contents of a first aid kit should be as simple as possible and the following items are suggested.

    1 pair blunt-ended scissors
    Assorted bandages
    Triangular bandage (sling)
    Adhesive plaster and dressings
    Sterilised cotton wool
    Sterilised gauze
    Mild antiseptic solution
    Safety pins
    Small forceps
    Eye bath.

m Laboratory stools or chairs are not normally required when pupils are engaged in experimental work. The design of the laboratory should therefore allow for the adequate and safe storage of stools.

n A school laboratory is normally designed for teaching experimental science and, as such, is not the proper place to keep stocks of chemicals. Most schools have more than two science laboratories and, in such a situation, separate storage facilities should be incorporated into the design of the laboratory complex. If the storage area is also to be used as a preparation room, it must be sufficiently large and so located that equipment and chemicals can be safely carried into the laboratories, preferably by means of trolleys the same height as the laboratory benches. The DES states that stock quantities of common volatile liquids like carbon disulphide, ethanol, propanone (acetone) and methylbenzene (toluene) 'must be stored in a specially constructed store outside the laboratory'. Moreover, if such a store is not a separate building from the school, 'it must be designed to have at least one hour's fire resistance, and is best planned in conjunction with the Fire Officer'. For further suggestions and recommendations see page 45 and the DES publication, *Safety in Science Laboratories* (1976 edition), page 4. For details of the design of an outside store, see Everett, K., Hughes, D. (1975). For storage regulations, see page 49.

Particular attention must be given to the storage and use of reagents known to deteriorate with time. Ethoxyethane (diethyl ether) sometimes contains explosively unstable peroxides after prolonged storage and the formation of such substances in alcohols or alcohol mixtures under similar conditions has also been reported. Any peroxide may be detected by the liberation of iodine on the addition of acidified, aqueous potassium iodide to the suspect alcohol or ether. Testing for peroxide formation should be a routine procedure before distilling any alcohol or ether which has been stored for a long time, i.e. more than six months. Note that some fluids used in wet copying processes contain a very high proportion of an alcohol such as propan-2-ol.

The storage of alkali metals also needs special attention. Potassium has been known to corrode the aluminium cans in which it is usually stored in schools.

o Few schools can now afford to set aside a laboratory exclusively for sixth-form work. In such work, experiments sometimes have to be left running for a long time and provision should therefore be made for the necessary apparatus to be left in a position where it will not be a hazard (or a temptation) to other users of the laboratory. In biology, where the use of living material has often to be planned weeks or months in advance, it is sensible to have animal and plant houses separate from the main biology laboratories. (See page 5.)

p An adequate level of lighting is particularly important in laboratories. The number and location of windows should not be such that the laboratory is an unreasonably bright and hot place in which to work on a warm summer's day. Bunsen flames can be very difficult to see in bright sunlight.

q Caution must be exercised in the use of plastics in laboratory fittings, e.g. in waste pipes, in bench tops, or as the working surfaces of fume cupboards. The chemical and thermal resistance of some plastic materials is summarised in Table 5.8 on page 103. See also the section on laboratory sinks, page 141.

The above safety aspects of laboratory design should be considered together with the other basic design principles discussed in Chapter 1.

The hazards of glassware

Glassware is produced and used in vast quantities and there is considerable variation in quality. Large glass containers must not be handled by the neck and Winchesters of reagents need special care in this respect. Reagents in glass bottles must not be stored in direct sunlight, near radiators or on warm floors, and it should be remembered that liquids in spherical containers can act as a lens and focus enough sunlight to cause a fire. Broken glass needs to be treated with particular care and should be cleared up as soon as possible. Small slivers of glass may be collected by means of a piece of plasticine and broken glass should be placed in a clearly marked container used exclusively for the purpose. Chipped or cracked glassware should not be used in laboratory experiments.

Glass tubing remains the cause of many unnecessary accidents. Lengths of glass tubing should be carried vertically, *not* horizontally and such tubing should not be stored above head height. When cutting glass tubing, a glass file or knife is used to make the initial scratch on the glass. Pressure is then applied to break the glass, during which time the hands should be protected by a cloth. The cut ends of glass tubing should always be fire-polished by rotating them in a bunsen flame.

When passing glass tubing or a thermometer through a cork or rubber bung it is particularly important to use the correct technique. A cork-borer, lubricated with glycerol (propane-1,2,3-triol) and of slightly greater diameter than the tubing, is passed through the bung or cork. The tubing is then inserted into the borer and the borer withdrawn.

The same procedure may be used, in reverse, to remove a seized tube from a bung. See Figure 5.11, page 109.

Pipettes and burettes can be dangerous if they are not properly used. Mouth-operated pipettes are inexpensive to purchase but there is a strong case for *never* pipetting any liquid using mouth suction. Pipette fillers are readily available and they overcome the hazards which arise when a pupil fails to keep the tip of a mouth-operated pipette fully immersed in the liquid being sampled. The use of pipette fillers in schools is therefore to be encouraged. Mouth pipettes must *never* be used for drawing up volatile liquids, aqueous ammonia, concentrated acids or alkalis, toxic liquids e.g. ethanedioates, bacterial broths or radioactive materials. Pipettes should be washed immediately after use and be stored in a proper rack. Laying a pipette on a laboratory bench may lead to contamination of the mouthpiece.

When properly clamped, the top of a burette is likely to be above the head of the operator. The burette should be released and brought down below eye level before any attempt is made to fill it using a funnel. Pupils should be forbidden to climb on stools to fill burettes.

Glass stoppers which have become jammed should be loosened by tapping *gently* with a wooden block wrapped in a soft cloth, or, as long as the contents are suitable, by running warm water over the neck of the bottle. In either case, the hands should be protected by gloves or a cloth.

Glass ampoules, e.g. of bromine, should be opened behind a protective screen with eye protection and gloves as appropriate. The heat of the hand can cause considerable pressure to build up in an ampoule of a volatile liquid so that they should not be handled more than is necessary.

It must be emphasised that glass is not a totally inert material. It should, therefore, be cleaned as soon as possible after use, using the mildest possible cleaning agent in the first instance, e.g. warm soapy water. Reagents such as 'chromic acid' should be avoided; modern cleansing agents are safer and more effective.

All glassware should be inspected regularly for flaws, and any that are detected should be repaired or the item rejected. On no account should laboratory glassware be used as drinking vessels.

For many routine operations carried out in school science teaching, glass apparatus is not essential and may be replaced by a less hazardous alternative, e.g. plastic measuring cylinders, beakers and tubing. Note, however, that plastic containers become very slippery in moist or wet hands and that, if dropped, they bounce rather than break so that the contents are ejected upwards.

The use of gas cylinders

Gas cylinders are widely used in schools and most accidents involving cylinders arise from inadequate storage or misuse. Gas cylinders must always be clamped in position, held in a trolley, or laid horizontally on the floor and wedged to prevent rolling. They must not be subjected to heat (e.g. radiator, sunlight) or to corrosive fumes.

A large gas cylinder is heavy but this does not necessarily imply great strength. Particular care should be taken to avoid bumping a cylinder, e.g. when taking it up or down stairs.

Before use, the valve of every cylinder should be checked. Cylinder valves must always be opened smoothly and slowly; a wrench or hammer must never be used to exert pressure on the standard key. If hand pressure is insufficient to open a main cylinder valve, the cylinder should be returned to the suppliers with an explanatory note. No attempt must ever be made to oil or grease a cylinder valve. Valves suspected of leakage should be tested with soapy water.

Cylinders, valves, pressure reducers and gauges for *combustible* gases (e.g. hydrogen), have outlets and fittings screwed with a *left-hand thread*. Those for non-combustible gases (e.g. nitrogen), are screwed with a right-hand thread. In all cases, the main cylinder key should be fastened to the cylinder or trolley so that it may be used to cut off the supply of gas from the cylinder in an emergency. A gas cylinder must never be connected to any apparatus without first establishing and controlling the rate of gas flow.

There is a British colour code for identifying the contents of a gas cylinder. This is not followed internationally, so that the printed word is a more reliable guide to the contents of a cylinder. Note that medical grade gases are stored in cylinders of a different colour from cylinders storing the same gas of industrial quality.

Some schools find it convenient to purchase 'lecture bottles', i.e. small cylinders, of some gases. These small cylinders and their fittings should be treated with the same respect as the larger cylinders. They are most safely stored in a rack which is designed to protect the valve and fittings from damage. Gas cylinders must not be stored in a closed space which would allow a high concentration of leaking gas to develop.

It should be noted that some Local Education Authorities forbid their schools to purchase certain gases in cylinder form.

Toxic reagents

The range of toxic materials is very great and includes many more substances than those which are scheduled officially as poisons. It is, therefore, sensible to keep scheduled poisons in a locked store, to label clearly all toxic materials and to insist on proper laboratory procedure and subsequent hygiene. Further, only minimum qualities of scheduled poisons should be stored in a school.

Toxic substances may enter the body by inhalation, ingestion or absorption. Inhalation is by far the most common mode of entry and, in some instances, e.g. carbon monoxide, the transfer from the lungs to the bloodstream may be very rapid. Table 3.1 quotes the maximum permissible concentrations of some materials found in school science laboratories. The values are taken from the Department of Employment, *Threshold Limit Values* and from the recommendations made by the USA National Institute for Occupational Safety and Health.

It is obviously important to use proper experimental technique, to have adequate natural and mechanical ventilation, and to give conscious thought to minimising or overcoming the hazards associated with a particular procedure. Toxic gases such as chlorine and nitrogen (IV) oxide should not be prepared in large quantities on an open bench, and experiments should be designed to minimise the amounts of toxic material consumed or produced. Wherever possible, toxic reagents should be replaced by less toxic alternatives.

Note that the use of benzene in schools and other educational establishments is the subject of a circular from the Department of Education and Science to Local Education Authorities. The advice given is that benzene should not be used at all as a solvent and that the use of benzene as a chemical reagent should be reduced to the absolute minimum (DES, M 72/0143, 2nd April 1974).

Mercury must also be used with care. The element may be absorbed through the skin but inhalation of the vapour is a more common mode of entry into the body. Everyone concerned with handling mercury in a school should be made aware of the hazards of its vapour and appropriate warning notices should be displayed. Where possible, an alternative should be used. But, if the use of mercury is essential, the hands must be protected and the procedure carried out under well-ventilated conditions. Mercury surfaces should not be exposed to the atmosphere for any length of time and the element should not be heated in contact with an open atmosphere.

Experiments involving mercury should be done on a tray to facilitate recovery of any spillage. Mercury from broken thermometers should be recovered immediately by means of a capillary tube attached to a flask and filter pump. Any which cannot be recovered from crevices or cracks should be sprinkled with sulphur. It is sometimes convenient to treat spilt mercury with a mixture of solid carbon dioxide ('dry ice') and propanone and then sweep up the solid mercury.

It is rarely worthwhile purifying mercury in schools. A supplier will generally take back impure mercury with an excellent allowance against a new purchase. A film of dirt may be removed by shaking the mercury gently in a strong bottle containing a few pieces of adhesive tape.

Some laboratory reagents, e.g. ninhydrin, are now available in aerosol form. Many of the substances used in aerosols are potentially hazardous. To prevent inhalation or skin absorption of these substances, aerosols should be used only in a fume cupboard or an alternative procedure, e.g. 'dipping', should be adopted. Aerosols should be

Substance	Threshold limit value (ppm unless indicated)	Nature of hazard	Notes
Ammonia	25	Attacks mucous membranes and eyes. '0.880' solution is corrosive	Wear eye protection when handling aqueous ammonia
Aniline	—	See phenylamine	
Benzene	10	Attacks liver and kidney tissue; destroys bone marrow tissue; a chronic poison	Do not use this substance if it can be avoided, e.g. replace by less toxic methylbenzene wherever possible
Bromine	0.1	Vapour attacks lungs, eyes, throat. Liquid causes serious burns and can inflame sawdust	Store and/or prepare minimum quantities. Eye protection necessary. Handle ampoules with care
Cadmium	0.05 mg m^{-3}	Extremely toxic	Do not use cadmium or its compounds
Carbon disulphide	20	Liquid absorbed through skin; vapour toxic and substance has a very low flashpoint	Label any bottle of this substance 'Highly flammable' and do not store more than 500 cm^3 in a laboratory. Use a less hazardous substitute wherever possible
Carbon tetrachloride	—	See tetrachloromethane	
Chloroform	—	See trichloromethane	
Chlorine	1	Attacks eyes, nose, lungs. Used as a poison gas in World War I	Prepare and use minimum quantities of gas. Fume cupboard
Hydrogen sulphide	10	Paralyses olfactory nerves and thereafter cannot be smelled	Use only in a fume cupboard
Iodine	0.1	Vapour is toxic, attacks eyes	More dangerous because it is not generally regarded as a poison
Lead	0.15 mg m^{-3}	A cumulative poison. Pupils are exposed to lead poisoning from other sources e.g. car exhaust fumes	Take all reasonable precautions to minimise likelihood of lead contamination
Mercury	0.05 mg m^{-3}	Cumulative poison of all tissues but especially kidneys	Avoid spillages. Use a mercury tray. Do not heat mercury or its compounds in an open container, use a substitute, where possible, in manometers
Phenol	5	Toxic and caustic. Absorbed through the skin	Handle with caution and with adequate protection
Phenylamine	5	Absorbed through the skin. Very poisonous	Use minimum quantities
Sulphur dioxide	2	Corrosive and toxic	Protect cylinders and siphons from damage. Prepare and use gas in minimum quantities
Sulphuric(VI) acid	1	Highly corrosive and dangerous material	Use acid of minimum concentration adequate for the task in hand. Use appropriate technique for diluting the concentrated acid
Tetrachloromethane	10	Toxic and absorbed through the skin	Use in well-ventilated conditions or replace by a less toxic alternative e.g. 1,1,1-trichloroethane
Trichloromethane	50	Absorbed through the skin. Toxic as well as anaesthetic properties	As for tetrachloromethane
Trichloroethene	100	Absorbed through the skin. Toxic	As for tetrachloromethane

Table 3.1 Threshold limit values

stored in a cool place and any details or instructions accompanying them should be scrupulously observed.

Chemicals are most likely to be ingested as a result of inadequate technique involving mouth pipettes. Pipette bulbs, filter pumps or some other pipette-filling device should be used wherever possible. A serious cause of accidents involving poisoning by ingestion is the practice of transferring toxic materials from the container in which they are supplied to a bottle normally used for storing an innocuous consumable material, e.g. weedkillers stored in an orange juice bottle. There is no known antidote to a weedkiller such as paraquat, and some weedkillers, fungicides, pesticides and herbicides are dangerous to use in concentrations exceeding those which are recommended. There is, therefore, a clear responsibility on the science teacher to ensure that such materials are securely stored, adequately labelled and used only under proper supervision.

Attention is drawn to the fact that exposure to a substance of biological or chemical origin, even in a dose as low as 10^{-9} g, may cause damage to body proteins. Such a small initial 'sensitising' dose usually produces no observable effects but subsequent exposure to an even smaller 'challenging' dose can cause a virulent reaction and severe tissue damage. In addition, 'cross-sensitisation' of the skin is possible, i.e. exposure to one substance may render the skin susceptible to challenging doses of a larger number of chemically related materials.

1-bromo-2,4-dinitrobenzene is a powerful contact sensitiser sometimes used in school experiments and, as with the corresponding chloro- and fluoro- derivatives, precautions must be taken against skin contact, inhalation and ingestion. The DES recommends that:

'teachers should not use substances with which they are unfamiliar, especially biochemicals, unless they have been given authoritative advice about them. Such advice can be obtained from the Regional Medical Services Adviser of the Department of Employment'.

Asbestos

The use of asbestos and asbestos products in schools and other educational establishments is governed by the following.

DES, Administrative Memorandum 7/76, (Welsh Office 5/76), *The Use of Asbestos in Educational Establishments*, July, 1976

SED, Memorandum 6/68, *Inhalation of Asbestos Dust*, May, 1968

In accordance with the advice given in these publications, asbestos wool or cord, asbestos gloves and tripod gauzes with asbestos centres, should not be used in school science teaching. Hard asbestos mats, while preferable to the soft variety, should be replaced as soon as practicable, and the DES advises that educational establishments 'discontinue the use' of asbestos tape and paper.

It should also be noted that accidents have been reported when hot glass or metal has come into contact with hard asbestos mats. Such accidents appear to arise from a reaction between the hot material and the binding agent used in the mat and/or from thermal shock.

It is sometimes necessary to cut, drill, file or otherwise treat sheets of asbestos in a manner likely to produce asbestos dust. Such operations should be kept to a minimum and conducted under efficient local exhaust ventilation with the exhaust suitably filtered before discharge. For details, see the appropriate Memorandum and the publications of the Asbestos Research Council (P.O. Box 18, Cleckheaton, West Yorkshire, BD19 3UJ).

It is clearly sensible to avoid the use of asbestos and asbestos products in a school as far as this is possible. Asbestos mats may be replaced by ceramic tiles, hardboard treated with fire-resistant paint, or cement boards reinforced with glass fibre. Ceramic fibre paper is an adequate substitute for asbestos tape or paper, and stainless steel tripod gauzes may be used instead of the conventional asbestos centred variety. Rocksil or other mineral/ceramic wool may be used instead of asbestos wool. Platinised asbestos should be replaced by platinised Kaowool or a similar product. Chromated cowhide or fibre glass provides a satisfactory material for use in the manufacture of heat resisting gloves.

Chemical hazard warning labels

Partly as a result of United Kingdom membership of the EEC, there has been a significant increase in the use made of chemical hazard warning signs in labelling laboratory reagents. These signs, intended to indicate the nature of the danger(s) associated with a given material, are based on the following six-fold, EEC classification.

Class I explosives and unstable substances not used as explosives, e.g. picrates, azides, methyl-2,4,6-trinitrobenzene (TNT).

Class II oxidising materials which give rise to a highly exothermic reaction when in contact with other substances, e.g. fuming nitric(V) acid, potassium chlorate(V) and potassium chlorate(VII).

Class III flammable substances. This Class includes five sub-classes of flammable materials.

very flammable solids
very flammable liquids
very flammable gases
spontaneously flammable substances
substances which emit flammable gases on contact with water

Many hydrocarbons fall into this category together with alkali metals, white phosphorus and calcium(II) dicarbide (acetylide).

Class IV toxic and harmful substances. Toxic materials are those which offer a serious risk of acute or chronic poisoning by any route. Harmful substances are those which, while less toxic, offer risks to those exposed to them.

The number of compounds which fall into this class is very large and includes pyridine, lead(IV) oxide, manganese(IV) oxide and phenol.

Class V corrosive and irritant materials. Corrosive substances are defined as those which destroy living tissue, and an irritant material is one which causes inflammation of the skin. Thus, phenol, potassium dichromate(VI), and concentrated sulphuric(VI) acid are all members of this class. 'Corrosive' materials carry a different warning symbol from those categorised as 'irritant' for which the same sign as in Class IV harmful substances is used.

Class VI radioactive materials.

The signs used to identify each of these classes are reproduced in Figure 3.1.

Many substances clearly belong to more than one of the six major classes. By convention, the number of warning symbols is restricted to two. When two symbols are used together, one is the symbol for Class I, II or III and the other is the symbol for Class IV or V.

Within member countries of the EEC, the authoritative source of information about the symbol(s) applicable to individual substances is the Council of Europe publication, *Dangerous Chemical Substances and Proposals Concerning their Labelling* (4th edition). By means of a system of serial numbers, summarised below, this publication also records the nature of the risk(s) associated with each of the substances classified, e.g. 'danger of cumulative effects'.

Class	Serial numbers
I	R1 to R5
II	R11 to R13
III	R21 to R35
IV	R51 to R72
V	R81 to R84

Where the serial number is preceded by the letter S, safety directions are also provided. These relate to storage, containers, inhalation, contact, personal protection, cleaning, fire, first aid etc. as appropriate.

For other sources of information about the hazards associated with laboratory materials, see the Bibliography at the end of this chapter.

Attention is also drawn to the following.

i The Association for Science Education publishes 'hazard labels' intended for use in school science

I Explosive substances

II Oxidising substances

III Flammable substances

IV Toxic substances

IV Harmful substances

V Corrosive substances

VI Radioactive substances

Figure 3.1 Warning signs recommended for use in labelling bottles of reagents

laboratories. These do not make use of hazard symbols but give explicit written details of the hazards associated with a given substance and of the appropriate emergency action to be taken in the event of an accident in which that substance is involved.

ii The symbols illustrated in Figure 3.1 differ from those found on many bulk-carrying vehicles and tankers, which are reproduced in Figure 3.2. The EEC labelling directives do not cover transport and the two systems will operate 'side by side' for the foreseeable future.

iii Section 6, paragraph 1(c) of the Health and Safety at Work Act, 1974, makes it the duty of any person who 'designs, manufactures, imports or supplies any article for use at work' to 'take such steps as are necessary to secure that there will be available in connection with the use of the article at work adequate information about . . . any conditions necessary to ensure that, when put to that use, it will be safe and without risks to health'.

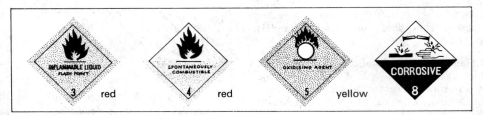

Figure 3.2   Warning signs used on road tankers and transport vehicles

Toxic plant material

The range of toxic plant material is much greater than is commonly realised and a school biology laboratory is likely to contain a considerable amount of such material, e.g. castor oil seeds. Also, the hazards associated with some garden plants are not as widely recognised as they should be. Table 3.2 lists some of the more common toxic species of plant material.

Common name	Botanical name	Poisonous parts
Garden flowers		
Aconite (winter)	*Eranthis hyemalis*	All
Christmas rose	*Helleborus niger*	All
Foxglove	*Digitalis purpurea*	All
Iris (Blue flag)	*Iris versicolor*	All
Larkspur	*Delphinium ajacis*	Foliage and seeds
Lily of the Valley	*Convallaria majalis*	All
Lupin	*Lupinus* sp.	All
Monkshood	*Aconitum anglicum*	All
Narcissus (daffodil, jonquil)	*Narcissus* sp.	Bulbs
Garden vegetables		
Potato	*Solanum tuberosum*	Green sprouting tubers and leaves
Rhubarb	*Rheum rhaponticum*	Leaves
Shrubs and trees		
Broom	*Cytisus (Sarothamnus) scoparius*	Seeds
Cherry laurel	*Prunus laurocerasus*	All
Laburnum (Golden Rain)	*Laburnum anagyroides*	All
Rhododendron	Species: *Azalea* American laurel Mountain laurel	Leaves and flowers
Yew	*Taxus baccata*	All; seeds lethal
Snowberry	*Symphoricarpus albus*	Fruits
Hedgerow plants		
Black nightshade	*Solanum nigrum*	All
Buttercups	*Ranunculus* sp.	Sap
Deadly nightshade	*Atropa belladonna*	All
Privet	*Ligustrum vulgaris*	Berries
Thorn apple	*Datura stramonium*	All
Marshland plants		
Hemlock	*Conium maculatum*	All
Hemlock, Water dropwort	*Oenanthe crocata*	All
Kingcup, Marsh marigold	*Caltha palustris*	Sap

House plants		
Castor oil plant	*Ricinus communis*	Seeds
Dumb cane	*Diffenbachia* sp.	All
Hyacinth	*Hyacinthus* sp.	Bulbs
Poinsettia	*Euphorbia pulcherrima*	Leaves and flowers
Woodland plants		
Cuckoo pint (wild arum)	*Arum maculatum*	All
Mistletoe	*Viscum album*	Fruits
Oak	*Quercus* sp.	Fruits and leaves
Poison ivy	*Rhus toxicodendron*	All
Toadstools	*Amanita muscaria* *A. pantherina* *A. phalloides*	All

Table 3.2   Some poisonous plants

Carcinogens

Some chemicals, notably some aromatic amines, are carcinogenic. The use and storage of such amines in schools and other educational establishments is forbidden, (see DES *Administrative Memorandum*, 3/70 and SED Circular No. 825).

The possibility of generating a carcinogenic or suspected carcinogenic substance as a by-product of a routine laboratory reaction should not be overlooked. Methanal (formaldehyde) and hydrogen chloride react rapidly in air to form an appreciable concentration of *bis*-chloromethyl ether, a substance known to induce tumour-formation in rats exposed to an inhaled air concentration as low as 0.1 ppm. This substance is also formed in aqueous and non-aqueous media and in Friedel-Crafts' reactions using methanal and a variety of metal chlorides. Hence, the possibility of its formation must be considered whenever methanal and hydrochloric acid come into contact either by design or fortuitously, e.g. cleaning biological glassware.

Note that a number of substances such as iodomethane (methyl iodide) and (chloromethyl)benzene, (benzyl chloride) have been shown to induce cancer in laboratory animals.

Many chemicals, e.g. benzene, phenylamine (aniline), and tetrachloromethane (carbon tetrachloride), are absorbed through the skin. If their use cannot be avoided, appropriate laboratory practice and hygiene is of particular importance.

Caustic and corrosive materials

Caustic and corrosive materials (which may also be highly toxic) are widely used in schools. Particular care is needed in handling the following categories of substance.

a Acids, e.g. nitric(V), methanoic (formic), phosphoric(V), and sulphuric(VI). Acids should be diluted by adding the acid to water and not vice versa. Protective clothing and an eye shield should be worn.

b Bases, e.g. sodium hydroxide, potassium hydroxide, and '0.880' aqueous ammonia.

c Substances which yield corrosive materials with water, e.g. acid chlorides, aluminium chloride, and sulphur dichloride oxide (thionyl chloride).

To these categories should be added bromine, phenol, phosphorus and sulphur dioxide.

Uncontrolled chemical reactions

The following list contains a number of categories of hazardous materials or mixtures which can react suddenly and violently.

Strong or concentrated acids with strong or concentrated bases.
Oxidising agents with metal powders or reducing agents.
Alkali metals or alkaline earth metals with water, acids or chlorinated solvents. Thus sodium or potassium must not be used for drying trichloromethane (chloroform) or tetrachloromethane (carbon tetrachloride).
Metal hydrides.
Hydrocarbons with halogens, chromic(VI) acid or sodium peroxide.
Concentrated nitric(V) acid with ethanol.
Propanone (acetone) with trichloromethane (chloroform). (Note that this mixture is often used to construct boiling point composition curves).

Note that explosions have occurred when small pieces of lithium have been heated in air, especially in conditions of high humidity. For further information about hazardous reactions, see the following.

Bretherick, L. (1975) *A Handbook of Reactive Chemical Hazards*, Butterworths.
Everett, K., Jenkins, E. W. (1976) *A Safety Handbook for Science Teachers*, John Murray.
National Fire Protection Association (USA) *Manual of Hazardous Chemical Reactions*, NFPA, Boston, USA.

It is important that science teachers control the amounts of materials used by pupils in experimental work and take all necessary steps to ensure that contamination of reagents does not occur. In general, liquid–liquid reactions are easier to moderate than solid–liquid reactions.

Plastics and polymers

The number of plastic and polymeric materials available to, and used in, schools has grown rapidly in recent years. The following points should be noted.

a Only small quantities of plastic materials should be heated for test purposes. Polystyrene produces toxic styrene vapour on heating and PVC yields hydrogen chloride. Strong heating of acetal resins leads to the evolution of methanal (formaldehyde) and acrylic materials decompose to give methyl methacrylate (2-methylpropenoate).

b Expanded polystyrene should be cut by a hot wire which is not sufficiently hot to cause the liberation of the toxic monomer. Expanded polystyrene should not be sanded or sawn except under conditions where the dust can be prevented from entering the lungs.

c Protective gloves must be worn whenever hot plastics, resins or adhesives are handled in the laboratory.

d Where polymerisation reactions are catalysed by peroxides, lauroyl peroxide should be used in preference to benzoyl peroxide.

e A number of organic reagents (e.g. trichloromethane, 1,2-dichloroethane, glacial ethanoic (*acetic*) acid) are commonly used to bond acrylic materials. The hazards associated with these materials are well known and appropriate precautions should be observed. Note that solutions of acrylic materials in ethanoic acid can cause severe and delayed blistering of the skin.

f Friction welding of acrylic materials should not be undertaken in school science workshops or laboratories.

g Poly(tetrafluoroethene), PTFE, should not be overheated as this may lead to the influenza-like symptoms of 'polymer-fume fever'.

h Methyl ethyl ketone peroxide, MEKP, is often used in conjuction with an accelerator of cobalt naphthenate to catalyse the setting of resins. MEKP and cobalt naphthenate together constitute an explosive mixture. Hence, it is advisable to use resins which incorporate the accelerator and require only the addition of the MEKP catalyst.

j The DES advises that polyurethanes should not be prepared in schools and states that the preparation of polyvinyls 'must not be attempted'.

Radioactive materials

The use of radioactive materials in schools in England and Wales is governed by Administrative Memorandum 2/76, May 1976 and the associated notes for guidance. Teachers in Scotland are referred to SED Circulars Nos. 689, 852 and 882. These documents are essential reading for any teacher who wishes to work with radioactive materials in a school, establishment of further education, or college of education. Where appropriate, teachers should also consult Williams, T. G. (1974) *The Use of Plutonium-239 Sources in Schools and Other Educational Establishments*, HMSO. It should be noted that Administrative Memorandum 2/76 cancels Administrative Memorandum 1/65.

In England and Wales, approval of the Secretary of State is not required for work with the following.

a The elements uranium, thorium, potassium and rubidium and compounds of these elements which are available through the normal chemical suppliers. It is, however, necessary for work with these substances to comply with the provisions of the Radioactive Substances (Uranium and Thorium) Exemption Order 1962 (Statutory Instrument No. 2710, 1962) and the Radioactive Substances (Prepared Uranium and Thorium Compounds) Exemption Order, 1962 (Statutory Instrument No. 2711, 1962). These orders relate to the amounts of these elements and their compounds which may be used and stored and to the amounts which must not be exceeded when arrangements are made to dispose of radioactive waste. The mass of all the uranium and thorium contained in all the waste disposed of in one day must not exceed 100 g.

b Equipment containing radioactive sources with activities of the order of 3.7 kBq (0.1 μCi) or less.

c Luminous painted surfaces which have protective coverings and which comply with the relevant parts of the Radioactive Substances (Luminous Articles) Exemption Order 1962. The related Statutory Instrument is No. 2644, 1962.

d Television receivers used in the normal way for viewing.

e Equipment in which electrons are accelerated by a potential difference not exceeding 5 kV.

f Other equipment containing cathode ray tubes in which electrons are accelerated by more than 5 kV, providing that the dose rate at a distance of 10 cm from the surface of the equipment does not exceed 0.1 millirem per hour. In general, teachers are advised not to purchase equipment of this type unless it is accompanied by a Certificate from the National Radiological Protection Board confirming that a test on an identical model showed it to conform with this dosage requirement.

The approval of the Secretary of State *is required* for all other work with radioactive and ionising sources. There are two categories of approval, A and B, the latter being for the less potentially hazardous work. Advice on category B work is given in the Notes of Guidance which accompany Administrative Memorandum 2/76. Teachers working with category A are referred to the *Code of Practice for the protection of persons exposed to ionising radiations in research and teaching*, HMSO. It is necessary to apply for approval in the category appropriate to the work to be carried out, and re-application is necessary if it is proposed to change the category. The relevant forms are IR(A) and IR(B).

Each of the two categories, A and B, contains three sub-categories, as follows.

Category A

A1 High-level work with radioactive sources and X-ray equipment.
A2 High-level work with radioactive sources and low-level work with approved X-ray sets.
A3 High-level work with X-ray equipment and low-level work with approved radioactive sources.

Category B

B1 Low-level work with approved radioactive sources—open and closed—and approved X-ray sets.
B2 Low-level work with approved radioactive closed sources only and approved X-ray sets.
B3 Low-level work with approved radioactive closed sources only.

Within each of these categories and sub-categories certain types of open and/or closed sources and work with certain types of equipment are permitted. Full details will be found in Administrative Memorandum 2/76, which also specifies the qualifications required by teachers who wish to work with radioactive sources and X-ray equipment.

Teachers should note the following points from the Memorandum.

a Where classes include pupils under sixteen years of age, the use of ionising radiations requiring the approval of the Secretary of State must consist of demonstration only.

b Pupils under the age of sixteen must not be allowed in laboratories where older pupils are conducting experiments with ionising radiations, except when specifically approved closed sources which have an activity of not more than 370 kBq (10 μCi) are being used.

c Responsibility for ensuring safe working with ionising radiations rests with Local Education Authorities in respect of maintained schools and with the governing bodies or proprietors of independent schools.

d The transport of radioactive substances by road e.g. between different sites of the same school or by student teachers on teaching practice, is governed by the provisions of the Radioactive Substances (Carriage by Road) (Great Britain) Regulations, 1970, Statutory Instrument No. 1826, 1970. In transporting substances of low activity, it may be possible to obtain exception from these provisions. Reference should be made to the relevant Statutory Instrument.

e The following provisions are made for student teachers using ionising radiations on teaching practice:

i the sources to be used must be closed sources relevant to work in category B3, unless the student teacher is qualified for work with ionising radiations in another category. The Memorandum states the kinds and levels of qualification required for work within each category;

ii the consent of the head of the establishment in which the teaching practice is undertaken is required before the work with ionising radiations is initiated;

iii if the work done by the student teacher on teaching practice is with sources or equipment brought by him or her to the practice school/college, the school or college may need to apply for permission to use such sources or equipment on an occasional basis;

iv a member of the science staff of the school must be present when a student teacher is using radioactive sources or X-ray equipment;

v all sources and equipment brought by a student teacher to a school or college must be returned by him or her to the training establishment.

f When work is undertaken with ionising radiations for which approval is required, the school or college must know the name and telephone number of a hospital from which advice may be sought or to which casualties of a radiation incident can be referred.

g Schools which comply with the Radioactive Substances (Schoools etc.) Exemption Order, 1963 may gain exemption from the requirement under the Radioactive Substances Act, 1960, to seek authorisation for the accumulation or disposal of radioactive waste, providing that certain conditions are satisfied. Details are provided in paragraph 3 of Appendix 2 of Administrative Memorandum 2/76.

The *Notes for Guidance* associated with Administrative Memorandum 2/76 give advice on many aspects of working with radioactive materials. The following points should be noted:

i the normal laboratory rules (no eating, etc.), are of particular importance;

ii a laboratory coat (or other suitable protective clothing) should always be worn. If possible, the laboratory coat should be kept solely for work with radioactive material and monitored as appropriate;

iii hands must be scrupulously washed and monitored after each practical class. Disposable towels should be used;

iv mouth operations, e.g. licking labels, are forbidden;

v protective gloves (disposable polythene) should be worn;

vi if possible, keep a set of glassware exclusively for work with radioactive materials;

vii label all radioactive materials clearly with an appropriate warning notice. The recognised symbol for radiation is a black trefoil on a yellow background;

viii part of a laboratory should be set aside for work with radioactive materials. If this is not possible, work with such materials should be confined to a set of large plastic trays lined with paper towels;

ix it is essential to check the quantity of radioactive materials, whether these are stored or being disposed of, at appropriate intervals;

x work should not be undertaken by anyone having a cut, abrasion or other open wound.

X-rays

The use of X-rays in educational establishments in England and Wales is governed by Administrative Memorandum 2/76. Details of the requirements for demonstration X-ray and other equipment operating at 5 kV or more are given on page 8 of the Memorandum, which should be read in conjunction with the relevant parts of the accompanying *Notes for Guidance*. However, X-ray equipment with approved designs is available commercially and this is likely to be adequate for almost all school work. Schools may submit proposals for the approval of apparatus of their own design and such approval will involve an examination of the equipment by the National Radiological Protection Board or other approved agency. Teachers in Scotland are referred to Circular No. 689.

Electrical hazards

The most obvious hazards associated with the use of electrical equipment are electric shock, burning, radiation damage and fire.

The lowest level of current which is detectable by the skin is about 1 mA at 50 Hz or 5 mA d.c. Any increase in current beyond about 20 mA at 50 Hz or 80 mA d.c. is a serious threat to life. However, it is important to stress that the electrical resistance of the human body varies very significantly from one person to another and is different in the same person at different times and under different conditions. Moreover, the electrical resistance of the skin depends markedly on whether it is dry or moist. Even with a skin resistance of 10 000 ohms, a 230 V supply could result in a current of 20 mA which could be lethal. For some particularly sensitive individuals, 60 V could produce a fatal current.

The above considerations apply when an electrical current passes through the body. In practice, a part of the body, such as a finger or hand, may short-circuit two conductors of different electrical potential. This may produce a less severe electric shock but it may also lead to severe, localised electrical burns.

Capacitors present particular hazards as they may recover a proportion (up to 10 per cent) of the original charging energy if left on open circuit after discharge. Capacitors should be kept individually short-circuited when being stored and all capacitors should carry a suitable warning notice adjacent to the terminals.

The risk of receiving an electric shock from electrical circuits or equipment can, of course, be minimised by adequate design, appropriate servicing and proper use. Switches should break the live circuit and equipment should be fitted with a distinctive on/off light. 'Fail-safe' devices should be incorporated wherever possible. The proper earthing of electrical equipment is essential. No attempt must ever be made to repair or adjust electrical equipment without disconnecting it from the mains supply.

All equipment, particularly if it is portable, should be systematically and regularly inspected. Worn grommets should be replaced and cables should not be allowed to swing about inside an instrument casing.

It is important to ensure that all mains equipment carries an adequate fuse rating and that the connections to the mains are correctly and securely wired. Since 1st July, 1970, Great Britain and 18 other European countries, have operated the following colour code.

Earth: Green and yellow stripes
Live: Dark brown
Neutral: Light blue

However, the obsolescent British colour code (earth, green; live, red; neutral, black), will be associated with older equipment for some years to come. Such equipment should be fitted with the new international coloured cable at the earliest opportunity. It should be noted that it is still possible to purchase items of foreign manufacture which do not conform to the new convention.

The teacher's example is of particular importance when working with electrical equipment. Use low voltage wherever possible and avoid long, trailing leads, worn flex and makeshift connections. 2- or 3-way adaptors and extension leads will be unnecessary if the laboratory has an adequate number of power points.

Lasers

Lasers designed for school use are generally of low power. However, experiments with animals show that retinal damage begins at about $10^{-2} \mathrm{W\,cm^{-2}}$ *received on the retina*. Since the lens of the eye focuses light, a much lower intensity of incident light represents a hazardous level of illumination and the DES recommends a maximum of $10^{-6} \mathrm{W\,cm^{-2}}$ for exposure of the pupil of the eye for continuous-wave operation. The use of lasers in schools in England and Wales is governed by Administrative

Memorandum 7/70 *Use of Lasers in Schools and Other Educational Establishments*. Teachers in Scotland are referred to SED Circular No. 766, October 1970.

It is important to remember that any light surface will reflect laser pulses. Work with lasers should be carried out in well-lit rooms to avoid enlarging the pupils of the eyes. Lasers should carry an appropriate warning notice. It is an added safeguard if the laser has to be switched on by using a removable key rather than by a simple switch incorporated in the instrument casing.

Ultraviolet radiation

Ultraviolet radiation is absorbed by the cornea and conjunctiva of the eye. The inflammation which subsequently develops (4–8 days) is exceedingly painful. Hence, ultraviolet lamps must be properly shielded and eye protection must be worn if the source is exposed. No source should ever be viewed directly. Note that an electric arc produces a large amount of ultraviolet radiation and is as dangerous as an open source.

Noise

The effect of noise on an individual depends to some extent upon the individual but primarily upon the frequency and intensity of the noise. Ordinary speech represents a sound level of about 60 dB. A motor cycle or heavy vehicle produces about 85 dB of sound and represents the approximate level at which prolonged exposure causes some damage to hearing.

In the school situation, everything should be done to maintain as low a level of background noise as possible, e.g. fume cupboards should be quiet in operation. It should be remembered that continuous exposure to low levels of noise may cause damage to the hearing, reduce efficiency or simply produce an irritable working environment. If communication with pupils is impaired by excessive background noise, this could constitute a hazard.

Microbiological hazards

Experimental work with micro-organisms is now quite common in schools and an increasing number and variety of such organisms are being used. For details of the relevant hazards and of appropriate safety precautions see Chapter 9, page 281.

Keeping animals in schools

See Chapter 8 for a detailed discussion of many of the problems associated with maintaining animals in schools.

Experiments involving pupils

The use of pupils as experimental subjects poses a number of problems. It is essential that teachers familiarise themselves with the hazards associated with any experimental procedure in which pupils provide the raw data. It is equally essential that pupils are not compelled to provide the data or perform the experiment and that they understand fully the need for all appropriate precautions. The following points are of particular importance.

a Care is needed in discussing personal data collected from the pupils, e.g. eye colour or blood group. Genetic studies may lead a child to realise that he/she must be adopted or illegitimate.

b Care is also needed in discussing publicly individual variations within a class, e.g. a pupil's reaction time, weight or height.

c Pupils must never be asked to swallow any chemical or drug as part of an experiment. The tasting of chemicals is not to be encouraged and, in cases of doubt, the advice of the School Medical Officer should be sought. The familiar experiment involving phenylthiocarbamide (PTC) must not be conducted in such a way that the toxic solution enters a pupil's mouth. The method using impregnated paper strips should be used and pupils must not be allowed to taste more than two strips. The stock of strips, solution, or solid PTC must be rigorously checked and, when not required, kept in the poison/store cupboard. If PTC solution or solid is ingested, medical advice must be sought as a matter of urgency.

d Blood sampling carries the risk of transmitting disease, e.g. infective hepatitis, if sterile procedures are not used. Sterile lancets must always be used to obtain the samples and these must be disposed of after use. Sampling work of this kind must not be undertaken until approved by a qualified medical practitioner. Pupils cannot be required to provide a blood sample and prior approval of the pupils and of their parents must also be obtained. Note, however, that neither pupils nor parents can be expected to be fully aware of the possible hazards involved.

e Teachers should be aware of any abnormal susceptibilities of individuals in their charge if these are likely to cause difficulties in some experiments. It is appreciated that such awareness raises questions of medical confidence, but experiments involving the measurement of lung pressure for example may pose particular risks for asthmatics or epileptics. Similarly, children may be sensitive to known allergies likely to be encountered in biological studies. The coats of laboratory animals, *Primula* and *Pelargonium* species, grasses and some fungi are well known sources of allergies.

The stem, leaves, flower heads and sap of daffodils, hyacinths, jonquils, narcissi and tulips may act as irritants, causing painful lesions on the fingers. The cowslip (*Primula veris*) has been known to cause dermatitis and the pollen and glandular hairs of the common houseplant *Primula obconica* contain an irritant poison. Many members of the ivy family, especially the poison ivy *Rhus toxicodendron*, are known to contain an irritant sap. If an individual has to work with biological material to which he is known to be allergic, protective gloves and/or a 'barrier' cream may offer sufficient protection.

A note on allergy to locusts is available from the Anti-Locust Research Centre (College Lane, Wright's Lane, London, W8).

f The Department of Education and Science 'strongly advises' teachers that they should not perform or encourage pupils to perform, experiments involving the use of an electroencephalograph to determine so-called biological feedback.

Field work

There is a wide range of activities organised by schools but conducted outside school premises, e.g. archaeological 'digs'; visits to exhibitions, museums, shops, factories, farms or overseas countries; involvement with community service; sponsored walks; and residence at recognised field centres. These activities raise a number of organisational and safety issues, which are related in the fundamental sense that safety is concerned with planning and with the anticipation and identification of possible hazards. Whatever the nature and duration of the work conducted out of school, it is essential to know the individual intellectual and physical capabilities of the pupils involved and to be familiar with all the hazards that the undertaking is likely to present. Teachers contemplating school outings will find the following publications useful.

Schools Council, (1972) *Out and About: A Teacher's Guide to Safety on Educational Visits*, Evans/Methuen Educational
DES, (1977) *Safety in Outdoor Pursuits*, HMSO

Science teachers are likely to be concerned primarily with field work and industrial visits. Field work conducted at a residential centre some distance from the school presents different problems from those associated with an afternoon visit or an occasional lesson 'out of doors'. Whatever the circumstances, the teacher must be familiar with any regulations concerning the conduct and insurance of school parties which may have been drawn up by the Headmaster and LEA or other employing authority. Recognition of a field trip or other excursion as an official school function has important legal implications (see page 51). It should be noted that while some LEAs provide detailed guidelines or rules about the conduct and composition of school parties, these guidelines or rules may vary from one Authority to another.

Parents should always be given adequate warning, preferably in writing, if their children are to be out of school for any length of time or if the school day is likely to be prolonged as a result of an occasional visit. Parents are sometimes asked to sign an indemnity form which at least ensures that they understand the nature of the school visit and the reasons for it. However, seeking such indemnities may cause unnecessary problems (see page 51).

The teacher must be aware of all the hazards presented by a field environment, in the broadest sense of that term. Particular consideration should be given to the following situations.

a Seaside locations. Tidal dangers should be considered when planning visits to seaside locations. Local advice or detailed local knowledge is often required. Information concerning the timing of tides, obtained from a tide timetable, should be regarded as subject to considerable local fluctuations.

Cliffs and caves pose hazards. Cliff walking must be rigidly controlled and climbing or caving forbidden without proper equipment and preparation. Swimming or bathing in the sea should only be allowed when conditions are suitable and when there is a sufficient number of qualified adults to mount a rescue operation if this becomes necessary. Again, local knowledge of tides, currents, depths of water, etc., is essential before a teacher allows pupils to swim in the sea.

b Forest and moorland areas. The country code should be followed at all times.

c Industrial visits. When taking pupils to factories, teachers should find out about the safety rules which apply in the factory and insist that pupils abide by them.

d Tents and caravans. There must be adequate sanitary arrangements and the teacher must assume responsibility for filling fire buckets and ensuring that they are properly located. A large party of schoolchildren living in tents or caravans will produce a considerable quantity of waste paper, empty bottles and tin cans. There must be proper provision for the collection and disposal of such waste.

e Rough country and steep slopes. Rough country, e.g. mountain and upland terrain, should be treated with respect. Field work conducted in such an environment requires comprehensive and careful planning. Everyone involved in the expedition must have an adequate knowledge of the field environment and, if a particular route is to be followed, each pupil should have a copy of that route and be able to follow it.

It is essential that the teacher is aware of the physical, intellectual and emotional demands that can reasonably be made on pupils in his charge. Young persons are, physiologically and psychologically, less resilient than healthy adults and, on no account, should muscular development be confused with physical fitness. Hence, an

adequate safety margin must be allowed in calculating walking times or the distances to be covered in a given time.

This safety margin must recognise that what appears to be an innocuous slope in warm sunshine can, within an hour, be converted by a sudden weather change into an extremely hazardous environment. Knowledge of weather forecasts is essential but inadequate when planning field trips to mountainous or exposed areas. Local advice must always be sought and all members of the expedition should have clothing and footwear which are adequate for the most adverse conditions. Each member of the group must also be aware of what he is required to do if the weather deteriorates seriously and rapidly. Fog, mist and snow are particularly common hazards in mountainous and upland regions. It follows that every party engaged in field work in a potentially hazardous environment must be supplied with a reserve of warm clothing, some emergency rations, a first aid kit, a compass and some means of summoning help in an emergency, e.g. a whistle and torch.

It is recognised that a scientific field trip is a somewhat different undertaking from such activities as caving, potholing, orienteering, etc. Nonetheless, the advice given in the Schools Council and DES publications referred to on page 43 should be considered by all science teachers who are planning field work of whatever duration.

Fire

The following are among the sources of fire hazard encountered in school science laboratories.

a Ignition of solvent vapours, e.g. ethoxyethane (diethyl ether), carbon disulphide. Some of these vapours ignite well below red heat and carbon disulphide vapour may be ignited by a hot radiator. Flammable, volatile liquids should not be poured from one container to another near a naked flame. If such liquids are to be heated, a water bath must be used and adequate precautions taken to ensure that the vapour is efficiently condensed. Such precautions include adjusting the rate of distillation to avoid liquid accumulating in the condenser.

On no account should flammable, volatile liquids be stored in a conventional, domestic refrigerator. Refrigerators should be modified when purchased, to remove spark sources from the cold chamber. The amount of vapour required to produce an explosive mixture in the confined space of a refrigerator is very small. Aerosols should also be stored away from sources of heat.

b Ignition by reactive chemicals, e.g. white phosphorus, alkali metals and their peroxides. Large quantities of such chemicals should not be stored in schools and the quantities which are kept should be stored in a fire resistant store. Such storage should also ensure that powerful oxidising agents such as chlorates(V), chlorates(VII), nitric(V) acid, peroxides and manganates(VII) are kept well separated from such substances as carbon, sulphur, paper or sawdust. Chloric(VII) acid, $HClO_4$ is an extremely dangerous oxidising agent and is unlikely to be used in schools.

c Uncontrolled chemical reactions. For a list of hazardous chemical reactions see the sources referred to on page 39. It is obviously essential that a teacher rehearses all the experiments that he and his pupils eventually undertake and that only the minimum quantities of materials are used. Particular care is needed in igniting a jet of hydrogen, in heating ammonium nitrate(V), in conducting thermit reactions, in preparing mixtures for analysis and in handling peroxides.

Hydrogen is involved in a large proportion of school laboratory accidents. *Before* lighting a jet of hydrogen, a sample must always be selected and tested to see that it is free from air. A similar testing procedure is necessary when town gas, rather than hydrogen, is used to reduce a metallic oxide. Where natural gas is supplied to laboratories, it may be necessary to use a hydrogen cylinder, although some schools have adopted the undesirable practice of bubbling the natural gas through methanol or ethanol to increase the reducing property of the gas flow. In these circumstances, it is essential that all air is removed from an apparatus before attempting to ignite the gas mixture.

For a summary of safeguards in the laboratory preparation of hydrogen, see *Education in Science*, November 1974, page 33.

d Inadequate storage and disposal techniques. Heating blocks, charcoal blocks, tripods, etc., must be thoroughly cooled in water before being put away. Phosphorus residues may be disposed of chemically (e.g. aqueous copper sulphate or bromine water) or by ignition outside the laboratory. White phosphorus must not be allowed to come into contact with plastic materials, e.g. in a bowl or sink. Volatile, flammable liquids must not be stored near radiators or in strong sunlight.

Peroxides pose particular problems of storage. Benzoyl peroxide (benzenecarbonyl peroxide) is explosive when dry. It should be replaced by the much safer dodecanoyl peroxide (lauroyl peroxide) in polymerisation experiments (see page 39). If benzoyl peroxide has to be kept in a school, minimum quantities only should be purchased and kept in a desiccator containing water in the lower compartment. For more general advice, see the *Code of Practice for the Storage of Organic Peroxides* published by Laporte Chemicals, Ltd., Warrington.

e Local heating due to electrical or other faults. If an electric motor is stopped mechanically with the current flowing, overheating will occur. Similarly, fires have resulted from the inadvertent obstruction of electric fan blades and the overheating of microscopes with built-in lamps. All electrical equipment used in schools must carry appropriate fuses and, wherever possible, incorporate a fail-safe mechanism. In temporary electrical circuits, it is important that no component is required to pass a higher current than that for which it is rated.

f Loose clothing or hair ignited by bunsen burners. Bunsen flames are often difficult to see in bright sunlight, but long flowing hair or an unbuttoned laboratory coat undoubtedly constitute fire hazards *whenever* pupils are working with naked flames. Laboratory coats should always be fastened and hair tied well back on the head.

g Misuse of gas cylinders. Gas cylinder controls should be opened slowly and smoothly and suspected leaks should be examined using soapy water and not a flame. Cylinders should not be stored near radiators, in warm sunlight or near major fire hazards.

h Inadequate maintenance, e.g. of electrical equipment or circuitry. All items of equipment used frequently in science teaching should be inspected regularly. Fires have been caused by using perished rubber tubing to connect a bunsen burner to the main gas supply.

j Inadequate laboratory design, e.g. it should not be possible for pupils to accidentally set fire to wooden shelving by placing a bunsen in the wrong place on the laboratory bench. Blackout curtains should not be free to blow into the working space when windows are opened.

Every laboratory should be scrutinised from the point of view of safety and, in particular, an assessment made of its degree of fire risk. Inadequate provision of waste bins, inadequate or inappropriate fire-fighting equipment, etc., may either cause a fire or allow a trivial incident to develop into a major accident.

k Static electricity. Electrical charges can accumulate on metal drums from which a non-conducting liquid, e.g. industrial ethanol, is being poured. This danger may be avoided by earthing the bare metal surface of the drum.

l Inadequate temperature control. The temperature in a solvent store or a laboratory should not rise unreasonably on a hot, sunny day. Note that a flask containing a liquid may act as a lens and thereby focus enough heat to cause a fire. Similar hazards may arise with large mirrors.

Before any fire can begin or be sustained, there must be a source of fuel, a support medium (usually oxygen), and sufficient heat to bring the fuel to a temperature at which combustion can be maintained. If any one (or more) of these three factors is removed, a fire will go out.

Fires are sometimes classified according to the nature of the material(s) that are burning and it is essential to use

Class of fire	Suitable extinguisher(s)	Notes
A Ordinary combustibles, e.g. wood, paper, cloth	Water from a hose reel or a CO_2 expelled extinguisher. Soda-acid may be used or CO_2 gas if water is not available	Soda-acid causes corrosion
B Flammable liquid fires, e.g. oil, fat, organic solvents	Depending upon the location of the fire, use CO_2, a vaporising liquid extinguisher (BCF), foam or powder extinguisher or a fire blanket	*Never* use water. Always a danger of re-ignition with this class of fire
C Electrical fires	CO_2 or vaporising liquid	Switch off power before tackling fire. Powder extinguisher may be used but may damage equipment
D Metal fires	Conventional extinguishers may be inappropriate. Use *dry* sand or an appropriate dry powder extinguisher	Such fires are more likely with the use of powdered zinc, magnesium etc., in schools

Table 3.3   Extinguishers suitable for different types of fire

an appropriate type of fire extinguisher. Table 3.3 summarises the position. A new classification of fires has been recommended and is being implemented. For details see BS–EN2: 1972, *Classification of Fires*, BSI.

Every school science laboratory should be equipped with the following fire-fighting appliances:

i *a fire blanket* made of woven fibre glass. The fire blanket should be accessible to pupils as well as staff and be easy to reach and release. Fire blankets should not be used on glass apparatus where there is a risk of spillage and a consequent spreading of the fire;

ii *a sand bucket and scoop*. Sand buckets should be clearly marked as such and be fitted with lids. *Small* fires involving pieces of phosphorus are conveniently extinguished with dry sand;

iii *an adequate number of suitable fire extinguishers*. For most general purposes in a school, carbon dioxide extinguishers are adequate. If a water reel is installed, it is likely to operate automatically as the reel is unwound. It is obviously important that the reel has an adequate range of operation.

Carbon tetrachloride (CTC, tetrachloromethane) fire extinguishers are obsolescent and should never be used indoors. Both tetrachloromethane and its decomposition products are highly toxic and on no account must a CTC extinguisher be used on burning sodium or phosphorus.

All fire-fighting equipment should be located in that part of a laboratory where the fire risk is lowest, e.g. near a door, but accessibility of equipment must be kept in mind. All fire-fighting appliances should be checked regularly and, if necessary, the science staff should take the initiative in seeing that this is done.

When fighting a fire, human safety is always of paramount importance. However, discretion is essential in deciding the lengths to which first aid fire-fighting

should be carried. Portable fire-fighting equipment is not designed to cope with extensive fires and a laboratory must be evacuated as soon as fire or smoke threatens the means of escape or the building structure.

Pupils and staff must know precisely what to do in such an emergency, since it is too late to find out very much when an emergency actually happens. Class control is more likely to be retained if the teacher acts methodically and quietly. Most schools have clear instructions for leaving a building during an emergency and pupils should be familiar with these instructions and with details of the assembly procedure. If possible, laboratory windows and doors should be closed after the pupils have been evacuated. If the Fire Brigade are summoned, they must be told of any special hazards in the fire area, e.g. radioactive materials, gas cylinders.

First aid procedures

The purpose of first aid is to make the patient secure and comfortable and to prevent deterioration in his condition until any necessary professional medical assistance is available. First aid procedures should, therefore, interfere as little as possible with the injured person. The following conditions require *immediate* attention:

i severe bleeding;
ii absence of breathing;
iii eye injuries;
iv shock.

On no account should an injured person be given anything to drink (unless this is an essential emetic or antidote). If a general anaesthetic is required later this could cause vomiting with serious consequences.

Minor cuts should be washed thoroughly with water and a suitable sterile dressing applied. Antiseptics should not be applied. If the cut was caused by a rough dirty object, e.g. a rusty nail, the injured person should be taken to, or advised to consult, a doctor for an anti-tetanus injection or a toxoid boost.

Severe bleeding from a severed artery is recognised by a spurting flow of bright red blood. The flow may be restricted by applying pressure at an appropriate point on an artery between the wound and the heart. Teachers should be aware of those pressure points where the flow of blood may be stopped by compressing an artery onto the underlying bone.

When an injured person is not breathing, artificial respiration must be started at once. All teachers and pupils should be able to undertake some form of respiratory resuscitation, e.g. the kiss-of-life technique. Such skills are of value in the community generally as well as in the school environment. Details of how to carry out artificial respiration will be found in *First Aid*, the authorised manual of the St. John and St. Andrew Ambulance Associations and the British Red Cross Society. Many local Ambulance Associations will provide demonstrations and instruction for schools.

Eye injuries should always be regarded as potentially serious. If corrosive materials have been splashed into the eye, the eye should be held open and washed with copious amounts of water. It is easier to do this if a laboratory tap is fitted with a short length of rubber tubing so that a stream of water may be directed at the eye. However, a person with corrosive material in his eye will be in acute pain and considerable physical force may be necessary before it is possible to treat the eye with water. All eye injuries should be referred to a doctor, however trivial they may appear initially. The wearing of safety spectacles by pupils is an obvious precaution against eye injury and should always be considered. See page 31.

Injury is always associated with some degree of shock. Shock may be recognised by faintness, giddiness, complaint of blurred vision, collapse, pallor, clammy or cold skin or the breaking into a sweat, and anxiety. Shock can be a serious and even fatal condition and it requires prompt attention. The patient should be laid down and, if possible, the feet raised slightly higher than the head. Reassurance is essential and the patient's anxiety should, as far as possible, be allayed. In cases of severe shock, medical advice is essential. In general, patients in a state of shock should not be moved unnecessarily nor should they be kept unduly warm.

Probably the commonest type of accident in school science laboratories involves burning or scalding. Chemical burns should be washed with copious amounts of water and no attempt should be made to carry out neutralisation reactions on the skin. Some substances, e.g. phosphorus and bromine, cause severe burns and medical advice must be sought as a matter of urgency. Heat burns are accompanied by the loss of fluid from the blood into the tissues causing blisters to form. Small burns should be treated by cooling the injured areas as rapidly as possible using running water or ice packs. A suitable sterile dressing should be applied, but lotions, ointments and oily dressings should be avoided.

If a person's clothing is on fire, it is imperative that the victim is put into a horizontal position immediately. This will limit the spread of the injury. The burning clothing should be extinguished by water or by means of a fire blanket. Note that any charred material should be left in contact with the burnt area as it will be sterile. If more than 10 per cent of the body surface is burnt the injury should be regarded as very severe and arrangements must be made to get the injured person to hospital as a matter of urgency.

Poisoning by ingestion should never occur in school science teaching. The most common causes are inadequate operation of a mouth pipette or drinking the contents of a bottle in error. Where known poisons are being used, e.g. ethanedioates (oxalates), preventive measures are essential, but it is a wise safeguard to have access to a suitable antidote.

Clearly, it is helpful if all science teachers have a simple but adequate knowledge of first aid. It is most important that a teacher be able to recognise a situation which needs professional medical attention and be able to act in those situations where delay could be fatal. The names of teachers properly qualified in first aid treatment should be known to all members of a school.

The first aid kit, kept in or adjacent to every laboratory, should be accessible to pupils and staff. For a suggested list of contents, see page 32.

Teachers should be aware of abnormal susceptibilities of individuals within their charge, e.g. those suffering from haemophilia, diabetes, hypersensitivity, colour blindness or epilepsy. Again, matters of medical confidence are involved in such information, but it is difficult to see how teachers can exercise full responsibility without the appropriate knowledge of individual pupils, which may, in any case, be available to the Head teacher.

A major epileptic fit can be a frightening experience, especially to anyone who has not seen an epileptic attack before. The first requirement is to prevent the victim from causing himself injury. Convulsive movements should not be restrained, but the victim should be prevented from knocking into hard objects, e.g. desks or stools. When the convulsions subside, cradle the patient's head in the arms and loosen any tight clothing around the neck. Do not insert a ruler or other hard object into the victim's mouth during the fit, although, if an opportunity arises, the corner of a clean, folded handkerchief may be inserted between the teeth to prevent the tongue and lips from being bitten. Such biting is relatively rare and recovery from an epileptic fit is normally rapid. On some occasions, confusion, involuntary micturition and consequent embarrassment are involved. If the epileptic convulsions are not over within a few minutes, a doctor should be called. After an epileptic fit, the victim should take some rest and, at an appropriate point, his parents should be informed. However, each case must, to some extent, be treated individually as each person will often have his own pattern of fits.

Teachers are more likely to encounter *petit mal*—which needs no first aid—or a *psychomotor attack*, than a major epileptic fit. A pupil who suffers from *petit mal* may partly lose consciousness for a few seconds while important instructions are being given by the teacher. Such loss of consciousness may not be complete and the pupil may simply fail to hear what has been said. Many attacks of *petit mal* may occur during a school day. *Psychomotor attacks* are manifest by behaviour which is suddenly irrational or inappropriate to the circumstances in which the person finds himself, e.g. a pupil may smack his lips repeatedly or get up and run around the laboratory. When the attack has passed, the individual will not know what has happened during the seizure.

Further information and advice about epilepsy is available from The British Epilepsy Association (3–6 Alfred Place, London, WC1E 7ED). It should be noted that pupils suffering from some forms of epilepsy may be following a prescribed course of drugs and that these drugs may be brought legitimately to school so that they can be taken at the proper time.

Finally, it should be emphasised that an accident within a school usually provides somewhat fewer problems than an accident which occurs on a field trip or expedition. Adequate preparation for out-of-school activities must recognise this fact and must include an appropriate procedure to be followed in the case of injury to a teacher.

Some legal issues

Waste disposal

Much of the legislation governing waste disposal, water pollution, atmospheric pollution and noise has been incorporated in the Control of Pollution Act, 1974. The provisions of the Deposit of Poisonous Waste Act, 1972 will be wholly repealed once the 1974 Act comes into force.

Under the 1974 Act, a Disposal Authority (a County Council in England) has wide responsibility for providing adequate arrangements for the disposal of 'controlled waste' (i.e. commercial, industrial, household or any such waste) and for preparing a plan for all aspects of waste disposal. Depositors of waste must be licensed by a Disposal Authority, which may impose conditions for disposal in granting the licence. There are severe penalties for contravention of the provisions of the Act.

In schools, the quantities of toxic, noxious or otherwise hazardous material which must be destroyed are usually small. Where possible, therefore, such materials may be rendered harmless by the science teacher. For information on chemical disposal see Gaston, P. J. (1970) *The Care, Handling and Disposal of Waste Chemicals*, Northern Publishers, Aberdeen. Where materials cannot be disposed of easily and safely in this way, appropriate arrangements should be made by the Local Education Authority. The disposal of radioactive, carcinogenic or highly toxic materials raises special problems and the advice of the appropriate Disposal Authority must be sought.

Local fire brigades will sometimes agree to collect and destroy flammable waste solvents. Any large teaching institution, e.g. a University, produces vast quantities of such waste and will have made provision for its regular collection and disposal. It is sometimes possible for a nearby school to gain the co-operation of such an institution in disposing of flammable solvents.

School science accommodation must include provision for the safe collection of several different types of solid waste. Broken glass or porcelain must be kept in a clearly marked container, reserved solely for the purpose. Bacteriological or fungal cultures should be collected in a special container if the unwanted cultures cannot be destroyed immediately by autoclaving or rigorous chemical sterilisation. Reactive chemicals, especially powerful oxidising agents, should not be placed in waste bins. In general, stocks of hazardous materials such as potassium or white phosphorus should be kept at the minimum necessary to ensure adequate supplies for routine teaching purposes. Dissection material or other tissues may be disposed of by incineration or by sterilisation followed by pulverisation and discharge via a waste disposal unit.

Illegal experiments

A number of experiments are forbidden by law. The making of explosives and, in particular, the mixing of sulphur or phosphorus with chlorates(V) is illegal. The DES (*Safety in Science Laboratories*, 1976, p. 39) also identifies experiments which are 'quite unnecessary in a school course'. These include experiments with rockets and rocket fuels; the preparation of phosgene, hydrocyanic acid, cyanogen, nitrogen tri-iodide and the oxides of chlorine; the explosion of hydrogen and oxygen; the explosion of acetylene and oxygen mixtures; and sealed tube combustions. The DES publication also lists a number of experiments which need particular care but which, with appropriate precautions, may be reasonably and safely demonstrated in a school.

Alcohol regulations

There are stringent regulations governing the purchase and use of industrial methylated spirit and duty free spirit. Such spirits cannot be obtained without a requisition order from the office of the Department of Customs and Excise. It is necessary to give an undertaking that the material will be securely stored, used exclusively for science teaching and be accounted for in a return of the amount used for teaching purposes. Special permission is required to purify industrial methylated spirit or duty free spirit. Mineralised methylated spirit is, of course, readily obtainable without special permission, but it should be noted that the use of a still for alcohol requires a licence whether or not permission to obtain duty free spirit has been sought.

Storage regulations

There are specific legal requirements governing the handling and storage of flammable materials, gas cylinders and other potential sources of hazard. The relevant Acts of Parliament and Orders in Council were not drafted primarily with schools in mind, but many of the provisions will have been incorporated in rules or a code of practice drawn up by the Local Authority or Fire Office. Science teachers must, therefore, be familiar with these rules and abide by them.

Substance	Flash point °C
Acetaldehyde (ethanal)	−38
Acetone (propanone)	−18
Acetyl chloride (ethanoyl chloride)	4
Amyl alcohols	19–32
Benzene*	−11
Butan-1-ol and butan-2-ol	24–29
Carbon disulphide	−30
Crotonaldehyde	13
Cyclohexane	−20
Cyclohexene	−60
1, 2-dichloroethane	13
Diethyl ether	−45
Dimethyl ether	−41
Ethanal (acetaldehyde)	−38
Ethanol (ethyl alcohol)	12
Ethoxyethane (diethyl ether)	−45
Ethyl acetate	−4
Heptane	−4
Hexane	−23
Methanol (methyl alcohol)	10
Methylbenzene (toluene)	4
Methyl acetate	−10
Pentane	−49
Petroleum ethers	below −17
Propanone (acetone)	−18
Toluene (methylbenzene)	4
Xylene	14

*See page 34.

Table 3.4   Some common liquids with flash points below 32 °C

The regulations relating to the storage and use of radioactive materials (see DES Administrative Memorandum 2/76 and SED Circulars 689, 852 and 882) and of highly flammable liquids are of particular importance. The Highly Flammable Liquids and Liquified Petroleum Gases Regulation, 1972 (No. 917) came into effect on 21st June 1973. A 'Highly Flammable Liquid' (HFL) is defined as 'any liquid, liquid solution, emulsion

or suspension which, when tested in a specified manner . . . gives off a flammable vapour at a temperature of less than 32 °C and supports combustion'. Some common liquids with flash points below this temperature are listed in Table 3.4.

Quantities up to 50 litres (approximately 20 full 'Winchester' bottles) of highly flammable liquids may be kept in a preparation room, *provided* that they are kept in a suitably placed cupboard or bin made of fire resisting material and fitted with retention sills to contain any spillage. Larger quantities must be kept in a separate structure away from the school laboratories. For a discussion of wooden fire cabinets, see Stark, G. W. V., *et al.* (1971) *Chem. and Ind.*, pages 1173–1174.

All storerooms, cupboards, bins and other containers used for storing highly flammable liquids must be clearly and prominently marked with the words HIGHLY FLAMMABLE or FLASH POINT BELOW 32 °C.

Not more than 500 cm^3 of any one highly flammable liquid should be kept on any shelf in a school laboratory.

Poisons legislation

There are several Acts of Parliament governing the purchase, storage and use of Drugs and Poisons. Again, these Acts are not directed primarily at schools, where a Court would probably accept that pupils must have reasonable access to some materials which are poisonous. However, every care should be taken to store and monitor adequately the stock of poisonous materials and to ensure that pupils are allowed access to only such materials as are appropriate to their age, ability and experience.

Work with living animals

Vertebrate animals are protected by the Cruelty to Animals Act, 1876 and the Protection of Animals Act, 1911. If there is any doubt about the interpretation or applicability of these Acts, as far as any school experiment is concerned, the advice of the Home Office Inspectorate should be sought.

Science teachers involved with or responsible for work with any living organisms are strongly urged to consult the publications referred to on page 52. See also Chapter 8, page 196.

Radioactive materials

The Administrative Memorandum 2/76 and its notes and appendices are essential reading for any science teacher who wishes to undertake work involving radioactive material in a school. See pages 40–41.

Contractual issues

In addition to the specific items of legislation referred to above, the teacher is an employee, usually of a Local

Education Authority, and as such is subject to the laws governing employer-employee relationships. In addition, a teacher is often regarded as acting *in loco parentis*, since he is assumed to have particular responsibilities towards children when they are in his charge. It should be noted that the implementation of the provisions of the Health and Safety at Work etc. Act, 1974, will impose additional and specific contractual obligations on both employers and employees (see page 51).

On appointment to a school, a teacher normally signs a written contract which defines the agreement between himself and his employer. The contract may also be assumed to incorporate any further particulars supplied when the post was advertised or quoted in a letter of appointment. In direct grant and maintained schools, every full-time teacher must be supplied with a copy of his contract and he has the right of access to any regulations referred to therein.

Science teachers—like other teachers—may be involved, by the nature of their profession, in legal processes which can be conveniently considered as falling into three categories.

a Arising from the law of negligence. A teacher is liable for the results of an accident to a pupil in his charge only if it is established that he has been negligent in some way. In this context, negligence could be taken to mean failure 'to take such reasonable care of his pupils as a careful father would take of his children, having regard to all the circumstances'. It is essential that all necessary precautions be observed by the science teacher when conducting experiments, and that pupils be asked to undertake only those experiments commensurate with their age, ability and experience. Pupils must have, and understand, copies of the laboratory rules and must be given *explicit* warning of any hazard associated with a particular experiment, technique or material. As always, it is the teacher's responsibility to prevent accidents by designing the teaching–learning situation in such a way as to minimise and, where possible, remove hazards.

If an accident involving a pupil does occur, it is important that medical advice be sought unless there is no doubt that this is unnecessary. Failure to summon medical attention when it was needed could constitute negligence.

It should be noted that a charge of negligence may arise out of any of the following three situations:

i a pupil injures himself;
ii a pupil injures another pupil;
iii a pupil behaves in such a way as to injure a third party.

Local Education Authorities normally provide indemnity for their employees by means of a Public Liability Insurance Policy. Non-LEA schools are likely to provide similar insurance cover.

An employing authority will have a standard procedure to be followed in the event of an accident to a pupil. However, an accident is an uncontrolled situation in which a teacher is emotionally, if not physically, involved. Some time for reflection and the consultation of any witnesses, e.g. laboratory technicians, is therefore essential before any formal report of an accident is submitted. Such a report should be confined to facts and should not attempt to account for the accident or to apportion responsibility.

b Wilful bodily injury to a pupil by a teacher. It is inappropriate to consider this matter in detail here. A parent may bring a civil action for damages, in which case the LEA or other employer could be enjoined as a co-defendant. A criminal action for assault could also be brought.

c Bodily injury to the teacher. If a teacher suffers injury from a pupil or a third party while conducting his normal teaching duties, he will be offered some protection by the LEA's insurance policy. If injury arises from some defect in the school building (or in equipment supplied by the LEA and being properly used), then the LEA is also liable for damages. Nonetheless, science teachers would be wise to report, in writing, any hazards which it is the responsibility of the employer to rectify, e.g. defective gas, water or electricity outlets and spent fire extinguishers.

The legal position of science teachers whose lessons are taken by a student teacher is sometimes a matter for concern. When a school, acting on behalf of the LEA, accepts student teachers for teaching practice, it does so recognising that they may have little or no experience of class management, of laboratory organisation, or of maintaining class discipline. All reasonable safeguards, consistent with allowing the student teacher to develop his self-confidence and professional competence, should therefore be invoked. Thus, regular staff should always be aware of what experiments a student intends to conduct in a given lesson and should indicate any hazards and suggest how these may be overcome or minimised. Experiments should always be thoroughly rehearsed by student teachers and the responsible member of staff should do all that is required to satisfy himself that the student is, or will become, a fit and competent person to take charge of a lesson involving experimental science.

In law, it appears that the student teacher on teaching practice is in the same position as a teacher who is an employee as far as insurance arising from the law of negligence or governing bodily injury to the teacher is concerned. In other words, student teachers are not distinguished from salaried employees as far as an LEA's Public Liability Insurance Policy is concerned. However, a student science teacher would be well advised to join a professional association as a student member, if he wishes to obtain some insurance against a claim for damages arising from the wilful bodily injury of a pupil while on teaching practice. No insurance can, of course, cover the consequences of any criminal action for assault, which may be successfully brought against the student teacher for inflicting such injury.

Out-of-school activities

Most LEAs have a clear policy and a set of regulations governing out-of-school activities. The LEA or other employer's responsibility for pupils and teachers is in no way diminished if the pupils and teachers are working on an approved out-of-school activity. The insurance taken out by an LEA normally recognises this fact but it seems wise that details of any out-of-school activity be submitted to the LEA for approval so that the adequacy of the insurance cover can, if necessary, be confirmed. In some instances, it is reasonable to ask parents to sign an indemnity against any claim made on behalf of their child in the charge of a teacher conducting an out-of-school activity, but such an indemnity may be unnecessary and, in some instances, may prejudice the rights of parents under common law or under an existing insurance policy. Indemnities should, therefore, be sought only after the advice of the LEA has been obtained.

When a pupil is injured at school, a teacher may offer or be asked to use his car to take the injured pupil home or to a hospital. If the teacher's motor car insurance is restricted in its coverage to social, domestic or pleasure purposes, it would be an offence to transport a pupil in this way. Moreover, it would be an offence with serious consequences if a further accident arose, involving the pupil and the car.

However, most LEAs have effected suitable Motor Contingent Liability Policies which indemnify teachers in respect of their legal liability for accidents to third parties arising from the use of a car in direct connection with voluntary participation in out-of-school activities and insofar as their own insurance arrangements are inadequate for the purpose. It is also a relatively inexpensive undertaking to modify an individual motor insurance policy to provide the necessary cover. Science teachers, like their colleagues, should therefore establish the legal position as far as such insurance is concerned before an emergency or other situation arises. It should be noted that some domestic insurance policies could be inoperative if the teacher used his car to transport apparatus or chemicals.

The Health and Safety at Work Act

The Health and Safety at Work etc. Act, 1974, applies to all persons at work, except domestic workers in private households, and is an enabling measure, superimposed upon existing health and safety legislation. The Act provides for the gradual replacement of existing health and safety requirements by revised and updated provisions. Existing Statutory Regulations under previous legislation remain in force until they are replaced by alternative regulations and codes of practice under the new Act. The Act is comprehensive in size and effect and its detailed implications for schools and other educational establishments are still being elaborated.

The Act became applicable to schools and colleges on 1st April 1975. Responsibility for administering the Act lies with the Health and Safety Commission and its Executive, which will appoint inspectors to carry out its enforcement functions. The inspectors have power to issue improvement or prohibition notices, or to stop an activity which causes risk of serious personal injury. Under the Commission's guidance, Local Authorities will enforce the legislation in 'non-industrial' areas of employment and regulations are to be issued after appropriate consultation.

The following points are of particular importance to school science teachers.

a The Act requires all employers to prepare and, where necessary, to revise written statements setting out their *policy* for the health and safety of employees and the procedures for effecting that policy.

b The Act requires that all reasonably practicable steps be taken to train employees in health and safety issues. This provision emphasises the need to ensure that technicians, as well as teachers, are properly trained in safety matters.

c Designers, manufacturers, importers and suppliers of articles or substances for use in schools or colleges must ensure that, as far as reasonably practicable, they are safe when properly used. This means that an article must be tested for safety in use and that details must be given of any conditions of use regarding its safety.

d Teachers have a duty under the Act to take reasonable care to avoid injury to themselves or to others by their teaching activities, and to co-operate with employers and others in meeting statutory requirements. The Act also requires employees not to interfere with or misuse anything provided to protect their health, safety or welfare in compliance with the Act. This provision emphasises the importance of adequate supervision of laboratories at all times.

e Section 37 of the Act makes it clear that if an offence against the Act is committed by a corporate body, not only the corporation but also the individual(s) who have been responsible by consent, neglect, connivance, etc., may be subject to legal action.

f An employer is required 'to give to persons (not being his employees) who may be affected by the way in which he conducts his undertaking, the prescribed information about such aspects of the way he conducts his undertaking as might affect their health or safety'. This sub-section of the Act could be considered as covering risks to pupils in schools and students in colleges and appropriate safety training should, therefore, be provided.

g The Health and Safety Commission, after appropriate consultation, will issue Regulations and approved codes of practice. These codes of practice will have a special legal status as they will not be statutory requirements, but may be used in criminal proceedings as evidence that a statutory requirement has been contravened.

h The Act provides for recognised Trades Unions to appoint safety representatives from among the employees and requires an employer, if requested by the safety representatives, to appoint a safety committee.

The Health and Safety Act places an obligation on teachers and administrators to develop and publicise appropriate safety-education procedures. Some LEAs already issue science teachers with a loose-leaf Safety Manual, produced by a Safety Committee charged with producing instructions and recommendations and with circulating these to appropriate individuals and organisations within the Authority. It is possible that this practice will become more widespread and be extended to all Local Authority employees.

Further information about the operation of the Health and Safety at Work etc. Act, 1974, may be obtained from the local offices of the Health and Safety Executive, from the pamphlets produced by the Health and Safety Commission or from the following publications:

Powell-Smith V. (1974) *A Protection Handbook—Questions and Answers on the Health and Safety at Work Act*, Osborne Books, Ltd.

DES, *Health and Safety at Work etc. Act, 1974*, Circular 11/74 (226/74 Welsh Office).

For details of how the Health and Safety at Work Act may affect school science teaching see:

Education in Chemistry, (1975) Vol. 12, No. 6.

Bibliography

Bretherick, L. (1975) *Handbook of Reactive Chemical Hazards*, Butterworths. Appendix 1 of this important book lists other specialised sources of data on chemical hazards.

Chemical Society (1975) *Code of Practice for Chemical Laboratories*, London.

Department of Education and Science, Administrative Memoranda:
3/70 *Avoidance of Carcinogenic Aromatic Amines in Schools and other Educational Establishments.*
7/70 *The Use of Lasers in Schools.*
2/76 *The Use of Ionising Radiation in Educational Establishments.* There are also accompanying notes for guidance.
6/76 *The Laboratory Use of Dangerous Pathogens.*
7/76 *The Use of Asbestos in Educational Establishments.*

Department of Education and Science (1976) *Safety in Science Laboratories*, HMSO.

—(1976) *Safety in Further Education*, HMSO.

Everett, K., Jenkins, E. W. (1976) *A Safety Handbook for Science Teachers*, John Murray.

Everett, K., Hughes, D. (1976) *A Guide to Laboratory Design*, Butterworths.

Gaston, P. J. (1965) *The Care, Handling and Disposal of Dangerous Chemicals*, Northern Publishers, Aberdeen.

Muir, G. D. (ed), (1977) *Hazards in the Chemical Laboratory*, Chemical Society, London.

Scottish Education Department:
(1968) Circular No. 689, *Ionising Radiations in Schools, Colleges of Education and Further Education Establishments.*
(1970) Circular No. 766, *Use of Lasers in Schools, Colleges of Education and Further Education Establishments.*
(1972) Circular No. 825, *The Use of Carcinogenic Substances in Educational Establishments.*
(1973) Circular No. 852, *The Temporary Use of Ionising Radiation in Schools, Colleges of Education and Further Education Establishments for Demonstrations by Visiting Lecturers or Student Teachers in Training.*
(1973) Circular No. 882, *Special Precautions for the Safe Handling of Radium 226 Closed Sources of an Approved type for Use in Schools, Colleges of Education and Further Education Establishments.*
(1968) Memorandum No. 6, *Inhalation of Asbestos Dust.*
(1970) *Safety in Science Laboratories.*

Weast, R. C. (ed), (1977) *Handbook of Chemistry and Physics*, 57th edition, Chemical Rubber Pub. Co.

Work with living organisms

ASE (1972) *Biology Teaching in Schools Involving Experiments or Demonstrations with Animals or Pupils.*

Schools Council, Educational Use of Living Organisms Project, 1974:
Animal Accommodation for Schools.
Small Mammals.
Micro-organisms.
Organisms for Genetics.
Plants.
Recommended Practice for Schools Relating to the Use of Living Organisms and Material of Living Origin, EUP.

Shapton, D. A., Board, R. G. (1972) *Safety in Microbiology*, Academic Press.

UFAW (1967) *Handbook on the Care and Management of Laboratory Animals*, Edinburgh and London.

Field work

DES (1977) *Safety in Outdoor Pursuits*, HMSO.

Schools Council (1972) *Out and About: A Teacher's Guide to Safety on Educational Visits*, Evans.

Legal issues

Barell, G. R. (1970) *Legal Cases for Teachers*, Methuen.

Cooke, A. J. D. (1976) *A Guide to Laboratory Law*, Butterworths.

DES (1974) Circular 11/74 (226/74 Welsh Office), *Health and Safety at Work etc. Act.*

Powell-Smith, V. (1974) *A Protection Handbook—Questions and Answers on the Health and Safety at Work Act*, Osborne Books, Ltd.

4 Audio-visual aids

Introduction

This chapter provides an introduction to the wide range of audio-visual resources available to the science teacher and to some of their many uses.

Most of the research findings relating to the value of audio-visual aids are highly specific and therefore difficult to summarise. However, the following general points are important and should be borne in mind when planning the use of audio-visual materials.

a Learning from visually presented materials is improved if the materials are carefully introduced to pupils who are aware of their purpose, and if there is an opportunity for pupils to become actively involved in their use.

b The effectiveness of learning from audio-visual sources is, not surprisingly, a function of the appropriateness of the aid for the purpose in mind. Appropriateness may refer to several independent criteria, e.g. whether the likely improvement in learning is worth the extra time and effort on the part of the teacher or pupils, and whether the material is best presented as a diagram, photograph, sketch, model or overhead projector transparency. Unfortunately, there is little research evidence to help the science teacher choose between the different methods; experience and common sense are probably the most important guides for selecting one visual aid rather than another or for choosing audio rather than video techniques.

c It is invalid to assume that photographs are more appropriate for younger pupils or for those of low educational attainment than for their older or more successful counterparts, who can cope more readily with diagrams and graphs. Such evidence as exists suggests that the ability to perceive and interpret visually presented information can, in large measure, be learnt.

It is not reasonable to assume that the perception by pupils of visual stimuli will be that intended by the teacher. This means that an accompanying verbal commentary is essential. Such a commentary helps pupils to interpret the visual images correctly and increases the likelihood that they will remember what they have been taught.

In recent years a number of commercial manufacturers have marketed sets of audio-visual materials in the form of 'multi-media packages'. These usually involve a set of slides or a film-strip with an accompanying commentary on a cassette tape, though other combinations are also encountered. These 'packages' are intended for use in an individualised learning situation or with small groups of pupils.

Chalkboards

A chalkboard is simply a board with a surface that allows chalk to rub off a stick on to the board. The commonest chalkboard is therefore the blackboard, but other colours, particularly green, are often employed.

The characteristics of a chalkboard surface are important. It must be sufficiently rough to allow the chalk to be transferred from the stick to the board but, if it is too rough, the board cannot be cleaned satisfactorily. If the surface is too smooth, the chalk tends to slip and drawing or writing on the board becomes impossible. The cheapest surface is wood coated with a suitable chalkboard paint. Other surfaces such as rubberoid or glass are more expensive but have several advantages. For example a specially prepared glass surface is particularly easy to write on and it will hold a greater amount of chalk than either rubberoid or wood.

A wooden or glass-covered board may be supported on an easel but more commonly, is fastened directly to the wall. The wall support can be designed to allow several boards to slide over one another vertically or horizontally, thus increasing the amount of writing area in a given location.

Rubberoid boards require roller supports which may be fastened to a wall or mounted on castors. The advantages of a roller-board are as follows:

i it is possible for the teacher to write at a constant and convenient height by rolling the surface up or down as required. Similarly, the height of the writing surface can be adjusted to meet the various needs of pupils;

ii more than one type of surface can be included on the roller board. In addition to a plain chalkboard surface, the roll may include a white section for projection and a section suitably marked for plotting graphs;

iii material prepared in advance of a lesson can be hidden from view until required. Also, material recorded on the board can be retained for longer periods of time than on other forms of chalkboard with a smaller surface area.

The problem of chalk dust

It is not possible to avoid the production of dust when using a chalkboard. So-called 'dustless' chalk is available and it has the advantage that it produces larger grains than conventional chalk. These fall directly downwards into a trough rather than forming an airborne dust.

The problems of dust can be minimised by:

i cleaning the board with downward strokes rather than horizontal sweeps;
ii ensuring that the eraser or duster is not saturated with dust;
iii using a proprietary brand of chalk dust control fluid, 'CDC', with its special applicator or a suitably impregnated cloth;
iv cleaning accumulated dust from the chalkboard trough so that it is not gathered by sticks of chalk resting there.

The maintenance of a chalkboard

A chalkboard is a standard item of equipment in almost all school laboratories and classrooms, and perhaps this is why there is a tendency to overlook the need to maintain the surface in optimum condition.

Chalkboards should be cleaned with a felt-block eraser or a dry cloth. Dusters dampened with water eventually make the surface shiny and a similar effect results from contamination of the surface with oil or grease (e.g. from the hands).

A shiny chalkboard will need a new coat of a suitable paint. The newly painted surface must be treated as a new chalkboard and be 'broken in'. The entire surface should be patted with a dusty eraser. When this dust is wiped off with a dry cloth, the board is ready for use. Failure to treat a new surface in this way is likely to make it difficult to erase chalk marks completely.

The siting of a chalkboard

The siting of a chalkboard in a laboratory or classroom is important. Lighting conditions change throughout the day and it is important to ensure that light reflected from the chalkboard surface does not prevent students from reading the material written on it. If the board is fixed in position, it may be necessary to arrange adequate shielding (e.g. blinds) of some of the windows. Chalkboards should never be placed adjacent to a window; the contrast in illumination so obtained makes reading the board very difficult.

Adequate illumination of the surface is essential. If artificial lighting of the board is necessary, the light must be so placed as to overcome the reflection problem already referred to, while ensuring an uninterrupted line of vision for each of the students.

Markerboards

White or off-white plastic-faced markerboards were first introduced in the 1960s. They have grown in popularity as felt tip pens have become widely available in a variety of colours. The principal advantage of markerboards is the elimination of chalk dust, and some surfaces are also suitable for projection.

It is *essential* to use the correct type of marker when writing on a markerboard. For ordinary purposes, where the writing is to be erased during or at the end of a lesson, water-based markers are used. The surface may then be cleaned with a damp cloth. For more permanent markings, spirit-based markers are recommended and the surface must eventually be cleaned with a solvent such as propanone (acetone) or a suitable material recommended by the manufacturers of the board.

Water-based and spirit-based markers are sold individually or in packs of assorted colours.

A later development is the markerboard made of enamel-coated steel. The enamel surface is tough and durable and requires only occasional cleaning, to remove dust and grease, to maintain it in optimum condition. These boards can be used with the usual range of markers but they have the additional advantage that they can be used as magnetic boards (see page 55).

Markerboards can be purchased in a variety of sizes, from 610 mm × 914 mm (2 ft × 3 ft) up to 3005 mm × 1225 mm (10 ft × 4 ft). They may be wall-mounted or held in a mobile stand. The latter has the advantage that a two-sided board, arranged to rotate about a horizontal axis, may be used. More specialised surfaces, e.g. a graph board, are also available.

Display boards

In designing a laboratory or classroom, some wall space should always be given to pin-board or other suitable surfaces for the display of charts, notices, photographs etc. Various types of composition board are available from builders' merchants but cork floor or wall tiles make suitable alternatives.

For information boards a system of 'plug-in' letters is often most satisfactory. Plastic letters or numbers are plugged into a tracked board which may be either wall-mounted or used in conjunction with an easel. Boards are available in assorted sizes and with a variety of frames. Some of these are intended for indoor use only. Letters and numerals are sold in boxed sets in a variety of sizes and colours.

For large displays of a temporary nature, a system of pin-board mounted on lightweight scaffolding is desirable. The flexibility of the system is increased if the scaffolding can be re-arranged to allow displays at different heights. Display boards of this type may be either covered in coloured hessian facing which should be fire-proof (BS 476) or made from 'peg-board'. Most manufacturers supply a range of sizes of single and double-sided panels in assorted colours and will also make display boards of any size to order.

When purchasing temporary display boards, consideration should be given to:

i the storage of the stands when not in use;
ii the range of accessories available, e.g. wires for book displays, shelves, etc.;
iii the mass and size of the screens if they are to be transported from one location to another;
iv the means of fastening charts to the screens;
v the stability of the stands when arranged for display; some are mounted on castors and others on plastic or metal feet.

Magnetic boards

Magnetic boards have a number of uses in science teaching and are particularly valuable in illustrating transfers or shifts of some kind. e.g. 'movement' of electrons in bonding, ions in solution, and chromosomes in cell division. For an account of the magnetic blackboard in chemistry teaching, see Maloney, M. J. *School Science Review*, 1968, 170, **49**, 126–127. See also Robertson, R. A. *School Science Review*, 1972, 186, **54**, 106–108.

A magnetic board is easily made from a suitable sheet of mild steel, which should be mounted in a wooden frame. Alternatively, one may be purchased from suppliers. Some manufacturers sell white, plastic-coated magnetic boards which may be either wall-mounted or used in conjunction with an easel. The surface of such boards permits them to be written on with 'dry marker' felt pens which can be erased easily with a soft cloth.

A wide range of magnetic items is available from general educational suppliers. Tape, self-adhesive on one side, may be cut into small pieces and attached to the back of cardboard 'cut-outs' for display purposes. Labelling tape, letters, numerals and a variety of specialised shapes and symbols are available in a range of colours. Magnetic string (usually 1 mm thick and sold in 10 m lengths) and magnetic tiles with self-adhesive backs are useful additions to the range.

Flannel boards

The flannel board or (flannel graph) is a cheap yet effective teaching aid which can serve a similar purpose to the magnetic board. Parts of a diagram may be removed, exchanged, added to and altered in other ways and members of the class may take part in the presentation of this 'mobile picture'.

At its simplest, the flannel board consists of a piece of fluffy material, such as an old beige blanket, mounted on a sheet of board. Cut-out shapes, words, letters and other items may be made from felt or even from thin card

backed with felt or coarse sandpaper. More elaborate commercially produced flannel boards, with accompanying 'cut-outs' are available from general educational suppliers.

Screens

The following are important points for consideration in the selection of screens for school use.

a Screen surface. A matt white surface allows little appreciable fall-off in brightness until the viewer is more than 40° on either side of the centre line of projection. Even at these angles fall-off is comparatively small. Matt white screens are available in heavy duty and normal grades of cotton up to 3.5 m high, but the surface, once dirtied, is difficult to clean satisfactorily. Such screens are therefore most suitable for use in permanent positions in lecture theatres or halls. A matt white, emulsion painted wall makes a perfectly satisfactory alternative and has the advantage that it can be renewed with further coats of paint. Plastic coated screens are also available in a variety of designs and sizes and have the advantage that they may be cleaned by wiping with a damp cloth. Plastic surfaces may be embossed with a regular pattern which allows light to be reflected from the screen at a wide angle. For smaller sizes a piece of white card has much to commend it. It is cheap and, once dirtied, is easily replaced. However, sizes larger than A1 (594 mm × 841 mm) are difficult to obtain.

A silvered surface may give a more brilliant picture than a flat white screen owing to its greater reflective properties. However, in some cases there is a considerable fall-off in brilliance as the viewer moves away from the centre line of projection until, at about 35° from the centre line, the degree of illumination may be unacceptably low. Screens made of plastic and other similar materials with silvered surfaces are available from suppliers. Aluminium paint, preferably sprayed on, makes an adequate alternative on a suitable surface.

The most expensive screens have glass beaded surfaces, but these are not often used in schools. The glass particles are attached to the screen surface by adhesive and this may discolour with time and with the attachment of dust particles. In some cases the brightness of the image on such screens falls off markedly as the viewer moves away from the centre line of projection.

It should always be remembered that no surface can reflect more light than that which reaches it from the projector. Although some surfaces are undoubtedly more reflective than others, manufacturers' claims about the reflective properties of their screens should be treated with caution. It is often the case that increased brilliance near the centre line of projection is achieved at the expense of decreased illumination at greater angles; indeed with screens whose reflective properties approach 100 per cent, this must be the case.

b Portability. An expensive screen should not be permanently mounted in a laboratory in which corrosive fumes are likely to be produced. In such instances a board or wall painted with matt finish white emulsion paint is probably adequate for most needs.

Portable screens are available in several designs. The simplest is a table- or bench-mounted model with top and bottom rollers, the top being held up by self-acting springs or stays. In more sophisticated versions, the stays may be folded to allow the screen to wrap around the lower roller which is enclosed in a wooden box. Free-standing tripod screens are generally less heavy than boxed models and, being independent of any other support, may be erected in any position. Several tripod-mounted screens incorporate a screen tensioning device and a means of tilting the top of the screen forwards. This reduces the 'keystone' effect, which may be particularly troublesome if an overhead projector is used on a vertical screen.

If a screen is to be moved frequently from one building to another, it may be convenient to select one which can be accommodated in the boot of a car. A canvas screen bag will help to keep the screen clean.

c Size. When selecting a screen size, the brilliance of the projector and its position in the room should be kept in mind. A doubling of the image width will reduce the illumination to one-quarter. Table 4.3 on page 60 indicates the relationship between image width, length of throw and focal length of the projection lens for slide projectors.

d Rear projection. In some situations where only a small viewing area is required, rear projection may be an advantage. Several commercially produced rear projection screens are available, generally of a small format. Translucent perspex or other similar plastic material may be used to improvise such a screen. Good quality tracing paper is also suitable, though very liable to damage.

Rear projection may be carried out in two ways. The image may be formed directly on the screen by a projector which is directed towards the viewer. Alternatively, the projector may be directed at an angle away from the viewer and the image reflected back onto the screen by means of a mirror. In the former case, slides and other projection material must be reversed left-to-right before loading into the projector.

A tea-chest or similar box with the bottom replaced by a suitably translucent material or an old television set with the television tube similarly replaced make excellent rear projection screens.

Wall charts

A wide range of wall charts is available from commercial suppliers. Some of these are sponsored by industrial and other manufacturers and may either be available free or carry a nominal charge. Wall charts may also be produced in school to meet a particular educational need. Many such charts are of value provided they are not over used. Pupils will ignore a chart if it is allowed to remain up long enough to become accepted as part of the furniture.

When planning or selecting a chart for teaching purposes, the following factors should be considered.

a Contents. The information given must be accurate and up to date and must be in a form appropriate for this means of presentation.

b Teaching situation. For what size of group and level of work is the chart intended? Are charts appropriate to the particular teaching situation? Charts may be used to provide initial stimulation, to demonstrate a particular skill, to provide a visual description, to organise knowledge and summarise information and for many other purposes. What purpose best fits the teaching situation in which the chart is to be used?

c Visual impact. Charts should not be overloaded with material but should have a simple arrangement with a balanced composition. Line and colour should be used effectively and consistently to give appropriate information and stress the main teaching points. An initial glance should make the purpose of the chart clear. All details should be visible at a distance of 8–10 m.

d Labelling. Letters which are to be seen at a distance of 10 m by normal vision should be at least 3 cm high. Lower case letters are easier to read than upper case as there is less similarity in letter height and shape, and hence there are more distinguishing features when a word is seen as a whole. Some teachers may be able to write clear, free-hand letters, but many will require some assistance. Stencil lettering using a drawing pen (such as *Rapidograph* or *Uno*) or dry transfer lettering (such as *Letraset* or *Lettapress*) are clear and attractive and add professionalism to the finished product. Label lines should be at least 3 mm thick and should preferably terminate in an arrowhead.

The simplest method of displaying a wall chart is on a pin-board, but a number of alternative methods of display are satisfactory. If a strip of mild steel is screwed to the wall in a suitable position, charts may be attached to it by means of magnets. Clipboards or 'bulldog' clips may also be screwed to the wall to hold charts. *Blutak* or even Plasticine may be attached to the corners of a chart which can then be secured to the wall with a little pressure. The use of transparent and other adhesive tapes should be avoided as tape can cause damage to both the surface of the chart and the surface to which it is attached.

The storage of wall charts presents problems. Small numbers may be kept between two sheets of thick card or hardboard held together with tape after the fashion of an artist's portfolio. For larger numbers the ideal system is to

Substance	Advantages	Disadvantages	Comments on use
Cardboard	Cheap; large models built quickly	Not very robust	Can be used to demonstrate optical isomerism/asymmetry. Very useful for complex structures such as zeolites or biochemical species, e.g. DNA
Cork spheres	Many sizes available. Easily cut and glued	Not easy to paint	Suitable for molecular and crystal structures. Can be joined with cocktail sticks
Marbles	Cheap, but limited range of sizes available	Can be cemented together, but must be held in a frame to do so. Not readily cut	Best used to demonstrate packing of species
Milk straws	Negligible cost, easily cut	Not suitable for permanent models	Used with pipe cleaners to make 'Dreiding' models
Plastic spheres	Cheap and available in many sizes. Brightly coloured	Some may be drilled already, but in general difficult to work with	Useful for 'packing' models or simple tangential models
Plasticine	Cheap, flexible, re-usable. Poor colour intensity	Limited rigidity: unsuitable for large molecules. Spheres must be 'rolled' from sticks to a uniform size	Useful for 'electron shapes' i.e. probability electron envelopes
Polystyrene	Available in blocks and as spheres from 12 mm ($\frac{1}{2}''$) to 50 mm (2'') in size. Easily cut, glued and painted	Mechanical weakness. Holes tend to wear and material dents easily	Perhaps the most suitable material for pupils to use to make models
Table tennis balls	Excellent sphericity. Low density	Only one size. Rather expensive but cheaper brands are available. Not easily glued together	Best used in conjunction with other materials
Wooden balls and beads	Beads are very cheap. Balls are obtainable from a handicraft shop in a wide range of sizes	Not easy to cut and painting required—more than one coat	Space-filling models should not be attempted with wooden spheres as a high degree of precision is required
Wire, plastic coated	Relatively inexpensive	Thicker wire not easy to bend. Repeated bending eventually causes fracture	Useful for 'Dreiding' models or polymer structures

Table 4.1 Relative advantages of materials for model making

use a plan chest with a series of shallow drawers. If a lath of wood is attached to the top of each chart a vertical hanging rack can be devised. Charts should not be folded and only those produced on linen should be rolled.

Samples and materials

Industrial and other organisations produce a large range of materials and samples for use in schools. Many of these items are obtainable free of charge. For others a nominal charge may be made.

Details of such aids are published from time to time in science teaching journals such as *The School Science Review* and *Education in Chemistry* and in the *Bulletin of the Schools' Information Centre on the Chemical Industry* (The Polytechnic of North London, Holloway Road, London, N7 8DB).

Queries should be addressed to the industrial organisation concerned and, where appropriate, to the Education or Schools Liaison Officer.

Models

Models are often used to convey an appreciation of three-dimensional structure. A wide range of biological models, some made from plaster of Paris and others from rigid or flexible plastic are available from commercial suppliers. Many of these models are expensive and in some cases their value must be questioned. However, skeletal and other anatomical models (see pages 142 and 75 have their place in the biology laboratory. Some models may also be improvised from relatively simple raw materials.

There are several different types of atomic and molecular models used to represent the arrangement of species in a lattice. *Ball and spoke* models are simple to make but the sphere sizes are usually arbitrary and no attempt is made at correspondence with the sizes of the species being represented. The 'spoke' lengths can, of course, be made to correspond to internuclear distances. The principal advantages of ball and spoke models are relative cheapness, good interior visibility even in a complex model, and versatility, as the models can usually be dismantled and re-arranged to build a model of another lattice. *Dreiding models* are essentially ball and spoke models without the spheres to represent the atomic species.

Space-filling models attempt to illustrate the overlap of electron clouds in space. Dimensions are based on van der Waals' radii. These models are suitable for organic and some inorganic structures but not for ionic lattices. They are particularly useful in predicting or explaining steric effects.

Tangential contact models are made from spheres of appropriate relative scale sizes in such a way that the spheres touch. They are based on ionic or covalent radii and are suitable for atoms and molecules of elements and molecules of inorganic and organic compounds. The direction of chemical bonds and the bond angles can be included in this type of model, which possesses a satisfactory degree of internal visibility except in very large structures.

Models may be made from a number of materials. Table 4.1 summarises the relative advantages of some commonly available materials.

Polystyrene, in the form of spheres, is perhaps the most commonly used material for making atomic and molecular models in schools. It may be cut with a saw, sharp knife or hot wire. If a hot wire is used, it should be slightly less than red hot in order to reduce the risk of producing the toxic styrene monomer. Note that the spheres may be reduced slightly in size by 'rolling' them beneath a firm wooden board.

Polystyrene spheres may be joined in a number of different ways. Adequate temporary 'bonds' may be made by means of pipe cleaners, thin dowel rod, or cocktail or match sticks, inserted into holes in the spheres. These holes can be made with a compass point or nail. More permanent 'bonds' are made by using a suitable glue, e.g. *Durofix, Evostik*. Polystyrene glue should not be used as this tends to dissolve the polystyrene.

Painting of polystyrene spheres is best done after any necessary holes have been made but before the model is finally assembled. Painting protects the spheres from minor damage and dirt and allows different species to be distinguished in the same model. A single coat of gloss paint is usually adequate, but an emulsion paint may be used to provide a dull, matt finish. Plastic 'enamel' paints produce the most attractive finish but are the most expensive.

If a large number of spheres is to be painted, the spheres may be fastened on thin nails hammered upwards through a thin board. There is no universally accepted colour code for painting spheres to represent the elements, although there is an international convention. It is clearly important to be consistent within a school, particularly between departments, but this does not, of course, preclude the use of 'special' colour features to highlight an important aspect of a particular structure. Table 4.2 summarises the codes suggested by the International Convention and by the Institute of Physics.

Element	International Convention	Institute of Physics
Bromine	Green	Blue-green
Carbon	Black	Black
Chlorine	Green	Grass-green
Fluorine	Green	Light-green
Hydrogen	Cream	White
Iodine	Green	Dark-green
Nitrogen	Ultramarine	—
Oxygen	Red	Red
Phosphorus	Purple	Purple
Silicon	—	Grey
Sulphur	Yellow	Yellow
Metals	Grey or brown	Brown (copper-coloured), silver, gold, in order of increasing co-ordination number

Note that it is possible to use different shades of the same colour to illustrate degrees of electronegativity, partial charge, etc.

Table 4.2   Colour codes for atomic models

Polystyrene spheres are best stored in polythene bags, either within cardboard boxes or on stackable trays. Each bag should contain spheres of a given size and it is sometimes convenient for distribution to a class if each bag contains a specified number of spheres, e.g. 36.

There is considerable literature on the production of atomic and molecular models using polystyrene spheres. The following sources will be adequate for all introductory purposes.

Bassow, H. (1968) *Construction and use of Atomic and Molecular Models*, Pergamon.

Beevers, C. A. (1965) Miniature Scale Models, *J. Chem. Ed.*, **42**, 273.

Burmlik, G. C. *et al.* (1964) Framework Molecular Orbital Models, *J. Chem. Ed.*, **41**, 221.

Feiser, L. F. (1963) Plastic Dreiding Models, *J. Chem. Ed.*, **40**, 457.

Hobson, D. C., Platts, C. V. (1966) Milk Straw Models, *SSR*, 163, **47**, 694.

Larson, G. O. (1964) Atomic and Molecular Models made from Vinyl Covered Wire, *J. Chem. Ed.*, **41**, 219.

Nuffield O level Chemistry Project (1967) *Handbook for Teachers*, Longman/Penguin Books, Chapter 14.

Platts, C. V. (1956) Models for Teaching Atomic Structures and Theories of Valency, *SSR*, 134, **38**, 96.

Sanderson, R. T. (1962) *Teaching Chemistry with Models*, van Nostrand.

Savory, C. G. (1976) A survey of crystal and molecular models, *Educ. Chem.*, **13**, No. 5, 136.

Tetlow, K. S. (1964) Modelling of Chemical Structures with Expanded Polystyrene Spheres, *Educ. Chem.*, **1**, 1, 7.

A detailed list of references to atomic and molecular models included in the *School Science Review*, *Education in Chemistry* and the *Journal of Chemical Education* will be found in:

Jenkins, E. W. (1974) *A Bibliography of Resources for Chemistry Teachers*, Centre for Studies in Science Education, University of Leeds, pages 67–71.

Where a large number of polystyrene spheres is to be used in making a model, marking and drilling holes in the correct position is a time-consuming activity. However, precise location of holes is important in constructing a large model. Angles around an equator of a sphere are easily located by means of a protractor cut away to fit the sphere.

For tetrahedral sites, a jig is a considerable advantage. Details of two such jigs will be found in:

Bentley, S. C. (1967) Simple Jig for Tetrahedral Carbon Atom Models, *SSR*, 165, **48**, 488.

Good, N. S. (1968) A Simple Jig for Tetrahedral Carbon Atom Models, *SSR*, 168, **49**, 48.

A wide range of atomic and molecular models is available commercially. Unfortunately, there is no agreement about the scale employed, although a degree of consensus about colour coding is evident. Some care is needed in purchasing sets of models as the set is unlikely to meet all the needs of a school science course. The possibility of augmenting the set by additional purchase is, therefore, important. In general, it is more economical to buy a complicated model, perhaps involving the drilling of many holes in each sphere, than to construct it in a school.

Episcopes and epidiascopes

The advent of the overhead projector has led to a decline in the use of episcopes and epidiascopes. With an episcope, the image depends upon reflected light and hence is weaker than that produced by transmitted light. Even with a tungsten-halogen lamp, an episcope must normally be used in a blacked-out room.

An epidiascope may be used as an episcope (e.g. for the projection of images of book illustrations) or as a direct projector of slides. The slide attachment is normally designed to take 8.25 cm × 8.25 cm ($3\frac{1}{4}$ in × $3\frac{1}{4}$ in) glass-mounted slides, but modifications for 5 cm × 5 cm and several other continental and American slides are available.

An epidiascope is normally a bulky piece of equipment which cannot easily be moved from one room to another unless it is mounted on a trolley.

The slide projector

The 5 cm × 5 cm slide projector has been widely used in schools for many years and has almost entirely replaced the earlier 'lantern slide' projector. Some slide projectors may, with appropriate modification, be used to project film-strips, but many better quality models can only be used for slide projection.

In its simplest form, the slides are fed by hand into a carrier which may be either the 'slide across' or the rotating type. This type of projector is perfectly adequate for viewing a few slides or in situations where pupils are themselves handling the equipment in a resources centre. Its simplicity has the advantage that there is very little that can go wrong with the mechanism. Also, it is usually this type of projector which can be modified for showing film-strips so there is the added advantage of flexibility.

For a more formal presentation or for the showing of a large number of slides, a good magazine-loading projector has much to commend it. Two basic types are available. The first takes rectangular box-shaped magazines each holding 36 or sometimes 50 slides, which are fed horizontally into the projection position. The second takes a circular magazine holding some 80 slides and rotates as each slide is dropped into the projection position. In many cases, magazine-loading projectors may be operated either manually or by remote control. In the latter case, the operator is free to face the class and to point out the salient features of individual slides. Some of these automatic projectors may also be used in conjunction with a tape-recorder in a tape/slide sequence in which the projection of the next slide is triggered automatically by a pulse recorded on the tape. Slides for magazine projection should be selected with some care as many magazines will only accept slides of less than a certain critical thickness. Thicker slides may lead to jamming of the magazine or, worse still, of the projector itself.

In selecting a slide projector for school use the following points should be considered.

a Flexibility. Is the projector also required for viewing film-strips? (see page 61 for a discussion of the relative merits of film-strips and slides).

b Automation. To what extent is a magazine-loading projector required and what degree of 'remote control' is desirable? In general, increased automation goes with decreased flexibility.

c Brilliance. In what sizes of room is the projector to be used? A powerful projector will be required for use in school halls if a sufficiently large and brilliant picture is to be viewed by all members of the audience. If the projector is only for classroom use or for viewing slides in a small resources area, a less powerful (and hence usually less expensive) model may be quite suitable.

d Projection lens. The size of picture formed by a projector at a given distance from the screen depends upon the focal length of the projection lens, those with larger focal lengths forming a smaller picture at a given distance from the projector. In many cases a choice of lenses is available and the ultimate selection will depend upon the size and nature of the room and the way in which the projector is to be used. Some more expensive projectors have 'zoom' lenses as optional extras, permitting the operator to 'home in' on the central portion of a slide, which thus becomes enlarged. The relationship between the focal length of the projection lens, the image width and the length of throw is shown in Table 4.3.

Focal length of lens		Image width in cm					
mm	in	75	100	120	150	200	250
75	3	1.5	1.8	2.8	3.0	4.0	5.0
85	$3\frac{3}{8}$	1.8	2.1	3.0	3.7	4.6	6.0
100	4	2.1	2.8	3.7	4.5	5.5	7.0
110	$4\frac{3}{8}$	2.4	3.0	4.0	4.5	6.0	8.0
125	5	2.8	3.4	4.5	6.0	7.0	9.0
150	6	3.4	4.0	5.0	6.5	8.0	10.5
180	7	3.7	4.5	6.0	7.0	9.5	12.0
200	8	4.0	5.0	6.5	8.5	10.5	14.0
250	10	5.0	6.0	7.0	10.5	14.0	17.0
400	16	8.0	10.0	13.0	17.0	22.0	28.0

Table 4.3   The length of throw in metres of a 5 cm × 5 cm slide projector giving a range of image widths and using a variety of projection lenses

e.g. A projector with a 150 mm focal length lens being used with 5 cm × 5 cm slides produces an image 100 cm wide at a projector/screen distance of 4 metres.

e Portability. If a projector is to be moved frequently from one room to another, this may be an important consideration.

f Heat. In some projectors, an excessive proportion of the energy from the lamp may be dissipated as heat, which may cause damage to projected slides. Any projector for school use should include a built-in cooling fan.

After use, a slide projector should be allowed to cool thoroughly before being moved. If a separate fan-cooling switch is included this may be left on to hasten the cooling. When not in use, a plastic or similar cap should be placed over the lens and the projector placed in its carrying case, well away from chalkboards to minimise the accumulation of dust on the lens.

Most 5 cm × 5 cm slides are formed from a single frame of 35 mm miniature film. This gives an actual transparency size of 36 mm × 24 mm (sometimes referred to as 'Leica-size'). In a few cases, the so-called 'half-frame' size may be used (24 mm × 18 mm). For a further discussion of this, see page 61. A very wide selection of slides is available from scientific and other suppliers, and teachers can easily prepare their own using a 35 mm camera. If this is of the single lens reflex type, then addition of a copying stand, which can be improvised from simple materials, makes it possible for slides to be made from almost any original material. The arrangement is shown in Figure 4.1.

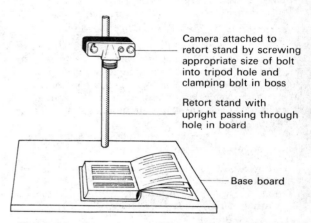

Camera attached to retort stand by screwing appropriate size of bolt into tripod hole and clamping bolt in boss

Retort stand with upright passing through hole in board

Base board

Figure 4.1   A copying stand

For work of this kind, the camera should be fitted with appropriate extension rings for close-up work. Care must be taken to prevent undue reflection from the surface of the material being photographed and in most cases natural light is preferable to artificial light. If tungsten filament lamps are being used, an appropriate colour film should be purchased as the standard 'daylight' film is unsuitable for use with artificial light.

Slides are most conveniently stored in transparent plastic *Viewpacks* which are available from many audio-visual aid suppliers. These hold 12, 20 and 24 slides in a single multi-pocket wallet in such a way that they may be viewed without removing them from the wallet. Some wallets are designed for storage in albums whilst others may be suspended from metal hangers in a standard steel filing cabinet. An alternative, but generally less convenient form of storage, is to place the slides in sets of plastic drawers of the type sold in hardware shops for the storage of screws, nuts, bolts, etc. Several similar designs have a 5 cm × 5 cm format and are therefore suitable for slide storage.

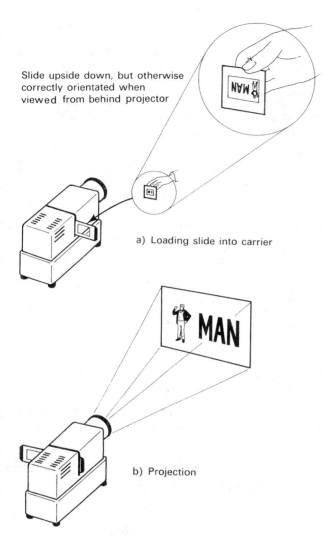

Slide upside down, but otherwise correctly orientated when viewed from behind projector

a) Loading slide into carrier

b) Projection

Figure 4.2 Loading slides into a slide projector

5 cm × 5 cm slides which operate a switch when they are inserted into the viewer. In this way the slide is seen against an illuminated, translucent plastic background. Most viewers of this type incorporate a lens to give a slight enlargement of the slide. A slide viewer is a convenient device which allows small numbers of pupils to examine slides in a resources centre or other similar learning situation.

The film-strip projector

As has already been suggested (see page 59), some slide projectors may serve for the viewing of film-strips as well as slides—using a film-strip carrier in place of the slide carrier. The main advantages of film-strips appear to be lightness, cost and sequencing of presentation. A film-strip bearing 36 frames weighs only a small fraction of that of a comparable number of slides—especially if the slides are glass mounted. The cost of cutting and mounting is such that commercial slides are also very much more expensive than the equivalent film-strip. The question of sequencing is one which is much debated, many teachers preferring the freedom to select their own order rather than depending upon that chosen by the film-strip maker. The result is that some schools buy film-strips and then cut them into their separate frames, mount them and use them as slides.

Two types of film-strip are available, according to the format of the individual frames (see Figure 4.3). The

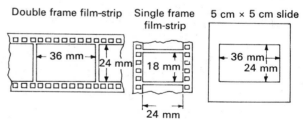

Double frame film-strip   Single frame film-strip   5 cm × 5 cm slide

← 36 mm →   24 mm   18 mm   ← 36 mm → 24 mm

24 mm

Figure 4.3 Two types of film-strip with a 5 cm × 5 cm slide for comparison

The life of individual slides is extended considerably if they are mounted in glass. However, this must be carried out with some care to prevent slides of excessive thickness being formed.

During projection, slides should always be fed into the magazine or projector in such a way that the slide appears to be upside-down but otherwise correctly orientated when viewed from behind the projector (see Figure 4.2).

Slide viewers

Several different designs of slide viewer are available from photographic dealers. Most of these take single

earliest type corresponds to single frames from 35 mm cine film and therefore each measures 24 mm × 18 mm with the longer axis running at right angles to the axis of the film itself. This type is sometimes called 'half-frame', though 'single-frame' is clearly a more accurate description. The other form resulted from the invention of the Leica camera at the beginning of this century and has frames measuring 36 mm × 24 mm with the long axis of each frame coincident with the long axis of the film itself. These 'double-frame' film-strips are gradually replacing the older single-frame format and the recent disappearance of the last available make of single-frame camera from the market can only accelerate the process.

Most film-strip projectors have twin, rotating conversion plates or some similar device on the film-strip carrier for use with either single-frame or double-frame films. However, a number of recent models employing rear projection onto a built-in screen have no such facility and can only be used with double-frame film. This factor should be borne in mind when selecting a projector if the school already possesses many single-frame film-strips.

Great care must be taken in loading and using a film-strip if the maximum length of service is to be obtained from the film. With most projectors, the film is prevented from cockling by passage between two glass plates during projection. These plates must be kept scrupulously clean and should be opened manually both during the loading of the film and when moving from one frame to the next during projection. Failure to observe any of these precautions will result in the film becoming scratched. Single-frame film-strips should be loaded onto the top spool, and transferred, frame by frame during projection, onto the lower spool. Double-frame strips are loaded onto the left-hand spool (as viewed from behind the projector) and transferred to the right-hand spool during projection. As with slides, each frame should be upside-down, but otherwise correctly orientated when viewed from behind the projector. Film-strips should only be handled by their edges; fingerprints on the frames themselves will attract dirt and may cause scratching of the film.

Recently a number of manufacturers have devised machines with a built-in rear-projection screen. These are of small dimensions and are only suitable for viewing by a few pupils in a resources centre or similar situation. Some models are also equipped with cassette tape play-back facilities and there is now an increasingly wide range of film-strips available with accompanying tape. Alternatively, teachers may prepare their own tapes to accompany film-strips already in the school. In this way, small teaching units can be built up which may prove useful in mixed ability groups and other similar teaching situations, where pupils are working individually or in small groups at different speeds.

Most film-strips arrive from the suppliers in metal or plastic drums. These are best stored in drawers which can be subdivided to provide one space for each strip. If each space and its corresponding strip are similarly numbered, missing items can be noted at a glance. The teaching notes and/or cassette tapes which normally accompany film-strips are best stored separately, with a suitable cross-referencing system for easy retrieval.

The overhead projector

In recent years the overhead projector has become a popular teaching aid in the classroom and laboratory. One of its great advantages over the earlier episcope and epidiascope is the brilliance of its image, which permits its use in full light; another is that its construction enables the operator to face the class while using the projector.

The overhead projector consists of a lamp and reflector situated below a horizontal glass table or stage—the latter being some 25 cm square. The projection lens assembly is placed above the stage and may be adjusted to bring about focusing. In most modern projectors, the light is refracted through the stage by means of a Fresnel screen consisting of a number of concentric rings which have a prismatic effect. In some cases, a curved reflector below the stage replaces the Fresnel screen. The arrangement of a typical overhead projector is shown in Figure 4.4.

One manufacturer markets an overhead projector which also serves as an episcope and may be adapted for use as a slide projector. However, in both these cases some degree of blackout is required as the brilliance of the illumination from the projector is considerably reduced.

It will be noted from Figure 4.4 that the screen used for the overhead projector is tilted forwards. This avoids the 'keystone' effect in which the top of the picture is wider than the bottom. The siting of screens and the choice of available surfaces is discussed on page 55.

One of the few disadvantages of the overhead projector is the comparatively short life of the lamps. However, this can be considerably extended by an efficient, fanned cooling system and in many cases the lamp cannot be switched on unless the fan is already in operation. If the fan is separately controlled, it should be left running after the lamp has been switched off, so that the lamp cools more rapidly. The projector must not be moved until the lamp has been allowed to cool completely. One projector has been produced with twin lamps which, in the event of one lamp failing, can be exchanged by moving a switch. Other models have two settings for lamp brilliance, use of the lower setting leading to greatly increased lamp life. When not in use, the projector should be kept free from dust by protecting it with a plastic cover usually available from the manufacturers. The stage should be kept clean at all times and should not be written on directly as this may lead to scratching. Projectors can be conveniently mounted on trolleys so that they can be moved easily from one room to another. Suitable trolleys are available from a number of suppliers or may be made in school from *Speedframe* or other similar angle iron.

Care should be taken not to leave projectors near windows where they may be exposed to direct sunlight for any length of time. In one recent case the sunlight falling on the projection lens assembly was focused onto the stage causing charring of transparency material and the consequent danger of fire.

Transparencies for use on the overhead projector can be produced in a number of ways. At its simplest level, the stage can be fitted with an acetate roll on which the operator writes as he is teaching the class. In this way, the projector fulfils the function of a chalkboard. Alternatively, transparencies can be produced in advance. This has the advantage that additional material,

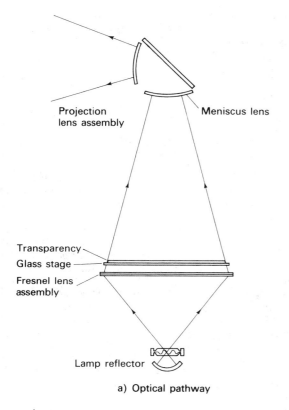

a) Optical pathway

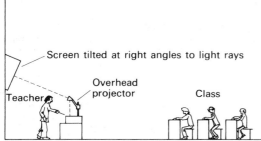

b) Position of screen in relation to the overhead projector

Figure 4.4   The overhead projector

e.g. labels or colours, may be added to the transparency as it is presented and other material may be superimposed on the basic transparency by means of 'overlays' carefully kept in register with the basic transparency. This technique is especially useful for building up diagrams of chemical and biological structures. Transparencies are normally prepared on acetate sheet, which is available in various thicknesses and two sizes (26 cm × 27 cm or

26 cm × 22 cm). The larger almost square format has the advantage that it makes full use of the square projector stage. Set against this is the fact that the larger transparencies do not fit conveniently into folders, files and briefcases. The much cheaper and less robust *Cellofilm* is not very suitable for elaborate transparencies, but makes a good alternative to the acetate roll.

The most convenient writing implement for use on acetate sheet is the *felt pen*. However, not all types of felt pen are suitable and care should be taken to ensure that only those recommended for use on an overhead projector are purchased. These may contain either water-based inks (e.g. *Ofrex Projectotip* or *Staedtler 'Lumocolor 315'*), or spirit-based inks (e.g. *Staedtler 'Lumocolor 317'*). In both cases a wide range of colours is available. Water-based inks can be removed with a slightly moistened tissue or cloth, but the spirit-based inks are more permanent and can only be removed with spirit. Sometimes a combination of both types of ink is an advantage; a permanent transparency made with spirit-based pens can have temporary additions made to it in water-based ink. These additions can then be erased with a damp cloth, leaving the permanent transparency undamaged. Black or coloured drawing ink may also be used for writing on acetate, but care should be taken to ensure that the surface is free from finger prints as any small patch of grease will repel the ink. For solid coloured effects a transparent, self-adhesive coloured film (e.g. *Transpaseal*) may be attached to the acetate sheet in appropriate areas. This should be cut to the correct shape before the self-adhesive backing is removed and ideally should be attached to the underside of the transparency.

Lettering on transparencies should always be of an adequate size. If the screen size is 1.25 m × 1.25 m, letters will need to be at least 4 mm high to be viewed at 8 m by a person with normal vision. Letters may be written directly on the transparency using felt pen or drawing ink with or without the aid of a stencil. Alternatively, they may be added by using dry transfer lettering such as *Letraset*.

Transparencies may also be prepared by making a master on paper and using a thermal copier or similar device for the production of the final tranparency (see page 70). Care should be taken to ensure that the ink used in preparing the master has a high carbon content (see page 70). Various types of film resulting in a black image on clear background, black on colour or colour on clear are available for thermal copying. Some publishers and manufacturers supply books of masters which are designed to be copied in this way.

Cardboard mounts are available to which transparencies may be attached with adhesive tape. These offer some protection to the material and are especially useful if a number of overlays are to be superimposed on a basic transparency as exact location of the transparencies is facilitated by the mounts. However, such mounts are expensive—usually more expensive than the transparencies themselves. As an alternative, an acetate sheet may be attached to a sheet of paper by a strip of adhesive tape along one edge. The reverse of the paper can be used for teaching notes, and the notes and transparency hinged in this way can be filed for easy storage. Care should be taken to ensure that the paper 'underlay' does not obstruct the air intake of the projector cooling fan as this may cause damage to the Fresnel screen or to the lamp.

A wide range of transparency material can be purchased in ready-made form. The majority of this is mounted in cardboard. However, prices are usually high and in many cases the material is not precisely suited to the educational needs of the class. The *Flipatran* system marketed by Transart consists of a metal frame or viewer which is fitted onto the projector stage. Transparencies and notes in the form of a spiral-bound book are clipped onto the viewer so that individual transparencies may be projected. A wide range of topics are covered by the Flipatran system.

The overhead projector also allows a number of models and demonstrations to be shown to a class. Cardboard 'cut-outs' of skulls and bones may be moved in relation to one another on the stage, thus producing a shadow puppet effect on the screen. Some X-ray plates make suitable transparencies. Bubble rafts in a crystallising or Petri dish make effective models of atomic or molecular structure or of the arrangement of cells in a piece of plant tissue. Three-dimensional models can also be produced in perspex or other similar transparent material and placed on the stage for projection, though the depth of focus of many projectors may not be very great. In some cases, a closer understanding of the relationship between a two-dimensional image and a three-dimensional structure may be obtained using the overhead projector.

Microprojectors

Several designs of microprojector are available from microscope manufacturers and general scientific suppliers. Most of these are high quality microscopes with the addition of a powerful light source to enable the image to be projected onto a screen. In some cases the microscope stage is horizontal, the optical pathway through the instrument is vertical and the final image is formed on a vertical screen via a mirror. In other instruments, the stage is vertical and the optical pathway horizontal, thus enabling direct projection onto a screen without a mirror. Many microprojectors can be fitted with a rear-projection screen which enables them to be used in normal lighting conditions; otherwise complete blackout is required.

Before committing a department to the purchase of a microprojector, serious consideration should be given to the use to which it will be put. If it is to be used mainly for the projection of prepared slides, then the alternative use of photographic slides and a 5 cm × 5 cm slide projector should be borne in mind. There is now a wide selection of excellent photomicrographs available in this format.

However, if living material (e.g. microscopic pond life) is to be examined by an entire class and individual microscopes are not available, there may be no alternative to the purchase of a good microprojector.

Motion picture projection

Motion film may be encountered in a number of sizes, the dimension referring to the overall film width. After some initial experiments, the earliest film used was in the 35 mm format and this is still the size preferred by most commercial film-makers and used in public cinemas. Occasionally even larger formats (e.g. 70 mm) are used for special purposes. Film used for educational purposes is normally of a narrower gauge—16 mm or 8 mm.

16 mm film is widely accepted and can be used for silent and sound projection, in colour or in monochrome, with a quality of image approximating quite closely to that obtainable with the larger 35 mm format. 16 mm silent film has perforations (sprocket holes for the claw mechanism) along both edges of the film, while that used for sound projection utilises one edge for the sound track, which may be optical or magnetic or both (see Figure 4.5). Sound film should therefore never be placed in a projector intended for use only with silent film. Another factor is that silent film is normally projected at 16 frames per second while sound film is shown at 24 frames per second. As a general rule, professionally produced films, which are normally available on hire to schools, are in the 16 mm format.

The original 8 mm gauge consisted of 16 mm film which was passed through the camera twice, the first time exposing only one half of the film and the second exposing the other half. The film was then processed in this 'double width' form and subsequently cut along its entire length to give the 8 mm format. The perforations on 8 mm film are therefore always along one edge only. In recent years a further development has been the introduction of the Super 8 format. This differs from Standard 8 film in a number of ways (see Figure 4.5). The picture area is increased by approximately 50 per cent in Super 8 film, thus giving a corresponding increase in brightness for the same size of screen. This increase in picture size has been brought about partly at the expense of the perforations, which are thus reduced in size. There is also a slight increase in the distance between perforations and Super 8 film has space left for a magnetic sound track. It is not possible to show Super 8 film in a Standard 8 projector or vice versa although dual-8 projectors will show both types of film. 8 mm film is the format normally used by amateur photographers and the price of film and equipment is such that its use is possible within the budgets of many schools. 8 mm film is also used for certain specialised teaching aids such as film-loops and casettes (see page 66).

Whatever gauge of film they show, all motion film projectors must complete four actions relating to the

a) 8 mm film

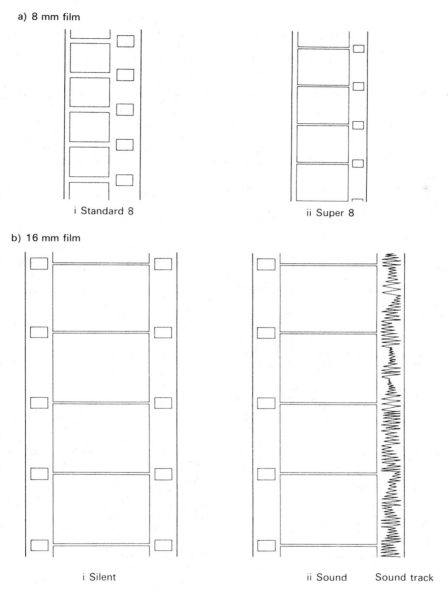

i Standard 8   ii Super 8

b) 16 mm film

i Silent   ii Sound   Sound track

Figure 4.5   Various types of motion picture film

showing of each frame. These are *holding*, *projecting*, and *obscuring* the light during the *transporting* of the film to the next frame. The film is *held* in a gate in such a way that the picture area itself is not touched, yet the film is maintained at right angles to the path of light so that the whole of each frame remains in focus. Gates vary in design, but can normally be opened, both for the insertion of the film and for cleaning. Cleaning is essential if film is to be maintained in good condition and should always be carried out with a small gate-cleaning brush before loading the film into the projector. During *projection* the lamp and its associated optical system concentrates sufficient light through the frame and onto the screen. Keeping in mind that the picture area of each frame of 16 mm film is approximately 10 mm × 7 mm and that this must be magnified to a screen width of some 2 m, it can be appreciated that every part must be of good quality and scrupulously clean for efficient projection. The light is *obscured* by the shutter while the film is *transported* to bring the next frame in line with the gate. This movement of the film is carried out by the claw mechanism which, in the better projectors, involves three claws. When threading a film into the projector, the manufacturers' instructions should be closely followed.

Different projectors offer a variety of facilities. The most useful of these is automatic threading in which the film leader (usually coloured white) is fed on rollers through the entire film path until it emerges near the take-up spool onto which it must be wound by hand. Another useful device is a powered, fast rewinding system by which the film is put back onto its original spool ready for the next showing. A reverse projection mechanism and one enabling the showing of single frames are other facilities which are sometimes available, though unlikely to be used much in the school situation.

Whereas 8 mm silent film is relatively cheap, commercially produced 16 mm sound film is very expensive. It is therefore most unlikely that individual schools will be in a position to purchase such films. However, a number of film libraries exist, e.g. those financed by the major science-based industries, from which schools and other organisations may hire films at a reasonable charge. Some Local Authority Teachers' Centres and Health Education Units also maintain small film libraries. Many good films are in considerable demand and several months' notice may be required to ensure that a film is available on a particular day.

Careless use of a projector can result in extensive damage to film, for which the school may be held financially responsible. Projectors must therefore never be used by anyone who is not fully acquainted with their mode of operation. Many Local Authorities do not allow teachers and technicians to operate projectors unless they have passed an appropriate test and they run short courses leading to such certification.

8 mm film-loops and cassettes

In recent years, two developments have had a considerable influence on the use of 8 mm film in school science teaching. The first of these was the introduction of the 'single concept' film-loop. These short films last for just 2–4 minutes and are housed in special plastic cassettes. The end of the film is spliced on to the beginning so that it can be shown repeatedly without rewinding. Film-loops are shown in special film-loop projectors, some of which incorporate rear-projection screens and can thus be used without blackout. A 'stop' button and remote control facilities are other features frequently included. This type of film received a considerable boost from the Nuffield Science Teaching Project which produced a large number of film-loops in connection with various courses. A very large selection of loops is now available from a variety of audio-visual-aid suppliers and from some publishers.

Some of the earlier loops were made in monochrome film, but most are now produced in colour. Loops may be in either the Standard 8 or the Super 8 format—in many cases a choice is available. Standard 8 loops cannot be shown in Super 8 projectors or vice versa. To reduce the possibility of a mistake being made, the cassettes are distinguishable by their shape (see Figure 4.6). The superior quality of Super 8 film and the fact that some manufacturers have recently begun to market some loops in Super 8 only, make it advisable for those contemplating the purchase of a projector for the first time to select one using the Super 8 format.

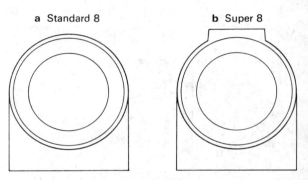

Figure 4.6   The shapes of Standard 8 and Super 8 film-loop cassettes

It is possible for schools to have their own loops made from suitable lengths of exposed 8 mm film. Norman Hemsley Productions of Richmond, Surrey, will enclose such film in cassettes for a small charge.

Partly because of the nature of the cassette itself and partly due to the frequency with which many loops are projected, the introduction of the 8 mm film-loop has not been an unqualified success. The splice joining the end to the beginning must be of the highest quality and, because there is some sliding between adjacent portions of film, correct lubrication during loading into the cassette is essential. In spite of these precautions, film-loops sometimes 'jam' in projectors and on other occasions they may 'jump'. In either instance the projector must be switched off immediately to prevent damage to the film. If the cassette is removed from the projector and the film 'inched' on manually inside the cassette, this will sometimes cure the fault.

Film-loops should be stored in a dust-free situation. Most manufacturers pack their cassettes in dust-proof plastic boxes and these are best stored in drawers which have been appropriately sub-divided and numbered or labelled so that missing loops can be quickly identified.

More recently, several manufacturers have marketed cassette-loading projectors in which threading of the film is fully automatic. The film is enclosed in plastic cassettes, which are made in a number of sizes giving running times from four minutes to 30 minutes. Fast rewind, reverse projection and still-picture facilities are also available. The extent to which cassette projectors replace film-loop projectors remains to be seen.

Television and radio

The BBC and IBA produce a large number of television and radio programmes for schools. Lists of the programme titles and a brief indication of contents and timing are circulated to all schools many months in advance of showing so that appropriate timetable arrangements may be made. Each series of programmes is designed for a specific age and ability range and notes for teachers and/or pupils are often available to supplement each broadcast or series. Programmes are normally 20–25 minutes in length and most are broadcast several times during the week and at different times during the school year.

The advent of relatively inexpensive videotape recording equipment has made it possible to record broadcast television programmes for use in schools at times other than those of transmission. However, such recording may only be carried out legally under certain conditions and teachers considering the practice should ensure that the following conditions are fulfilled:

i the school must be in possession of a current television receiving licence;
ii the BBC only allows the recording of 'Educational Broadcasts' which are specifically put out for use by schools and colleges. Other programmes which may have an educational content (such as nature programmes, *Horizon*, etc.) may not be recorded;
iii the IBA also allows the recording of Educational Broadcasts, but only if the school has obtained a licence for such recording. Licences are obtainable from the Independent Television Companies Association;
iv Open University broadcasts may only be recorded if the institution is licensed for the 'right to record' and a fee is charged for every programme so recorded. This fee only covers the retention of such a recording for one year.

All other recording of radio or television broadcasts and of gramophone records is strictly illegal.

Where several departments in a school require videotape recording facilities, it is important to develop a reliable system of booking and distribution of equipment. This should allow time for routine maintenance and servicing. VTR facilities purchased by different departments within a school should always be compatible.

Some television programmes are converted into 16 mm films which are then available for hire or purchase. Details are available from the appropriate television companies.

In addition to television programmes transmitted by the BBC and IBA, a number of Local Education Authorities have established closed-circuit television systems for relaying educational programmes to their schools.

A television camera, in conjunction with a monitor, may be useful in several situations in school science teaching, e.g. in demonstrating dissection technique or observing thermometer readings with a large class of pupils. It is clearly economic if a monitor used as part of a closed-circuit television network can also be used to receive broadcast transmissions.

VTR equipment is available in both reel-to-reel and cassette form, the latter being generally more simple to operate in that no threading of the tape is required. Tapes are also available in various widths ($\frac{1}{4}$ inch, $\frac{1}{2}$ inch and 1 inch), the narrower formats being generally more convenient for portable use, though the quality of recording is normally better with wider tape.

For an account of television and the teaching of science see:

Harris, B. R. (1971) Broadcasting and the Teaching of Science, *SRR*, 182, **53**, 5.

Broadcast radio programmes are easily recorded on a conventional reel-to-reel audio tape recorder. Note, however, that relatively inexpensive and portable cassette recorders are available and that these may have other uses in science teaching, e.g. the development of audio or audio-visual programmes for use with pupils of low educational attainment. For an example of the use of tape recorders in this way see:

Townsend, I. J. (1971) Science for the Special Child, *SSR*, 181, **52**, 668 and (1972) *SSR*, 184, **53**, 475.

Worksheets

Recent changes in the methods of teaching science and the widespread introduction of mixed-ability classes have resulted in an increased use of worksheets as a means of individualising the learning activities of pupils. Ideally, a worksheet should carefully structure the activities required of a pupil and check that the pupil understands the work which he has undertaken. Diagrams are frequently employed in worksheets and, in certain circumstances, photographs, film-loops, charts, slides and other aids may also be incorporated. Worksheets are also frequently used in work with pupils of low educational attainment.

Worksheets have a number of inherent advantages and disadvantages:

i they enable pupils to work at their own pace, but *may* minimise the personal contact between pupil and teacher;
ii they provide a record of the work done by a pupil in a standard and easily recognisable form, but they may

fail to encourage self-expression and creativity on the part of the pupil;

iii they are time-consuming to produce, especially if audio-visual material is incorporated; they may also need re-writing and up-dating from time to time;

iv if well constructed, they can motivate pupils, especially those of low educational attainment whose skills at reading and writing may be poorly developed;

v they may be helpful in providing a means by which pupils who have been absent from school can catch up on work which has been missed;

vi badly written worksheets do little to raise the level of literacy of the pupils.

When constructing worksheets, the following points should be borne in mind.

i the amount of written material should be appropriate for the pupils for whom the worksheet is intended;

ii the work should be attractively presented and laid out in such a way that pupils can easily find their way through it. The use of different colours for different sections (instructions, questions, apparatus requirements, etc.) should be considered;

iii flow diagrams may sometimes be used to summarise instructions, though care should be taken to ensure that the pupils are able to follow such presentation;

iv diagrams, cartoons, etc., may sometimes be used to convey practical instructions;

v a generous allowance of space should be made if the pupils are expected to write their answers on the worksheet itself;

vi worksheets should be produced to a standard size and adequate provision made for the pupils to file the sheets for easy reference;

vii worksheets should be tested, modified and re-written from time to time;

viii a variety of means of duplication of worksheets is available. This is discussed in the following section.

Stencil duplication

This is one of the commonest methods of reproducing typed or written material and most schools possess a stencil duplicator. The stencil consists of a fibrous, plastic-coated sheet which is impervious to ink until it is 'cut' (see below). This is attached to a thin strip of card by means of which the cut stencil is fastened round the drum of the duplicator. Rotation of the drum forces ink through the cut areas of the stencil onto sheets of absorbent paper which are passed through the machine. The drum may be rotated by hand, but in recent years such machines have been largely replaced by electrically operated duplicators. A well-cut stencil can produce several thousand copies and, as the cost of the stencil is fairly low, it is also an economic method of duplication for small numbers of copies (see Table 4.4).

Different makes of duplicator vary in the means of attaching the stencil to the drum and those fitting one model are unlikely to fit conveniently onto another. In the event of it being necessary to use a cut stencil on a machine of another make, the thin card by which it is normally attached to the drum should be removed and replaced by one fitting the machine. This can be cut from an old stencil, and then secured to the required stencil with adhesive tape.

The most common method of cutting a stencil is with a typewriter, which must be used on the stencil cutting

Method of duplication	Number of copies					
	1	5	10	30	100	500
Typing with carbon papers						
Excluding time for typing	.3	.3	—	—	—	—
Including time for typing	40	8	—	—	—	—
Stencil duplication						
Typed stencil excluding time	6	1.3	.75	.40	.27	.22
Typed stencil including time	46	9.4	4.8	1.7	.67	.30
Stencil cut electronically	10	2.3	1.2	.53	.31	.23
Stencil cut thermally	16	3.4	1.8	.71	.37	.24
Spirit duplication						
Typed or written master excluding time	3	.90	.60	.38	.31	—
Master cut thermally	4.7	1.2	.73	.44	.33	—
Electrostatic copying (Xerox)	1.8–2.0*	1.8–2.0*	1.8–2.0*	1.1*	.78–.80*	.72–.80*

* N.B. Costs do not include the charge for hire of the machine which may be from £15–£50 per month depending on the nature and versatility of the particular machine. There is also usually a basic minimum charge in addition to the hire fee; the volume of work being duplicated must therefore be carefully considered before contemplating this means of duplication.

Table 4.4   The relative costs of duplication by various means shown as price (in p) per copy. The costs do not include labour except for typing. (1976 prices)

setting so that the ribbon of the typewriter does not come between the keys and the stencil. Stencils may also be cut by hand using a special stylus and a plastic backing sheet. Some manufacturers produce special stencils for hand cutting which give better copies than those designed for typewriter cutting. Stencils may also be cut by means of an electronic scanning machine (see page 71) or a thermal copier (see page 70). These stencils are rather more expensive to purchase than the conventional variety, but they enable complicated illustrations and even half-tone photographs to be reproduced.

The principal disadvantage of a stencil duplicator is that it is not possible to reproduce different colours on the same copy without using different inks. This requires the use of several drums—one for each colour—and the cutting of a separate stencil for each colour. Care must then be taken to keep the copies in register during their several passages through the machine.

Cut stencils may be stored after copying and used to run off further copies on a subsequent occasion. Some slight deterioration of quality may take place, however, owing to some of the cut portions of the stencil becoming blocked with ink. Several manufacturers produce storage cabinets in which stencils may be hung from special cardboard hangers, which are purchased separately. Suitable cabinets may also be made in the school workshop. It is advisable to attach one run-off copy to the stencil for identification purposes and also to devise an efficient indexing system for easy retrieval of the appropriate stencil.

Spirit or hectographic duplication

This is another common method of reproducing typed or written material, but only in short runs of up to a hundred copies. The spirit copying machine is quick and simple to operate and is found in many school staffrooms and laboratory preparation rooms. A master copy is produced on a sheet of glazed paper which is placed on top of a sheet of hectographic carbon paper of appropriate colour. The material is typed, written or drawn on the matt side of the master so that a layer of carbon is transferred onto the glazed side to give a mirror image of the material. A hard, flat writing surface is required as a backing during this process and special backing sheets are available, though a melamine-surfaced table or a sheet of glass are equally suitable. The hectographic carbon may be exchanged for one of another colour at any stage during the production of the master, thus producing a master made up of several colours. Seven colours of carbon are available (black, purple, blue, red, green, brown and yellow), though in practice the last two usually show up less well than the others.

To produce copies, the master is wrapped around the drum of the spirit duplicator. The absorbent run-off paper is then passed under a felt pad soaked in spirit and brought into contact with the master. The hectographic carbon, being soluble in spirit, is transferred to the paper which thus receives a positive print of the original material. The number of copies obtainable from one master is limited by the fact that each print removes a small amount of the hectographic carbon deposited on the master, and so copies will become gradually fainter. Careful adjustment of the pressure with which the paper is applied to the master and moistening of the paper itself are necessary to produce the maximum number of copies. Too great a pressure in the early stages, or paper which is too moist, will remove too much carbon on the first few copies thus reducing the number which can be satisfactorily produced. The number of copies obtainable also depends on the quality of the hectographic carbon employed. Under ideal conditions over a hundred copies may be produced from a single master.

Spirit masters may also be produced using a thermal copier (see page 70). In this case only one colour of carbon, usually purple, is employed.

Its speed and convenience and especially its ability to produce copies in several colours make the spirit duplicator a very popular machine. It is particularly suitable for the production of maps and diagrams where colours give added clarity. Both hand and electrically operated machines are available.

Spirit masters may be stored for a short period and then re-used, provided the deposit of carbon has not been completely removed during the first duplication. However, in general, the storage properties of spirit masters compare unfavourably with those of stencils. The relative cost of duplication using different methods and for various numbers of copies is shown in Table 4.4.

Dyeline or diazo duplication

This process utilises paper impregnated with light-sensitive diazonium salts. When exposed to ultraviolet light these salts decompose so that the paper becomes white. Unexposed portions of the paper become black. The master is drawn on tracing paper or any other similar transparent or translucent material. Best results are obtained if this is done with drawing ink, but pencil or black felt pen make adequate alternatives. The master is then laid over the diazo paper which is then exposed to a light source with a high ultraviolet component. All those parts of the diazo paper which are unprotected by the lines on the master become bleached. The paper is then 'fixed' in a cabinet containing ammonia gas or in an appropriate fixing fluid. Unfixed diazo paper will bleach all over with exposure to natural light containing ultraviolet radiation. A selection of coloured diazo salts enables a variety of colours to be obtained by this process, which can also be used to produce coloured transparencies for the overhead projector. Copies can also be made directly from an overhead transparency thus ensuring that members of a class have an exact replica of a transparency viewed on the screen.

This method of duplication is widely used for reproducing plans and technical drawings on a larger format than those frequently copied by the processes mentioned above. However, this process of duplication is more time-consuming and generally more costly (unless very small numbers of copies are required) than those already outlined.

Offset duplication

The range of offset duplicators is extensive and there is a corresponding range in price. In general, offset duplication is more expensive both in terms of capital and of running costs than the simpler methods described above. The process is only likely to be used in those institutions where large numbers of good copies are required on a regular basis.

The offset process depends upon the immiscibility of oil and water. The master, or offset, is prepared by one of a number of different processes (typing, photography, printing, etc.), and is fastened round the cylinder of the offset duplicator. The master is then dampened so that those parts which are not to be reproduced—the so-called non-image areas—acquire a thin layer of water. The master is then passed through inking rollers and the image to be reproduced is coated with a grease-based ink. This image is transferred to the copy paper via a third 'blanket' cylinder.

The differences between one offset duplicator and another relate mainly to the degree of sophistication of the machine—the complexity of the inking/printing system, the operating capacity, the quality of reproduction, etc. These factors are mainly responsible for the wide price range and should be carefully considered before selecting a machine for school use.

Electrostatic duplication

Electrostatic transfer is the basis of the widely used *Xerox* system. Light-sensitive selenium is deposited on a revolving drum which carries an electrostatic charge. An image of the original is focused onto the drum and where the light makes contact with the drum, the charge is reduced or removed. The drum is then coated with a fine powder so that when the copy paper is brought into contact with it, the image is transferred to the paper. This image is 'fixed' by heating the paper so that the powder is fused onto the paper.

There are a number of variations of this electrostatic copying process. For example the light-sensitive coating may form the top surface of the actual copy paper. In general, the method is simple to use and a wide range of machines using the principle is available. Some models are only available for hire and in this case the anticipated

frequency of use should be carefully considered before entering into a hiring agreement. There is no limit to the number of copies which can be prepared from a given original and the quality of reproduction is generally good, although it is not always possible to reproduce photographs, solid areas or colours by this method. Overhead transparencies may also be produced and copied by electrostatic duplication.

Electrostatic copiers are available with a variety of additional facilities. Some produce copies at a variety of reduced sizes, while others can use both sides of the copy paper. Most machines take a variety of paper sizes.

A careful record should be made of all uses of an electrostatic copier, so that charges can be made to the appropriate departments. Also, a member of staff should be charged with the responsibility of checking that the machine is kept supplied with paper.

Thermal copying

In the simplest form of thermal copying, a copy paper is placed in contact with the original and infra-red radiation is used to turn the sensitive coating of the copy paper black in those areas where the image absorbs heat. The white or 'non-image' areas simply dissipate the heat.

In the slightly more complex heat transfer system, the copy paper is placed between the original and the transfer sheet. The infra-red radiation affects the coating of the transfer sheet and this is then transferred to the copy paper. This method gives a better quality of reproduction than the simpler 'direct' system.

Thermal (or thermographic) copiers are simple to use and many desk top and portable models are available. It is not possible to produce colour by this method (except for overhead transparencies). Paper and plastic offset masters may be produced thermographically, but the quality of reproduction is not generally very high. It is also possible to produce spirit masters and to cut stencils using a thermal copier.

Heat copying machines are excellent for preparing overhead projector transparencies. These may be made directly using the appropriate heat sensitive film, which is available in a range of forms including those producing a coloured or black image on a clear background, or a black or white image on a coloured background. High quality transparencies may also be produced by an indirect method in which the coating of a transfer sheet is transferred to the surface of the original. On a second passage through the machine this in turn affects the heat sensitive transparency. A considerable range of line thickness and tone may be achieved by this method.

The original for thermal copying must be produced with a medium based on carbon or metal, e.g. a typewriter ribbon, carbon paper, drawing ink or pencil. Ball-point pens and felt-tipped pens are not normally satisfactory. In an attempt to overcome this, a dual process based on the use of ultraviolet and infra-red radiation has been developed. A thin paper is placed underneath the original and ultraviolet radiation is passed through the paper onto the original. The radiation reflected back from the white areas of the original causes appropriate parts of the light-sensitive coating on the paper to decompose. This paper is then placed in contact with the copy paper and treated with infra-red radiation so that the coating on the copy paper decomposes to produce a positive image.

Because of their versatility, thermal copiers can be used with a wide range of heat-sensitive papers and films. It is therefore necessary to select the appropriate 'software' for the task in hand with some care. Full instructions, including details of the range of materials available, are normally supplied by the manufacturers.

Diffusion transfer copying

This widely used technique is capable of producing offset masters, spirit masters or single copies. Sensitised paper is placed face downwards on the original. Light is then passed through the back of the sensitised paper; the white areas of the original reflect the light back through the sensitised paper, while the black areas cause no such reflection. A 'latent negative' is thus formed. This may then be placed face downwards on the final copy paper and fed into a developing machine. As the negative develops, the chemicals so formed diffuse onto the transfer sheet to give a positive image. The negative and the positive are left in contact for about 10 seconds to allow sufficient diffusion to produce a good image.

In an alternative method, the latent negative is developed to produce a visible negative of the original. A second sheet of sensitised paper is placed face to face with the negative and exposed once more to the light. The second sheet of paper is then developed to give a positive image of the original.

Detailed operating instructions are usually provided with each model, but the quality of duplication is dependent on exposure time and is to some degree susceptible to variation in temperature or to changes in voltage.

An extension of the technique is the so-called *gelatin transfer process*. The original is exposed to light and the reflected light passed through an emulsion-coated paper. The emulsion contains developing and dyeing agents and the pattern of reflected light is a function of the detail on the original. The exposed emulsion is immersed in a solution which softens the 'image areas' of the emulsion and releases suitable dyes. The paper carrying the emulsion is then passed through rollers while in contact with a sheet of copy paper. The soft, dyed image is thus transferred to the copy paper. Once again, correct exposure is critical for good results.

Electronic scanning stencil cutters

The range of material for stencil duplication may be greatly increased with the use of an electronic scanning stencil cutter. In recent years there has been a marked increase in the use of such machines in schools and in Teachers' Centres. The original is fastened round one end of a horizontal cylinder and a special copy stencil is located at the other end. An offset master may be used instead of a stencil. The cylinder is then rotated electrically at high speed and illuminated and scanned by a photoelectric cell. The signals received by the cell control the movement of a stylus in contact with the copy stencil or offset master. This stylus burns through the top layer of the stencil and thus 'cuts' it. The cut stencil is then used on a conventional stencil duplicator which may produce several thousand copies of the original.

The advent of electronic scanning stencil cutters has enabled half-tone photographs and other complex illustrations to be reproduced. Coloured copies may be obtained in the normal way by using a coloured ink cylinder on the stencil duplicator. If multiple colours are required, separate originals must be drawn and stencils cut for each colour. Each colour is then transferred to the copy paper by separate passage through the duplicator, using a different coloured ink cylinder for each passage. Accurate registration is essential on each passage through the machine.

Preparation of originals for duplication

The quality of copies obtained by any of the means of duplication described above is, to a very large extent, dependent on the quality and means of production of the original. In most cases, the highest quality of reproduction will be achieved if the original is prepared on white paper and copies made from this by appropriate means (e.g. by electronic scanning stencil cutter or thermal copier and stencil duplicator). In this way, the original is undamaged and may be retained for further use. It may, therefore, be worth taking rather more care than would be appropriate in preparing a spirit duplicator master which would only serve to produce one hundred or so copies. However, as in many situations in science teaching, a compromise must be struck which ensures a reasonable quality of production without excessive investment of time.

In almost every case the best originals are those prepared with black drawing ink. A number of makes of drawing pen (e.g. *Rotring*) are available in a range of nib widths; they may be purchased from good stationery shops or drawing office suppliers. These pens may be used in conjunction with stencils of appropriate size to produce a professional look to the lettering. Alternatively, transfer lettering of the *Letraset* type may be employed for this purpose, though its use is rather more time-consuming than a pen and stencil. It should be remembered that such originals may be used not only to produce stencils and hence copies, but also to produce overhead transparencies. In this case the size of the overhead projector projection stage should be kept in mind when producing the original.

Reprography and the copyright laws

The Copyright Act, 1971 deals with the legal aspects of copyright, which also apply to material used for copying for educational purposes. As a general rule, publishers, authors and artists permit the making of *single* copies of illustrations or *small* portions of text for use solely in the course of instruction in an educational establishment. This does not mean that teachers and others are free to make multiple copies of illustrations or text for class use. The Society of Authors and the Publishers' Association have produced a pamphlet entitled 'Photocopying and the Law' and teachers are recommended to consult this for further guidance on this matter.

The selection of reprographic equipment for school use

No single item of reprographic equipment is likely to be adequate to meet the diverse demands of a school. Most schools possess spirit and stencil duplicating facilities, though the use of the latter may be restricted to the office staff. Thermographic and electronic scanning facilities are available on an increasing scale. More sophisticated and expensive items of copying equipment are sometimes found in Teachers' Centres, and this obviates the need for an individual school to purchase such items.

When selecting equipment for school use, the following factors are likely to be important.

a Capital and running costs. The latter are not simply those of paper and ink, but must include the financing of any maintenance and repair facilities. The comparative costs for single copies and for short, medium and long runs are shown in Table 4.4.

b Servicing. The adequacy of servicing, repair and replacement facilities should be established.

c Portability. The movement of expensive items of equipment of this sort from one location to another is generally to be avoided. However, if copiers are to be moved the size and weight are significant factors.

d Versatility. The facility to produce single or multiple copies and to prepare overhead transparencies, offset and other masters should be considered.

e Flexibility of use. Some copiers require the original to be passed through the machine, hence necessitating the use of single sheet originals. Other machines will accept the

material in book form. As copy paper is expensive, it is economic to be able to use different sizes of paper as required.

f Hire or purchase. The relative merits should be considered. In some cases machines are only available for hire, though most copiers must be purchased.

g Multiple users. The use of copying facilities by several members of staff causes organisational problems. Records should be kept of the amount and sizes of paper used so that charges may be made to appropriate departments. Alternatively, each department should hold its own stocks of paper and other 'software'. If portable equipment is used, stock control is more difficult than if the facilities are maintained in a central location.

h Range of copying facilities and their interdependence. Individual items of reprographic equipment may increase the range of, or depend upon, other items of equipment in the school. A thermal copier, for instance, may not only serve as a copier, but also produce spirit masters, stencils and overhead transparencies. The full versatility of the machine may only be realised if all the other items of equipment are also available.

j Storage. Not only must the equipment itself be given space, but there must also be sufficient and appropriate storage space for the consumable materials associated with the equipment. It will also be necessary to provide storage facilities for those stencils and/or transparencies produced.

k Maximum exposure area. The maximum size of original which may be copied is important and may sometimes be less than it appears.

l Speed of copying. The speed of producing multiple copies should be considered where appropriate. The mode of operation (hand or electric) is a significant factor here, but it also has an important influence on the cost. If large numbers of copies are to be produced frequently, electrical operation is clearly desirable as it releases the operator to carry out other tasks as well as reducing the time taken for the running off. A good stencil duplicator, operated electrically, may produce as many as 160 sheets per minute.

Educational games

In recent years, there has been a marked increase in the number of games and simulation exercises designed for use in school science teaching.

The simulation exercises that have been developed include case studies, role-playing activities, 'in-tray' exercises, and decision-making 'kits' based upon models of actual systems. For most of these simulation exercises, specially prepared materials are necessary and many also require considerable preparation and practice if they are to be used to maximum advantage.

Educational games are usually intended to develop specific aspects of a pupil's competence e.g. the ability to write or manipulate chemical formulae, to use the concept of the mole, or to understand the principles governing the flow of electric current in a circuit. Some games are played in accordance with the rules of games likely to be familiar to the pupils e.g. dominoes, bingo, Monopoly. Others, for which materials are often manufactured commercially, require that pupils learn the rules of a new game before they can begin to take advantage of the educational benefit the game claims to offer. Occasionally, an educational game uses readily available materials such as dice or playing cards in a novel or unconventional manner. See, for example Eloosis, *J. Chem. Ed.*, (1974), 532, 51, a game played with an ordinary pack of cards and intended to provide an insight into the nature of scientific inference.

Games and simulation exercises may be used for a variety of purposes, e.g. to provide practice in the use of a skill already acquired, to engender learning of facts or principles, to increase motivation or to provide a novel means of revision. Empirical evidence about the effectiveness of games and simulation exercises as learning devices is scanty but there seems little doubt that they are techniques which many pupils find acceptable and, in many instances, enjoyable.

Details of new games and simulation exercises are reported from time to time in the science teaching journals and in publishers' catalogues. For reviews of the field see:

Gibbs, G. I. (1974) *Handbook of Games and Simulation Exercises*, Spon.

Tansey, P. J. (ed.) (1971) *Educational Aspects of Simulation*, McGraw-Hill.

For details of games of particular interest to science teachers see:

J. Biol. Ed. (1973) **7**, 5, pp. 1–2.

Daniels, D. J. (ed.) (1975) *New Movements in the Study and Teaching of Chemistry*, Temple Smith, pp. 177–180.

Attention is also drawn to the *Society for Academic Gaming and Simulation in Education and Training* located at the University of Loughborough.

Bibliography

Archenhold, W. F., Jenkins, E. W., Wood-Robinson, C. (1977) *Addresses for Science Teachers*, Centre for Studies in Science Education, The University, Leeds, LS2 9JT.

Atkinson, N. G., Atkinson, J. N. (1975) *Modern Teaching Aids*, Macdonald and Evans.

British Film Institute, *Film and Television in Education*, BFI.

Butler, H. C. (1966) *8 mm in Education*, NCAVAE.

Cable, R. (1965) *Audio-Visual Handbook*, ULP.

Coppen, H. (1964) *Wall Sheets, Their Design, Production and Use*, NCAVAE.

Coppen, H. (1965) *Aids to Teaching and Learning*, Pergamon.

Dale, E. (1969) *Audio-Visual Methods in Teaching*, Holt, Rinehart and Winston.

Davies, I. (1971) *The Management of Learning*, McGraw-Hill.

Kidd, M. K., Long, C. W. (1963) *Projecting Slides: Practical Aspects of Slide, Filmstrip and Episcope Projection*, Focal Press.

Leggatt, R. (1970) *Showing Off or Display Techniques for the Teacher*, NCAVAE.

NCAVAE (1965) *A Guide for the Production of Wall Charts*, NCAVAE.

—*Visual Aid Year Book*, NCAVAE and *Visual Education* (monthly).

Powell, L. S. (1966) *Guide to the Overhead Projector*, BACIE.

Romiszowski, A. J. (1968) *The Selection and Use of Teaching Aids*, Kogan Page.

Vincent, A. (1965) *The Overhead Projector*, NCAVAE.

5 Apparatus for school science teaching

Air-tracks

Linear air-tracks allow virtually friction-free motion for investigating dynamic processes such as uniform velocity, uniform acceleration and momentum conservation. By restricting the movement of masses to one dimension, the air-track is particularly suitable as an *introduction* to elastic and inelastic collisions.

The following features are of importance when considering the purchase of an air-track.

a Design. A robust construction minimises the possibility of sagging. Suitable designs include a triangular extruded section or a square tube section. The track is normally mounted horizontally on two supports positioned an equal distance from each end, with an edge of the track uppermost. A catapult arm, with slots for elastic bands, is required at each end of the track, and the supports should allow for fine adjustment of the position of the track.

b Material of track and supports. To permit the use of magnets in elastic collision experiments, the various components making up the track must be made of a non-magnetic alloy, e.g. aluminium alloy.

c Length. Air-tracks are available in lengths from about 1.5 m to over 2 m. Factors such as storage facilities, the length of tables in the laboratory and relative cost should be considered before deciding on a particular length of track.

d Air holes. Holes drilled alternately on opposite sides of the track, with a spacing of about 2.5 cm, increase the stability of stationary 'vehicles'.

e Vehicle design. The vehicles should be of non-magnetic alloy construction and have facilities for the attachment of various types of buffer at each end. A vertical hole in the centre of the vehicle for a straw or white plastic rod is useful for experiments involving stroboscopic photography, and a vertical slot in the top of the vehicle enables a white card to be positioned for photoelectric timing experiments.

f Number of vehicles. The minimum number for a normal range of experiments is three—two of equal mass when any two buffers are attached, and a third vehicle having half that mass when any two buffers are attached. Momentum calculations are simplified if the masses are in a ratio 2:1, say 0.4 kg and 0.2 kg respectively.

g Buffer attachments. These include:
 i small catapults with slots for an elastic band;
 ii magnetic buffers incorporating a ceramic magnet;
 iii a holder for loading with plasticine.

h Air blower. This can be any conventional cylindrical vacuum cleaner used in reverse. For safety, the air blower should be electrically earthed via a three-core electric cable, and the flexible air hose should terminate in an adaptor making an airtight seal with the inlet of the track. It is possible to service more than one track from a single air blower by using a suitable Y-piece adaptor.

Ammeters (see Meters)

Amplifiers

Amplifiers operated from the mains (usually 240 V a.c.) or from batteries (usually 9 V d.c.) have come into common use in school physics courses, particularly with the increasing emphasis on electronics and its applications.

Before purchasing an amplifier, it is extremely important to check its specification to ensure that it will meet the requirements of the course. The cost of replacement batteries or of a charger module to recharge 'rechargeable batteries', needs to be considered when buying battery-operated amplifiers.

In common with other electronic equipment, amplifiers should incorporate 4 mm sockets for input and output signals. A clearly marked 'on-off' switch is particularly important for battery-operated amplifiers to conserve battery life.

Power amplifiers

Mains-operated, general purpose power amplifiers are often included in an instrument that also contains a loudspeaker or a signal generator, which may be used separately and so save possible expenditure. Such amplifiers should respond to frequencies from zero (d.c.) to about 50 kHz and have a variable gain control with an output of several watts into a 3–8 Ω load. Battery-operated power amplifiers give a lower output, usually in the region 50–300 mW. A high impedance voltage output should be available for experiments involving a.c.

D.C. amplifiers

These may be used to measure currents in the range 10^{-8} A to 10^{-11} A and charge in the range 10^{-7} C to 10^{-9} C. Built-in calibration is desirable.

Operational amplifiers

These amplifiers may be used to magnify both steady and varying inputs by a very large factor (say 10 000), and to perform mathematical operations such as integration. Operational amplifiers are always used with feedback, so that the gain of the arrangement is largely independent of the amplifier's own characteristics and depends mainly on the values of the resistors and capacitors connected in the circuit.

Circuits including operational amplifiers can be devised to solve differential equations and consequently these circuits can be used as analogues of any physical system which obeys such equations. When purchasing operational amplifiers, it is important to obtain from the suppliers suitable circuits, which include bias and stability arrangements, for various applications.

Pre-amplifiers

These are battery-operated, transistorised voltage amplifiers, which usually function over a particular frequency range, either audio (up to say 50 kHz) or high frequency (up to say 10 MHz). They are used as an intermediate amplifier prior to the signal being passed to a power amplifier or an oscilloscope.

Anatomical models

A wide range of anatomical models is available from commercial suppliers. The types most likely to be useful in schools are discussed below.

a **Small scale models.** These are constructed of polystyrene and are available in male or female forms. A transparent outer 'skin' encloses the principal organs of the body and bones of the skeleton. The models are approximately one-fifth life size (linear scale) and are normally supplied in separate parts which need to be assembled and glued with polystyrene cement before use. Alternative parts are available for the female to simulate the internal condition during advanced pregnancy.

b **Approximately life-size models.** These include torsos, hearts, heads, brains, etc. The separate parts, which are realistically coloured, may be detached for a closer examination and then re-assembled. Such models may be constructed of rubber latex which is slightly flexible and effectively unbreakable, or they may be made of plaster of Paris which is easily chipped.

c **Large scale models.** Individual parts, such as ears, eyes, kidneys and larynx are available on an enlarged scale, usually four or five times life size (linear scale). Like the life-size models they may be constructed of either rubber latex or plaster of Paris, though certain parts, such as the lens and cornea of the eye, may be made of other suitable plastic material. These models are realistically coloured.

d **Skeletal models** (see Skeletons, page 142).

Aquaria

An aquarium is one of the most economical and useful ways of maintaining a large number of living plants and animals in a semi-natural way in the laboratory. Once established it requires little maintenance. Algae, aquatic angiosperms, protista, planaria, aquatic snails and insects, crustaceans, water mites, amphibia and fish can all be kept in suitable freshwater aquaria. The range of types of organism is even larger in the case of marine aquaria, though there are greater problems of maintenance. A carefully prepared aquarium can also provide an excellent example of a miniature ecosystem exhibiting a delicate balance between the organisms inhabiting it and their environment. Even if no other living organisms are being maintained in a laboratory, an aquarium will prove invaluable as a source of organisms for simple physiological work. (See page 281 for details of setting up aquaria.)

Various types of aquarium are available from biological suppliers and high street aquarists. Suitable aquatic plants and animals may be obtained from the same sources or even collected in the field. There are three basic types of aquarium.

a **Metal-framed rectangular tanks**

Iron, stainless steel or chromium-plated models exist with slate or plate-glass bases and drawn-glass sides. Larger models may have plate-glass sides. Iron frames should be either painted with a non-toxic paint or, preferably, covered with non-chip plastic to resist corrosion. A wide range of sizes from 45 cm × 25 cm × 25 cm to 95 cm × 37.5 cm × 37.5 cm is available from suppliers. Larger or non-standard sizes may be specially ordered. Aquarium stands are also available.

The main advantages of metal-framed aquaria are that they can be relatively easily repaired and the glass provides good visibility. Set against this, they are relatively heavy and costly compared with plastic tanks.

b Moulded glass tanks

These are generally of a smaller size than the metal-framed models. A range from 25 cm × 17.5 cm × 17.5 cm to 35 cm × 22.5 cm × 22.5 cm is available from suppliers. The glass should be taped around the top edge to reduce the risk of chipping. Although these tanks provide greater all-round visibility, the quality of the glass is inferior to that used in metal-framed tanks and this results in considerable visual distortion. Once damaged—unless through relatively minor chipping—they are irreparable. Moulded glass tanks are not very suitable as general aquaria, but are quite adequate for breeding and other specialised purposes.

c Transparent plastic tanks

These are available in a wide range of sizes from 30 cm × 20 cm × 20 cm to 90 cm × 37.5 cm × 37.5 cm. Some models also have bow fronts for greater visibility. For the smaller sizes, plastic tanks are generally cheaper than glass models. However, the surface of the plastic is liable to scratching and hence a reduction in its already limited transparency.

For most general purposes, metal-framed tanks are recommended, and smaller sized plastic or moulded glass versions are best reserved for breeding or other specialised functions.

AQUARIUM ACCESSORIES

Aerators

In a natural aquatic environment, the respiratory activity of the animals and plants is balanced by the photosynthetic activity of the plants and the gas exchange taking place at the surface of the water. As a result the organisms in the environment do not normally suffer from anoxia or carbon dioxide poisoning. However, in many aquaria it is not possible to achieve such a natural balance. This is mainly due to the fact that the concentration of animals in an aquarium is generally very much higher than would be found under natural conditions. The problem can be overcome by the use of an aerator which will break up the water surface and hence speed up the rate of gas exchange. It will also increase the circulation of water, so distributing heat and respiratory gases more evenly through the tank. An aerator is also essential if any form of filter is to be employed.

A variety of aerator pumps is available, of diaphragm or piston type, the latter being generally quieter in operation. Some models are of the 'pressure and vacuum' type and are more versatile than those which only provide air at increased pressure in that they can be used for experimental work where recirculation of air is required.

Aerators are normally used in conjunction with a diffuser stone or a filter. Models are available which will supply from one to 18 diffusers and the choice must depend upon the number and siting of aquaria within the laboratory.

N.B. *Any exposed metal part of an aerator pump must be correctly earthed.*

Artificial sea water and pond water

Various manufacturers market artificial aquarium media in powdered or tablet form. These are simply diluted with an appropriate volume of water as described in the manufacturers' instructions. Artificial media may also be made up in the laboratory (see page 191).

Canopies

An aquarium canopy is recommended for many purposes. It minimises water loss through evaporation—an especially important consideration with tropical aquaria. It also prevents the entry of dust and reduces the risk of unwanted interference with the tank. Some fish are able to jump from the surface of the water and in these cases a canopy is obviously essential. A canopy also provides an anchor point for aquarium lights. Aluminium and plastic canopies are suitable for either tungsten filament or fluorescent lamps.

Alternatively, a simple sheet of glass serves many of the purposes mentioned above. Two corners should be cut from the glass to allow the entry of heater flex and of air lines, and the edges of the glass should be taped to prevent chipping and to reduce the risk of fingers being cut while handling the canopy. Flat glass canopies should be slightly raised above the aquarium edge to permit a rapid exchange of gases. This may be done with small pieces of rubber or plasticine. If a glass canopy is used, lighting must be supplied externally.

Diffuser stones

If an aerator pump is being used in the absence of an undergravel filter, the air line should be fitted with a diffuser stone. This is placed on the surface of the aquarium gravel and serves to break up the bubbles of air, thus permitting a more efficient gas exchange and preventing excessive turbulence of the water.

Feeders

If an aquarium is being left unattended for several days, food may be provided through an automatic feeder. This should not be necessary for adult fish, which will survive up to a week without feeding. The principle of the automatic feeder is illustrated in Figure 5.1. The central disc rotates once every 24 hours. Each time an aperture passes beneath the food hopper it picks up a measured volume of food which is delivered to the tank some hours later when the aperture is positioned above the delivery tube. The frequency of feeding may be reduced by

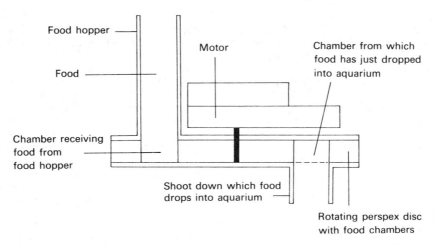

Figure 5.1   An automatic aquarium feeder

blocking off some of the apertures and the size of feed may be varied by using sleeves to alter the size of the aperture. Automatic feeders incorporate a time switch so that aquarium lighting may also be controlled automatically.

Filters

A filter removes some of the waste products which might otherwise prove injurious to the inhabitants of the aquarium. One type of filter is sited beneath the aquarium gravel at the base of the tank. Used in conjunction with the aerator pump it draws detritus down into the gravel where bacterial decomposition releases salts back into the water. Different designs of undergravel filter are used in marine and freshwater tanks and care should be taken to ensure that the correct one is employed. These types of filter are relatively inconspicuous and there are no filter materials to be changed or renewed.

A more elaborate filter involves the use of a polymer wool to remove suspended solids from the water and activated charcoal to remove dissolved gases. Both substances must be changed regularly otherwise serious pollution can occur. This type of filter is generally fitted outside the tank and hence tends to be more conspicuous than the undergravel type. It is not suitable for use if fish fry or other small animals are present as these are liable to be drawn into the filter along with the water.

Gravel and stones

Aquarium gravel serves a number of functions: it acts as an anchorage for rooted plants; it allows for the effective removal of detritus from the clear water of the tank where it would be stirred up by the aerator or the animals; it harbours bacteria which will decompose the falling detritus; and finally it adds to the appearance of the aquarium.

Aquarium gravel should be chemically inert. Seashore gravel, shells and artificially dried material are best avoided. The particles should be smooth and fairly uniform in size and should make for an open texture so that detritus can easily pass through it. Larger stones in an aquarium should be employed with considerable caution as many will adversely affect the pH of the water or add injurious chemicals to the tank. Gravel should always be thoroughly washed before use (see page 281).

To calculate the amount of gravel required for a given aquarium allow 1.5 kg gravel for every 10 litres of water. Hence a $60 \, cm \times 30 \, cm \times 30 \, cm$ tank holding approximately 50 litres of water will require about 7.5 kg of gravel. This will give an average depth of about 3 cm which is usually best distributed by sloping the gravel from a depth of about 2 cm at the front of the tank to about 4 cm at the rear.

Heaters

Thermostatically controlled heaters are required for all tropical tanks. The power of the heater will depend on the size of the tank and upon the difference between the required temperature of the water and the minimum temperature to which the surroundings are likely to fall. As a general rule, 0.2 watt per litre for each degree Celsius difference in temperature should prove sufficient. For example, a $60 \, cm \times 30 \, cm \times 30 \, cm$ tank (holding approximately 50 litres of water) which is to be maintained at 25°C in a room which may drop to 15°C will require a 100 W heater. (Volume of water × temperature differential × 0.2, i.e. $50 \times 10 \times 0.2 = 100 \, W$). Heaters are normally sealed within a Pyrex glass tube or a plastic-coated aluminium alloy and are doubly protected by an inner sheath of mica or other insulating material. This double insulation is required by the Electrical Equipment (Safety) Regulations, 1975, and the use of other forms of heater is not recommended. The advice of the LEA Science Adviser or other suitable authority should be sought before installing heaters of an older design.

Heaters should always be used in conjunction with thermostats. These are available either as separate units or in combination with the heater itself (see Figure 5.2). The latter are normally more convenient as they require fewer electrical connections. However, should a fault develop, the entire unit—heater and thermostat—must be replaced.

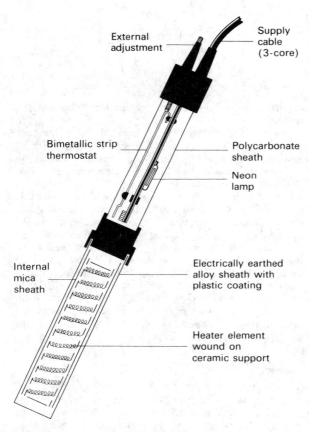

Figure 5.2   Aquarium thermostat and heater

The heaters and thermostats described above are intended for immersion in the water of the aquarium. Some thermostats are designed to allow the control knob to protrude from the surface for ease of adjustment. In other cases the entire unit is immersed, which has the advantage that it makes unwarranted interference much more difficult. In either case they are usually provided

with a screw adjustment for setting to the appropriate temperature. However the actual temperatures are not normally indicated on the thermostat itself and some experimentation is necessary to determine the setting which will maintain the water temperature in the tank within appropriate limits. Most thermostats are fitted with a pilot light, which aids this procedure by indicating when the heater is operating. Details concerning the wiring, adjustment and operation of the pilot light should be supplied as part of the manufacturer's instructions accompanying the thermostat and/or heater.

The position of the heater in the tank is not usually significant. Convection currents and movement of water by the aerator and animals should result in a fairly even distribution of heat throughout the aquarium. No part of the heater or thermostat—except in some cases the control knob—should protrude above water level or be buried in the gravel at the bottom of the tank; this can lead to overheating, cracking of the glass tube or failure of the heater with unfortunate and possibly dangerous consequences.

When an aquarium is being set up (see page 281) organisms should never be introduced to heated water unless it has been established that the heater-thermostat is operating efficiently and maintaining the water temperature within acceptable limits.

N.B. *Under no circumstances should heaters be used unless they are correctly earthed and fitted with an appropriate fuse. They should also be designed to ensure that, in the event of a cracked glass tube, the entry of water causes the fuse to blow.*

Lights

Fluorescent or tungsten filament lamps may be used. The initial cost of installation is considerably lower for tungsten filament lamps, but they need replacing more frequently. They may also give rise to high temperatures in the top few centimetres of water in the tank. A 15 W fluorescent tube or a 40 W tungsten filament lamp should provide sufficient light for a 60 cm × 30 cm × 30 cm tank if it is left on for 6–8 hours each day. However, some experimentation is necessary for best results. If too much light is provided, the tank may develop a heavy growth of algae; if the illumination is insufficient, the plants will not grow.

N.B. *All aquarium lights, especially if used with a metal canopy, must be correctly earthed.*

Nets

A net must always be used when transferring fish from one tank to another. The overall length of the net should be several centimetres more than the depth of water in the tank; the width should be about one-third of the minimum dimension of the tank. The sides and end of the net should form right angles, otherwise fish can lurk in the corners of the aquarium. The net mesh should be made of nylon and be fine enough to prevent fins and scales from getting snagged yet coarse enough to allow easy movement through the water. Biological suppliers and aquarists usually provide a suitable selection to choose from.

Scrapers

An aquarium scraper is simply a razor blade in a holder attached to a long handle. It is used for scraping algae from the sides of the tank. Purpose-made scrapers are also available from biological suppliers and aquarists.

Sealing compounds

Various types of aquarium glazing compounds are available for fixing the glass into metal-framed aquaria. If small leaks develop they can sometimes be cured by the addition of 'Aquastop' to the water. The compound is drawn into the leaking area, where it forms a gel and hence seals the leak. It has the advantage that it can be used without removing the aquarium inhabitants. Larger leaks may be repaired with aquarium putty, but if this is not successful the whole plate may have to be removed and resealed in position. Decorator's putty is toxic to aquarium inhabitants and, therefore, is not suitable for sealing leaks or for fixing the glass in metal-framed tanks.

Siphons

Siphoning is usually the most convenient method of emptying a tank. The siphon tube should be of flexible plastic or rubber and have an internal diameter of about 1 cm. If a piece of rigid plastic or glass tubing is fitted to one end, this can be used for removing small pockets of debris and uneaten food if it should become necessary.

Alternatively, a dip tube can be improvised using a length of glass or stiff plastic tubing fitted with a rubber bulb.

Stands

Metal-framed stands, constructed of angle iron, may be used as supports for metal-framed aquaria. Stands are available from suppliers in a restricted range of standard sizes, but may also be manufactured to order through local aquarist shops. In schools, it is normally more convenient to place tanks directly on a laboratory bench but stands may form a useful alternative in some circumstances.

Thermometers

A thermometer should always be included in an aquarium, and is especially important if heating is being employed. Aquarium thermometers generally float vertically in the water and are fitted with a suction disc with which they can be attached to the side or bottom of the tank. The bulb of the thermometer should always be free in the water and not immersed in the gravel. Aquarium thermometers purchased from high street aquarists are frequently calibrated in degrees Fahrenheit, while those from scientific suppliers generally have both Fahrenheit and Celsius scales.

Thermostats (see section on heaters, page 77).

Asbestos

The use of asbestos in schools is governed by advice from the appropriate government departments. For details, see page 36.

Atmometers

Atmometers are used for measuring the rate of evaporation from a porous surface under different environmental conditions. Some commercially manufactured atmometers are combined with a potometer so that the rates of evaporation may be compared with the rates of uptake of water into a cut leafy shoot. A laboratory atmometer may be improvised from a length of glass tubing, a porous pot and a rubber bung (see page 156), and a field atmometer (or evaporimeter) may be made from a length of capillary tubing, a small disc of filter paper and a scale (see page 159).

Autoclaves

An autoclave is an essential piece of apparatus if any microbiological work is to be undertaken, as sterilisation under pressures of approximately $100\,kPa$ ($15\,lb\,in^{-2}$) above atmospheric is necessary. Domestic pressure cookers may be employed for the purpose but their small size restricts the amount of equipment which can be sterilised at one time. However, if pre-sterilised plastic Petri dishes are used a pressure cooker may be sufficiently large for the sterilisation of media and other small accessories. Domestic pressure cookers which are used for microbiological work should be kept solely for that purpose.

Larger, but portable, autoclaves are available from commercial suppliers. These may be fitted with electrical heating elements or may have to be heated on a gas burner. Autoclaves and domestic pressure cookers are sealed with a rubber gasket and care should be taken to ensure that this is cleaned and dried after use to prevent it from perishing. Autoclaves are normally provided with an adjustable pressure controller which may be set at either 70 or $105\,kPa$ (approximately 10 or $15\,lb\,in^{-2}$) above atmospheric pressure. Pressures much in excess of these levels will automatically operate a valve thus preventing dangerous pressures from building up. It is therefore important to ensure that the control is set to the appropriate pressure before use as it is difficult to adjust once it is hot. Autoclaves are also normally provided with a pressure indicator. Some portable autoclaves incorporate a siphon tube, which allows the removal of the water and steam after sterilisation so that the contents are dried and maintained under near-vacuum conditions. Autoclaves (and domestic pressure cookers) are normally provided with inner containers for holding the apparatus being sterilised. Stainless steel autoclave drums are also available which enable apparatus to be dry or steam sterilised.

Vessels of media being sterilised should be plugged with cotton wool and covered with metal foil or paper held in place with string or elastic bands. This will prevent the cotton wool becoming sodden with water. If it is necessary to sterilise a large batch of media, rectangular bottles with aluminium screw caps (medical flats) may be used. These can then be stored for class use at a later date. Media should not be shaken as they are removed from the autoclave as this may cause them to boil and spill. Details of sterilisation by autoclave can be found on page 278.

Balances

The general adoption of SI units throughout the world has accentuated the desirability of calibrating measuring instruments in these units. In particular, the physical quantities of mass, weight and force should be measured in the appropriate SI units.

Certain measurements, for example those required for density and calorimetry calculations, clearly involve the concept *mass*, and should be made in kilograms or grams. A lever balance calibrated in mass units usually has sufficient accuracy for this type of work.

However, if a *force* is required to act on a spring for a Hooke's Law experiment or on a beam to investigate the principle of moments, then it is necessary to calculate the gravitational force on the mass used, i.e. the *weight*, which is measured in newtons. To obtain the weight, the mass is multiplied by the Earth's gravitational field strength at the particular place, which may generally be taken as $10 \, N/kg$ to a first approximation. In simple terms, a gravitational force of 1 newton acts on a mass of 100 grams. It is clearly preferable to refer to 'masses' and 'masshangers' rather than 'weights' and 'weighthangers', in the above experiments.

Similarly, forces exerted by a spring balance or force-meter should be measured directly in newtons, and conversion scales, made of thin, self-adhesive metal tape, are available from some manufacturers for certain spring balances calibrated in grams. A 'home-made' paper scale, showing markings in newtons and stuck over the old scale with clear adhesive tape is cheap and quickly made.

Lever balances

This type of single-pan balance is particularly useful for obtaining rapid readings of mass. The versatility of the balance is increased if it has two ranges, $0–250 \, g \times 1 \, g$ and $0–1000 \, g \times 5 \, g$ or $10 \, g$, the change-over being made by rotating a hinged arm. If the zeros on the two scales correspond, no re-levelling is necessary when changing the range of the balance.

The base should be fitted with a levelling screw for adjusting the zero setting, and a hook fitted to the underside of the pan support will extend the use of the balance to include density work on air pumped into a large plastic container.

Spring balances (force-meters)

The most common types of spring balance are as follows:

  i tubular barrel;
 ii flat barrel;
iii dial type;
 iv compression type;
  v personal weighing machine.

For individual class experiments, the tubular barrel spring balance has become popular. The barrel should be corrosion- and scratch-resistant, and the movement of the inner scale is made more reliable if nylon bushes are provided as guides inside the barrel. All balances must have an adjustable zero and be clearly calibrated in newtons.

The most popular tubular or flat barrel spring balances used in school physics teaching have a range of $0–1 \, N \times 0.01 \, N$ and $0–10 \, N \times 0.1 \, N$ (generally figured every newton), the latter measuring the gravitational force on masses up to 1 kilogram. Higher capacity tubular or flat barrel balances are available, but dial type demonstration balances, with a large and easily-read scale and a range of $0–50 \, N \times 0.25 \, N$ or $0–100 \, N \times 0.5 \, N$ are probably more useful. A popular range for a compression type spring balance is $0–50 \, N \times 0.25 \, N$, the more recent models having a rigid plastic load pan. The platform type personal weighing machine should be fitted with a magnifier over the scale opening, and have a range of $0–1250 \, N \times 5 \, N$ or $0–1000 \, N \times 5 \, N$.

Chemical balances

The purchase of a chemical balance is probably one of the most expensive single outlays of capital expenditure by a school science department. General information on balances, weights and weighing will be found in any comprehensive text of quantitative chemical analysis, e.g. Vogel's *Quantitative Inorganic Analysis*.

The following factors will govern the choice of a balance for school use:

  i size of instrument;
 ii weighing capacity;
iii degree of accuracy;
 iv robustness;

v ease of use, i.e. number of controls and their position, left- and right-handed access to the balance pan, ease of reading (including parallax errors in optical scales, etc.);
vi ease of loading the balance pan(s);
vii ease of cleaning;
viii servicing facilities;
ix use to which the balance is put;
x price.

In recent years, top-loading balances have become very popular in schools. In many instances these balances are portable and do not require the purchase of boxes of weights. Top-loading balances are ideal for main school work but may not always be sufficiently accurate and/or reliable for some sixth-form exercises. One of the largest Examination Boards requires that candidates be able to weigh to within ± 2 mg and recommends that aperiodic beam balances be used for the purpose. 'Top-pan' balances which are sufficiently accurate to weigh within this limit are, of course, available, but they generally require adequate shielding from draughts and should not also be used for other quantitative work.

Balance pans invariably deteriorate with use. Spillages are almost inevitable and, in the case of solids, damage can be minimised by using filter papers on the pan or by using the paper or plastic weighing boats which are now commercially available. Grease can be removed from a balance pan by warm, soapy water—abrasive or strong chemical reagents should not be used. In the case of a top-pan balance, the cleaning process may be so severe that it is impossible to re-set the zero on the balance. With a beam balance, the adjustment of the balance is readily destroyed. It follows that spillage is best prevented rather than treated. Hence, volatile and deliquescent or highly corrosive substances must always be weighed in a closed container.

The location of a balance is an important consideration but one in which some compromise is usually necessary. Portable, top-loading balances render a balance room unnecessary, but for accurate beam balances a reserved location is essential. The atmosphere must be free from corrosive materials, including excessive moisture and warmth, and the balance must rest on a vibration-free bench. If a beam balance is kept in a balance case, the atmosphere within the case should be kept dry by means of a suitable desiccant, e.g. calcium oxide or 'self-indicating' silica gel.

If it is necessary to move a balance, considerable care must be taken. Top-loading instruments usually have a screw-lock which fixes the beam and this should be turned to the lock position before the balance is moved. If a beam balance has to be moved, the pan(s) and support(s) should be adjusted so as to reduce the load and minimise any potential damage to the knife edge.

A detailed record should be kept of the servicing received by each balance. The intervals between servicing should depend on the amount of use and transportation of the balance, rather than upon some arbitrary period of, for example twelve months.

Barometers

An accurate value of air pressure is required occasionally in most science courses. Whereas an aneroid barometer gives an approximate reading, the *Fortin barometer* enables air pressure to be determined with great accuracy. A *barograph* provides a continuous record of air pressure over a seven-day period and can be used to illustrate changes in pressure with different weather conditions.

Fortin barometer

The following points require attention when the purchase of this instrument is being considered.

a Quantity. One instrument per school should be sufficient. Departments other than science, e.g. geography, may wish to take occasional readings.

b Type. A typical instrument has a scale marked from 675 to 820 mm of mercury ($\times 1$ mm) (91–111 kPa $\times 0.135$ kPa) with a vernier reading to 0.05 mm (but see note **f** on altitude). A screw device is the most accurate for adjusting the mercury level in the cistern to touch the tip of the fixed pointer. It is useful to have a thermometer $-10\,°C$ to $+50\,°C$ ($\times 0.1\,°C$) mounted on the hardwood board to which the barometer is attached.

c Case. Serious consideration should be given to the desirability of purchasing a lockable, hardwood case with glazed front and side panels as protection for this delicate and expensive item of equipment.

d Transportation. Prior to installation, the barometer must be carried upside down, i.e. with the cistern uppermost, to avoid possible damage to the cistern.

e Site The science preparation area or a 'quiet corner' in a laboratory are suitable sites, bearing in mind that the barometer should not be exposed to strong direct sunlight or heat radiation. It should be fixed in a vertical position using a plumb line, so that the top of the mercury column is at a convenient height for observation—approximately 150 cm above floor level.

f Altitude. If the school is situated at a high altitude, i.e. above 500 metres, the height above sea level must be stated when purchasing the barometer, in case any special scale has to be provided. Correction and conversion scales are provided in BS 2520 for readings taken at temperatures other than $0\,°C$ and at heights other than sea level.

g Insurance. The suppliers recommend the overseas customer to insure against damage in transit. The supplier will usually arrange such insurance upon request.

Barograph

The seven/eight-day recording barograph is based on an aneroid system covering a pressure range from approximately 700 mm to 810 mm of mercury (95 kPa to 110 kPa). A magnifying lever movement is attached to a recording arm which leaves an ink trace on a calibrated barometer chart wound on the slowly revolving cylinder. Periodically the barograph readings should be checked against the pressure determined on a Fortin barometer and suitable adjustments made if necessary.

Some barographs have a bimetallic element incorporated in the same case so that temperature can also be recorded on a continuous basis. The chart paper ordered for such a 'barograph/thermograph' must have markings in pressure and temperature units.

Batteries (heavy duty)

With the increasing availability in recent years of variable low-tension power-packs run from the mains supply (see power supplies, page 137), the use of accumulators or storage batteries has decreased somewhat. However, for laboratories not fitted with ring mains and for certain experiments in 'modern physics' courses, the heavy-duty battery may still be the only suitable source of direct current.

When purchasing heavy-duty batteries, the choice rests

	Lead–acid battery	Nickel–cadmium battery
Electrolyte	Liquid	Liquid (solid for export)
Battery voltage	12 V from 6 cells	6 V from 5 cells
Voltage steps	× 2 V	× 1.2 V
Typical capacity	40 Ah at 10 h rate	40 Ah at 5 h rate
Other capacities	20 Ah, 30 Ah, 60 Ah, 80 Ah	10 Ah, 16 Ah, 23 Ah, 32 Ah
Robustness	Cells in polystyrene containers in tough moulded-rubber case	Cells in polystyrene containers in wooden case
Laboratory use	Rechargeable, high discharge current but easily damaged by short circuits. Should be regularly charged	Rechargeable, can be overcharged, inadvertently short-circuited and left uncharged without deterioration

Table 5.1   Characteristics of lead–acid and nickel–cadmium batteries

between the lead-acid type car battery and the nickel-cadmium alkaline battery. Both should be fitted with 4 mm sockets on each cell so that it is possible to tap off intermediate voltages. The maintenance/charging instructions supplied with the batteries should be followed strictly and stored safely for future reference. The acid or electrolyte is normally ordered separately. The major features of the two types of battery are summarised in Table 5.1.

MAINTENANCE

a Lead-acid batteries

Lead-acid batteries should never be allowed to stand in a discharged condition. If the voltage falls below 2.0 V per cell (on open circuit), it must be fully recharged as soon as possible. The open-circuit voltage of a fully charged battery is 2.1 V per cell.

Many commercial chargers provide a *constant voltage charge*, with a high initial current falling to a low value at the end of the charge. This method has the advantage of charging the battery reasonably quickly without risk of serious overcharging. If the battery is charged on a constant current *trickle charger*, the charging current should not exceed 2.5 mA per ampere hour of the battery's rated capacity. When charging, the positive of the supply must be connected to the positive of the lead-acid battery.

Batteries should be checked regularly and topped up with distilled water. If a cell is overfilled, a syringe may be used to remove the excess liquid. After topping up, the terminals and inter-cell connections should be smeared with battery grease. It is good practice to keep a record of the dates of charging.

b Nickel-cadmium batteries

The nominal voltage of a nickel-cadmium cell is 1.25 V per cell, although the voltage may rise to 1.45 V per cell on maximum charge. To prolong the life of a battery, the cells should not be allowed to discharge below 1.0 V per cell.

The constant voltage charging method must *not* be used with nickel-cadmium batteries because of the very low internal resistance of the cells. It is essential that nickel-cadmium cells are charged by the *constant current* charging method. Some commercial chargers may be set at the appropriate charging rate (as specified in the maintenance/charging instructions supplied with the battery) for a given time period, at the end of which the mains supply to the charger is automatically disconnected.

The cylindrical cells, dimensionally equivalent to 1.5 V 'U2' cells and having a capacity of 4 Ah each, may be recharged using a steady charging current of 400 mA for about 10 hours.

Beakers

Beakers are available in a variety of shapes, sizes and materials. The common shapes are illustrated in Figure 5.3.

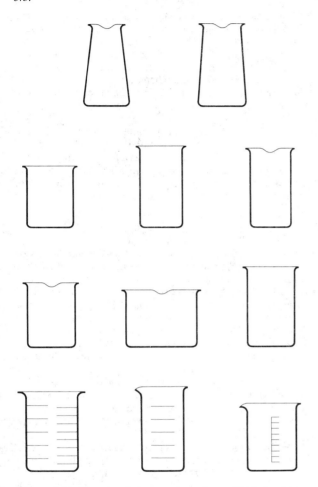

Figure 5.3   Shapes of some commonly available beakers

In schools, the type of beaker most commonly used is the squat variety. Many different sizes are available with a nominal capacity ranging from $5 \, cm^3$ to $5000 \, cm^3$ (5 litres). Most of these sizes are available with or without graduations.

Conical beakers are marginally more expensive than the squat variety but they are less easily knocked over and can often be used as substitutes for conical flasks, e.g. in carrying out titrations.

Beakers are manufactured in a number of materials, including aluminium, nickel, stainless steel, borosilicate glass, PTFE, polycarbonate, polypropylene and poly(methylpentene). Plastic beakers, which may be transparent or opaque, have the particular advantage that they do not break. Further, they are usually cheaper than corresponding glass beakers. The resistance of plastic beakers to heat and chemical reagents (especially organic solvents) is surprisingly high. For example, the heat distortion point of poly(methylpentene) is approximately $80 \, °C$. Plastic beakers are undoubtedly satisfactory for many school laboratory operations.

Consideration should be given to the sizes of beakers purchased by a school. Large beakers, i.e. 2 litres and above, are expensive and are used relatively infrequently. Similar considerations apply to the purchase and use of very small beakers (5 or $10 \, cm^3$). On the other hand, there seems little point in having a stock which includes beakers of 50, 100, 250, 400, 500, 600, 800, 1000, 1500, 2000, 3000, 4000 and $5000 \, cm^3$ capacity. In many schools, it will be satisfactory to have sufficient numbers of beakers with capacities of 50, 100, 250, 600, 1000 and $2000 \, cm^3$ capacity. Beakers are often sold in *shelf* packs (of 12, 10, 6 or 4 items), or *standard* packs containing a larger number of beakers, the precise number depending on the capacity of the beaker concerned. As is the case with many items of laboratory glassware, considerable savings arise from the single purchase of a large number of standard packs. However, the discount which is offered, usually from 5 per cent to 20 per cent, needs to be considered in the light of the very considerable storage problems which may arise. It is for this reason that some Local Education Authorities purchase glassware and other equipment in bulk quantities and make the items available to their schools as required.

Beakers are among the most heavily used items of laboratory glassware, and, as such, can become very dirty. Beakers and other glassware which are superficially dirty may be cleaned using hot water and a detergent, with a suitable brush if required. Once washed in this way they should be rinsed with clean water and left upside down to drain dry. It is easier to wash and dry items of laboratory glassware if a *large* sink, supplied with hot and cold water, is available together with appropriate draining boards and drying racks.

Most school beakers can be satisfactorily cleaned using the above procedure. However, some stains and residues may be resistant to the action of detergents and warm water. In general, beakers should be cleaned in the first instance with the mildest cleaning agent. If this is ineffective specific chemical reagents and/or mild abrasives should be tried. Thus manganate(VII) (permanganate) stains may be readily removed using aqueous sulphur dioxide or dilute nitric acid. Where the action of such specific chemical reagents is unsuccessful, the so-called 'chromic acid' mixture is often used. However, this practice is to be deprecated. A mixture of concentrated sulphuric acid and sodium dichromate(VI) is highly corrosive and an extremely powerful oxidising agent. Explosions have been reported when 'chromic acid', from a stock bottle, has been used as a cleaning agent. A number of powerful and quick-acting detergents have been produced for laboratory use and these should be employed in preference to 'chromic acid'.

Care should be taken when using abrasives to clean beakers as glass is not a completely inert material and it is very easily scratched. If it is known that a reaction to be carried out in a beaker is likely to leave a residue which is difficult to remove (e.g. the action of concentrated sulphuric acid on sucrose to produce carbon), consideration should be given to carrying out the reaction in a vessel which may be discarded afterwards.

Beakers which are chipped, cracked or very badly scratched should not be used.

Blood-grouping apparatus

Various types of blood-grouping apparatus are available from commercial suppliers. 'Eldon cards' bear small labelled areas which have been treated with the appropriate sera and have then been dried *in situ*. The addition of a few drops of tap water reconstitutes the sera, which are then ready to receive small drops of blood. The agglutination of the red cells is easily seen and remains visible after the blood has dried. The cards may thus be stuck in students' notebooks as a permanent record. Eldon cards have the advantage that they avoid the use of bottles of liquid sera, which may be spilled. They may also be stored for at least two years at room temperature.

An alternative method incorporates the use of self-adhesive 'Type Tabs' which again can be kept as a permanent record. These are used in conjunction with appropriate liquid anti-sera.

Undoubtedly the cheapest method, but one which does not provide the student with a permanent record, is to obtain phials of the appropriate anti-sera from a local hospital or blood bank. These may be obtained more readily if the teacher and/or technician is prepared to donate blood!

Strict precautions should be observed when pupils are obtaining samples of their own blood (see page 43) and no pupil must be compelled to produce a sample. The area from which blood is to be obtained should first be sterilised with a medical swab soaked in alcohol or surgical spirit. Disposable sterile lancets should be used to prick the skin and these must be disposed of immediately after use.

Blood lancets

These are made of stainless steel and are essential for the safe collection of small blood samples. They are individually packed and sterilised and for school use are best obtained in bulk quantities of 250 or more. *Blood lancets must never be re-used.*

Bottles

a Dropping bottles

Two types of dropping bottle may be encountered in the biology laboratory. The 'TK' pattern bottle has a ground-glass stopper bearing two vertical grooves of different size and a conical projection from which the drops of liquid are dispensed. To obtain liquid from the bottle the stopper is rotated until the grooves are aligned with complementary grooves in the shoulder of the bottle. Drops will be delivered when the bottle is tilted. TK pattern bottles are available in nominal capacities from $30 \, cm^3$ to $100 \, cm^3$. Some sizes are also available in amber glass.

Polystop bottles are also made of glass but have a dustproof polythene stopper fitted with a glass dropper and a vinyl teat. They are available with nominal capacities from $30 \, cm^3$ to $500 \, cm^3$. The larger sizes are fitted with a vinyl suction bulb for dispensing larger volumes of liquid and the glass dropper may also be calibrated.

b Reagent bottles

Reagent bottles may be purchased with ground-glass or interchangeable plastic stoppers. They may be of clear or amber glass and have a wide or narrow mouth. Narrow-mouth reagent bottles are preferred for liquids and these are available in capacities of 30, 60, 125, 250, 500 and $1000 \, cm^3$. Smaller sizes should be used for expensive reagents or for reagents which deteriorate with storage.

Amber coloured glass reagent bottles are used for substances, such as aqueous silver nitrate, which undergo photochemical decomposition.

It is sensible to purchase a set of reagent bottles bearing permanent labels, identifying the common laboratory reagents such as dilute sulphuric acid or sodium hydroxide. These permanent labels are highly resistant to chemical attack.

Where large bottles or jars are required to store solid reagents, such as sulphur or copper(II) sulphate, which are used in considerable quantities in a large school, square bottles are often more convenient than round. Sweet jars are adequate substitutes for the large bottles available from laboratory suppliers.

The correct and clear labelling of reagent bottles is extremely important. Self-adhesive labels bearing the names of reagents are available from the usual suppliers and, in some instances, labels can be made satisfactorily from 'Dymo' tape. The ASE produces sets of labels bearing safety and other information in addition to the names of the reagents. For chemical hazard warning labels using pictograms see page 37.

Buffer solutions

Buffer solutions are required for a variety of different

purposes in school chemistry teaching, e.g. the 'setting' of a pH meter and the control of pH in complexometric titrations. However, for most chemical reactions studied at school, very strict control of pH is not required.

Buffer solutions are most conveniently prepared by means of 'buffer tablets'. These are obtainable from a number of commercial suppliers and one tablet dissolved in 100 cm³ of distilled water will provide a solution of a stated pH, at a specified temperature, often 20 °C.

Buffer tablets are purchased in packs of 50. However, these tablets are available for certain pH values only, e.g. 4.0, 7.0 and 9.2. Where other pH values are required it is necessary to make up the particular buffer solution needed. This is most easily done using a 'Universal Buffer Mixture'. For example, the BDH Universal Buffer Mixture may be used to make 1 litre of buffer solution with a pH of from 2.7 to 11.4. Other buffer mixtures are also available from commercial suppliers, together with full details of how to use the buffer mixture to produce a solution of a given pH.

Where buffers are required for more specialised purposes, e.g. in the study of biological systems, they are best purchased directly from the suppliers. Details will be found in the appropriate catalogues.

Burettes

Burettes are available in two grades, A and B. The specification for each of these grades is given in the British Standards publication BS 846 (1962) entitled *Specification for Burettes and Bulb Burettes* and its amendment dated 17th January 1964. The differences between the two grades are illustrated in Table 5.2.

Nominal capacity burette *	Sub-divisions	Tolerance on capacity (ml)		Delivery time			
				A		B	
		A	B	Min. s	Max. s	Min. s	Max. s
25	0.05	±0.03	±0.05	120	170	85	170
25	0.1	±0.05	±0.1	70	100	35	70
50	0.1	±0.05	±0.1	105	150	75	150
100	0.2	±0.1	±0.2	100	150	65	130

* ml = unit used in BS 846

Table 5.2 Differences between Grade A and Grade B burettes

Tables for use in the calibration of volumetric glassware are given in BS 1797 (1968).

Grade B burettes—which are less expensive than Grade A—are adequate for almost all volumetric work done at school. They are available in several sizes ranging in nominal capacity from 1.0 cm³ to 100 cm³. Unlike Grade

A burettes, where the jet must be an integral part of the burette, Grade B burettes may be fitted with a variety of jets, subject to certain conditions. Those most commonly found in schools are as follows.

a The pinch valve. These valves, made of rubber, open when the centre is pressed with the fingers and are fastened directly on to the end of the glass burette. The valves are interchangeable and are used with glass jets, which are purchased separately from the burette. Pinch valves are adequate for most elementary volumetric work but they are not suitable for use with some reagents. The pinch valve is more satisfactory than a piece of rubber tubing held by a screw, spring or Mohr's clip, because a finer degree of control is possible.

b Stop-cocks. These may be made of glass or PTFE and it is now possible to purchase burettes with interchangeable stop-cocks. The stop-cock is more satisfactory for accurate work than the pinch valve and it is correspondingly more expensive. In the hands of students, stop-cocks tend to either come loose during use or become jammed during storage. The former difficulty may be resolved by employing the proper technique and by using burettes in which the stop-cock is held in place by means of wire, a plastic spring or some other retaining device on the side opposite the tap. Jammed stop-cocks are best prevented by thoroughly cleaning burettes after use and the problem arises less frequently with PTFE stop-cocks than with glass ones. If a stop-cock does become jammed it may be loosened by *gentle* tapping or by placing the tap in warm (not hot) water until sufficient expansion has occurred to allow the tap to be turned. However, glass contracts exceedingly slowly to its original volume and calibrated volumetric glassware must not, therefore, be heated.

Students should be warned of the dangers of filling burettes above head height and be taught the correct techniques for using them. They should be cleaned thoroughly and regularly. Grease is a particular nuisance because it causes the contents of a burette to collect in drops on the wall of the burette as it is drained, and therefore leads to inaccuracy in a volumetric estimation. Grease can be removed with a suitable detergent and water but several rinsings are necessary to ensure that all traces of the detergent are removed. For more thorough cleaning, a burette may be filled with the detergent and left to stand overnight.

Burettes are more likely to stay clean if they are properly stored away from laboratory fumes and dust. After being thoroughly washed and allowed to dry, they may be stored vertically in a glass-fronted cupboard by means of 'Terry' clips fastened to the back wall. It is important that the tip of the jet of the burette is not allowed to bear the weight of the burette by resting on a shelf or the base of the cupboard. This may be avoided by inverting the burette or by placing the 'Terry' clips in suitable positions.

Burettes are easily broken during school use. If the top of a burette has been chipped or broken above the graduated scale, this may be repaired by 'squaring off' the burette and then fire polishing the cut end. A very slightly chipped glass jet may be smoothed by means of carborundum or other abrasive, but care must be taken to limit the widening of the jet, as too fast an outflow will make the burette unusable.

Cages

A INSECT CAGES

Several designs of insect cage are available from commercial manufacturers. The simplest consists of a cylindrical aluminium or sheet-steel base onto which fits a cylinder of transparent plastic. The lid, made from the same material as the base, has a perforated section to permit ventilation. This type of cage is quite suitable for a variety of insects which can be maintained at room temperature. However, if regular access is required (e.g. for feeding) a cage with a more convenient lid is preferable.

Vivaria made from enamelled steel, with a sloping perspex panel which slides up and down, are especially suitable for the maintenance of stick insects. Ventilation is provided by two perforated panels on the sides.

Locusts are increasingly kept as experimental animals in schools. For the completion of their life cycle, temperatures considerably higher than those normally found in school laboratories are required. Some additional form of heating is therefore necessary and is most conveniently supplied by an electric lamp. Several manufacturers supply locust cages based on a design suggested by the Nuffield Biology Project. These cages are made from sheet aluminium with a glass or perspex front. A false floor of perforated zinc allows the faeces to fall through. A number of egg tubes is also included. Some cages have a single lamp to provide both heat and light and this is sited in the main cage. An alternative design has a second lamp below the false floor, thus allowing heat to be provided with a minimum of light. (See page 232 for details of maintenance of locusts.)

Environmental chambers are also available, the temperature of which can be maintained to within 1 °C of that desired. These are basically similar to the locust cage described above, though slightly smaller and with the addition of a thermostat. The close control of temperature permits the chamber to be used for the incubation of eggs as well as for other purposes requiring a constant temperature.

Many other forms of insect cage may be improvised from materials available in school laboratories (see page 160).

B SMALL MAMMAL CAGES

The size and type of cage will depend upon the species of mammal and the numbers that are to be kept. Table 5.3 provides a rough guide to size.

Species	Height cm	Floor area cm²
Mouse (*Mus musculus*) breeding pair or trio	13	500
Rat (*Rattus norvegicus*) breeding pair	25	1000
Gerbil (*Meriones unguiculatus*) breeding pair	20	1000
Hamster (*Mesocricetus auratus*) pair or female plus litter	20	1000
Guinea-pig (*Cavia porcellus*) breeding pair	30	2500
Rabbit (*Oryctolagus cuniculus*) single animal	45	5500

Table 5.3 Recommended cage sizes for six species of small mammal

Materials for construction

Wood, chipboard and other similar materials may be used for cage construction, but they are absorbent and should be painted with several coats of polyurethane varnish. They are also subject to gnawing and this should be borne in mind at the design stage. Wood is also liable to attack by wood-rotting fungi and wood-boring beetles and cannot by autoclaved. However, for the larger cages (e.g. for guinea-pigs and rabbits) wood is particularly suitable. It is not recommended for the other species.

Cement-asbestos sheeting may be used in certain circumstances provided it too is covered with several layers of polyurethane varnish to reduce its absorbency. The sheet may be cut by scoring and breaking, and a cage constructed by attaching it to a wooden frame. However this is suitable for the larger animals only and so the cage is likely to be rather heavy. (N.B. *Care should be taken not to inhale the dust from cement-asbestos sheet and for this reason, it should not be sawn.*)

Cages constructed entirely from metal (e.g. aluminium) are suitable for guinea-pigs and rabbits and a number of commercial suppliers stock a suitable range for these animals (see page 88). Also, metal grid lids are frequently used on plastic base boxes (see below). These lids may be made from galvanised or stainless steel, the latter being more expensive.

A variety of plastics is commonly used for the construction of the base boxes of cages for the smaller species. Polythene has the disadvantage that it cannot be autoclaved; it is also opaque and rather too flexible for cage construction. However, polythene washing-up bowls, with a lid constructed from wire mesh attached to a wooden frame, make perfectly adequate temporary cages. Many commercially produced cages have polypropylene base boxes. Polypropylene is more rigid than polythene and may also be autoclaved. Being translucent it is a suitable material for breeding cages, where excessive visibility may be a disadvantage. Polycarbonate is a transparent material which may be autoclaved and although it is a little more expensive than polypropylene, the greater visibility that it provides has considerable advantages for all but breeding cages.

Points for consideration in selecting small mammal cages

a Security. Cages should be perfectly secure against the escape of stock and particular attention should be paid to the possibility of young animals escaping. Lids and doors should be easily opened by those looking after the animals, but should fit in such a way that they cannot be left unsecured. The animals should also be accessible to the handlers so that they can be removed easily when required.

b Safety. There should be no rough-edged metal parts or bad joints which may injure animals or people. The cages should also be constructed in such a way that the animals' feet, heads, teeth, etc., cannot be trapped.

c Cleaning. It must be possible to clean the cages easily and they should withstand immersion in disinfectant and/or sterilisation by autoclaving.

d Provision of food and water. Pelleted food should always be provided in chute or wire mesh hoppers and not in open dishes which are liable to spillage and contamination. The hoppers should have a sufficient capacity to last over weekends and the animals should be able to feed from them easily. In the case of rabbits and guinea-pigs, this means having a chute hopper with an opening width of 8–10 cm (see Figure 5.4). Wire mesh hoppers, which may form an integral part of the cage lid, should have a mesh width of approximately 8 mm.

Water should always be provided in drinking bottles and not in open dishes. The bottles should be wide-necked to facilitate cleaning and may have plastic, metal or rubber lids and spouts of sufficiently hard material to withstand gnawing. Stainless steel is ideal for this purpose, but home-made spouts may be improvised from 8 mm internal diameter glass tubing, reduced at the tip to a 2 mm diameter by partially closing in a bunsen flame. The bottles should be suspended in the cage at an angle of about 45° and at such a height that all, including the smallest animals, can drink, and yet be sufficiently high off the floor to avoid the bedding piling up against the spout as this will cause the bottle to leak and wet the bedding. For the smaller species (mice and hamsters), bottles of 250 cm³ capacity are suitable. For the larger mammals

a) Rat pellet hopper

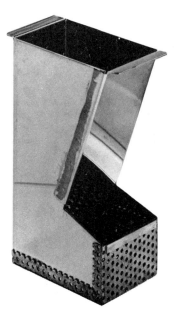

b) Guinea-pig hopper

Figure 5.4   Food hoppers for small mammal cages

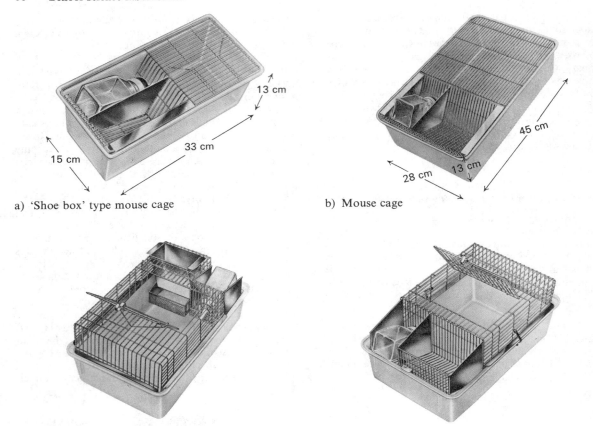

a) 'Shoe box' type mouse cage

b) Mouse cage

c) The same base as cage b) but fitted with alternative lids for guinea-pigs (left) and gerbils, hamsters or rats (right)

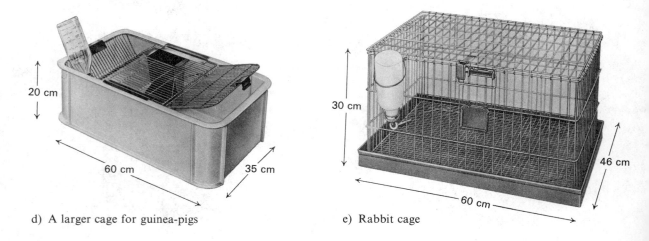

d) A larger cage for guinea-pigs

e) Rabbit cage

Figure 5.5   Examples of small mammal cages

(rats, guinea-pigs and rabbits), 500 cm³ bottles will be required. Mongolian gerbils do not require water if sufficient fresh fruit and vegetables are being provided, but in the absence of such a diet, a 250 cm³ water bottle should be used.

e Visibility and ventilation. It is advantageous to have a clear view of the animals except in the case of breeding specimens. Guinea-pigs and more especially rabbits, should be provided with a sheltered sleeping compartment in which they are not normally visible. Ventilation should be adequate but not excessive: cages made entirely from wire mesh usually provide too much ventilation, while deep aquaria with mesh lids may provide too little.

f Labelling. There should be adequate provision for labelling.

A variety of types of cage is illustrated in Figure 5.5. See also page 243 for details of maintenance of small mammals.

Capacitors

When purchasing capacitors—sometimes still referred to as condensers—it is necessary to state the type (e.g. electrolytic, paper, tuning), capacitance value, working voltage and tolerance required. Manufacturers' catalogues should be consulted to ascertain the values available and to compare prices. Capacitors with a tolerance of ± 20 per cent are quite satisfactory for most purposes and cheaper than those manufactured to a tolerance of ± 5 per cent.

Small double-ended capacitors can be mounted temporarily on a clip component holder consisting of a wooden or plastic base with two crocodile clips, each visibly connected to a 4 mm socket.

Permanently mounted capacitors are available from the manufacturers, sometimes in *Substitution Boxes*, which either cover a range in equal steps, such as 1–10 μF in 1 μF steps, or offer a larger range, covered in steps of constant ratio, e.g. 0.22–1000 μF in eight steps. The maximum working voltage should be marked on the box. A cheaper alternative is to purchase the capacitors separately and mount them in a plastic box connected to a rotary selector switch. Circuit connectors can then be made via 4 mm sockets placed at the usual minimum separation of 19 mm. Capacitors must be discharged after use by short-circuiting and be stored in an uncharged condition.

The following types are the most commonly used in school physics experiments.

a Electrolytic capacitors

In this type, two strips of aluminium are separated by paper strips impregnated with ammonium borate. A thin film of oxide, which has a high insulation strength, is formed on the anode and acts as the dielectric. The anode end (or positive terminal) is clearly marked on the case and must always be connected to the positive side of the circuit.

The most popular sizes for school use are in the range 50–1000 μF, typically 50 μF, 100 μF, 250 μF, 500 μF and 1000 μF, with a nominal working voltage of 50 V. Larger value capacitors, such as 10 000 μF or 25 000 μF, with a working voltage of 30 V, are suitable for illustrating the storage of charge in a capacitor. Such a capacitor should, for safety reasons, be mounted in a plastic box with 4 mm sockets (placed at a minimum separation of 19 mm) for connections to the charging and discharging circuit.

It should be noted that as the thickness of the dielectric film depends on the average voltage at which the capacitor is working, the stability of such a capacitor is not particularly high. However, electrolytic capacitors are very useful for low voltage work and as smoothing capacitors in rectifier circuits, where this drawback is of no consequence.

b Paper capacitors

Two long strips of metal foil form the two electrodes. These strips are interleaved with two similar strips of paper impregnated with paraffin wax or oil to improve insulating qualities, and the four strips are rolled into a tight cylinder.

Capacitors of this type are fixed in value, and vary from small capacity, for example 1 pF, to several microfarads. For large capacitances the roll is often packed into a rectangular metal box with two connecting terminals on the top.

c Tuning capacitors

Tuning capacitors are used in experiments illustrating the principles of radio reception. The capacitance is varied by altering the area of overlap of two sets of interleaved metal plates which are insulated from each other. One set of plates is fixed to the frame while the other set is fixed to a central shaft turned by the 'tuning knob'.

The maximum capacitance of these variable air capacitors is typically 365 pF or 500 pF.

d Capacitor plates

For a variety of experiments, including the measurement of the permittivity of air and basic investigations on capacitors, a pair of sturdy flat aluminium alloy plates is most appropriate. The size of each plate should be 25 cm × 25 cm × 6 mm thick, with 4 mm sockets in one edge of each plate. Polythene spacers should be available for placing at the four corners to produce a nearly constant separation between the plates.

For experiments illustrating charge transfer or ionisation in a candle flame, thin circular metal plates of 12.5 cm diameter are suitable. Each plate should have a

4 mm diameter metal contact pin and be mounted on an insulating plastic rod which may be held in a laboratory clamp.

Capillary tubing

This is frequently sold in nominal 1.5 m lengths with a bore diameter from 0.5 mm ($\pm$0.25 mm) to 3.0 mm ($\pm$0.4 mm). It should be cut and stored in the same manner as conventional glass tubing.

Small tubes for use in melting point determinations or in surface tension experiments can be conveniently made by drawing out clean glass tubing until the tube produced has a sufficiently narrow diameter. Alternatively, melting point tubes (open at both ends) may be purchased in packs of one gross.

Cells

Dry 1.5 V cells or batteries of four of these cells are required for much class practical work in low voltage electricity.

When ordering the basic 1.5 V 'U2' type cell, diameter 34 mm and height 61 mm, it should be noted that code letters vary from one manufacturer to another.

Although the leak-proof type of battery—either 1.5 V or 9 V—is more expensive, purchase of them can save money in the long run if accidental damage from a leaking battery is a realistic possibility.

Cell holders

The most common dry-cell holder is designed to hold four 'U2' type cells to provide voltage steps of 1.5 V, 3.0 V, 4.5 V and 6.0 V by plugging into 4 mm socket terminals.

Rechargeable cells

Non-leak nickel-cadmium cells, dimensionally equivalent to 1.5 V 'U2' cells, have a nominal voltage of 1.25 V and a capacity of 4 Ah. A useful life of several hundred charge-discharge cycles is claimed with proper care. However, rechargeable cells are expensive; they can be forty times more expensive than an equivalent leak-proof non-rechargeable 1.5 V cell. The recharging of cells is described in the section on batteries (page 82).

Cellulose tubing

Cellulose tubing is also known as dialysis tubing or 'Visking' tubing. It is made of seamless, transparent cellulose and is available in a range of sizes from 6 mm to 80 mm diameter when inflated. The larger sizes may be opened out and used in sheet form. The tubing, which is supplied in rolls, should be soaked in water before use,

whereupon it swells slightly permitting separation of the two sides. Once wet it may be knotted securely.

Cellulose tubing may be used several times provided it is kept wet between one occasion and the next. If it is being stored for more than a day it should be placed in screw-top jars containing water to which a little methanal (formaldehyde) solution has been added to prevent attack by micro-organisms.

Centrifuges

In general, school centrifuges operate at speeds of 2000–3000 r.p.m. Hence, the lid of a centrifuge should not be opened until rotation has stopped and no attempt should be made to slow down or stop the rotation by means of the hands or fingers.

Centrifuges require relatively little maintenance if they are properly used. Regular inspection, paying particular attention to the switches, and lubrication of the motor is all that is normally required and this should be done at six- or twelve-month intervals. The brushes in the centrifuge motor should not normally need attention.

The following points should be considered if the maximum working life is to be obtained from a centrifuge:

i avoid an uneven loading of the rotating head. Always use an even number of heads and counter balance the specimen tube with another containing water and placed directly opposite. Uneven loading is easily detected by means of the excessive noise or vibration it produces;

ii avoid spillage in the centrifuge. If accidental spillage does occur, insist that it is reported to the appropriate person and take the necessary steps to deal with the situation;

iii avoid using a centrifuge in a corrosive atmosphere. Corrosive vapour from spilled liquids can seriously corrode and weaken the rotating head or even damage the motor;

iv as with any piece of mains-operated equipment, a centrifuge should be occasionally but regularly inspected for worn or loose electrical wiring;

v use the appropriate sizes of centrifuge tube in the sockets in the rotating head.

Chart recorders

In recent years, a number of schools have introduced their pupils to automatic methods of recording experimental data. The chart recorder is one such method and a number of chart recorders, designed especially for school use, are now commercially available.

A chart recorder has two essential components.

a A method of converting the change(s) taking place in an experiment into the mechanical movement of a stylus or recording pen. In the simplest case, the motion of the piston of a gas syringe connected to a reaction vessel may be transmitted directly to cause the corresponding motion of a recording pen attached to the end of the piston. More generally, the input into a chart recorder is an e.m.f. and, in some instances, the pointer of a moving-coil meter is attached directly to the recording pen. In more sophisticated instruments a servo amplifier is used to control the movement of the pen. This allows a much greater voltage sensitivity and servo-operated recorders therefore respond much better to small changes in the input signal.

b A drive mechanism which allows the recorder paper to travel at a steady rate when in contact with the recording pen. Precise control of the chart speed is necessary and this is best obtained by a gearbox arrangement rather than a belt or friction drive from the electric (or, rarely, clockwork) motor.

The following features are of particular importance when considering the purchase of a chart recorder.

a Input sensitivity. Most chart recorders have several calibrated ranges.

b Chart speed. A variety of chart speeds should be available and switching from one to another should be a straightforward operation.

c A zero shift. The presence of a zero shift control enables the recording pen to be 'placed at zero', at any point on the recording paper. Data may, therefore, be recorded from left to right or vice versa. In addition, it becomes possible to 'zero' the recording pen at an appropriate point on the chart so that 'positive' and 'negative' inputs are recorded as movements to either side of the arbitrary point.

d Event marker. Although by no means essential, it is convenient if a chart recorder has a switch or push-button arrangement which allows 'an event' (e.g. a particular point in an experiment) to be recorded on the chart.

e Chart characteristics. The main issues here are the chart width (usually much greater than the writing area) and the cost of chart roles. In general, chart rolls are no more expensive than graph paper of the corresponding type.

f Servicing and repair facilities

g Size and storage characteristics

Choice chambers

Several designs of choice chamber are available from commercial suppliers. A number of types may also be made in the school workshop (see page 157). The Waterhouse pattern consists of a circular base 10–15 mm in depth and divided in two by a vertical partition. A lid of similar dimensions, but without the vertical partition, is placed over the base, the two being separated by a sheet of fine nylon gauze on which the animals under investigation are placed. In a modification of this design, the base is divided into four equal segments, rather than two.

It is important that the partitions in the base do not protrude as far as the mesh or the experimental animals may sense them through the nylon. The animals (e.g. woodlice, *Tribolium*, *Calliphora* larvae) are introduced through holes in the lid which may be sealed with cellulose tape. If wet and dry conditions are being investigated, it is important to allow the conditions inside the choice chamber to reach equilibrium before the introduction of the animals; this will take at least 15 minutes. A measure of the relative equilibrium humidity of the various parts of the chamber may be determined using cobalt thiocyanate paper inserted through the holes before the introduction of the animals.

Chromatography

Many chromatography items are now manufactured for use in schools. Some may be purchased in the form of kits suitable for paper or for thin layer chromatography (TLC). These kits are normally supplied with an instruction sheet or booklet containing details of appropriate experiments. The convenience of any kit needs to be considered in the light of its cost compared with that of individual items.

There are four main methods of separating compounds by *paper chromatography*:

i descending paper chromatography;
ii ascending paper chromatography;
iii radial paper chromatography;
iv reverse phase paper chromatography.

It is unlikely that any one kit will cater for all of these techniques.

For simple illustrative experiments, it is possible to use filter paper for separation by paper chromatography. However, for many experiments in schools, chromatography paper is highly desirable. Chromatography paper is manufactured in four grades and sold as sheets, as nominal 100 m reels of 3-cm-wide strips, or as strips of a standard size (e.g. 36 cm × 5 cm).

The four 'Whatman' grades are as follows:

Grade 1 (Chr)	For general use; medium flow rate
Grade 2 (Chr)	For electrophoresis of amino acids, proteins, peptides; slow rate
Grade 3 (MM Chr)	For inorganic chromatography
Grade 4 (Chr)	For amino acids and sugars; fast running

Sheets of chromatography paper are also sold with

punched holes at the corners, slotted edges, etc. Packets of chromatography sheets are normally marked to indicate the direction in which the flow rate is at a maximum.

Thin layer chromatography is a highly versatile technique capable of separating a wide variety of substances. The best results are obtained by using the most appropriate adsorbent. The common adsorbents for TLC are silica gel, alumina, cellulose powder, and kieselguhr. These may be purchased with or without a binder (usually calcium sulphate) and with the optional addition of a fluorescent indicator for the easier identification of chromatogram spots under ultraviolet light. The adsorbents vary in the speed at which they allow the chromatogram to 'run'. Examples of the uses of these adsorbents will be found in the standard texts on TLC.

Coating a glass or polyester sheet with an adsorbent is best done with a TLC 'spreader' or 'coater'. However, such devices are expensive to purchase. It is possible to purchase pre-coated TLC plates; the smaller sizes are relatively inexpensive and should be used as occasion demands in a school.

Once prepared, chromatograms need to be dried and then, in some cases, 'developed'. Depending upon the nature of the chromatogram, an oven and/or an electric dryer will therefore be needed.

Many inexpensive techniques have been improvised for carrying out basic chromatographic techniques with the simplest of apparatus available in almost all schools. Some of these improvised techniques are referred to on page 157. For electrophoresis, see page 99.

Colorimeters and spectrophotometers

The general arrangement of the components of a colorimeter/spectrophotometer is represented diagrammatically in Figure 5.6. The optical filter confines the light

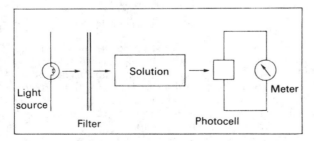

Figure 5.6   General arrangement of the components of a colorimeter/spectrophotometer

beam to a narrow band width at which maximum absorption will occur in the sample. In expensive instruments, the optical filter produces a monochromatic beam. Light intensity is controlled by the shutter and the

photocell produces an electric current which is directly proportional to the intensity of the light which has passed through the solution in the sample tube. The meter records this electric current, usually in μA. The meter may be calibrated by using standard dilutions of a known solution in the sample tube and plotting a graph of concentration against meter reading. This graph enables other readings on the meter to be interpreted quantitatively.

For school use, it is possible to purchase the optical part of a colorimetric unit separately from the meter. This has some educational advantages and is, of course, much cheaper than purchasing a complete colorimeter containing a built-in meter calibrated in optical density. Whatever form a colorimeter takes, it is necessary to purchase a set of optically matched tubes or cuvettes for use in the instrument. If the colorimeter does not have a direct wavelength control it is also necessary to purchase a set of optical filters, one of which is placed at the appropriate point in the optical path.

Colorimeters/spectrophotometers must be kept clean and free from dust. Great care must be taken not to spill liquids in the chamber that holds the sample. It is sensible to purchase spare bulbs together with several spare sample tubes.

If properly used, the colorimeter/spectrophotometer is a reliable and versatile instrument which may be used in a variety of school experiments. Some of these experiments have been incorporated in recent curriculum reform and many are described in the booklets supplied by some equipment manufacturers or retailers.

Since the output from a colorimeter/spectrophotometer is a small electric current, it is possible to use these instruments in conjunction with a suitable chart recorder. The recorder must, of course, be capable of operating from the output of the colorimeter and may be used to study the kinetics of a reaction in aqueous or other solution in which there is a suitable colour change, e.g. the iodination of propanone (acetone).

Conical (Erlenmeyer) flasks

These are manufactured in borosilicate glass with nominal capacities from $25 \, cm^3$ to 5 litres. Some sizes are available in poly(propene) and many conical flasks are graduated at suitable intervals. Like most items of common laboratory glassware, conical flasks may be purchased more cheaply in bulk quantities.

The conical flask most commonly used in school chemistry teaching has a nominal capacity of $250 \, cm^3$ although many of its functions, e.g. in titration, can be performed equally well by a conical beaker. Conical flasks are useful for overnight drying of ethereal or other extracts as they can be stoppered, but a conventional flat-bottomed flask will serve this purpose equally well.

Corks

Natural corks for use as stoppers are available in 'short' or 'long' forms and are manufactured in a range of diameters. They may be purchased in packs of individual or of assorted sizes.

Composition corks, made from cork pieces, are cheaper than natural corks but they are unsuitable for boring. All corks are porous and cannot, therefore, be used as stoppers for vessels containing highly volatile liquids. However, they may be made impermeable by coating them with (or immersing them in) molten wax.

When boring a hole in cork, the borer should have a fractionally smaller diameter than that of the tube which is to be passed through the cork. The cork should be held between the forefinger and thumb and bored from the narrow end, rotating the borer steadily in one direction. Cork borers are purchased in sets of six, twelve or eighteen borers with diameters of from 4 mm to 27 mm. They need to be sharpened from time to time and special sharpeners are available from the usual laboratory suppliers. Alternatively, borers may be sharpened using a file or hone.

When passing glass tubing or a thermometer through corks, the technique described on page 109 should be followed.

Corks are best stored in a unit with a large number of small drawers, or in a stackable tray system with suitably small trays. Each drawer or tray should contain corks of only one size. Alternatively, corks may be kept in labelled cardboard boxes stored in a drawer.

Crucibles

A school is likely to possess several different types of crucible.

Porcelain crucibles are the cheapest and the most commonly used. Porcelain is a complex heterogeneous mixture of an aluminium silicate and the oxides of aluminium and potassium, which is hard and does not fracture when strongly heated. The porcelain is rendered non-porous by covering with an appropriate glaze. If this glaze is to be permanent, it must not soften on heating and must have a coefficient of expansion which is very close to that of the porcelain itself. It must also be impermeable to those chemical reagents with which it is likely to come into contact.

The purchase of inexpensive, low quality porcelain crucibles is a false economy: they readily become stained and crack if unevenly heated. There should be no difficulty about cleaning a glazed porcelain crucible if it has been properly used and strongly abrasive materials should, in any case, be avoided.

The metal crucibles found in a school laboratory are likely to be made of nickel, stainless or mild steel or, more rarely, platinum. Metal crucibles are robust and will not crack on heating, but the contents should not include compounds likely to be reduced by the metal of which the crucible is composed. Similarly, materials which may fuse with the metal crucible should be avoided, e.g. caustic alkalis, the salts of lead and of the alkali metals, antimony, arsenic, bismuth, lead, silver, tin, zinc, sulphur, phosphorus and silicon.

Silica crucibles have several advantages. Silica is available in a transparent variety as well as in the more common translucent or opaque form and the substance has an extremely low coefficient of expansion. As a result, silica crucibles can be subjected to sudden and severe changes in temperature without cracking. Also, silica is a relatively inert material, unaffected by the common laboratory mineral acids. It is, however, readily attacked by hydrofluoric acid and by cold concentrated tetra-oxophosphoric(V) acid (orthophosphoric acid) above 200 °C. It is also attacked by those metallic oxides which, at high temperatures, combine with silica to form glass.

The principal disadvantage of silica crucibles is the additional expense of purchase although there are situations in which the extra cost is undoubtedly justified. Silica devitrifies on very strong heating, but the necessary temperatures are not likely to be reached in a school.

Similar considerations apply to crucibles made of alumina. This inert, impermeable material is more expensive than silica, but it is unaffected by reducing gases and by many oxides. Alumina crucibles are therefore suitable for fusing metals and can be used at temperatures up to about 1900 °C.

Fireclay crucibles are extremely hard and will withstand temperatures greater than 2000 °C. They are brittle and in schools are generally used for conducting the 'thermit' reaction. However, 'Morgan' fireclay crucibles are available in a variety of sizes and shapes, each designed to serve a particular function.

Crucibles may be cleaned by the action of dilute or concentrated mineral acid. It is sometimes helpful if contact between the crucible and its contents is minimised by means of a small piece of asbestos. The slow combustion of magnesium ribbon contained in a crucible is a common school experiment. Note that there have been reports of mild explosions when nitric acid has been subsequently added to the crucible to clean it.

Crucible lids are easily lost or broken and, in a school, it is generally necessary to ensure that there is an adequate supply of lids for the number of crucibles to be used. Lids are not usually universally interchangeable and some system of associating a particular lid with a given crucible is often helpful.

Crucible tongs

Crucible tongs may be made of mild or stainless steel, brass or nickel. More specialist tongs, e.g. with platinum tips, are also available. Nickel tongs are generally the most expensive, but they are more resistant to attack by acids, whereas steel or iron tongs, although cheaper, rapidly rust in the corrosive atmosphere of the laboratory.

Crucible tongs should not be used to handle red-hot metal crucibles as some degree of alloying may occur.

Deionisers

The chemical composition of a laboratory water supply varies from one part of the country to another but, in most cases, the water contains too many dissolved salts for it to be suitable for use in some scientific experiments. A supply of reasonably pure water may be obtained by distillation (see page 144) or by deionisation.

Laboratory deionisers make use of ion-exchange resins. There are two types of resin—anion and cation exchangers—and most school deionisers contain an ion-exchange *cartridge* which is packed with both of these types. When tap or 'raw' water containing dissolved solids is passed through the cartridge, the dissolved cations, e.g. $Na^+(aq)$, are removed in exchange for $H^+(aq)$ released by the cation-exchange resin. Similarly, the anion-exchange resin exchanges $OH^-(aq)$ for dissolved anions such as SO_4^{2-} (aq). The net result is the substitution of H^+ (aq) and $OH^-(aq)$ ions, i.e. water, for other dissolved cations and anions respectively. The electrical conductivity of tap water thus decreases as it passes through an ion-exchange column.

Most deionisers are therefore fitted with a battery-operated conductivity meter which indicates the extent to which the water has been deionised. If water is flowing through the resin at the appropriate rate, the meter reading is also an indication of the degree of consumption of the ion-exchange cartridge. Sometimes the scale of the meter is divided into colour zones, e.g. green, amber, and red. A meter reading in the green zone indicates that the cartridge is behaving as if it were new. An amber zone reading indicates the production of water of slightly lower purity and a reading in the red zone indicates that the ion-exchange column is spent and that the cartridge should be replaced. The differences in the quality of water in the amber and green zones can be considerable as Table 5.4 indicates.

As the conductivity meter in an ion-exchange apparatus is usually battery operated, it is helpful if the same meter can also be used to test the 'life' of the battery.

	Green zone	Amber zone
Carbon dioxide ppm	0	As in raw water
pH	6.5 to 7.0	4.0 to 5.0
Silica ppm	0.05	As in raw water
Electrical conductivity $\mu\Omega^{-1} m^{-1}$	100	10^3 to 1500
Residual dissolved solids ppm	1	*c.* 15

Table 5.4   Differences in the quality of water available from a deioniser

The replacement of ion-exchange cartridges is a straightforward matter; spent cartridges are simply sent to the supplier for regeneration. It is helpful if a school is able to purchase a spare cartridge at the same time as the deioniser. However, spent cartridges are usually returned quickly and most suppliers operate a system whereby five or six cartridge exchange vouchers are supplied at the time of purchase of the deioniser. These vouchers enable a new cartridge to be obtained at the further cost of postage only.

The relative merits of deionisation and distillation processes for water purification include the following:

i distillation of water is an expensive process which is almost certain to be more costly than deionisation unless the raw water is very hard;

ii if the raw water contains much dissolved material, ion-exchange processes become expensive and inconvenient in that the ion-exchange cartridge needs to be renewed frequently;

iii it is possible to purchase portable deionisers. A still has to be fixed to the wall in one room;

iv although the rate of production of distilled or deionised water from raw water varies from one piece of equipment to another, deionisation is generally a much more rapid process than distillation;

v distillation always removes dissolved gas as well as dissolved solids from tap water. *Some* deionisation processes may be unsatisfactory in that they may allow a considerable quantity of $CO_2(aq)$ to remain dissolved in the water, particularly if the ion-exchange resin is partly spent.

As with distilled water, deionised water will absorb atmospheric gases—especially carbon dioxide—on prolonged contact with the air, so deionised water should not be stored for long periods of time if such dissolved gases are likely to have a significant effect upon its use.

Desiccators

A school will need a small number of conventional desiccators and perhaps one vacuum desiccator. At least one large desiccator is also useful so that several specimens may be allowed to dry at the same time. Such specimens must, of course, be clearly labelled.

Desiccator lids should be removed by sliding them sideways and this is facilitated by thinly greasing the ground-glass surfaces—excess grease must be avoided. Air should always be introduced slowly into a vacuum desiccator so as to minimise the risk of disturbing or spilling the contents.

If a hot object is placed in a desiccator, the air should be allowed to expand for a few seconds before the lid is replaced. This ensures minimum disturbance of the contents when the lid is removed—provided it is removed slowly.

Probably the most frequently used desiccant in a school

is granular, fused calcium chloride. This substance is cheap, but it is not a particularly effective drying agent. It is, however, very useful for drying substances which have been washed with ethanol. The relative efficiencies of some common laboratory drying agents are given in

Drying agent	Residue of water mg litre^{-1} of air
P_4O_{10}	0.00002
BaO	0.0007
$Mg(ClO_4)_2$	0.0004
H_2SO_4, concentrated	0.003
$CaSO_4$	0.005
Al_2O_3	0.005
KOH, (fused, sticks)	0.015
SiO_2 gel	$c.$0.015
NaOH, (fused, sticks)	0.6
$CaCl_2$ granular	1.0
$ZnCl_2$ (fused, sticks)	0.1
$CuSO_4$	2.5

Table 5.5 The relative efficiencies of some drying agents

Table 5.5, which confirms that phosphorus(V) oxide is the most powerful common desiccant. When choosing a desiccant for a particular purpose, consideration should be given to its chemical suitability, the degree of desiccation required, the cost of the desiccant and the possibility of its regeneration. Silica gel has the advantage that it is available with a little added cobalt chloride, the colour of which indicates the extent to which the desiccant should be replaced.

In humid, tropical areas, large desiccators may be used to store items of sensitive optical equipment.

Disposable laboratory ware

A wide range of items is manufactured in disposable form—largely for medical use. Some of these may be usefully employed in the school situation.

a Aprons. White polythene aprons (usually available in bulk packs of 10 or 25) are useful when particularly messy work is being undertaken. They may also be used by a technician or teacher engaged in large amounts of microbiological work. Polythene aprons should be disposed of after use and if they have been used for microbiological work they should first be autoclaved in a disposal bag (see below and page 280).

b Gloves. Disposable gloves are available in both polythene and rubber form. The latter are intended for surgical use and are suitable for autoclaving, but they are relatively expensive and are unlikely to have much use in the school situation. Thin polythene gloves are very cheap

and are useful for both microbiological and chemical work. Those used for microbiology should be autoclaved in a disposal bag (see below) before disposal. See also page 41.

c Petri dishes. Perspex dishes are available from suppliers in both 5 cm and 9 cm diameter sizes. There is very little difference in cost, but it should be remembered that the former use very much less medium (approximately one-third if filled to the same depth) and may therefore prove considerably cheaper in the long run. They are normally supplied pre-sterilised in packs of ten. Plastic Petri dishes are intended to be disposable items and must be treated as such, especially after microbiological work. They should be autoclaved in disposal bags prior to disposal (see below and page 280).

d Syringes. Pre-sterilised syringes are available in 1 cm^3, 2 cm^3, 5 cm^3, 10 cm^3 and 20 cm^3 sizes. In each case the graduated volumes are indicated. The 1 cm^3 size has finer graduations and is relatively more expensive. Disposable syringes have a wide variety of uses in science teaching and, provided they are not being used for microbiological work, may be used on a number of occasions. Most disposable syringes have a Luer fitting and should only be used with Luer fitting needles. Disposable needles are available in bulk packs (100 or 125) in a variety of sizes as indicated in Table 5.6. They are supplied individually packed and sterilised.

Needle size	Gauge	Diameter mm	Length mm
1	21	0.8	38
12	23	0.6	28
15	23	0.6	25
17	25	0.5	24
20	25	0.5	16
Serum (standard)	19	1.1	38
Serum II	19	1.1	51

Table 5.6 Disposable syringe needles

Needles should not be used more than once as they may become blocked and fly off if excessive pressure is applied to the plunger. All syringes and needles used for microbiological work must be used only once and autoclaved in disposal bags prior to disposal (see page 280). Used syringes from hospitals, surgeries and clinics must *never* be used in schools.

e Disposal bags. Autoclavable disposal bags are intended for the disposal of contaminated plastic ware such as Petri dishes, gloves and syringes. They are available in bulk packs with wire ties which should be loosely attached during autoclaving to permit penetration of steam. After autoclaving the bag should be placed in the waste bin without removing the contents (see also page 280).

Dissecting apparatus

Boards

Larger specimens such as rats, rabbits and dogfish are normally attached to boards for dissection. The wooden boards, which are usually at least 30 cm × 45 cm in area should be at least 2 cm thick to reduce the risk of warping. They are normally made of several sections of wood which should be joined with a waterproof adhesive and be soft enough to permit steel pins to be pushed in by hand. It is advisable to have some means of preventing liquid from the dissected specimen running off the board and on to the bench. This may be achieved by cutting a groove or channel round the edge of the board to collect the liquid or by fitting a rim which protrudes approximately 1 cm above the board surface. After use, boards should be scrubbed clean with a stiff brush, detergent and hot water and be allowed to dry before storage.

In some circumstances, it may be convenient to dissect insects and other small animals mounted on cork boards. These are available from suppliers but may easily be improvised from pieces of cork tile available from shops specialising in flooring material.

Chain hooks

It is sometimes necessary to stretch tissues during dissection. This may be done with chain hooks consisting of three hooks each attached to a central ring by means of a short length of chain. Chain hooks should be manufactured from stainless steel or other suitable rust-proof material.

Dishes

Smaller specimens (e.g. mice, amphibians, insects and earthworms) are normally dissected in wax-based dissecting dishes to permit the additions of water, saline or Ringer's solution to the dissection. Dissecting dishes are best made from white-enamelled steel and are available in two forms. Vertical-sided dishes, also called instrument trays, are available in a range of sizes from 15 cm × 20 cm to 35 cm × 45 cm. They are not easily stacked, and are more expensive than stackable 'pie' dishes which are available in various sizes from 17 cm × 24 cm to 26 cm × 36 cm.

Molten dissecting wax should be poured into the dishes to a depth of approximately 1 cm so that specimens may be easily pinned out for dissection. The correct black dissecting wax should be used and not paraffin or candle wax which fails to adhere to the sides of the dish and thus floats to the surface during dissection. Candle wax is also more likely to crack with the use of pins.

Forceps

A wide range of types of forceps is available from biological suppliers and an even greater selection may be provided by those specialising in the supply of instruments to the medical profession. For most school purposes, two types, with blunt points and fine points, will suffice. Both should have serrated points and be provided with a guide pin on one shank and a corresponding hole in the other to ensure the alignment of the points. Forceps are available in nickel-plated steel and, at a somewhat greater cost, stainless steel.

For more delicate dissection, forceps with very fine, unserrated points (e.g. watchmakers' forceps) will be required. Fine pointed, curved forceps are also available but are more liable to damage resulting in non-alignment than those with straight points. Nylon forceps are useful for younger pupils who are not involved in dissection but who require a means of handling small, delicate objects; they are very much cheaper than those made of nickel-plated or stainless steel.

A number of other more specialist types of forceps may be required in certain circumstances. Entomological forceps have curved, flattened ends enabling a good grip to be gained on the mounting pins of set specimens. Specimen forceps are rather larger instruments with flattened, serrated ends and are useful for removing preserved specimens from alcohol or other preservative. A school is unlikely to need more than one or two pairs of these. Artery forceps have shanks resembling scissors which may be locked in the closed position, thus clamping a vessel or length of rubber tubing. They are available in steel or polypropylene, the latter being very much cheaper.

Needles

Steel needles are available mounted in either wooden or aluminium handles. The former have a tendency to become loose as the wood dries out, but this may be overcome with a little strong glue such as 'Araldite'. Needle holders are also available with slotted heads and screw collars designed to hold replaceable needles of various sizes. These have some advantages for more advanced work, but are not suitable for junior use where detached needles may prove dangerous. (See also seekers on page 97.)

Pins

Dissecting pins are used for pegging specimens to the dissecting board. They consist of long steel spikes mounted in a wooden, metal or plastic handle. As it is sometimes necessary to hammer the pins into the boards, it is important that their construction should withstand such treatment. Dressmakers' pins may also be used for pinning out smaller specimens to wax or cork and for the display of structures during dissection.

Scalpels

Dissecting scalpels are of two basic types. The first consists of a steel blade permanently set in a plastic or steel handle. Blades are available in various patterns and in lengths of from 30 mm to 50 mm. Such scalpels need frequent resharpening on a fine carborundum stone if they are to be maintained in a suitable condition for dissection. In the school situation they are best suited for junior work where razor-sharpness is not required.

For much advanced work, the fixed-blade scalpel has been superseded by instruments with replaceable blades. The best known of these (*Swann-Morton*) is made in three handle shapes for which a dozen or so shapes of blade are available. Blades are normally sold unsterilised in packs of five and care should be taken to ensure that appropriate blades are ordered for a particular handle as they are not fully interchangeable. The smaller blades (Numbers 10–15 inclusive) fit the No. 3 and the longer No. 5 handles while the larger blades (Numbers 20–25) are made for the No. 4 handle. Blades may be resharpened if required.

Scissors

Although a wide range of scissors is available, especially from shops specialising in surgical instruments, they are mostly of two basic types. Scissors with closed shanks have handles which lie alongside each other as they leave the pivot. Those whose shanks diverge as they run from pivot to handle are known as open-shank scissors. For normal school use, open-shank scissors are recommended and are available in two sizes, the larger size being more suitable for junior work. Fine dissection scissors with curved points and micro-scissors with sprung handles for more delicate work are also available, but these are not normally necessary at school level. Most forms of scissors are made in both nickel-plated and stainless steel, the latter being more expensive. Bone scissors (bone forceps) are a useful addition for advanced work.

Section-cutting razors

These are similar to the old-fashioned 'cut throat' shaving razor except that one side is ground flat. Special razors are available for left-handed users. Unless they are being very regularly used, a safety razor blade mounted in a suitable holder (available from suppliers) will serve the purpose equally well—perhaps better—at a greatly reduced cost and without the continual sharpening necessary for a section-cutting razor.

Seekers

A seeker is an invaluable piece of dissecting apparatus permitting the movement and investigation of organs without damage to tissues. Seekers (sometimes called blunt needles) are available in straight or bent forms mounted in wooden or metal handles. There is a tendency

for those mounted in wood to become loose; for this reason they should not be left soaking in water for any length of time.

Storage tanks

Dissected specimens which are being retained for further work and have been preserved in methanal (formaldehyde) solution may be stored in fibreglass tanks available from suppliers in various sizes. They should have close-fitting lids to prevent the escape of methanal fumes into the laboratory.

The care of dissecting instruments

Good dissecting instruments are expensive, but with proper care and attention can last for many years. For junior work, it is generally convenient to store instruments of the same type together. 'Cut-out trays' are recommended for this purpose. These not only protect the instruments by keeping them apart, but also permit a rapid check for missing items (see Figure 5.7).

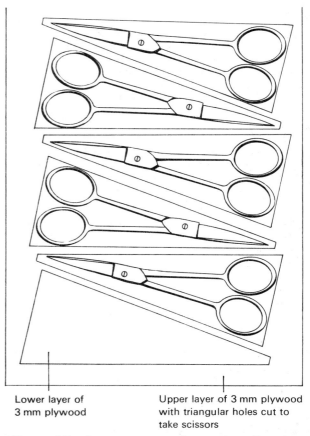

Lower layer of 3 mm plywood

Upper layer of 3 mm plywood with triangular holes cut to take scissors

Figure 5.7 A cut-out tray for storing dissecting instruments

At sixth-form level, it is preferable to issue each pupil a set of instruments for which he is responsible throughout the course. Such a set might include large and fine scissors, large and small scalpel, fine and blunt forceps, seeker, mounted needles, paint brush, hand lens and teat pipette. Sub-divided canvas rolls are probably the most convenient means of storing each set. Other items are best stored centrally and issued as required.

All instruments should be thoroughly cleaned before being put away. They should also be lightly oiled to prevent corrosion. A cloth reserved for this purpose may be conveniently stored with each set of instruments.

Dynamics trolleys and runways

The study of Newton's Laws and dynamics in general has been transformed from teacher demonstrations using, e.g. 'Fletcher's Trolley', to pupil experimentation with dynamics trolleys.

Trolleys

The essential requirements are robustness and low-friction bearing wheels. A hardwood or metal construction provides rigidity for the trolleys, which may be constructed from a kit of parts. A quick-release plunger spring is required for momentum experiments, and it is convenient to have two dowel pegs at each end of the trolley to stack trolleys of equal mass one on the other. Such pegs can also serve as anchor points for the rubber accelerating cords which have eyelets at each end.

The number of trolleys required in multi-mass experiments can be kept to a minimum by stacking one (or two) trolley mass equivalents on the base trolley. A steel mass equivalent may be located firmly on the base trolley by rubber locating pads fixed to the steel plate.

Runways

Runways should be constructed of 2 m or 2.5 m lengths of warp-resistant blockboard with metal angle side rails to ensure rigidity. Alternatively, a track can be assembled from a kit consisting of twin plastic track on metal angle side rails.

If 8, 12, or 16 runways are provided, the problem of safe storage and of transport to and from the work bench must be considered. The shorter 2 m boards have advantages in this respect.

Runways should generally be carried vertically until the bench is reached. A suitable store consists of a rack with dividing bars screwed to a wall, so that the runways stand vertically between the bars. A retaining bar or strap should be fitted to ensure that the runways do not fall over.

Runways may be tilted to compensate for friction by placing one end on adjustable paper pads or using continuously variable feet held in a metal angle plate and

bolted to each side of the runway. A block of softwood, possibly padded with rubber, and held firmly across the bottom of the runways, avoids the possibility of damage to trolleys which run off the base of the track.

Ecology apparatus (see also: Environmental comparators; Nets, collecting; Soil apparatus)

Point frames (point quadrats)

A point frame is generally used where vegetation is short (e.g. on a lawn) to estimate the percentage cover by plants. It consists of a horizontal bar firmly supported either by spikes at each end or by a central spike driven into the ground. A number of probes which can be raised and lowered either individually or together, are situated at regular intervals (often 10 cm) along the bar. Point frames are available from commercial suppliers, but can easily be improvised. Lengths of 'speedframe' or similar tubular steel are ideal for the horizontal bar (metal legs from broken laboratory stools make an adequate substitute). A central support can be made by screwing a length of steel rod (say 5 mm) into the middle of the bar, and the probes can be made from knitting needles or 3 mm steel rod. (See also Nuffield A level Biological Sciences (1970) Laboratory Book, Penguin Books, page 125.)

Quadrats

A quadrat usually takes the form of a square frame and is used as a sampling device in ecological work. The size of the quadrat used in a particular situation will depend upon the nature of the terrain and the species being sampled. The distribution of algae on a tree trunk may be investigated using a small quadrat the sides of which measure a few centimetres. At the other extreme, an examination of the distribution of trees in a wood may necessitate using quadrats measuring 10–20 metres along each side. Selection of the correct size of quadrat is important and is too often neglected in ecological work. If the area sampled is too large then a great deal of unnecessary work will be involved and very little, if any, extra information will be obtained. If too small an area is used, the information will be incomplete. A useful exercise is to use quadrats of various sizes in the same habitat and then examine the results to see which gives sufficient information for the minimum amount of work.

For small-scale work, plastic quadrats measuring 12 cm × 12 cm and subdivided into 1 cm × 1 cm squares are available from suppliers. These have the advantage that they are flexible and hence can be used on surfaces which are not flat, such as tree trunks. Metal-framed quadrats measuring 50 cm × 50 cm or 100 cm × 100 cm are also available. In the latter case, it is advisable to use a folding model; rigid 100 cm × 100 cm quadrats are cumbersome and difficult to transport. Whatever size of quadrat is being purchased or made, it is strongly recommended that the sides be subdivided so that the same frame may be used for sampling smaller areas. Thus a 50 cm × 50 cm frame can be marked at 10 cm intervals along each side. String or straight rods may then be used to enable 10 cm × 10 cm, 20 cm × 20 cm, 30 cm × 30 cm, 40 cm × 40 cm as well as the full 50 cm × 50 cm squares to be used.

Quadrats are very easy to improvise in the field from a metre rule and a length of string. They may also be made in a school workshop from metal rod or wood.

Electromagnetism

An electromagnetic kit, containing 542 items of 27 different components was designed originally for the Nuffield O level Physics Project to enable 32 pupils to work in pairs on a series of experiments involving magnets, magnetic fields, simple electromagnetic induction, the construction of a simple motor, dynamo and galvanometer, and simple alternating current work.

When purchasing such a kit it should be noted that the price quoted is usually for a *one-quarter kit*, suitable for *four pairs* of pupils.

An important consideration is the provision of spares. If, for example, one of the axle-rods becomes bent or lost, then one pair of pupils will be unable to construct the model motor unless a spare or substitute is available. It is clearly unnecessary to order spares for all components in the kit. However, it is useful to list those items which are likely to sustain damage or which are consumable, and to order spares of such components at the time of ordering the kits. Such a 'spares order' would probably include some Ticonal and Magnadur magnets, split pins, axle-rods, M.E.S. lamps (2.5 V, 0.3 A), as well as consumables such as reels of cotton, adhesive tape, valve rubber and reels of 26 SWG covered copper wire.

Electrophoresis

As with chromatography (page 91), electrophoresis equipment may be purchased in kit form. However, most schools will already have a power supply unit which can provide the necessary voltage and current for electrophoresis so that there is no need to purchase a power source specially for electrophoresis experiments. If the power source has to be connected to the sheet(s) or strip(s) of paper in a tank containing a solvent, it is important to ensure that the appropriate electrical connections are available or may be safely improvised.

Electrostatics

The various parts included in a Malvern electrostatics kit are specifically designed to enable pupils working in pairs to carry out a large number of electrostatics experiments.

The materials are modern, such as polythene and cellulose acetate strips, which can be charged by rubbing with a polishing cloth. The polythene strip gains a negative charge; the acetate strip gains a positive charge. These materials retain their charges rather better than the 'traditional' ebonite and glass rods, particularly in damp conditions. The effect of high humidity cannot, however, be completely eliminated, and the kit should be stored in a dry and preferably warm place prior to the lesson. In those parts of the world where the humidity is particularly high at certain times of the year, the study of electrostatics should be planned for the dry season.

A most important instrument in electrostatics is the *gold leaf electroscope*. A number of different designs of electroscope are available for purchase. The following are desirable features:

i *a detachable plate* makes the capacity of the electroscope very low and reduces leakage of electric charge to the atmosphere;

ii *a hook electrode*—the electroscope can be charged by lifting the hook from the case with a charged rod;

iii *a good insulating bush* to support a metal rod to which a single gold leaf can be attached;

iv *an earth socket*—4 mm terminal provided on the case;

v *windows*—front window is of clear glass and rear window of ground glass for projection purposes when demonstrating to a class.

Environmental comparators

Several types of environmental comparator are now available from commercial suppliers. These can be used to take measurements of physical conditions in a range of environments. Conductivity, light, oxygen, pH, sound and temperature are among the factors which can be measured in this way. A comparator consists essentially of a power output (usually provided by batteries), a meter and an appropriate environmental probe. Variations in the level of the particular factor being measured affect the resistance of the probe and hence the reading on the meter.

When selecting a comparator for purchase, the following points should be considered.

a Versatility. Some comparators can only be used to measure one factor while others are available with a range of modular attachments enabling measurements of several different factors. It may be worth considering the purchase of a more versatile instrument with only one or two attachments, and to acquire additional modules as the need arises and appropriate finance becomes available.

b Portability and durability. As most measurements are likely to be taken in the field, a comparator must be transported easily by hand or in a rucksack. The use of printed circuits in the modules reduces the likelihood of damage, even in the event of the modules becoming wet.

c Aquatic use. It is often desirable to take measurements in aquatic as well as in terrestrial environments. Care should be taken to ensure that the instrument is suitable for this purpose.

d Choice of scale. In some models, readings may be taken directly for some factors (e.g. temperature, oxygen level). In others it may be necessary to refer to a calibration graph in order to obtain a reading of the precise level of the factor being measured.

e Sensitivity. A selection of sensitivity levels is an added advantage, especially for factors such as temperature which may show very slight, but significant, variations from one micro-habitat to another.

f Use as a colorimeter. In some models the light module may also be utilised as part of a colorimeter.

g Field probes. If measurements are to be taken in ponds and rivers, up trees and in other places where the meter itself cannot be placed easily, it is important to ensure that appropriate extension probes are available.

Ergometers

Bicycle ergometers

Bicycle ergometers are used to measure the power output of the legs and to determine the relationship between metabolic rate and exercise. They normally consist of a welded angle-iron frame (or an old bicycle frame) with a bicycle pedal and sprocket wheel mounted at one end, linked by a bicycle chain to a grooved wheel of 1 m circumference mounted at the other. Work is done against friction which is provided by a rope passing over the grooved wheel connected to a dynamometer (force-meter) at one end and a hanging load (or a second dynamometer) at the other. When the grooved wheel is rotated the force applied by the rope on the wheel is indicated by the difference between the dynamometer reading and the hanging load and so the work done in a given time can be calculated.

As the circumference of the wheel is 1 m, the force applied is moved through 1 m for every revolution of the wheel. Most bicycle ergometers are supplied with an adjustable revolution counter. Appropriate hanging loads (e.g. sandbags) must be purchased separately. (See Figure 5.8).

Arm ergometers

Arm ergometers are used to measure the power output of the arm. The principle of operation is similar to that for

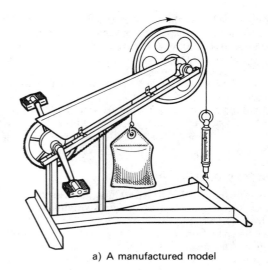

a) A manufactured model

Force-meters hooked on to adjustable cross-piece

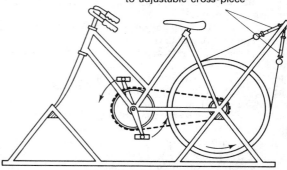

b) A home-made version using a bicycle frame and rear wheel

Figure 5.8   Two designs of bicycle ergometer

the bicycle ergometer except that the grooved wheel is turned directly by the hand. Arm ergometers are normally attached to the laboratory bench by means of a pair of large G-clamps, which should be purchased separately. (See Figure 5.9).

Filter flasks

These are manufactured in borosilicate glass in accordance with BS 1739. Some sizes are also available in poly(propene).

Filter papers

Various types of filter paper are available. For general filtration purposes, *qualitative* grade filter papers are

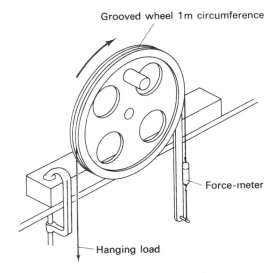

Figure 5.9   An arm ergometer

used. These should be distinguished from the ashless and hardened papers required for accurate, quantitative work. *Ashless* papers are produced by removing the mineral components of the paper by treatment with hydrochloric and hydrofluoric acids. For this reason, such papers are also sometimes referred to as 'acid-washed' filter papers. Both 'single' and 'double' acid-washed grades of filter paper may be purchased. *Hardened* papers are designed to withstand the force of suction at a filter pump and to be resistant to concentrated acid or alkaline solutions. They are made of almost pure cellulose and have considerable strength even when wet. Hardened papers are particularly useful in accurate gravimetric analysis.

It follows that it is important to select the appropriate type of filter paper for the purpose in hand. The principal manufacturers of filter papers produce leaflets or pamphlets summarising the characteristics of the filter papers they produce. Such information normally indicates the retentivity of the filter paper together with the filtration speed. Table 5.7 provides such information for a number of qualitative grade filter papers.

Filter papers are, of course, manufactured in a variety of diameters. In a school, only a restricted range of sizes is necessary and ashless or hardened papers will be needed relatively infrequently. All filter papers should be stored in a clean, dry atmosphere, and any expensive grades of paper should be clearly distinguished.

It is occasionally helpful to use fluted filter papers. Fluting the papers generally increases the rate of filtration and this may be important in some circumstances, e.g. if a hot solution is to be allowed to crystallise, more rapid filtration reduces the changes of crystallisation occurring in the filter funnel.

For *gravimetric* work at school level, Green's 807 or Whatman's 589/3 grades will normally be adequate. Such papers will retain fine precipitates such as barium sulphate and their slow filtering speed can be overcome by use of a filter pump. Both grades are acid washed, low-ash papers.

Manufacturer's number		Characteristics
Green	Whatman	
904	15	Rapid filtration; suitable for viscous liquids and juices
$788\frac{1}{2}$	13	Medium filtering speed; general qualitative work
702	5	Slow filtering speed; very retentive; suitable for fine precipitates
704	4	Fast speed; use for coarse and gelatinous precipitates
798	3	Medium speed; highly retentive; suitable for viscous liquids
797	2	Slow speed; thick papers; suitable for general qualitative analysis
795	1	Medium speed; general purpose, thin papers but will not retain fine precipitates

Table 5.7   Characteristics of some qualitative grade filter papers

Filter pumps

These are manufactured in borosilicate glass, moulded poly(propene) or stainless steel. (The latter are the most durable but also the most expensive.)

In some plastic filter pumps the inner cylinder and jet can be removed for cleaning. There is usually little to choose between filter pumps in terms of maximum pumping speed (4 litre min^{-1}). The important factors, therefore, are (i) cost, (ii) durability, and (iii) degree of vacuum which can be produced under a water pressure usually of the order of 150 kN m^{-2}. Some filter pumps have a threaded 'male' or 'female' tap connection which is of considerable help if the laboratory taps are correspondingly threaded.

Force-meters (see Balances)

Freezing mixtures (see page 182)

Fume cupboards

Science teachers are becoming increasingly involved in the design and planning of fume cupboards, so it is important that the principal aspects of fume cupboard construction should be more widely known.

A fume cupboard is a container designed to prevent contamination of the laboratory, so that under all normal conditions air should flow into the container and not out of it. In a badly designed and/or sited fume cupboard, this may not be the case.

The following considerations are important in designing or selecting a fume cupboard for installation in a laboratory.

a Face air velocity

The velocity of the air flow into a fume cupboard must be sufficiently high to overcome the normal flow (and its disturbance) in the laboratory. There seems to be no general agreement on a minimum air velocity but it is suggested that *at least* 30 m min^{-1} is appropriate for general purpose fume cupboards. If a fume cupboard is aerodynamically designed, a face velocity as low as 20 m min^{-1} might be adequate. When designing a fume cupboard for school use it is desirable to work to an average face velocity of the order of 45–60 m min^{-1}; just sufficient to blow a paper tissue across a bench.

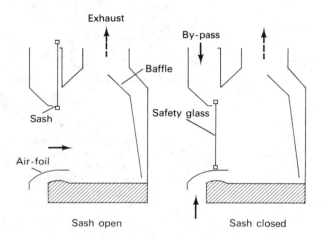

Figure 5.10   Fume cupboard design incorporating a bypass

A fume cupboard may be used with its window in any number of positions from fully closed to fully open. Accordingly, a fume cupboard must be so designed that working with the window almost closed does not produce an excessively high velocity across the working surface in the fume cupboard. This may be done by a suitable 'bypass' system (see Figure 5.10).

For a review of laboratory ventilation see Everett, K., Hughes, D. (1975) Chapter 6.

b Working aperture

It is obviously helpful if the maximum working aperture corresponds to the maximum opening of the window of the fume cupboard. However, this is not the case with many fume cupboards and the height and width of the working aperture should be considered in relation to the height of those who are likely to make most use of this laboratory facility.

c Extraction efficiency

This involves rather more than simply the volume of air shifted in unit time. The fume cupboard should be so designed that eddying is minimised and heavy vapours, e.g. bromine, are effectively extracted from near the base of the working surface.

d The control and siting of services

In general, services should be positioned towards the *front* of the working surface and operated by controls placed outside the fume cupboard. These controls are often more conveniently placed on the side of the fume cupboard as this avoids the necessity of perforating the base of the cupboard for drainage and other services.

e Flood-proofing

If the front of the working surface of a fume cupboard is fitted with a small raised lip, this will prevent spilled liquids from running out of the fume cupboard on to the laboratory floor.

f Construction

It should be remembered that a fume cupboard is a container in which hazardous reactions are carried out. Combustible construction materials should, therefore, be kept to a minimum and the internal surfaces of the cupboard must be resistant to heat and chemical action. The range of corrosive, toxic substances used in school science teaching is, of course, relatively restricted but the selection of an appropriate material for the working top is important. Some fume cupboards have interchangeable working surfaces but the additional expense of this feature is unlikely to be justified in a school.

Suitable materials for use in a school fume cupboard are slate, lead, glazed brick, stainless steel, formica, melamine resin plastics and tiles fixed with acid-resistant cement. Table 5.8 illustrates the resistance of some materials to chemical action. The working top of a fume cupboard is normally subject to hard wear and spillages so that stains quickly appear. The appearance of the working top is generally more easily maintained if the surface is coloured black or some other dark colour.

The glass portions of a fume cupboard should be constructed of toughened safety glass such as that used for car windscreens. Toughened safety glass is available as sheet or float glass; the former is usually adequate for glazing fume cupboards.

Fume cupboards should not be used for storing hazardous materials. If these are not removed when the

	Formica and melamine resin plastics	PVC	Poly (propene)	Stainless steel
Maximum working temperature/°C	154	60	145	—
Fire retardant	—	yes	no	—
Benzene*	—	3	2	—
'Chromic acid'* (80 per cent)	2	1	1	—
Dimethylbenzene	—	3	3	—
Ethanoic acid (glacial)	1	2	1	1
Ethanol	1	1	1	—
Ethyl ethanoate	1	3	2	—
Ethoxyethane	1	3	2	1
Hydrochloric acid (conc.)	1	1	1	3
Nitric(V) acid (50 per cent)	2	1	2	1
Propanone	1	3	1	1
Sodium hydroxide	1	1	1	3

KEY to Chemical Resistance: 1 = satisfactory at 20 °C
2 = slight attack
3 = attacked
— = no information

* These substances should not, in general, be used in schools. See page 34 and page 33.

Table 5.8   The chemical and thermal resistance of some materials
adapted from: Hughes, D., Design of radionuclide laboratories, *Chem. Brit.* **4**, 2, pp. 63–66, 1968

fume cupboard is to be used, they constitute an additional hazard and may well interfere with the carefully designed flow of air into/out of the fume cupboard. Flammable solvents should be stored in a proper store and toxic materials kept in a more secure location than the back of the fume cupboard.

Funnels

a Buchner funnels

These may be of porcelain, borosilicate glass or rigid polythene and are manufactured with nominal capacities in the range $2 cm^3$ to $1000 cm^3$. Some Buchner funnels are made in two or three sections which can be dismantled to allow thorough cleaning or replacement of the filter disc. Other varieties have an integral disc which may be made of porcelain or sintered glass. Where Buchner funnels have fused-in sintered glass discs, the discs are normally manufactured to conform with BS 1752.

Laboratory sintered or fritted filters. The British Standard Publication identifies four grades of filter, the porosity and principal uses of which are given in Table 5.9.

Buchner funnels should always be thoroughly cleaned immediately after use. Where sintered glass filters are involved, cleaning is most easily done by reversing the flow of water through the disc, using mild suction only. Discs must not be cleaned with concentrated alkali solutions.

Porosity	Maximum pore size µm	Use
1	100–120	Filtration of coarse precipitates, or of mercury
2	40–60	Filtration of medium precipitates or of mercury under vacuum
3	20–30	Extraction of fine grain material
4	5–15	Analyses of very fine precipitates, e.g. barium sulphate

Table 5.9   Porosity and uses of different grades of filter

b Dropping funnels (Tap funnels)

These are normally made of borosilicate glass and in nominal capacities of 25, 50, 100 and $250 cm^3$. Their use instead of thistle funnels is to be encouraged (see page 104).

c Filter funnels

These are manufactured in glass, polythene or polypropylene and may be plain or fluted. Glass filter funnels—which may be made of borosilicate or of soda glass—

normally have sides inclined at 60° with the stem ground at 45°.

Plain filter funnels are available in several sizes, normally described by reference to the maximum diameter of the filter circle. Commonly available sizes are 50, 75, 100, 150 and 200 mm. In a school, it is sensible to stock no more than two or three sizes in any quantity. (The sizes of filter papers will need to be chosen accordingly). Fluted funnels—glass or plastic—are available in a much more restricted range of sizes and tend not to be widely used in schools.

Plastic filter funnels are generally less expensive than glass funnels of the corresponding size and they have an obvious advantage as far as breakages are concerned. They are, therefore, of considerable use in carrying out those routine filtration operations which form part of any main school chemistry course. However, plastic funnels are subject to discolouration as a result of prolonged or repeated contact with deeply coloured solutions or organic solvents.

d Hirsch funnels

Hirsch funnels are available in porcelain or borosilicate glass. The filtration disc is always an integral part of the funnel and it may be made of perforated porcelain or of sintered glass. The use of Hirsch funnels is governed by similar considerations to those described above for Buchner funnels.

e Separating funnels

These are made in several shapes (cylindrical, spherical, conical) and sizes, normally 50, 100, 250, 500, 1000, 2000 and 5000 cm^3. The stop-cocks and stoppers may be made of borosilicate glass, PTFE or polypropene (polypropylene). Interchangeable stop-cocks and stoppers in each of these materials are readily available and the stoppers are often purchased separately from the funnels themselves. It should be noted that the very large (1 litre) separating funnels are normally available only in a spherical shape and that it is possible to purchase cylindrical separating funnels which are graduated.

As with any item of glassware fitted with a stopper or stop-cock, it is important that separating funnels are thoroughly cleaned as soon as practicable after use. A procedure for dealing with 'seized up' stop-cocks is described on page 33 and the same technique may be applied to loosen glass stoppers.

f Tap funnels (see Dropping funnels)

g Thistle funnels

Thistle funnels are made with or without a safety bulb and with either a long, uniform or short, tapering stem. However, the introduction of a safety bulb into the stem of a thistle funnel increases the cost of purchase very considerably and negates one of the principal advantages of the thistle funnel—its cheapness. It is also possible to purchase thistle funnel heads made of PVC. These 'heads' have a short stem into which glass tubing of approximately 6 mm diameter may be fitted directly.

One of the principal disadvantages of the thistle funnel in a gas preparation apparatus is that the addition of further reagent via the funnel usually means that air is introduced into the reaction vessel. In the case of some gases, particularly hydrogen, this constitutes a hazard. Hence, on safety grounds, tap (or dropping) funnels are to be preferred to thistle funnels for use in this way. The thistle funnel has no equivalent in the range of 'Quickfit' apparatus. If such apparatus is being used for a gas preparation, the function of the thistle funnel is fulfilled either by a plug-type dropping funnel or by a tap funnel.

Fuses

The increased use of electrical equipment in school physics has led to a greater measure of standardisation over sockets, plugs and fuses.

The 20 mm long, 5 mm diameter cartridge fuse, in which the wire link is visible in a glass tube and contact is made through metal end caps, is becoming standard. The specification is covered by BS 4265 (1968), and the most common ranges are 50 mA, 100 mA, 200 mA, 500 mA, 1 A, 2 A and 5 A, although other ranges are available.

The 25 mm long, 6 mm diameter cartridge fuse, covered by BS 1362, is used in British three-pin fused plugs, the ranges being 2 A, 3 A, 5 A, 10 A and 13 A.

Surge-resisting fuse-links are more expensive than ordinary fuses and are not required for standard school physics equipment.

Fuse wire, designed to melt when the electric current exceeds a given value, typically 1 A, 2 A, 5 A, 10 A or 15 A, is usually sold in 100 g reels, though shorter lengths may be available if specified.

From the safety point of view, it is essential that the fuse in an electrical instrument corresponds to the design requirements. In a typical Low Tension Power Supply the primary mains circuit is protected by a 2 A fuse, while the secondary output circuit is protected by a 10 A fuse, possibly of the magnetic cut-out type. As a general rule, to calculate the size of fuse required, the maximum specified power input from the mains in watts should be divided by the mains voltage in volts to give the maximum allowed current in amperes. For example, a maximum power input of 500 W and mains voltage of 250 V, gives a maximum mains current of 500 W/250 V = 2 A, hence a primary mains fuse of 2 A is required.

Galvanometers (see Meters)

Gas analysis apparatus

Three relatively simple methods are available for the analysis of the respiratory gases (oxygen and carbon dioxide) in air samples. The method employed will depend upon the volume of air to be analysed and upon the manipulative and other skills of the pupils. All three methods employ the reagents potassium hydroxide (for the absorption of carbon dioxide) and potassium pyrogallate (for the absorption of oxygen). The reagents should be used in this order as potassium pyrogallate also absorbs carbon dioxide. The apparatus required for the three methods is described below.

a Gas burette

This is a graduated glass tube fitted at one end with a rubber stopper and a plastic stop-cock. The other end is sealed with a Suba-Seal stopper. The graduations normally run from 0–14 cm^3 in subdivisions of 0.1 cm^3. An allowance is made for the volume of the Suba-Seal stopper at the appropriate end of the burette. This apparatus is suitable for use in analysing air samples of approximately 10 cm^3 or larger volumes from which a suitable sample can be extracted. The method involves the injection of the reagents through the Suba-Seal stopper using a disposable plastic syringe and it is therefore essential that the stoppers be replaced frequently to prevent the leakage of part of the air sample from the burette. The method also involves 'levelling-off' to equalise internal and external pressures.

b Capillary analysis tube

This consists of a length of capillary tube fitted at one end with an air-tight brass screw and collar. The sample of air for analysis is drawn into the tube, which is sealed at each end with a drop of water by adjusting the brass screw. The reagents are drawn into the tube in a similar way. The capillary tube may be bent into a J-shape to facilitate the extraction of an air sample for analysis. This method is suitable for the analysis of very small volumes of air (considerably less than 1 cm^3). For successful operation of the capillary analysis tube, it is essential that the bore of the tubing remains clean and free from grease. After use, dilute sulphuric acid should be drawn into the tubes which are then left to soak overnight. They should then be thoroughly washed in hot water and detergent, rinsed in running water and dried by passing clean, dust-free air through the bore. It is also essential that the brass screw and collar remain airtight. The thread of the screw should be lightly greased with 'Kilopoise' or a similar grease. As an alternative to the brass screw and collar, the air sample may be moved along the tube by means of a 1 cm^3 disposable syringe fitted to the tube by a short rubber connector.

c Gas syringe (see also page 108)

This is a 100 cm^3 ground-glass syringe and is therefore rather more costly than either the gas burette or the capillary analysis tube mentioned above. It is used in conjunction with plastic squeeze bottles of 500 cm^3 capacity, in which the reagents are placed. The bottles are connected to the syringe by means of a short length of rubber or vinyl tubing fitted with tubing or artery forceps. This method requires an air sample of 100 cm^3 but has the advantage that percentages of oxygen and carbon dioxide may be read directly without the necessity of a calculation.

See the following for a more detailed account of these methods of gas analysis.

Nuffield O level Biology Project (1966) *Text and Teachers' Guide III*, Longman/Penguin. Gas burette and capillary tube methods.
Nuffield Secondary Science (1971) *Theme 3—The Biology of Man*, Longman/Penguin. The gas syringe method.
Nuffield O level Biology Project (Revised) (1975) *Teachers' Guide 2*, Longman. The capillary analysis tube and gas syringe methods.

Gas cylinders

Most school chemistry laboratories use a range of gas cylinders. The contents of a cylinder may be gaseous (O_2, N_2, CO, CO_2); liquid (NH_3, SO_2, Cl_2, C_2H_4); or in solution (C_2H_2 in CH_3COCH_3).

Purchase

Cylinders may be purchased together with the appropriate fittings so that the only additional expense is the cost of refilling the cylinder. Alternatively, gas cylinders may be hired and replaced by full cylinders as they become empty. The choice between these two methods is largely one of personal preference. Out-right purchase may mean that the school is without a source of gas while the cylinder has been sent away to be recharged. The alternative method obviates this particular difficulty but is likely to be more expensive in the long term.

Cylinders are available in several sizes. Large cylinders are heavy and need a cylinder trolley if they are to be frequently moved within a school. Many gases are available in cylinder form as small 'lecture bottles'. These are excellent for demonstration purposes, but are more expensive and require relatively frequent refilling. The gases commonly found in cylinder form in schools are oxygen, nitrogen, sulphur dioxide, carbon dioxide, hydrogen and chlorine, although some Local Education Authorities do not allow their schools to stock chlorine in cylinder form for laboratory use. It is generally convenient to purchase the larger size of cylinder for oxygen, hydrogen, nitrogen and carbon dioxide. Carbon dioxide is also used for physics and biology teaching and biologists are likely to use oxygen for food combustion and other experiments. The usual source of sulphur dioxide in schools is a 'siphon' or a disposable metal container which is not rechargeable.

Cylinder fittings

Every cylinder should be fitted with a reducing valve and pressure gauge before use. Cylinder valves, pressure reducers, gauges, etc., for combustible gases (e.g. hydrogen, ethyne) have outlets and fittings screwed with a left-hand thread. Those for non-combustible gases (e.g. nitrogen, oxygen) are screwed with a right-hand thread.

When using a gas cylinder, all relatively permanent connections should be made secure by the use of such items as 'Jubilee' clips. The cylinder master key should be permanently attached to the cylinder so that it is possible to cut off the gas supply rapidly in an emergency.

Stiff valves must be treated very cautiously. It should be possible to open a cylinder valve by hand pressure on the standard key. On no account should a hammer or wrench be used. A gentle tap with a piece of wood is permissible on the wing nut which screws the pressure regulator into the cylinder head. Cylinder valves should always be tested in the open air or in a well ventilated area before being taken into a laboratory for use. Dust or other foreign matter must be blown from the cylinder outlet before the regulator valve is fitted.

Identification of gas cylinders

The only legally recognised means of identifying a gas cylinder is the written word and the colours of cylinders are merely a secondary guide. Cylinders of gas for medical use are painted differently from those containing the gas in a form intended for industrial use.

Where American cylinders are in use, special care is needed as the USA colour code is different from that recommended in the United Kingdom. The complete colour code for United Kingdom gas cylinders is governed by BS 349 (1932). The following are likely to be relevant to the school situation.

Air	Grey
Carbon dioxide (non-medical)	Black
Chlorine	Yellow
Coal gas	Red (name stencilled on cylinder)
Ethyne (acetylene)	Maroon
Hydrogen	Red
Oxygen	Black
Nitrogen	Grey with black shoulder

Storage

Cylinders are most conveniently stored in manufactured trolleys, although similar trolleys can be made readily from one of the scaffolding materials which are commercially available. It is essential to purchase a trolley which corresponds in size to the cylinder.

Alternatively, cylinders may be clamped or chained firmly to a wall or bench.

Cylinders must be kept away from sources of heat, e.g. steam, radiators, underfloor heating elements and direct sunlight. If empty cylinders are to be stored temporarily in a horizontal position, they must be wedged to prevent rolling or slipping.

Use

Most accidents with gas cylinders are due to maltreatment. Cylinders containing liquid or dissolved gases must be used in a vertical position.

More than half the accidents associated with compressed gases involve oxygen. Oxygen-enriched atmospheres increase the fire risk enormously and on no account should *any* cylinder valve or outlet be oiled or greased. Also, if a valve of a cylinder containing a flammable gas, e.g. hydrogen, is opened too rapidly there is a danger of ignition of the gas from a spark generated by static electricity.

Cylinders should therefore be securely held when in use and all valves should be opened slowly and cautiously. The rate of flow of gas from a cylinder should be established before connecting to any apparatus. Suspected leaks should be tested out of doors using soap solution.

A cylinder should never be emptied completely. It is best left with slight positive pressure and the valve closed to prevent diffusion of air into the cylinder. This is particularly important with flammable gases. Empty cylinders should be marked clearly and unambiguously.

With cylinders of gases such as oxygen and nitrogen, the pressure indicates the quantity of gas remaining in the cylinder but with liquefied or dissolved gases the enormous variation of pressure with temperature makes weighing the best guide. The withdrawal of dissolved gases, e.g. ethyne, at a very rapid rate entails the risk of contamination of the gas by solvent vapour.

Geiger-Müller tubes

Two types of G-M tubes are generally used in physics investigations involving the detection of radiations in radioactivity experiments:

i halogen-quenched tube, particularly sensitive to α and β radiation, but also detecting γ radiation. This type of tube has a metal body and a mica end window, protected by an open-mesh plastic guard;

ii halogen-quenched all-metal tube sensitive to γ radiation only.

Table 5.10 shows typical data for tubes manufactured by Mullard Ltd.

A third type of tube manufactured by Mullard, Ltd., MX142, is designed for use with radioactive liquids, and has a capacity in the range 5–7 cm^3.

When purchasing G-M tubes, it is advantageous to buy ones which are interchangeable both electrically and

Mullard Type No.	MX 168	MX 180
Radiation detected	α, β, γ	γ only
Threshold voltage (max)	370 V	370 V
Plateau length (min)	100 V	100 V
Operating temperature range	$-55\,°C$ to $+75\,°C$	$-55\,°C$ to $+75\,°C$
Dead time	100 µs	100 µs
Window thickness	2.5–3 mg/cm^2	not applicable
Wall thickness	not applicable	375 mg/cm^2

Table 5.10   Comparison of MX 168 and MX 180 G-M tubes

mechanically. A holder with a G-M tube socket should be linked to a coaxial cable for connection to a PET socket on a scaler or ratemeter.

Simple tube stands are available which hold the G-M tube and holder in a horizontal position on the bench for inverse square law and absorption experiments. For the latter, sets of absorbers containing a graduated range from tissue paper and very thin aluminium foil in plastic mounts, through thicker aluminium and lead of various thicknesses, are commercially available. Bench stands to hold the absorbers vertically should be fitted with a 4 mm diameter horizontal socket to accept a mounted radioactive source.

Solid-state detectors

Solid-state detectors are smaller than the conventional G-M tubes and may be mounted in a glass tube for magnetic deflection experiments of α particles in the absence of air. A solid-state detector, attached to a pre-amplifier, enables a scaler or ratemeter to be used for counting.

Safety aspects (see also page 40)

The use of ionising radiations in schools and establishments of further education in England and Wales is fully discussed in Administrative Memorandum 2/76 issued by the Department of Education and Science and the Welsh Office. Teachers must consult this document (or a similar one issued by an appropriate authority in other countries) to ascertain the conditions under which work involving radioactive substances may be conducted, and how approval for such work may be obtained.

The Memorandum is accompanied by *Notes for the guidance of schools, establishments of further education and colleges of education on the use of radioactive substances and equipment producing X-rays*. A summary of the more important points in the Memorandum and these Notes will be found in Chapter 3, pages 40 and 41.

Teachers in Scotland are referred to the appropriate SED *Circulars*; see page 40.

Generators

Signal generators

The following features are of importance when considering the purchase of a versatile signal generator for school use:

   i the possibility of a sine and square wave output;
  ii a continuously variable amplitude control;
 iii a frequency range from 1 Hz (for vibration generator to 100 kHz (for a.c. experiments) preferably in five decade ranges;
  iv a clearly calibrated scale which can be read easily although it is non-linear;
   v a full power output at all frequencies from low impedence terminals for vibration generators or loudspeakers;
  vi full output voltage from high impedance terminals for a.c. experiments;
 vii an attenuator control, e.g. 0, $-20\,dB$, $-40\,dB$;
viii an on-off switch;
  ix portability—size and weight characteristics.

Van de Graaff generator

The hand- or motor-driven self-exciting van de Graaff generator has now replaced the Wimshurst machine for the generation of high voltages to illustrate electrostatic phenomena, such as sparking and 'electric wind'. The high potential difference produced can also drive a kinetic motion simulator.

If a van de Graaff generator is placed in series with a 1 MΩ resistor and a light beam galvanometer, the current flow is sufficient to establish the identical nature of static and current electricity. A kit of accessories which includes a 'head of hair', an electric whirl and a point discharger can be purchased or improvised.

The use of modern materials has made the generator more reliable, but in particularly humid or dusty conditions it is good practice to dust the various parts prior to use and to dry the belt.

The following factors should be considered when purchsing a van de Graaff generator

   i *drive*. Hand-operated models may appear to be more versatile, but a good Earth connection is still required. This Earth connection is provided via the *Earth lead* in the 3-core cable of motor-driven generators;
  ii *variable or fixed speed*. Motor-driven models have either a fixed speed induction motor or a variable speed control;
 iii *size*. The larger the diameter of the metal collecting dome, the larger is the voltage reached before discharge. A dome diameter of 25 cm can reach a voltage of 750 kV under good conditions, corresponding to a spark discharge of over 10 cm in length. However, a sphere of half the above diameter, i.e. 12.5 cm, can reach a voltage of 250 kV which is high enough for all normal applications;

iv *discharge sphere*. A discharge sphere is essential and is normally supplied with the generator;

v *storage*. Most models can be dismantled after use;

vi *accessories*. The collecting dome should have a 4 mm socket in the top to accept various accessories and to facilitate electrical connections;

vii *spares*. The belt is the most likely component to require replacement.

Vibration generator

A vibration generator is designed to produce mechanical vibrations when connected to a suitable signal generator. It is desirable for the full stroke movement of the piston to be maintained over the entire frequency range—typically 1 Hz to 100 kHz—and for the generator to respond to sine and square waves. The maximum applied voltage allowed, generally about 4 V a.c., should be clearly stated on the instrument to minimise the risk of accidental damage.

Glass rod

Glass rod is available in soda or borosilicate glass and is purchased in bundles containing nominal lengths of approximately 1.5 metres. The number of lengths in the bundle depends upon the diameter of the rod. Glass rod is available in diameters ranging from 3 mm to approximately 15 mm.

Glass rod may be cut in the manner described for glass tubing (see page 109) and it should be stored in a similar way. When making stirring rods from a length of glass rod, it is important to either fire polish the cut ends or to flatten one of the cut ends by pressing the heat-softened glass vertically downwards on to a metal surface.

It should be noted that polypropene (polypropylene) stirring rods may also be purchased.

Glass syringes

Glass syringes usually have a capacity of 100 cm³. They are precision instruments and therefore relatively expensive to purchase. The barrel and the piston are ground to very close tolerances and each usually carries a number or other identifying mark as the pistons are not interchangeable. The risk of breakage can be minimised by storing glass syringes in the pre-formed polystyrene containers in which they are supplied or in a drawer containing a thick layer of cotton wool.

It is advisable to tie each piston to its barrel with a piece of string. This should be long enough to allow the piston to be withdrawn slightly beyond the 100 cm³ mark but short enough to prevent the piston from being pulled or dropped out of the barrel. Glass syringes should not be clamped directly but be mounted on a board by means of two 'Terry' clips and then clamped in the usual way.

Glass syringes will give satisfactory service only if they are thoroughly cleaned after use. In particular, fine solids such as dust or copper powder must not be allowed to enter the syringe. The barrel and piston should be separated, washed with water, then ethanol and finally with propanone (acetone) or ethoxyethane (ether). Alternatively, the syringe parts may be washed with a mixture of ethanol and distilled water and the excess solvent removed by means of an air jet. If the latter method is adopted, a piece of filter paper should be used to reduce the amount of dust being drawn or passed into the syringe. When a syringe is clean, the piston will run smoothly and evenly. On no account should the piston be lubricated, e.g. with Vaseline.

Plastic syringes sometimes have a smaller capacity (50 cm³) than the glass counterparts. They are much cheaper to purchase but are rarely airtight and therefore have very limited use in quantitative work. Plastic syringes should be cleaned with cold water as organic solvents (particularly propanone) are likely to attack the polymer of which the syringe is made.

It is sometimes necessary to use a glass syringe at temperatures above 100 °C, e.g. in a molecular weight determination. In such circumstances, sudden changes of temperature should be avoided. At temperatures above 150 °C most glass syringes begin to leak quite badly.

Glass tubing

Most of the glass tubing used in school science teaching is made from soda glass, which softens at approximately 700 °C and is, therefore, easily worked with a bunsen flame. However, it is possible to purchase borosilicate glass tubing, which softens at a slightly higher temperature (*c*. 820 °C) and which has the properties generally associated with 'Pyrex' brand glassware.

Glass tubing is normally purchased in bundles of nominal 1.5 metre lengths. Manufacturers' catalogues specify the number of lengths in a bundle; the larger the diameter of the tubing, the fewer the lengths which make up the bundle. When ordering glass tubing, it is important to take note of both the external diameter of the tubing and the wall thickness. Glass tubing is available in external diameters of from 3 or 4 mm to about 55 mm.

Whenever possible, glass tubing should be stored horizontally. Plastic guttering fastened to the wall of the preparation room can be used for this purpose. If space does not permit horizontal storage, a suitable rack should be constructed or purchased which will store the tubing vertically. Whatever form of storage is used, the different sizes and types of glass should be kept separate from one another and be appropriately labelled.

Many laboratory accidents arise from bad technique or carelessness in the use of glass tubing. Lengths of glass tubing must always be carried vertically and it is important to ensure that the type of glass being used is suitable for the task in hand.

Glass tubing of not more than about 15 mm external

diameter may be cut by means of a glass knife, a triangular glass-cutting file or a rotary glass cutter. The glass is first scratched with the knife or file, the scratch being made at right angles to the long axis of the tubing. Then, protecting the hands with a clean cloth, the thumbs are placed on either side of the scratch and close to it. Application of firm but not excessive pressure by the thumbs in a direction away from the body should be sufficient to break the glass tubing cleanly. If the glass does not break readily, it may be necessary to deepen the original scratch. The cut ends of the glass tubing must then be fire polished by rotating the ends in a hot bunsen flame.

Cutting glass tubing of external diameter wider than about 15 mm requires a different technique. A glass file, knife or rotary cutter is used to make a *continuous* scratch around the circumference of the tubing. The tube can then be made to break at the scratch in one of the following ways:

i a loop of resistance wire is fastened around the scratch. The wire is then heated electrically (12–24 V d.c.) and the heat causes the tube to crack. If the glass tubing has thick walls, more than one application of the heated wire may be necessary;

ii the glass tubing is *gently* warmed at the scratch either by means of the wire as in (i) or in a low bunsen flame. A drop or two of cold water released from a teat pipette on to the scratch usually causes the tubing to break cleanly;

iii the scratch may be touched in several places with a hot glass rod. Repeated application of this procedure causes a clean crack to develop.

When glass tubing is to be bent, it is essential to heat the area to be bent uniformly, and a batswing burner (or a flamespreader) is often used for this purpose in a school. The diameter of the glass tubing must be maintained throughout the bend; flat or distorted bends should not be used. If the tubing is to bend to a particular angle other than 90°, it may be helpful to have the angle drawn on a piece of paper to act as a guide.

The threading of glass tubing through corks or bungs continues to cause many accidents. One of the safest and simplest techniques uses a cork borer and is illustrated in Figure 5.11.

The softening of soda glass in a bunsen flame is indicated by the appearance of an intense yellow colouration. Under these conditions, soda glass tubing may be drawn out to produce tubing of a narrower diameter or be tapered to produce a teat or a dropping pipette. Tubing which has been drawn out to a thin diameter may be cut and then sealed by heating strongly while rotating the end to be sealed. The sealed tube may be blown into a bulb by alternate heating and blowing until a bulb of the desired size is obtained. For details of these and other operations involving glass, the reader is referred to the sources quoted in the Bibliography on pages 164–165.

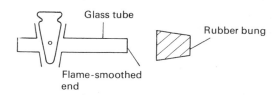

1   Glass tube   Rubber bung   Flame-smoothed end

2   Choose the correct size cork borer.

3   Lubricate the cork borer with, e.g. glycerol, teepol, soft soap.

4   Bore the hole in the normal manner so that the operation leaves the borer in the bung with the hand-grip on the same side as that from which it is wished to insert the glass stop-cock.

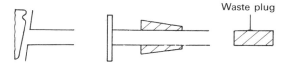

Waste plug

5   Select and lubricate the next largest cork borer and slide it over the first until it has passed through the bung.

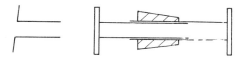

6   Withdraw the smaller cork borer and slide the glass tube into position.

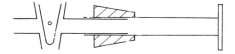

7   Withdraw the cork borer.

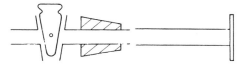

Figure 5.11 Technique for threading glass tubing through a bung or cork

Gloves

Several different types of gloves should be available in a school science block.

i *Insulated gloves*, where protection from heat is required. Such gloves should not contain asbestos.

ii *PVC gloves.* For general cleaning duties. Like in-sulated gloves these are purchased as individual pairs. PVC gloves should be of laboratory, rather than standard, grade.

iii *Polythene 'disposable' gloves.* These are purchased in packs usually of 100 gloves. They are useful for a wide variety of laboratory tasks, including work with radioactive materials. See also page 95.

Goggles (see Safety spectacles)

Haemacytometers (Blood cell counting chambers)

A haemacytometer is essentially a microscope slide bearing a small well of known depth, the base of which is marked with squares of known dimensions. During use the well is covered with a special coverslip (usually 0.4 mm thick) so that an exact and known volume of liquid is enclosed within the well. There are various designs of

haemacytometer, the two commonest of which are described below. They are available with single, double or quadruple counting chambers so that single, double or quadruple counts may be made of the same or different samples. Although designed for counting the numbers of red and white blood corpuscles in a blood sample, haemacytometers may also be used for counts of algae, yeast cells, protozoa and other similar suspensions.

For blood counts, the haemacytometer is used in conjunction with haemacytometer dilution pipettes (Thoma pipettes). This is necessary because the cells are far too numerous for direct counting. Two types of pipette are required. One (for red cells) has graduations at 0.5, 1 and 101. Blood is drawn up to either the 0.5 or the 1 mark and then diluted with an appropriate isotonic blood anticoagulant (e.g. Hayem's solution—see page 183). A dilution of 200 or 100 times is thus obtained (200 will be the most convenient in the ensuing calculation). The second pipette is for white cell counts and is marked at 0.5, 1 and 11 and is used in a similar way to the first,

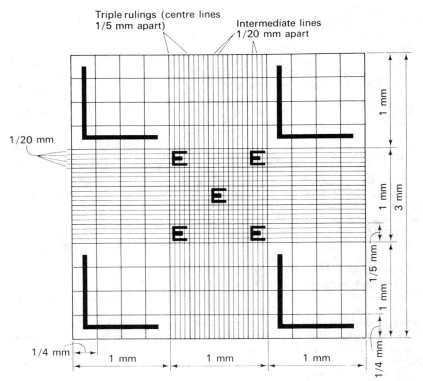

The five blocks of sixteen small squares (each 1/20 mm square) marked E are used for counting red blood cells.

The four blocks of sixteen larger squares (each 1/4 mm square) marked L are used for counting white blood cells.

Figure 5.12   The improved Neubauer haemacytometer

enabling dilutions of 10 or 20 times. Detailed instructions for the use of haemacytometers are normally supplied by the manufacturers, but may also be found in Bradbury (1973).

The Improved Neubauer haemacytometer

This is the type most commonly encountered. The base of the chamber is marked in the manner shown in Figure 5.12. When the coverslip is in position, the chamber is 0.1 mm deep and the volume above the 3 mm square is thus 0.9 mm³ (i.e. 3 mm × 3 mm × 0.1 mm). The volumes above the various smaller sub-divided areas may be calculated in a similar way. The red cells are normally counted in five groups of 16 of the smallest areas, each of these measuring 0.05 mm × 0.05 mm as shown in Figure 5.12. Thus the total area counted is 0.05 mm × 0.05 mm × 16 × 5 and the total volume 0.05 mm × 0.05 mm × 16 × 5 × 0.1 mm or 0.02 mm³. When counting cells, all those touching the lines delineat-ing the top and left-hand side of the square are included; all cells touching the lines delineating the bottom and right-hand side of the square are excluded. (See Figure 5.13).

The number of red blood cells in 1 mm³ of blood is calculated in the following way.

Total volume = 0.02 mm³.

Let the total number of red cells counted in this volume be R.

If there are R cells in 0.02 mm³ diluted blood,

there are therefore R/0.02 red cells in each mm³ of diluted blood.

The number of red cells in 1 mm³ of the original sample of undiluted blood is therefore R/0.02 × dilution.

If a dilution of × 200 has been used, the calculation is made easier.

No. of red cells per mm³ of blood = R × 200/0.02 = R × 10 000.

The white cells are normally counted in the four corner

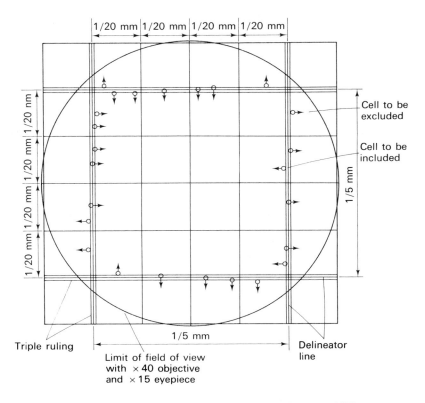

Arrows pointing inwards from the cells towards the group of sixteen 1/20 mm squares indicate those cells to be included in the count; arrows pointing outwards indicate cells to be excluded from the count.

Figure 5.13   The method of counting blood cells in a haemacytometer

squares, each with an area of $1\,mm^2$ and therefore $0.1\,mm^3$ in volume. These are indicated in Figure 5/12. The total volume is therefore $0.4\,mm^3$.

Let the total number of white cells counted in this volume be W.

If there are W cells in $0.4\,mm^3$ of diluted blood, there are therefore $W/0.4$ white cells in each mm^3 of diluted blood. The number of white cells in $1\,mm^3$ of the original sample of undiluted blood is therefore $W/0.4 \times$ dilution.

If a dilution of $\times 20$ has been used,

No. of white cells per mm^3 of blood $= W \times 20/0.4 = W/2 \times 100.$

The Fuchs Rosenthal haemacytometer

This has a larger chamber than the Neubauer haemacytometer, of side $4\,mm$ and depth $0.2\,mm$ when the coverslip is in position. The total volume over the graduated area is thus $3.2\,mm^3$ (i.e. $4\,mm \times 4\,mm \times 0.2\,mm$). The square is divided into sixteen $1\,mm \times 1\,mm$ squares each of which is further subdivided into sixteen $0.25\,mm \times 0.25\,mm$ squares in the manner shown in Figure 5.14. Appropriate adjustments must be made to the calculation given for the Neubauer haema-

cytometer in order to determine the number of red or white cells in a blood sample.

Incubators

An incubator is an essential piece of apparatus for microbiological work, for the culture of *Drosophila* and for the incubation of chick eggs. It may also serve as a drying oven in some circumstances. The design of a standard incubator is basically the same as that of an oven (see page 134). However, it is important to realise that incubators should have a narrower temperature fluctuation than is normal in most ovens—fluctuations of more than $0.5\,°C$ should be avoided. The overall temperature range is also important and should run from $5\,°C$ above ambient to $80\,°C$ or $100\,°C$. Most drying ovens only operate efficiently above $40\,°C$ and this is too high for many biological purposes. In selecting an incubator the following points should be borne in mind:

  i the maximum temperature fluctuation which can be tolerated (generally not more than $0.5\,°C$);

  ii the temperature range over which the incubator operates efficiently;

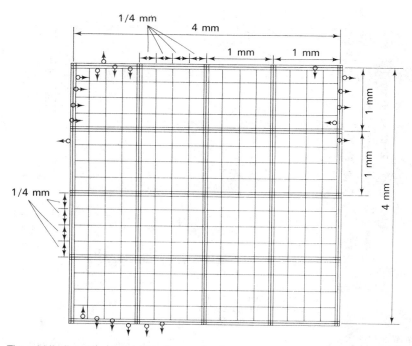

The middle line of each triple ruling marks the boundary of each 1 mm square.
The same rules for counting apply as with the improved Neubauer haemacytometer.

Figure 5.14   The Fuchs Rosenthal haemacytometer

iii the size required, including number of shelves, ease of adjustment, etc.;

iv the presence of an independent safety thermostat which will ensure the 'cut-out' of the heating element in the event of failure of the main thermostat;

v the presence of an inner toughened-glass door permitting observation without affecting the inside temperature;

vi the ease with which the interior may be cleaned;

vii the efficiency of the ventilation system. The shelves should permit free circulation of air and there should be an adjustable ventilator in the 'roof' of the incubator. Some models are fitted with fan circulation which ensures uniform temperature distribution;

viii the presence of an indicator light to show whether or not the incubator is operating;

ix the ease of servicing (e.g. replacement of element, thermostat, etc.).

Some models are designed specifically for incubating chick eggs and are fitted with an observation dome. These incubators are not as versatile as the standard 'cupboard' models but may be desirable if chick eggs are being incubated regularly and in large numbers.

Indicators

The acid-base indicators most widely used in schools are litmus, phenolphthalein and methyl orange. They may be purchased ready for use or made up in accordance with the instructions given in Chapter 7. Table 5.11 summarises the pH ranges and colour changes of a number of acid-base indicators.

Redox indicators are organic compounds which show contrasting colours in their oxidised and reduced states and which readily and reversibly undergo conversion from one of these states to the other.

The 'neutral' point corresponds to equal con-

centrations of the oxidised and reduced forms at the mid-point of the appropriate oxidation curve. As with the pH of an acid-base indicator, the electrode potential of the redox indicator must be such that it will undergo a colour change at the region of the marked change in potential at the end-point of the titration in which it is to be used.

Table 5.12 indicates the colour changes and $E^{\ominus}$ values of some common redox indicators.

Indicator	Colour change	$E^{\ominus}/V$ (293 K and pH = 7)
Diphenylamine sulphonic acid (in M H_2SO_4)	violet—colourless	+0.83
Diphenylamine	blue—colourless	+0.76
Thionine	violet—colourless	+0.06
Methylene blue	blue—colourless	+0.01
Neutral red	red—colourless	−0.32

Table 5.12   Colour changes and $E^{\ominus}$ values of redox indicators

Interchangeable laboratory glassware

Interchangeable laboratory glassware such as the 'Quickfit' range is now widely used in schools. This glassware is manufactured in resistant borosilicate glass and is available in conventional and semi-micro sizes. It is more expensive than ordinary laboratory glassware, principally because of the interchangeable ground-glass joints fitted to each item.

These joints are made in accordance with the ISO and the appropriate BS specifications (BS 572 and 2761) and are of two kinds:

i conical joints, designated by reference to the nominal diameter of the wide end and length of engagement of the cone. Thus a 29/32 joint has a maximum diameter of 29 mm and is engaged for 32 mm of its length;

ii spherical joints, designated simply by reference to the appropriate spherical diameter in millimetres. Thus, an S/29 joint has a diameter of 29 mm.

'Quickfit' glassware should be cleaned as soon as possible after use. Ground-glass joints should not be greased or scoured with abrasives. If a joint cannot be loosened readily, it should be placed in warm, running water.

In some instances, new interchangeable cones or sockets may be fitted to glass apparatus but considerable glass-working skills are required and these may not be readily available in a school.

Indicator	Colour change	pH range
Thymolphalein	colourless—blue	9.3–10.5
Phenolphthalein	colourless—violet	8.2–9.8
Thymol blue	yellow—blue	8.0–9.6
Cresol red	yellow—purple	7.0–8.8
Phenol red	yellow—red	6.4–8.2
Bromothymol blue	yellow—blue	6.0–7.6
Bromophenol red	orange—purple	5.2–6.8
Bromocresol purple	yellow—purple	5.2–6.8
Methyl red	violet—orange	4.4–6.2
Bromocresol green	yellow—blue	3.8–5.4
Methyl orange	pink/red—yellow	3.1–4.4
Congo red	violet—orange	3.0–5.2
Bromophenol blue	yellow—violet	3.0–4.6
Thymol blue	red—yellow	1.2–2.8

Table 5.11   pH ranges and colour changes of some common acid-base indicators

Kymographs

In recent years there has been a considerable increase in both the extent and the sophistication of the experimental physiology performed at school level. As a result, the kymograph is now used quite commonly for sixth-form work. At university or research level, great sensitivity and accuracy are required and it is therefore necessary to use smoked paper which is varnished after the trace has been made; for this purpose a metal cylinder (usually brass or aluminium) is essential. However, at school level less sophistication is required and a number of manufacturers have produced relatively simple models which are quite adequate for a range of experimental work including nerve-muscle preparations, studies of the frog heart-beat and in conjunction with a spirometer for investigations of human breathing (see page 144).

The following points should be considered before selecting a suitable kymograph for purchase.

a Construction. The instrument and its associated equipment should be compact and of a rigid construction. The controls should be easily manipulated and clearly labelled. Access to the motor should be simple for ease of repair and servicing.

b The cylinder. Unless smoke traces are being made (see above) a rigid plastic cylinder is quite adequate and considerably cheaper than the metal versions. There should be a lip at the bottom of the cylinder to prevent the chart paper from slipping off. Adjustment of the cylinder height should be an easy matter.

c The writing attachment. The mode of operation of the pen should be simple and the writing smooth and continuous. The reservoir of the pen should be emptied after use and the pen thoroughly washed. It should be washed again before refilling. Should the pen become blocked, it can usually be cleaned with a piece of very thin wire, which should be supplied by the manufacturers (a piece of thin fuse wire or a single thin strand from a length of electrical flex are suitable substitutes).

d Cylinder rotation. At least three speeds will be required for a range of experimental work:

i approximately 1 mm per second (e.g. for work on breathing when used with a spirometer);
ii approximately 10 mm per second (e.g. for frog heart-beat traces);
iii approximately 500 mm per second (e.g. for nerve-muscle preparations).

e Attachment of accessories. There should be an adjustable arm attached to the instrument so that a perspex bath may be fitted to contain the preparation under investigation. In some cases, kymographs are supplied with a full range of accessories (perspex bath, stimulator electrodes, heart lever with clamp, writing level and pen, hangers and masses for loading, etc.). Other manufac-

turers supply some of these as extras. The range of accessories accompanying the instrument should therefore be carefully checked as it has a considerable bearing on the cost.

f The stimulator. It is a considerable advantage to have the stimulator housed in the same case as the kymograph mechanism itself. It should be possible to elicit a single pulse either manually through a small switch, or by using a switch mounted on top of the kymograph case which is triggered by the rotation of the cylinder. It should also be possible to obtain repeated pulses over a wide frequency range (e.g. one pulse per 5 seconds to 50 pulses per second). The pulse itself should be of approximately 1 ms duration and the output should be variable from 0–20 V.

g Chart paper. A variety of papers is available and the method of recording the time base differs from one kymograph to another. In the most convenient forms the paper includes a printed time base so that intervals may be read off directly. If this facility is not provided, the manufacturers' instructions should be followed for recording the time base.

Lamps

It has become more common in recent years to use lamps (bulbs) of a given type, particularly with regard to voltage. This makes it easier to plan for spares, as lamps are less likely to be 'blown' by the accidental application of too high a voltage. In spite of this, particular requirements of voltage, wattage and current mean that a variety of lamps has to be stocked. Lamps should be carefully labelled in a 'tidy box' and an adequate stock should be maintained. It may prove helpful to colour-code bulbs (particularly the MES type) with a small dot of paint, using a different colour for each voltage.

Bayonet cap (BC) lamps

Bayonet cap lamps are generally of a high wattage and powered by mains supply, e.g. clear or pearl gas-filled lamps, 240 V, 60 W; 240 V, 100 W. A 240 V, 40 W pearl lamp in a BC holder and reflector is suitable for projecting shadows on a translucent screen.

For experiments illustrating the action of the pinhole camera, carbon filament lamps provide an easily recognisable and bright image. The resistance of the carbon filament decreases as the temperature of the filament increases, so that power dissipation can vary between 120 W and 200 W.

BC lampholders are most suitably mounted on a base with at least 2 m of flex linked to an appropriately fused mains plug.

Small bayonet cap (SBC) lamps

Two sizes of SBC lamp are generally available:

i small wattage lamps, 18 mm in diameter, e.g. 12 V, 6 W;

ii large wattage lamps, 38 mm in diameter, e.g. 12 V, 36 W.

These gas-filled bulbs generally have an axial line filament. Commonly used 38 mm diameter lamps include the 12 V, 24 W and 12 V, 48 W lamps.

Both size lamps fit into an SBC lampholder, which is most conveniently mounted on a wooden block or metal base so that electrical connections can be made via 4 mm sockets. Units with one, two or three lampholders per base may be purchased from the major suppliers.

Miniature Edison-Swan (MES) lamps

These very low voltage, low wattage, screw-fitting lamps are used in experiments with dry cells, for example with circuit boards in current electricity. The most popular size is the 1.25 V, 0.25 A round spotlight bulb. Other widely used lamps include the 2.5 V, 0.3 A type in the electromagnetic kit.

MES lampholders are usually mounted on a Bakelite base, but for circuit-board work, they are mounted on special spring connectors.

Lasers

A laser emits a high intensity, low divergent, monochromatic coherent beam of light which may be used to demonstrate a wide range of optical phenomena, such as diffraction, interference, polarisation and the properties of holograms.

For school use, it is most convenient for the laser to be supplied with an integral power-pack, and to be designed for use on an optical bench or table. A typical Helium-Neon gas laser suitable for school use has a power output of about 1 mW, and provides an emergent beam of unpolarised red light of wavelength 632.8 nm. An accessories kit should be made up from existing stock or may be purchased from the usual suppliers.

Safety aspects

Before any experimenting is done, it is essential that teachers read through the code of practice to be followed when using lasers in schools. The appropriate Administrative Memorandum, obtainable from the Department of Education and Science is No. 7/70 *Use of Lasers in Schools and Other Educational Establishments.*

The Scottish Education Department's regulations are contained in circular No. 766 *Use of Lasers in Schools, Colleges of Education and Further Education Establishments.*

Under no circumstances should the laser beam be viewed directly, and careful checks must be made that there is no possibility of specular reflections. It is advisable to use an optical alignment system before the laser is switched on, and to wear laser safety goggles while setting up a demonstration. For further discussion of safety aspects see page 42.

Leads

Flexible leads with 4 mm plugs, to fit 4 mm sockets, have become firmly established because of the speed and convenience with which electrical apparatus and components can be connected.

Older equipment can be 'converted to 4 mm' by fitting insulated 4 mm terminals or plug sockets. Temporary conversions on certain apparatus, such as a metre bridge, may be carried out using hexagonal, screw-threaded brass connectors, ending in a 4 mm socket, or a spade terminal fitted with a 4 mm end socket, or crocodile clips whose shanks accept 4 mm plugs. For d.c. work, red and black should be used for the positive and negative terminals respectively, while green is reserved for the Earth terminal. Another colour, often yellow or white, is suitable for a.c. terminals.

The most versatile—but also the most expensive type of lead—is a length of well insulated, flexible lead of copper conductors which can be safely used with 5 kV power-packs and with low voltage apparatus, and which is firmly secured to an insulated and stackable 4 mm head at each end. Such leads may be purchased in different colours, typically red, black and green, and in different lengths, from the 15 cm or 25 cm lead for short connections to the 1 m lead for long connections.

Stackable, insulated 4 mm socket terminals and suitably insulated connecting wire can be purchased separately to make up leads of different lengths and colours.

Leads can be stored by plugging the ends into the 4 mm holes in a sheet of pegboard fastened to battens on the wall.

Lenses

Lenses for school use should be purchased with ground edges and should preferably have been manufactured from optically worked glass. Figure 5.15 illustrates the

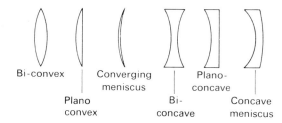

Figure 5.15 Shapes of lenses

various shapes available, either in spherical or cylindrical form.

There has been a welcome attempt in recent curriculum development projects to standardise on certain sizes of spherical lenses, i.e. 25 mm diameter for small lenses as used in the model microscope and the large 38 mm or 50 mm diameter lenses for most other purposes.

When ordering lenses, it is important to specify the shape, the diameter of the spherical lens or dimensions of the cylindrical lens, and the focal length or power.

$$\text{Power in dioptres (D)} = \frac{1}{\text{focal length in metres}} \text{ or } \frac{100}{\text{focal length in cm}}$$

Hence, a converging lens of focal length $+40$ cm has a power of $+2.5$ D. Similarly, a concave meniscus lens of focal length -33 cm has a power of -3 D.

Magnetic stirrers

A magnetic stirrer consists of a magnetic stirring bar which is made to rotate by the revolution of a powerful magnet attached directly to an electric motor or a water-driven turbine. The following points are important in evaluating magnetic stirrers.

a Speed of rotation. A variable speed is desirable although the maximum speed of rotation of any given magnetic bar will be a function of the size of the bar and of the quantity and viscosity of the medium in which it is used.

b Size of the working area. The area over which the rotating magnet can be made to drive the stirring bar is usually much less than the top surface area of the stirrer.

c Electrical power supply. In the case of electrically driven magnetic stirrers, it is important that the motor is designed to run off the voltage normally supplied in the laboratory, e.g. 220–240 V at 50 Hz.

d Material of construction. It is important that the electric motor be protected from corrosion and that the stirrer be made from a corrosion-resistant material which can be easily wiped clean.

Some magnetic stirrers are fitted with a boss at the base so that a rod may be supported vertically and used to carry a clamp. It is also possible to purchase a combined hot-plate/magnetic stirrer.

Magnetic stirring bars are available in various sizes, e.g. 15, 25, 40, 55 mm length and may be coated with PTFE or poly(propene). The PTFE-coated bars are usually more expensive. It is helpful to purchase or improvise a magnetic stirring bar retriever. This consists of a glass or other rod, perhaps 30 cm long, to one end of which a small magnet (or another stirring bar) has been

fastened. Such a device enables a stirring rod to be retrieved easily from the bottom of a conical flask or beaker.

Magnets

With the demise of school magnetometry, the need for the 10 cm or 15 cm steel bar magnet has largely disappeared. Shorter bar-shaped or cylindrical magnets, selling under names such as Ticonal or Alcomax, are very satisfactory for simple magnetic-field plotting.

Anisotropic ceramic ferrite, e.g. *Magnadur* magnets—5 mm thick with the poles on the 50 mm × 19 mm large faces—have come into use as pole pieces on a steel yoke, as in the Westminster electromagnetic kit.

Horseshoe-shaped magnets made of Alcomax are very strong, for example the Eclipse type C which has a height of about 3 cm and a gap of 2.2 cm between the pole pieces. The larger Eclipse Major is one of the most powerful permanent magnets available for school use and will deflect β particles from a radioactive source through 90° or more.

The keepers must always be kept in place when magnets are not in use, to provide a path for the magnetic flux.

If iron filings have become 'stuck' on a magnet, they may be removed with soft plasticine or 'Blue-tak'.

While the timing mechanism of some watches is not affected by magnetic fields, it is good practice to remove wrist watches when handling 'strong magnets' to avoid the possibility of damage.

Magnifiers

A variety of types of hand magnifier is available from commercial suppliers. Two types are recommended for school use.

a Folding magnifiers (hand lenses)

These will be found invaluable at all levels of school work. The simplest consists of a single lens (usually giving a magnification of × 5) housed in a folding plastic or metal casing. A lens doublet will reduce chromatic and spherical aberration somewhat and is probably worth the small extra cost involved. For most purposes a × 8 lens will be sufficient, but on occasions there may be a need for a magnifier providing a greater magnification and a better quality image. In this case an achromatic triple-folding lens magnifier is recommended. × 10, × 15 and × 20 versions are available but they will cost about five times as much as the doublet magnifiers mentioned above.

b Dissecting magnifiers

These are larger than folding magnifiers with a lens diameter of up to 10 cm (c.f. 2 cm for the largest of the folding magnifiers). Such lenses are normally encased in a metal surround and may be mounted on tripod legs or a flexible arm. It is useful to have a number of these instruments in the biology laboratory as they free the hands for manipulation or dissection of the specimen being examined. Dissection magnifiers normally have a magnification of $\times 3$ to $\times 5$.

Masses (weights)

For a discussion on mass and weight, see Balances, page 80.

Although Imperial masses are still available for purchase, they have become obsolete with the international changeover to SI units.

Iron masses with a lifting ring are available in 50 g, 100 g, 200 g, 500 g, 1 kg, 2 kg, 5 kg and 10 kg sizes, and this range will satisfy all likely requirements for experiments on inertia and 'the Earth's pull on given masses'.

For sonometer and spring extension experiments, the most convenient form is a hanger with a hook to which slotted masses may be added, the mass of the hanger having a mass equal to that of one of the slotted masses. The masses should be designed to slot firmly onto the holder and to stack without sliding. Sets of masses are available which have a total mass of 100 g ($\times$ 10 g), 1000 g ($\times$ 100 g) and other combinations to a total of 20 kg for sonometer work.

A mass of 100 g hung vertically from a steel spring in the Earth's gravitational field exerts a vertical downward pull of 0.98 N, which may be taken as 1 N for most practical purposes. Similarly, a force of 9.8 N, usually taken as 10 N, is exerted by the Earth's gravitational field on a mass of 1 kg. The geographical variation of the gravitational field strength is absorbed in the mass tolerance of ± 1 per cent.

Measuring cylinders

Measuring cylinders are manufactured in glass or plastic and with capacities in the range 5 cm³ to 2 litres. Glass cylinders may be made of soda glass or of borosilicate ('Pyrex') glass, the latter being considerably more expensive. Plastic cylinders may be made of poly(propene) or poly(methylpentene) and are adequate for almost all purposes in school science teaching.

In addition, measuring cylinders may be purchased *either* with a spout *or* with a glass/plastic stopper, and some are manufactured with a detachable plastic base. Measuring cylinders should be made in accordance with the relevant British Standard, BS 604, *Graduated Measuring Cylinders*.

Glass measuring cylinders are easily knocked over and broken. If only the top is chipped, the cylinder may be repaired using the technique for cutting wide-bore glass tubing described on page 109. Much breakage can be prevented by fitting glass measuring cylinders with sponge-rubber ring 'protectors'. These may be purchased to fit cylinders of capacity 50 cm³ to 500 cm³.

Measuring cylinders may be stored in a special rack which allows the stem but not the base of the cylinder to pass through a holding frame. The cylinders are thus held in an inverted position which means that the racks may also be used for draining the cylinders after washing and rinsing.

Measuring cylinders made of soda glass are susceptible to thermal shock. When cleaning such cylinders or when diluting concentrated reagents in them, care should be taken to avoid sudden large changes of temperature.

Mechanical stirrers

Most stirrers are designed to be mounted on a retort stand or on laboratory scaffolding by means of a bosshead. When selecting an electrically driven stirrer for school use, the following points should be considered.

a Range of speed available and the degree of speed control. The cheapest stirrers have a fixed speed which is constant over a wide load range. This speed may be altered by means of a variable transformer. However, a stirrer with a direct variable speed control, although more expensive, is a much more versatile instrument. In some models, speed-reduction gears are fitted, but a high degree of speed stability is not normally necessary in the school laboratory.

b Range of stirring rotors. It is helpful if the stirrer can be adjusted to accommodate a range of stirring rotors. These rotors may be made of glass, stainless steel or stainless steel coated with polythene. The adjustment is normally made by means of a chuck and key arrangement. In those circumstances where a direct drive is impossible, it is an additional advantage if a flexible drive arrangement can be used in conjunction with the stirrer.

c Motor characteristics. The motor may be of the induction type and it is important that it is not damaged if the stirring rotor is somehow prevented from rotating. More commonly, the motor may be of a series wound universal type and fan-cooled. In such cases the brushes of the motor should be examined periodically and replaced as necessary.

It is helpful if the motor mounting can be adjusted to give flexibility in the orientation of the drive shaft.

Mini-stirrers, driven by a 12 V battery or via a suitable transformer, are designed to be held in the hand or used in conjunction with standard ground-glass joints.

	Supply for 16 working groups	Possible advantages and disadvantages
Single range meters	16 ammeters 0–1 A 16 ammeters 0–5 A 16 voltmeters 0–5 V 16 voltmeters 0–15 V 16 galvanometers	Single range meters with two terminals cause least confusion to young or less able pupils
Dual range meters	16 ammeters (0–1 A, 0–5 A) 16 voltmeters (0–5 V, 0–15 V) 16 galvanometers	Dual range meters have three terminals, which could cause confusion; however, a total of only 48 meters is required
Basic galvanometers with shunts and multipliers	32 galvanometers 16 shunts 0–1 A 16 shunts 0–5 A 16 multipliers 0–5 V 16 multipliers 0–15 V	Only 32 meters required, with choice of shunts for advanced work. However, conversion factors have to be applied to the scale reading, unless replaceable scales are available

Table 5.13   Provision of meters for classes up to age 16

Meters

The provision and maintenance of measuring instruments for current and voltage in a school physics laboratory can be very expensive and requires careful consideration.

Up to the age of sixteen, i.e. for pre-sixth-form work, there is usually no need for a.c. meters on a class basis. Any a.c. requirements at this stage can be met by demonstration meters, for which a large range of interchangeable scales are available. The d.c. ranges required on a class basis are 0–1 A, 0–5 A, 0–5 V and 0–15 V, as well as centre reading galvanometers. The three main methods of meeting this particular need are shown in Table 5.13, with a list of possible advantages and disadvantages. Combinations of the three methods are, of course, possible.

For advanced-level work, the $100 \mu A$ basic movement with a dual scale marking of 2–0–10 and 1–0–5 is most useful. With a coil resistance of 1000Ω, a voltage of $100 mV$ gives a full scale deflection, so for most purposes there is negligible voltage drop across the meter. The modern meter is diode protected against accidental overloads up to 12 V d.c., and this protection should be looked for.

The d.c. current ranges most commonly required at a more advanced level are 1 mA, 10 mA, 100 mA, 1 A and 10 A, and the d.c. voltage ranges 1 V, 10 V and 100 V.

The a.c. shunt ranges available for the 1000Ω moving-coil galvanometer generally include the following: 10 mA, 50 mA, 100 mA, 1 A, 5 A, 10 A; 5 V/10 V, 20 V/50 V and 100 V/500 V.

It should be noted that some moving-coil galvanometers, with shunts for conversion into ammeters and voltmeters, have been designed for use on the overhead projector.

For certain measurements, the *internal light beam galvanometer* is most suitable. Typically, such a galvanometer has a voltage sensitivity of $1.8 mm \mu V^{-1}$, a current sensitivity of $25 mm \mu A^{-1}$ and a charge sensitivity of $75 mm \mu C^{-1}$, with a rotary switch to reduce sensitivity in steps of $10 \times$. The scale of such galvanometers is from 150 mm to 180 mm long, and a bright spot is used as the 'pointer'. Protection against overload should be a feature of these sensitive galvanometers.

Multi-range meters

A versatile multi-range meter, having both direct and alternating current and voltage ranges, as well as a series of resistance ranges, is generally considered essential for 'fault finding' in electric circuits and for certain advanced and demonstration experiments.

The following ranges meet the likely needs of most school physics experiments:

i	*direct current*	0–50 A to 0–10 A	in either 6 or 7 ranges
ii	*direct voltage*	0–0.25 V to 0–5 kV	in 8 ranges
	or	0–0.1 V to 0–3 kV	in 9 ranges
iii	*alternating current*	0–10 mA to 0–10 A	in 4 ranges
iv	*alternating voltage*	0–2.5 V to 0–5 kV	in 6 ranges
	or	0–3 V to 0–3 kV	in 7 ranges
v	*resistance*	0–2 Ω, 0–200 kΩ 0–20 MΩ	

Modern meters are provided with overload protection circuits or an automatic cut-out as a safeguard against accidental overload.

A number of cheaper $100 \mu A$ movement 'Multi-minor' instruments, some of which do not measure alternating current and which generally have fewer ranges than the multi-range meter described above, are now on the market and provide an alternative to meter provision for sixth-form students.

Internal batteries are usually supplied with the meter. It is advisable to keep a record of the types of batteries required and to maintain a small stock of spares.

Microbiological apparatus

Autoclaves (see page 79)

Bottles—media

A number of designs and sizes of media bottle are available from suppliers. Narrow-mouthed Bijou bottles have a rubber-lined aluminium screw cap and are available with nominal capacities of $7\,cm^3$ and $15\,cm^3$. They are normally purchased in bulk packs, already sterilised. McCartney bottles are basically similar to Bijou bottles but have the larger nominal capacity of $27\,cm^3$. They are available both in the sterilised and unsterilised form. Universal containers are similar in size to McCartney bottles, but have a wider mouth. Spare caps, with rubber liners, are available for all three types.

Medical flats are used for sterilising culture media in bulk and are available with nominal capacities of from $50\,cm^3$ to $500\,cm^3$.

Colony counters

A colony counter normally consists of a circular translucent platform illuminated from beneath. A Petri dish is placed on the platform so that the transmitted light enables some colonies to be seen more clearly. Other species may be more conspicuous if the background is black and this should be borne in mind if a colony counter is being purchased. Some models are fitted with a magnifying lens and a squared guide plate which is placed beneath the Petri dish to facilitate counting. More expensive models have a digital counter which may be operated by a hand-held needle probe that completes a circuit when the culture medium is touched. Alternatively, the colonies may be counted through the base or lid of the dish using a fibre-tipped pen incorporating a sensitive microswitch. In the school situation it is unlikely that a sophisticated colony counter will be required unless a great deal of microbiological work is being undertaken.

Discs

A wide range of impregnated discs is available from commercial suppliers of microbiological materials. Some of these discs are impregnated with micro-organisms and, when incubated in a suitable medium, will develop into flourishing cultures. It is essential to maintain the discs under sterile conditions and to employ sterile technique in transferring and setting-up cultures. Failure to observe these precautions will result in contamination of the cultures.

Discs impregnated with penicillin and other antibiotics are also available. Some of these take the form of *Multodiscs* consisting of six or eight separate discs each impregnated with a particular antibacterial agent and joined to one another by narrow strips. Various combinations of antibacterial agents are available in multodisc form and others may be made up as special orders.

Disposable plastic ware (see page 95)

Incubators (see page 112)

Inoculating loops

These are used for inoculating and streaking cultures. The loop is made from 26 or 28 SWG Nichrome wire and should be 2–3 mm in diameter. The formation of the loop should be such that there are no sharp edges, which will tend to dig into the agar. The free end of the wire is attached to a wooden or metal handle. Inoculating loops may be purchased from commercial suppliers or made in the school (see page 160).

Petri dishes

Details of disposable plastic Petri dishes are given on page 95. Soda glass Petri dishes are available at approximately ten times the cost of plastic ones, but they have the distinct advantage that they can be sterilised and re-used many times. The most commonly used standard size is 9.5 cm in diameter and 1.5 cm deep, with a slightly larger, shallower lid. Borosilicate glass dishes are also made in a variety of diameters from 5.0 cm to 14 cm. The standard 9.5 cm size is between two and three times the price of the soda glass version. Stainless steel sterilising boxes are available which will take ten 9.5 cm diameter dishes.

Spreaders

A spreader is a length of glass rod bent in the manner shown in Figure 5.16. It is used to achieve an even spread (or 'lawn') of bacteria over the surface of an agar plate. Spreaders may be purchased in pre-sterilised packs or made in the laboratory. The rod should be bent in such a way as to ensure that no part will dig into the surface of the agar when the spreader is used.

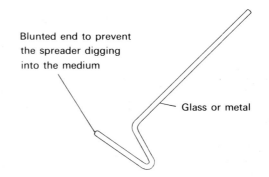

Blunted end to prevent the spreader digging into the medium

Glass or metal

Figure 5.16 A microbiological spreader

Transfer chambers

In schools specialising in microbiological work, teachers may favour the use of a transfer chamber for plating out, sub-culturing and other forms of transfer work. The basic requirements of an efficient transfer chamber are the following.

a The inner surfaces should be made of a smooth material which can easily be wiped over with a suitable antiseptic cleaner.

b The chamber should be fitted with a burner for the sterilisation of inoculating loops, etc., and to provide a flow of convected air in at the front of the chamber, over the flame where any spores are killed and out through the chimney. In some models a filter is included in the chimney for trapping the spores. (See Figure 5.17). The burners are available for Town or Natural gas and the appropriate type should be specified when ordering.

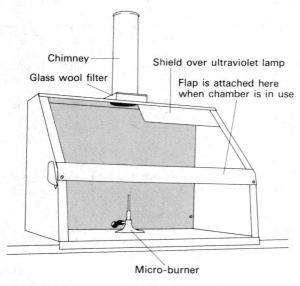

Figure 5.17   A microbiological transfer chamber

c The chamber should be fitted with a bacteriocidal ultraviolet lamp with an emission at about 260 nm. This should be adequately shielded so that ultraviolet radiation does not reach the eyes or the skin. The lamp should be positioned so that its radiation reaches all parts of the chamber and for effective sterilisation should be switched on some 20 minutes before the chamber is to be used. *The lamp must be switched off before the chamber is opened.* The ultraviolet source may also be used to irradiate cultures of bacteria or yeasts to induce mutation, the lids being removed from such cultures during the irradiation period.

d The chamber should permit the easy manipulation of apparatus through the front and clear visibility through a glass or perspex window. For details of improvising a transfer chamber see page 163.

Microscopes—monocular

Microscopes are among the most expensive items of equipment required in the biology laboratory. Considerable care should be taken therefore in selecting suitable instruments for purchase. Monocular microscopes have a single eyepiece (or ocular) through which the specimen is viewed with one eye. There are several basic designs and the relative advantages and disadvantages of each are partly a matter of personal preference.

TYPES OF MICROSCOPE

The standard microscope (see Figure 5.18)

The standard microscope is a well established design for which a wide range of objectives and eyepieces are available from the manufacturers. Appropriate combinations of lenses can give a range of linear magnifications from $\times 20$ to $\times 1500$.

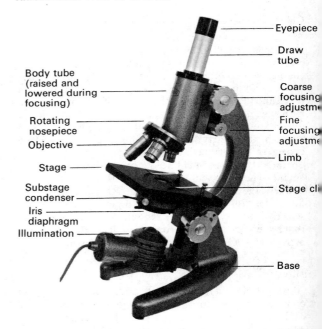

Figure 5.18   A standard microscope

The main body or *stand* of the microscope consists of a *base* to which the *limb* is attached; this may be inclined in relation to the base by rotation about a hinge joint. The *body tube*, which contains the *lens system*, is attached to the limb by means of an adjustable *focusing system* which normally comprises coarse and fine control knobs. The upper end of the body tube bears a *draw tube* housing the *eyepiece*, while the lower end supports one or more *objectives* frequently attached to a *rotating nosepiece* so that the lenses may be interchanged easily. The horizontal *stage* is also attached to the limb so that when the latter is inclined, the stage is tilted from the horizontal.

Beneath the stage there is frequently a *substage condenser* with its own focusing system, which serves to bring the incident light to a focus on the *specimen* which is mounted on a glass slide and held to the stage by *stage clips*. The diameter of the cone of light illuminating the specimen is controlled by an *iris diaphragm* situated beneath the condenser. Cheaper instruments may lack a condenser and instead utilise a concave mirror to focus light onto the specimen. They may also have—in lieu of an iris diaphragm—a circular plate with a series of holes of different sizes, which can be rotated to control the diameter of the cone of light reaching the specimen. Finally, beneath the condenser and iris diaphragm, is a means of illumination. This may consist of a built-in light source, powered from the mains directly or via a transformer, or from a low-voltage source, or alternatively a plano-concave *mirror* used to reflect rays from an external source onto the specimen. The plane side of the mirror should always be used in conjunction with a condenser and the concave side only as an alternative means of focusing light onto the specimen.

In spite of its long-standing use, the standard microscope is being replaced by the fixed inclined-limb instrument (see below). The principal disadvantage of the standard instrument would seem to be its relative instability—especially when in the inclined position—and the fact that inclination of the limb is not possible when examining freshly mounted specimens because the fluid mountant and the specimen itself gradually slide downwards under the force of gravity. With external sources of lighting, the mirror must be readjusted after inclination. Also conversion to a binocular instrument is impossible with the standard microscope.

There are a number of modifications of the standard instrument. In the most common of these the stage is maintained permanently in the horizontal plane so that the limb cannot be inclined at an angle to the base (see Figure 5.19). This type of microscope is quite commonly used for work with junior forms in secondary schools, but is not suitable for prolonged use when inclination of the limb is essential in the interests of comfort.

The fixed inclined-limb microscope (see Figure 5.20)

In recent years the standard microscope has to a large extent been replaced by the fixed inclined-limb microscope. In this type of microscope the stage is maintained permanently in the horizontal plane and the body tube is inclined towards the user. In some models rotation of the body tube in the horizontal plane is possible. This may be found useful in the teaching situation as the teacher can view the specimen without moving the instrument and without exchanging places with the pupil. With this type of microscope, focusing is achieved by raising and lowering the stage with its attached condenser and iris

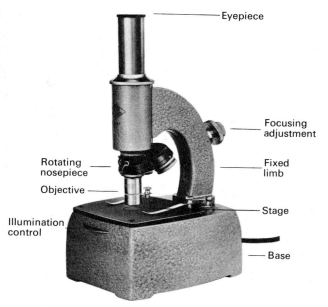

Figure 5.19   A standard microscope with a fixed limb

diaphragm. A wide range of eyepieces and objectives are obtainable to give a range of linear magnifications from × 20 to × 1500. In some cases it is possible to replace the monocular body tube with a binocular version.

In recent years, two further designs of microscope have become available.

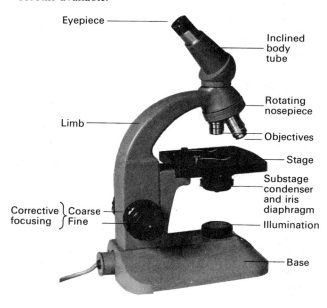

Figure 5.20   A modern microscope with a fixed inclined limb

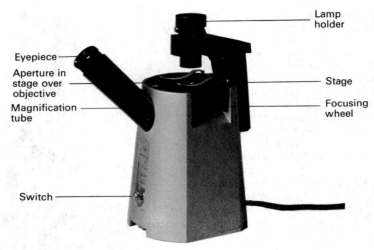

Figure 5.21   An inverted microscope

The inverted microscope (see Figure 5.21)

The inverted microscope has its light source uppermost. The specimen on its slide is placed on the stage and a serrated wheel is turned to move the single objective up and down to achieve a focus. Different degrees of magnification are obtained by inserting one, two or three magnification tubes between the eyepiece and the objective. In comparison with the two instruments already described, the inverted microscope operates at lower magnifications, from × 60 to × 200, and therefore may be suitable only for junior work. But it does have the added advantage for junior work in that it is impossible to bring the objective into contact with the specimen on its slide.

The McArthur microscope (see Figure 5.22)

This microscope was designed to meet the particular needs of the Open University, where equipment is sent through the post for students to use in their own homes.

The instrument is therefore compact (measuring only 13 cm × 7.5 cm × 2.5 cm) and of a lightweight plastic construction. A sliding stainless steel mirror permits the use of an external source of light, which may either be reflected onto the specimen by the mirror or allowed to strike the specimen directly. There is also provision for an internal light source powered by two small dry cells housed within the instrument. The objectives are supported on a small sliding bar which may also be moved up and down by turning a knurled wheel to bring the specimen into focus.

The principal advantage of the McArthur microscope is its compactness, which enables it to be taken on field work visits where other types of instrument would be too heavy and cumbersome. Set against this is the fact that considerable practice is required to manipulate the instrument, though this is made easier by mounting it on a tripod which is available from the suppliers. The quality of the lens system is also considerably inferior to those in other types of microscope.

Objective			Total linear magnification with various eyepieces					Approximate diameter of field of view with various eyepieces in mm				
Focal length		Primary magnification	Eyepiece magnification					Eyepiece magnification				
mm	in		× 6	× 8	× 10	× 12	× 15	× 6	× 8	× 10	× 12	× 15
40	$1\frac{2}{3}$	4	24	32	40	48	60	6.7	5.0	4.0	3.3	2.7
16	$\frac{2}{3}$ (Low power)	10	60	80	100	120	150	2.7	2.0	1.6	1.3	1.1
8	$\frac{1}{3}$	20	120	160	200	240	300	1.3	1.0	0.8	0.7	0.5
4	$\frac{1}{6}$ (High power)	40	240	320	400	480	600	0.7	0.5	0.4	0.3	0.3
2	$\frac{1}{12}$ (Oil immersion)	80	480	640	800	960	1200	0.3	0.3	0.2	0.2	0.1

Table 5.14   Details of the fields of view and linear magnification obtained with various combinations of eyepiece and objective. Assumed tube length 16 cm

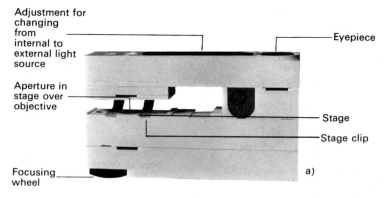

Adjustment for changing from internal to external light source

Aperture in stage over objective

Focusing wheel

Eyepiece

Stage

Stage clip

a)

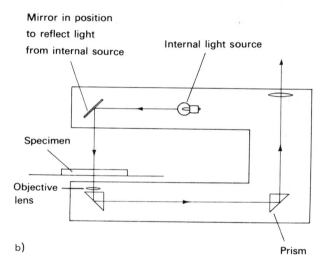

Mirror in position to reflect light from internal source

Internal light source

Specimen

Objective lens

b)

Prism

Figure 5.22   a) A McArthur microscope and b) the optical pathway through it

Microscope components

Lens system

The magnification of a microscope and the diameter of its field of view depend upon the precise combination of eyepiece and objective and also upon the tube length of the instrument. Many microscopes have a tube length of 16 cm. Table 5.14 shows the magnifications and the diameters of the field of view resulting from a variety of combinations of eyepiece and objective, calculated on the assumption that the tube length is 16 cm. Microscopes with tubes of greater length will have a correspondingly greater magnification and reduced field of view.

Objectives

The most commonly found objectives on instruments for school use are those with a focal length of 16 mm and 4 mm, though lenses with a focal length of about 40 mm will be found invaluable for overall views of large specimens and for junior work. For sixth-form work it is useful to have one or two very high power objectives with focal lengths of about 2 mm and designed to be used with a layer of cedar-wood oil (oil immersion) or water (water immersion) filling the space between the bottom of the objective and the top of the coverslip. This practice ensures that refraction is considerably reduced as the light passes from the coverslip to the objective thus minimising optical aberration and increasing brightness. Oil immersion objectives are normally marked 'OIL', though 'OEL' and 'HI' (homogeneous immersion) markings may also be encountered. Water immersion objectives are normally marked 'WI' or 'Was'. Immersion lenses should never be used without the appropriate immersion fluid.

Objectives are normally available with achromatic, apochromatic or fluorite (semi-apochromatic) lenses.

Achromatic lenses are corrected for chromatic aberration as far as red and green light are concerned, but are only partly corrected for spherical aberration. Apochromatic lenses (usually marked 'Apo') correct chromatic aberration for blue light as well as for red and green and they also have a wider range of correction for spherical aberration. However, they are considerably more expensive than achromatic objectives and will not normally be needed in the school situation. Part of the lens system of fluorite objectives is made from the mineral of that name and the effect is intermediate between achromatic and apochromatic lens systems. Most standard school microscopes are supplied with achromatic objectives.

Microscope objectives are normally marked with either the focal length or the primary magnification and both of these are shown in Table 5.14. Older objectives may have the focal length indicated in British units. The terms 'low power' and 'high power' are also frequently encountered, the former referring to objectives with focal lengths of 16 mm or more and the latter to those with focal lengths of 4 mm or less.

Some manufacturers produce high-power objectives in which the lenses are spring-loaded within the outer casing. Should the lens come into contact with the coverslip (a common fault with inexperienced microscopists) damage is reduced by the lens being pushed up against a spring located inside the lens housing. Spring-loaded objectives are recommended for junior use.

Objective lenses are usually mounted on a rotating nosepiece so that they may be conveniently interchanged. Nosepieces may be double, triple or quadruple according to whether they take two, three or four objectives. The objectives are said to be 'parfocal' if the nosepiece can be rotated from one lens to another while maintaining the specimen approximately in focus. Parfocal objectives are strongly recommended for school use where focusing may be a difficult exercise for those not experienced in microscopy.

Eyepiece

An eyepiece normally consists of two lenses, an eye-lens nearest to the eye and a field-lens nearer the objective. Between the two, at the focal plane of the eye-lens, is a metal annulus called the field diaphragm. Eyepieces are normally optically self-correcting in that the relative position of the two lenses tends to cancel out any chromatic aberration. It is not, therefore, normally necessary to have achromatic or apochromatic lenses. For advanced work in which apochromatic or fluorite objectives are used it is usual to have eyepieces which are optically corrected for use with such objectives. These are described as compensating eyepieces and are normally marked 'COMP'.

Eyepieces are available in a range of magnifying powers from $\times 4$ to $\times 15$ and this is usually stamped on the metal ring supporting the eye-lens. For most school purposes a $\times 10$ eyepiece is suitable, though a lower ($\times 5$ or $\times 6$) and a higher ($\times 15$) magnification may be useful on occasions. For teaching purposes a pointer eyepiece is invaluable. This contains a needle, the position of which may or may not be adjustable, so that it can be directed to a particular part of the specimen by the teacher or pupil. If all the members of a class use similar instruments, the teacher can carry around a single pointer eyepiece and insert it into each miscroscope in turn, as he assists individual pupils.

Focusing system

Most microscopes are provided with both a coarse and a fine focusing knob. In the case of standard instruments, the rack and pinion mechanism enables movement over some 5 cm when the coarse adjustment is used and over about 2 mm when the separately located fine adjustment is operated. Fixed inclined-limb models normally have a shorter travel distance and frequently have co-axially placed coarse and fine controls. For school work, especially at junior level, it is advisable to have a focus stop which prevents contact between the objective and the coverslip. This precaution is less necessary if spring-loaded high-power objectives are used (see above).

Illumination of the specimen

However good the optics of the instrument, maximum resolution and definition will not be obtained if the specimen is incorrectly illuminated. For most biological work with the monocular microscope at school level, specimens are viewed in transmitted light. The source of light may be either built into the instrument itself or provided from a bench or microscope lamp designed specially for the purpose (see page 131). In the latter case a 60-watt pearl bulb placed at approximately 20 cm from the mirror and at the same level as the mirror will provide suitable illumination.

If no substage condenser is fitted to the microscope the concave side of the mirror should be used to focus the light on the specimen. The plane side of the mirror should always be used in conjunction with a condenser. The function of the iris diaphragm is to control the diameter of the cone of light illuminating the specimen. This can also be matched to the numerical aperture of the objective being used. (The numerical aperture of the objective, NA, is a measure of its resolution and is commonly stamped on the casing of the objective. A 4 mm objective with NA 0.70 will give sufficient resolution for most purposes. Higher numerical apertures indicate greater resolution).

Focusing of the incident light onto the specimen is normally brought about by the condenser. This is provided with its own focusing system enabling it to be raised and lowered as desired. Many students and even some more experienced biologists frequently fail to realise the importance of the correct use of the condenser and diaphragm. The procedure outlined below should be followed at all times if the optimum use is to be made of the optical system of a microscope.

PROCEDURE FOR SETTING UP A MICROSCOPE

1 Place the microscope on the laboratory bench, well away from the edge of the bench itself and with the body tube—inclined or otherwise—towards the user.

2 If an in-built light source is being used, switch it on. If a bench light is being used, place this approximately 20 cm from the mirror and at the same level as the mirror. The use of a daylight blue filter in the filter ring, located beneath the iris diaphragm will improve the results.

3 Open the iris diaphragm to the maximum size of aperture.

4 Rotate the nosepiece to bring the 16 mm objective into operation.

5 Remove the eyepiece and look down the body tube from a distance of 10–20 cm. Adjust the plane side of the mirror until the back of the objective lens is fully and evenly illuminated. Replace the eyepiece.

6 Secure a microscope slide with a well-stained specimen to the stage by means of the stage clips and bring the substage condenser to the top of its range of travel.

7 Focus the 16 mm objective on the specimen.

8 To centre the condenser, close the iris diaphragm to its minimum size and move the condenser down until the image of the iris aperture appears. If this image is not in the centre of the field of view, the centring screws supporting the condenser should be adjusted until it is. Re-open the aperture of the iris diaphragm. (It is only necessary to check the centring of the condenser occasionally if the microscope is in continuous use).

9 Hold a pencil or other sharply pointed object against the surface of the lamp and move it until its image comes into view. Focus the condenser until this image is sharp. Check that both the specimen and the image of the pencil point are sharply focused. If the lettering or grain on the surface of the bulb is visible, *slightly* adjust the condenser to remove this.

10 Remove the eyepiece once again and look down the body tube from a distance of 10–20 cm. Close the diaphragm until about three-quarters of the diameter of the back lens of the objective is illuminated. Replace the eyepiece. The microscope is now set up for use with the 16 mm objective and care should be taken not to move the mirror or change the focus of the condenser.

To change to high power

1 Look at the slide and the objectives from the side and carefully rotate the nosepiece so that the high-power objective is brought into the optical axis of the instrument. Make sure that the objective does not touch the stage clips, slide or coverslip while it is being swung into position.

2 Bring the specimen into focus. If the objectives are parfocal (see p. 124) only a small adjustment of the fine focusing knob should be necessary. If the objectives are not parfocal, the high-power objective should be lowered until it almost touches the coverslip. *Watch the coverslip and objective from the side whilst doing this to ensure that the two do not come into contact.* Now look into the eyepiece and slowly rack upwards using the fine adjustment until the specimen comes into focus.

3 Remove the eyepiece and look down the body tube from a distance of 10–20 cm. Adjust the iris diaphragm until about three-quarters of the diameter of the back lens of the objective is evenly illuminated. Replace the eyepiece. The microscope is now set up for use with the high-power objective and care should be taken not to move the mirror or change the focus of the condenser.

To change to oil immersion

1 Carry out the above procedure and make sure that the specimen is central in the field of view when under high power.

2 Rotate the nosepiece to bring the oil immersion objective into the optical axis of the instrument, once again making sure that it does not make contact with the stage clips, slide or coverslip.

3 Carefully raise the objective a few millimetres and then place one drop of immersion oil either on the tip of the objective itself or on the coverslip exactly over the area to be examined. While viewing from the side, lower the objective till the oil fills the space between it and the coverslip. Continue to lower the immersion lens until it *almost* touches the coverslip. Look into the eyepiece and slowly rack upwards using the fine adjustment until the specimen comes into focus.

4 When examination under oil immersion is complete, the oil should be very carefully wiped from the surface of the lens using lens tissue barely moistened with xylene followed by clean, dry lens tissue. The surface of the coverslip should also be cleaned in a similar way.

Some workers prefer to use golden syrup diluted with an equal volume of water in place of immersion oil in cases where xylene is liable to damage the preparation during cleaning. The golden syrup is removed with a lens tissue dampened with water.

CARE OF MICROSCOPES

In the school situation it is often convenient to purchase a number or even all of the microscopes from the same

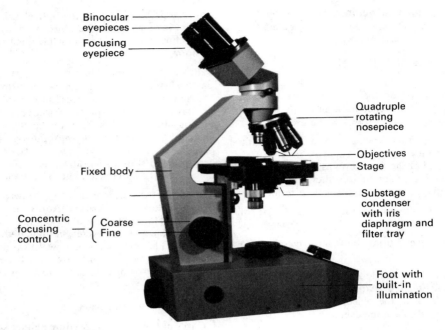

Binocular eyepieces

Focusing eyepiece

Quadruple rotating nosepiece

Objectives

Stage

Fixed body

Substage condenser with iris diaphragm and filter tray

Concentric focusing control — { Coarse / Fine

Foot with built-in illumination

Figure 5.23   A binocular microscope

manufacturer. Servicing can then be carried out on a regular basis by one of the maker's service engineers.

a Repairs should not be undertaken by anyone other than a skilled instrument mechanic, as this can often cause additional damage to the microscope.

b When not in use, all microscopes should be stored in the cases designed for them and in a dust- and fume-free atmosphere. Microscope cupboards should not be sited near chalkboards as the chalk dust can easily cause damage to the instruments. For short-term storage (e.g. for an hour or so) the microscope may be enclosed in a polythene cover.

c The eyepiece should always be left in position in the draw tube to prevent dust settling on the back lens of one of the objectives.

d All lenses and mirrors should be cleaned with lens tissue, though larger specks of dust may be removed with a clean camel-hair brush. *Handkerchiefs or dusters must never be used* to clean lenses as they will almost certainly scratch the surface of the lens. Once used, a piece of lens tissue should be discarded. (Lens tissue is available from scientific suppliers and also from photographic dealers). If an objective becomes greasy or dirty with Canada Balsam, it should be gently wiped with a clean lens tissue barely dampened with xylene and then immediately polished with clean, dry lens tissue. Excess xylene will dissolve the lens cement. Alcohol must never be used for cleaning lenses. The condenser lens should be cleaned in the same way as the objectives and eyepieces.

e Keep the stage clean by wiping with a piece of clean linen. Always ensure that slides are firmly held by the stage clips.

f Do not exchange eyepieces between microscopes of different design and/or make and never exchange objectives. Always ensure that objectives are fully screwed into the nosepiece and that eyepieces are properly seated in the draw tube.

g Always provide adequate support for the instrument if it is being carried from one place to another and never use it on a surface which is not stable and horizontal.

h After use, ensure that the low-power objective is left in the optical axis of the instrument and that there is no slide on the stage. Never carry the instrument with a slide in position on the stage.

Microscopes—binocular

Prolonged work with a monocular microscope can result in eye strain. This strain may be considerably reduced by the use of a binocular instrument (i.e. with two eyepieces). Binocular microscopes can have either a single or double objectives. In the former case an arrangement of prisms deflects light from the objective into the eyepieces, which normally lie parallel to one another. Such instruments usually have the two eyepieces inclined at an angle of approximately 45° to the vertical (see Figure 5.23). In

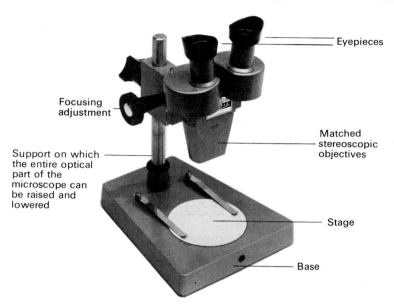

Figure 5.24   A binocular stereoscopic microscope

some cases it is possible to convert an instrument from a monocular to the binocular state and vice versa.

When selecting binocular microscopes for purchase it is wise to choose those with independent focusing of the eyepieces. Few people have eyes which are optically identical and it is therefore essential to be able to bring sharply focused images on to both retinas. The stage of many low-power binocular instruments takes the form of a ground-glass plate which is illuminated from below, either by a built-in electric lamp or by means of a mirror.

For some low-power work, such as dissection, it is necessary to view the specimen through two objectives and two eyepieces so that a stereoscopic effect is obtained. (See Figure 5.24). Such instruments have additional prisms so that erect images are formed. A range of objectives is normally available for these microscopes. For work of this kind it is often desirable to use reflected light, which may be provided by a separate bench light or be built into the instrument itself. Some stereoscopic microscopes have a movable light fitting so that they may be used with either reflected or transmitted light.

For details of microscope components and the care and maintenance of microscopes, see page 123.

Microscope accessories

Stage micrometer

This consists of a glass microscope slide with a scale engraved on it. The normal scale is 1 cm in length and has graduations every 0.1 mm (100 µm) with an extended line

every tenth graduation. The engraved part of the slide is protected by a coverslip. A stage micrometer is used to calibrate the graduations on an eyepiece graticule (see below).

Eyepiece graticules

These are available with either horizontal or vertical scales and also with simple cross lines at 90° to one another. Each consists of a glass disc, usually 21 mm in diameter, with the scale engraved on one surface. The scale normally covers 1 cm and is subdivided into 100

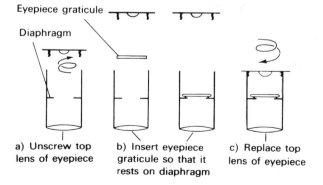

a) Unscrew top lens of eyepiece

b) Insert eyepiece graticule so that it rests on diaphragm

c) Replace top lens of eyepiece

Figure 5.25   Inserting an eyepiece graticule into the correct position within an eyepiece

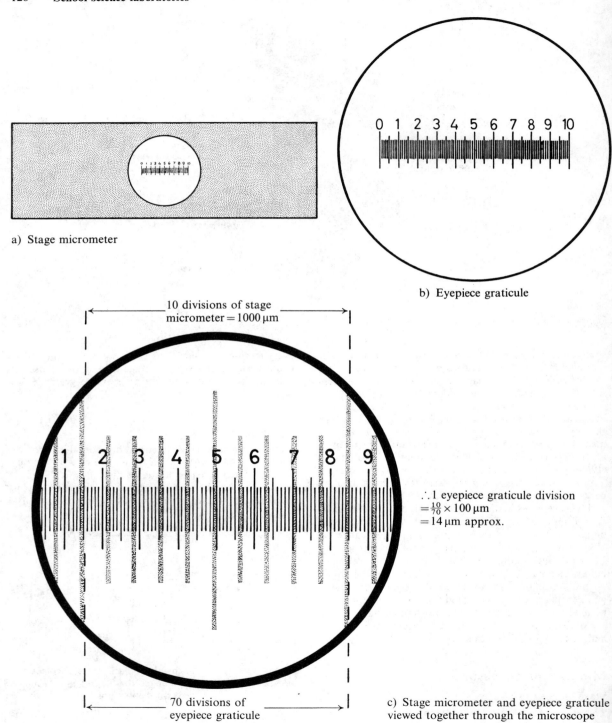

a) Stage micrometer

b) Eyepiece graticule

10 divisions of stage
micrometer = 1000 µm

$\therefore$ 1 eyepiece graticule division
$= \frac{10}{70} \times 100$ µm
$= 14$ µm approx.

70 divisions of
eyepiece graticule

c) Stage micrometer and eyepiece graticule
viewed together through the microscope

Figure 5.26   Calibrating an eyepiece graticule with a stage micrometer

graduations at 0.1 mm intervals with an extended line every tenth graduation (i.e. every millimetre). However, the exact dimensions of the divisions are not critical as the graticule must be calibrated for each microscope and for the particular combination of objective and eyepiece being used. The eyepiece graticule is placed inside the eyepiece, resting on the field diaphragm. Its insertion will necessitate removal of the field-lens of the eyepiece (see Figure 5.25).

To calibrate an eyepiece graticule:

i place a stage micrometer on the microscope stage and secure it with the stage clips;

ii insert the eyepiece with the eyepiece graticule in position and bring the stage micrometer into focus;

iii alter the position of the stage micrometer so that its initial graduation coincides with the initial graduation mark on the graticule;

iv count along both scales until a point is reached where there is again coincidence between a graduation mark on the graticule scale and one on the stage micrometer. (See Figure 5.26);

v as each division on the stage micrometer is 100 μm, it follows that:

One division on the graticule scale (in μm) =

$$\frac{\text{Number of divisions on stage micrometer scale}}{\text{Number of divisions on eyepiece graticule scale}} \times 100$$

Mechanical stage

This device enables a slide to be moved both transversely and backwards and forwards by mechanical means. This not only permits a convenient method of examining an entire slide and ensuring that no area is omitted, but it also enables exact measurements to be made as the transverse and forward movements are provided with vernier scales. Mechanical stages may be fitted to most standard or fixed inclined-limb microscopes, provided the microscope stage is square. The maximum distance of transverse movement is normally in the region of 50–60 mm, while the maximum forward movement is usually 25 mm.

A mechanical stage is also a convenient means of locating a particular part of a prepared slide. The slide is placed on the stage and the two knobs controlling transverse and forward movement adjusted until the 'ideal' portion of the slide is visible. The readings on the two vernier scales are then noted in pencil on the slide label together with an indication of the orientation of the slide on the stage. On a subsequent occasion the slide may be placed on the stage in a similar position and the knobs adjusted till the same readings are obtained on the two vernier scales. The 'ideal' portion of the slide should now be visible once again.

Phase contrast equipment

When a stained specimen is examined microscopically, different parts may be distinguished because they are of different colours or because they have different refractive indices. With an unstained specimen, the different parts can only be distinguished if they have different refractive indices. The light rays passing through areas with a high refractive index will be retarded relative to those passing through parts with a lower refractive index. Although the rays emerging from the two parts will have the same intensity, they will be out of phase and, when combined, will interfere with each other. It is this interference which enables the different parts to be distinguished. If a specimen under examination is very thin and transparent an ordinary microscope will not provide sufficient contrast and the different parts of the specimen will appear almost empty and structureless. In this case the use of phase contrast techniques may considerably help differentiation.

In phase contrast microscopy, the interference effects mentioned above are deliberately increased by passing different light rays through two different thicknesses of glass. Those passing through the thicker glass are retarded in relation to those passing through the thinner glass. The interference between the two sets of rays on recombination is therefore increased and results in a change in intensity.

The principle of phase contrast miscroscopy is illustrated in Figure 5.27. Light rays from an external source are directed into the microscope by a plane mirror, but before they reach the condenser they pass through an annular diaphragm, which replaces the normal iris diaphragm. A hollow cone of light is therefore focused on the specimen. Above the back lens of the objective is placed the *phase plate*. This consists of a glass plate with an annulus of thinner glass corresponding exactly in size with the image of the annulus in the diaphragm. Hence, rays passing through this thinner area will be out of phase with those passing through the thicker area of the phase plate. Direct rays will therefore pass through the annulus and on through the thinner portion of the phase plate. Those rays, however, which have been diffracted from the specimen and which have been consequently retarded by about one-quarter of a wavelength, pass through the thicker portion of the phase plate and hence are further retarded, emerging about half a wavelength out of phase with those passing through the system directly.

There are a number of variations in the techniques of phase contrast microscopy, but the basic principles remain the same. Kits are available from a number of microscope manufacturers, which enable an ordinary microscope to be converted to a phase contrast instrument. Essentially, such kits consist of a substage condenser assembly with annular diaphragms, phase contrast objectives and an auxiliary or focusing magnifier.

Phase contrast is particularly useful for the exam-

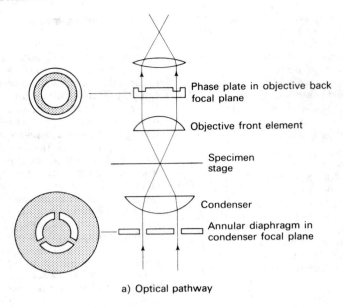

Phase plate in objective back focal plane

Objective front element

Specimen stage

Condenser

Annular diaphragm in condenser focal plane

a) Optical pathway

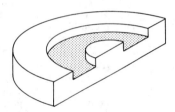

b) A cut-away diagram of the phase plate. The annular groove has a layer of vacuum deposited metal on its floor to reduce the intensity of the direct light.

Figure 5.27   The optical pathway through a phase contrast microscope

ination of protozoa and other transparent living cells where cytoplasmic structures are being observed. It is most unlikely that phase contrast microscopy will be used below sixth-form level.

Microscopy apparatus

Balsam bottles

These are made in clear glass and have a nominal capacity of 30 cm³. They are supplied with a glass rod for applying the balsam to the slide and with a domed loose cap which fits over the rod as well as the neck of the bottle.

Cavity blocks (Also known as staining wells, solid watch glasses and embryo cups)

The cavity block consists of a polished glass block 4.0 cm × 4.0 cm × 1.0 cm, hollowed out on one surface with a cavity 3.0 cm in diameter. They are normally supplied with a 4.0 cm × 4.0 cm cover and are used for staining procedures and various other techniques. They are also available in black glass which renders some specimens more easily visible.

Coverslips (Also known as coverglasses)

These are made in a number of shapes, sizes and thicknesses, the most commonly used being square or circular. Square coverslips are cheaper than circular ones, but the latter are perhaps easier to use when making permanent preparations. Larger rectangular coverslips may be used for specimens covering a large proportion of the slide (e.g. blood smears, mounted trematodes, etc.). Although there is now a British Standards Specification on coverslip thickness (Thickness No. $1\frac{1}{2}$ or 0.160–0.190 mm), a number of other thicknesses are available as outlined in Table 5.15.

Thickness number	Thickness in mm	Shapes available	Cover sizes in mm
0	0.085–0.130	Square	18×18 22×22 24×24
		Circular	16 mm diam. 19 mm diam.
1	0.130–0.160	Square	18×18 22×22 24×24
		Circular	16 mm diam. 19 mm diam. 22 mm diam.
		Rectangular	22×40 22×50
$1\frac{1}{2}$	0.160–0.190	Square	18×18 22×22 24×24
		Circular	16 mm diam. 19 mm diam. 22 mm diam.
		Rectangular	22×40 22×50
2	0.190–0.250	Square	18×18 22×22 24×24
		Circular	16 mm diam. 19 mm diam. 22 mm diam.
3	0.250–0.350	Square	18×18 22×22 24×24
		Circular	16 mm diam. 19 mm diam. 22 mm diam.

Table 5.15   Common sizes of coverslip

Labels

All slides of a permanent or semi-permanent nature should be clearly labelled. Self-adhesive labels in rolls or on a rectangular backing sheet are suitable. A maximum size of 24 mm × 24 mm should be used.

Lights

Many modern microscopes have built-in illumination, either as a standard fitting or as an optional extra which must be specified on purchase. Frequently there is a choice of mains or low voltage models. For most purposes, especially if the instruments are to be used by pupils below the sixth form, the low voltage models are preferable on safety grounds, but care should be taken to ensure that there are suitable low voltage outlets in all the laboratories in which the microscopes are to be used. The provision of such outlets must be borne in mind in laboratory design (see page 14).

Microscopes without built-in illumination are normally used in conjunction with a separate bench light. This has the added advantage that the light is not restricted in its use and may be used independently for other purposes, such as dissection. Bench lights are available in both mains and low voltage versions. Once again, for work below the sixth form the low voltage models are recommended on safety grounds, assuming suitable outlets are provided in the laboratories.

Opal lamps should always be used in microscope lights, 60 W being recommended for mains lights. Most low voltage lights are supplied with 24 W lamps. If a condenser is fitted to the microscope and correctly focused, the trade mark of the lamp may appear in the field of view (see page 125); this may be removed either by slightly refocusing the condenser or by carefully removing the mark from the lamp itself using metal polish or a very fine household abrasive.

Stability and ease of adjustment are two important features to be considered when purchasing microscope lights. Those with a metal base plate and a vertical steel rod to which the lamp is attached by means of a fitting similar to a bosshead are commended as they allow easy rotation of the lamp in both the vertical and the horizontal axes. Models with long folding supports of the 'anglepoise' type are not usually suitable for microscope work. Optional extras available with microscope lights are filter holders and iris diaphragms.

Pith

Some support is required when delicate tissues are being sectioned with a section-cutting razor. Generations of microscopists have used elder pith for this purpose, though carrots and other similar structures are preferred by some workers. In recent years styrofoam has been found particularly suitable as it is more uniform in quality than the natural elder pith.

Slides

The standard size of microscope slide is 75 mm × 25 mm and is normally 1.0–1.2 mm thick with ground edges. Such slides are available in boxes of 100 and there is often a considerable reduction in price for bulk purchase.

Larger slides (75 mm × 38 mm and 75 mm × 50 mm) are also available for mounting unusually large specimens. For some purposes, thinner slides are desirable (e.g. 0.8–1.0 mm) but these are usually almost twice the price of the standard slide.

For thick specimens, hanging drop cultures and various other purposes, cavity slides are required. These may take the form of a slide with a round polished cavity or with a fused-on glass ring or cell in the centre.

Slide drying plate

This is a useful piece of apparatus if permanent slides are being made on a regular basis in school. It takes the form of a steel plate which can be electrically heated to a thermostatically controlled temperature. Slides are then placed on the plate so that the heat dries out the mountant more rapidly than would be the case at room temperature.

Slide boxes

Prepared microscope slides should always be stored in suitable boxes. Two basic designs are available. In the first, the slides are stored horizontally in cardboard, wooden or plastic trays which in turn are placed, one above the other, in a stout cardboard or wooden container. Boxes holding about 100 slides or cabinets taking well over 1000 slides are available. In the second type, the slides are stored vertically in slotted boxes or trays. Boxes holding as few as 2–5 slides or cabinets taking several thousand are available.

Staining racks

A staining rack consists of a block of wood with a number of cylindrical depressions to hold dropping bottles. They may be purchased from suppliers, but can easily be made in the school workshop from varnished hardwood.

Staining troughs

These are glass containers used to hold slides during staining and associated stages of microtechnique. The slides are kept apart from one another by grooves in the sides of the trough. In one form (Coplin jars) the slides are held back to back with their long axes vertical, each jar holding ten slides. In an alternative design the same number of slides is held horizontally.

Mortars and pestles

Conventional laboratory mortars are made of agate, glass or porcelain. Agate is by far the most expensive and brittle of these materials and small agate mortars are used only in special circumstances. Agate is unsuitable for grinding alkaline materials because of its high silica content.

Most school mortars are made of porcelain rather than glass, which is more expensive. The porcelain may be entirely unglazed or glazed only on the non-grinding surface. Unfortunately, unglazed porcelain mortars often retain the colours of substances ground in them. They may be cleaned by grinding in them a mixture of sand and aqueous sodium hydroxide but thorough rinsing with water is essential. Mortars—of whatever material—must not be subjected to sudden changes in temperature and so care should be taken not to produce a highly exothermic reaction in a mortar in an attempt to clean it.

In general, one-piece pestles are preferred to those which have a wooden handle. Two-piece pestles are often more expensive and, with use, the wooden handles tend to shrink and loosen. Loose handles may be fixed by means of an appropriate 'impact' adhesive.

Nets, collecting

A small number of nets is essential for the collection of living material for use in biological work. Any school involved in ecological work will require a greater range and number of nets. When selecting nets for purchase, the following points should be given close consideration.

The anticipated frequency of use

Nets vary greatly in both quality and price. Those used for work in fresh water, especially collection of material in fast flowing streams or water with large growths of weeds, are subjected to considerable wear. Cheap nets, frequently used under these conditions, have a very short life.

The need for detachable net-bags

In some designs the fabric may be removed from the metal frame. This is useful, not only for replacement, but also for washing. The life of a freshwater net may be considerably extended if it is thoroughly washed after use. Detachable net-bags are normally secured to the frame by rustless fasteners which may be quickly unclipped. Some designs of frame are also built to accommodate a variety of net-bags of different mesh size or different shape.

The handle

Many nets are fitted with a standard length handle (e.g. 1 m) bearing a female thread to which extensions may be added. Nets to be used in aquatic environments should have thick hardwood handles which will withstand the considerable strain put upon them by the resistance of the water as the net is dragged along. This should especially be borne in mind if an extension handle is being used. Handles which are longer than 1 m may be difficult to transport.

The nature of the work for which the nets are to be used

There are four general purposes for which nets are likely to be used in the school situation. In most cases, there are several different types of net designed for each purpose. These are discussed below. The Freshwater Biological Association have been responsible for pioneering methods of collection and sampling in fresh water and if any quantitative sampling is to be carried out in this environment, FBA nets should be used.

a Air nets for the collection of terrestrial and airborne insects should be of lightweight construction (e.g. made with an aluminium frame). The diameter of the opening should be not less than 35 cm and the length of net-bag sufficient to enable it to be turned over to prevent the escape of any insects. (For example, a net of diameter 35 cm should be 70–80 cm deep). Air nets, sometimes called butterfly nets, may have a circular or a 'pear-shaped' opening. Some designs have a collapsible frame for easy carrying in the field. For most work a very short handle will be sufficient, but there are occasions when a longer handle (rarely as long as 2 m) may be an advantage. Long handles make rapid movement and manoeuvering difficult. The net-bag should be made of terylene or other similar material with sufficient mesh size to enable rapid movement through the air and good visibility of the insects inside when the net is held against the light.

b Freshwater nets for bottom collection have a square or D-shaped frame (see Figure 5.28a) so that a straight side of the net opening may be scraped along the bottom of a stream or pond. As the edge of the net-bag is constantly being rubbed against stones and bottom debris, it is subjected to a great deal of wear and should therefore be constructed of heavy gauge canvas or similar material. Net-bags are available with various mesh sizes, but for most bottom collecting a large size (e.g. 0.9 mm), trapping only the larger fauna, is desirable. Net-bags should be of the replaceable type. Bottom collection nets may also be used for general collecting, as an alternative to circular section nets, and thus serve a dual function. If only one type of freshwater net is being purchased, this is the best to get..

c Freshwater circular section nets for general collection should be used as dip nets and not for bottom collection. They are normally fitted with a conical dip net-bag (see Figure 5.28b) but a parallel-sided bag similar to that used with bottom collection bags may also be used. Conical dip net-bags are also available with an open tip to accept a rubber funnel or a sample tube for the collection of small animals and plants. Conical dip net-bags are available with a range of mesh sizes. Those with a mesh size of 0.1 mm are suitable for unicellular algae and protozoa while those with the larger mesh size of 0.3 mm

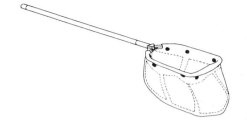

a) Square-framed bottom net

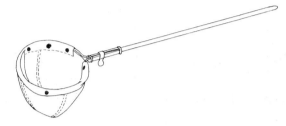

b) Circular-framed net for general collection

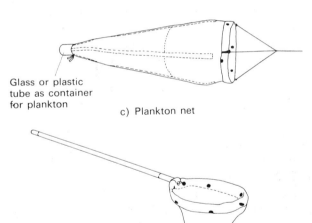

Glass or plastic tube as container for plankton

c) Plankton net

Glass or plastic tube as container for animals

d) Collecting net for small freshwater animals

Figure 5.28   Different designs of collecting nets

may be used for small crustacea and the smaller aquatic insects.

d Plankton nets consist of a circular frame some 30 cm in diameter with three points for the attachment of a tow line

	Class oscilloscope	Single beam demonstration oscilloscope	Double beam demonstration oscilloscope
Screen	Minimum diameter 7 cm, sometimes larger	10 cm × 8 cm	10 cm × 8 cm
Persistence	Medium	Medium	Medium
Y-amplifier	Direct connection to Y plates		
Sensitivity	$1\,V\,cm^{-1}$ to $5\,V\,cm^{-1}$	$100\,mV\,cm^{-1}$ to $50\,V\,cm^{-1}$	$100\,mV\,cm^{-1}$ to $50\,V\,cm^{-1}$ on each trace
Frequency response	d.c. to 10 kHz	d.c. to 1 MHz	d.c. to 5 MHz on each trace
X-amplifier	Direct connection to X plates	Continuously variable	Continuously variable
Sensitivity at maximum gain	Not applicable	$20\,mV\,cm^{-1}$	$100\,mV\,cm^{-1}$
Time base	1 sweep s^{-1} to 1000 sweeps s^{-1}	$100\,ms\,cm^{-1}$ to $1\,\mu s\,cm^{-1}$	$100\,ms\,cm^{-1}$ to $1\,\mu s\,cm^{-1}$ on each trace
Trigger	Synchronisation should be automatic	Auto or trigger level selection	Auto or trigger level selection

Table 5.16   Design characteristics of school oscilloscopes

(see Figure 5.28c). The net-bag itself is conical in shape with a mesh aperture of 0.3 mm for zooplankton and 0.1 mm for phytoplankton. The tip of the bag is designed to accept a plastic or rubber funnel or a sample bottle for the collection of the specimens. Plankton nets of this type are only suitable for towing behind a boat and may therefore only be used in larger freshwater lakes or, more usually, in the sea. Freshwater plankton samples may be collected using a circular net frame fitted to a hardwood handle and supporting a conical dip net-bag with an appropriate size of mesh and a sample bottle (see p. 134).

Oil baths

An oil bath may be fitted with a series of removable concentric metallic rings, allowing flasks of different sizes to be immersed up to about half-way in the oil. The remaining surrounding rings help to reduce fumes and evaporation of the oil. Alternatively, an oil bath may simply consist of a deep metallic pan fitted with handles at each side.

Oil baths are used in the range 100–180 °C and become unpleasant, and even dangerous, at higher temperatures. The 'oils' used are liquid paraffin, dibutyl esters (e.g. dibutyl benzene-1,2-dicarboxylate) or mixtures of concentrated sulphuric acid and one of its salts. Mulliken's mixtures can be used to produce temperatures in excess of 100 °C. Details are given in Vogel, A.I., *Qualitative Organic Analysis*, Longman, 1966, page 6.

For most purposes, oil baths have been superseded by electrothermal heating mantles. It is also possible to purchase electrothermal heating tapes. In general, electrically heated mantles allow higher temperatures to be reached and a finer degree of temperature control to be employed.

Oscilloscopes

Three types of oscilloscope are desirable in a school physics laboratory.

a Class oscilloscope. A simple and inexpensive oscilloscope—the maximum number of pupils should be four per oscilloscope.

b Single beam demonstration oscilloscope. A general purpose demonstration oscilloscope fulfilling general requirements up to age sixteen.

c Double beam demonstration oscilloscope. This is required for more advanced work, including phase comparisons, simultaneous current and voltage displays, and investigations of integrating and differentiating circuits.

The control panels of school oscilloscopes should be as uncluttered as possible to avoid confusion. The basic controls should be clearly marked and will include power on/off, brilliance, focus, sweep speed (coarse and fine), Y-shift and Y-gain (and X-gain where available). A transparent graticule, marked in 1 cm squares, is helpful when measuring the heights of waveforms.

A comparison of typical design characteristics for the three types of oscilloscope is made in Table 5.16.

Ovens and drying cabinets

An oven or drying cabinet is useful in a school science laboratory and may be used for drying glassware, for the slow evaporation of solutions, for the drying of chromatograms or as an incubator.

When selecting an oven for use in school science teaching, the following considerations are important:

i the temperature range over which the oven operates and the degree of temperature control required and available;

ii the incorporation of a circulating fan, allowing the oven to be used for rapid drying;

iii the size of the oven;

iv the number of oven shelves and the ease with which they can be adjusted and/or removed;

v the ease with which the oven thermostat can be reset or replaced;

vi the ease with which the heating element(s) may be replaced;

vii the possibility of using a drying oven as an incubator;

viii the incorporation of a 'cut-out' safety device;

ix the presence of an indicator light to show whether or not the oven is operating.

In general, ovens require little maintenance. They should be wiped clean after use. The interior may be made of stainless steel or aluminium so that corrosion should be minimal. If the motor of the circulating fan requires lubrication, this must be attended to from time to time. If the thermostat is actuated by the making and breaking of points, these should be cleaned and reset at appropriate intervals.

pH meters

Advances in solid state electronics have led to a revolution in the design and operation of pH meters. Meters accurate to ± 0.02 pH units can now be purchased at a reasonable price.

However, a pH meter is generally only as good as the electrodes used in conjunction with it. For school purposes, a combined glass and reference electrode is generally satisfactory. The reference electrode, which is often silver/silver chloride, is sealed inside an insulating glass stem which is fused to a glass membrane. Contact between the reference electrode and the glass membrane is made by means of a standard solution, e.g. 0.1 M hydrochloric acid. When the glass membrane is in a solution containing hydrogen ions, it develops an electric potential, the magnitude of which is dependent on the hydrogen ion concentration. The pH meter therefore measures the variation of the potential with hydrogen ion concentration using the reference electrode as a standard.

A large range of pH meters and electrodes is available and there is a corresponding variation in price. When considering the purchase of a pH meter for general school use, the following points are important:

i *portability*. It is helpful if the pH meter can be used outside the laboratory, e.g. in field work;

ii *power source*. pH meters may have a mains and/or a battery source of power. The latter is obviously essential if the instrument is to be of general use 'in the field';

iii *range*. Most pH meters have a range of 0–14 but this is not always the case;

iv *accuracy*. The measurement of pH to ± 0.02 units is generally adequate for school experiments. However, much more accurate (and expensive) instruments may be purchased;

v *temperature adjustment*. This may be manual and/or automatic;

vi *ease of standardisation*. Before a pH meter is used it has to be calibrated or standardised by immersing the glass (or other) electrode in a buffer solution of known pH. Such buffer solutions may be purchased ready for use or in an easily made-up form. After correcting for temperature as necessary, the pH meter is then adjusted to show the pH of the buffer solution;

vii *use as a millivoltmeter*. A pH meter measures electric potential, usually in mV. Thus some pH meters have a scale calibrated in mV as well as in pH units. In more expensive meters, the mV output may be fed directly to an automatic chart recorder, so that variations in pH may be continuously recorded;

viii *maintenance*. pH meters require little maintenance other than the replacement of batteries. Glass electrodes are also relatively robust and provided they are washed thoroughly with distilled water after use and kept in a clean atmosphere, will provide prolonged and satisfactory service. Detailed operating and maintenance instructions are supplied with pH meters and electrodes;

ix *interchangeability of electrodes*. It is obviously desirable to be able to use a given pH meter with a number of different electrodes if the need arises.

Photography (see Strobe photography)

Pipettes

Pipettes are available in two classes, A or B (BS 1583). For normal school use, class B pipettes are entirely satisfactory.

a Bulb pipettes. These have one graduation mark, are calibrated for delivery and are the most commonly used variety in schools. Bulb pipettes are available in capacities from $0.5 \, cm^3$ to $100 \, cm^3$. $25 \, cm^3$ pipettes are most commonly used in schools, but it is desirable to have a small number of other sizes, e.g. 5, 10 or $50 \, cm^3$.

b Capillary pipettes. These deliver a small quantity of liquid from a calibrated capillary tube. The commonly available sizes are 0.1 ($\times 0.005$), 0.2 ($\times 0.01$) and 0.5 ($\times 0.02$) cm^3.

c Dropping pipettes. These consist of simple glass or PVC tubes fitted with rubber or PVC teats. They are readily made in the school laboratory and usually have a capacity of approximately 2 cm³, although larger sizes are available.

d Graduated pipettes. These are commonly available in the following sizes.

Capacity cm³	Sub-divisions cm³
1	0.01
2	0.02
5	0.05
10	0.1
25	0.2

e Micropipettes. These are available with capacities from 10 mm³ to 250 mm³ and in a school are of specialist interest only.

f Teat pipettes. See Dropping pipettes.

g Safety pipettes. A 'safety' pipette is one in which the stem of the pipette is fitted with a sliding glass sleeve which acts as a suction pump.

In recent years, there has been a trend towards the use of disposable pipettes. These are generally slightly less accurate than conventional glass pipettes but they have are involved. An ordinary 25 cm³ bulb pipette can be filled by means of a filter pump and a length of rubber tubing or, more conveniently, by means of one of the commercially available pipette fillers. (Micropipette fillers are also manufactured). The use of rubber pipette fillers is safe and hygienic and should be encouraged in schools. In all cases, it is essential to ensure a sufficient depth of liquid in the vessel from which the substance is being taken.

In recent years, there has been a trend towards the use of disposable pipettes. These are generally slightly less accurate than conventional glass pipettes but they have the advantage of convenience and are sufficiently accurate for school use. How far this trend is encouraged in schools is likely to be a function of cost which is, of course, substantially lowered by bulk purchase. At the moment, disposable, graduated *capillary* pipettes are not manufactured on a large scale in the United Kingdom, although they can be imported.

It is also possible to purchase fast-flow pipettes and 'blowout' pipettes, in which the last drops of liquid must be expelled from the jet by blowing. At the present time, these useful types are not covered by a BS specification, but it is likely that appropriate changes will be made in the specification.

It is important that pipettes are thoroughly cleaned after use. A biodegradable detergent and cold water are normally sufficient although the pipette may be stood in a measuring cylinder full of detergent if it is particularly greasy. Whichever cleaning method is used, the pipette must be thoroughly rinsed afterwards. Pipettes may be stored in a rack or by means of the method suggested for burettes (page 85). It is important that a relatively dust-free atmosphere, away from laboratory fumes or vapours, is chosen. Particular care must be taken not to damage the delivery tip of a pipette.

Plant growth apparatus

Auxanometers

An auxanometer is designed to measure the growth rate of the apex of a plant. It consists of an adjustable, counterbalanced arm which is attached by thread to the growing point of the plant—a runner bean in a plant pot being a suitable specimen. Several forms of auxanometer are available from suppliers and they differ from each other mainly in their means of recording the growth. In the simplest form the arm travels on an arc-shaped scale and readings are made by observation at regular intervals. In some mechanical forms the arm bears a bristle which makes a trace on a smoke disc rotating at a regular speed (usually one revolution per hour). The distance between adjacent circular—or more correctly spiral—lines indicates the hourly growth magnified by a built-in magnification factor. In other models, the recording is made on a rotating drum to which a sheet of paper is attached. One auxanometer, for demonstration purposes, is usually quite sufficient for a school.

Clinostats

A clinostat consists of an electric or clockwork motor which turns a shaft at a constant speed (usually three or four revolutions per hour). A thick cork disc, enclosed by a perspex cover to retain moisture, can be attached to the shaft so that the growth of seedlings pinned to the disc can be studied in situations where gravity is no longer acting in one direction only. It is sometimes useful to be able to attach an entire plant pot to the shaft and it is therefore advisable to select a clinostat which has this facility. Electric motors are generally more convenient than clockwork for powering a clinostat; however the latter are quite satisfactory if mains electricity is not available.

Germination chambers

Several designs are available from suppliers, but a germination chamber is easily improvised in the laboratory.

Plugs (see Leads)

Pneumatic troughs

Pneumatic troughs are used much less frequently in schools nowadays than in the past. They are available in glass, polythene, or stoneware, although stoneware troughs are obsolescent. Clear glass troughs have obvious advantages for demonstration purposes. Plastic troughs are almost unbreakable but it is often cheaper to use a plastic washing-up bowl instead.

Polystyrene spheres

Polystyrene spheres are available in a variety of nominal sizes ranging from 12.5 mm to about 50 mm diameter. The density of each sphere depends upon the method of manufacture, but it is usually possible to compress a sphere to an intermediate size by rolling it between wooden boards. Spheres are usually purchased in packs of 100 and they are best kept clean by storing them in the original polythene containers.

Polystyrene spheres can be cut by a saw, a razor blade or a sharp knife. However, if large numbers of spheres are to be cut, it is sometimes more convenient to use a hot wire, but this wire must *not* be red hot (see page 39). A length of Nichrome wire (SWG 26)* carrying a current of about 2 A from a 'labpack' is adequate.

Spheres may be painted using conventional gloss paints and if a large number is to be painted, spraying is likely to be the easiest method. The spheres may be held on a board of thin nails. An appropriate colour code should be followed to represent atoms, ions or molecules, or different elements. Painting also protects the spheres from minor damage.

Polystyrene spheres can be used to make tangential-contact or space filling structural models. To do so, it is often necessary to hold the sphere firmly so that successive holes may be drilled at particular angles, e.g. the tetrahedral angle. This is best accomplished by means of a jig. Details of a jig for tetrahedral holes are given in the *School Science Review* 1967, 166, **48**, 810. For a more general review of the use of polystyrene spheres in making models, see Bassow, H. (1968) *Construction and Use of Atomic and Molecular Models*, Pergamon.

Potometers

Many different forms of potometer are available for school use. Some of these are illustrated in Figure 5.29. Others may be improvised from materials available in most school laboratories (see page 162). The most important feature of any potometer is the junction between the apparatus and the base of the plant shoot whose rate of water uptake is being measured. Any leak at this point will prevent satisfactory results being obtained. The junction should be made from flexible rubber tubing which fits tightly over the base of the cut shoot; the latter must therefore be of circular cross-section and preferably *slightly* larger than the internal diameter of the tubing. The use of such materials as Vaseline to reduce leakage is rarely satisfactory.

* For approximate metric equivalents to SWG, see page 149.

For class experiments, a piece of straight capillary tubing fitted with a rubber junction (see Figure 5.29a) makes a perfectly adequate and very cheap potometer. One or two more elaborate models may be required for demonstration purposes. A potometer fitted with a graduated syringe enables accurate measurements of water uptake to be made without knowing the dimensions of the capillary tubing.

Power supplies

Voltage supplies are required in three ranges for school physics experiments, generally specified as Low Tension (LT), High Tension (HT), and Extra High Tension (EHT).

Low tension, 0–25 V a.c. and d.c. up to a loading of 8 A

In older laboratories, the LT requirements were met by a centralised supply with distribution units—sometimes with individual fuse carriers—fixed to the benches. The development in the 1960s of portable LT supplies operated from the electrical mains supply removed the need for a centralised supply and so increased flexibility.

Low Tension variable voltage supply units are generally designed to give a continuously variable or stepped output, from 0 to above 24 V, a.c. or d.c. The output voltage drops as the current drawn increases—typically a 25 V unsmoothed d.c. output at no load reduces to a 22 V unsmoothed d.c. output at a load current of 8 A. It should be noted that the unsmoothed d.c. must be fed into a smoothing unit if ripple is to be reduced—an essential requirement in certain advanced experiments.

Some LT power-packs have separate 12 V a.c. output terminals, a useful addition for fixed voltage requirements, as, for example, in ray lamp experiments.

A cost-saving device on some LT supplies is a magnetic cut-out which trips at a particular load current, typically 9 A. This enables the unit to be reset quickly when the overload has been removed and avoids the cost of replacement fuses.

High Tension, 0–300 V d.c.

An HT power supply should give a continuously variable 0–300 V d.c. output with an additional supply continuously variable from 0 to 25 V d.c. (negative bias) at a maximum current of about 60 mA. A voltmeter with dual range should be provided to indicate output, with a toggle switch to alter the range on the voltmeter. Two sets of 6.3 V a.c. outputs for heating filaments are an extremely useful addition.

Extra High Tension, 0–5 kV d.c.

The maximum voltage allowed for the operation of discharge tubes and spark counters is 5 kV (DES 2/76).

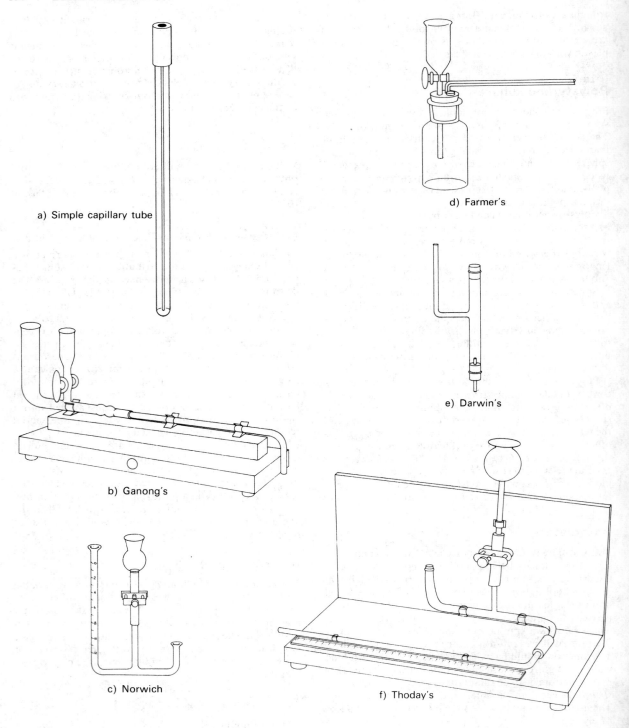

a) Simple capillary tube

b) Ganong's

c) Norwich

d) Farmer's

e) Darwin's

f) Thoday's

Figure 5.29   Various designs of potometer

It is preferable to purchase EHT supplies which are so designed that either the negative or positive side of the output can be connected to Earth or to the filament supply. A voltmeter should be incorporated in the unit, and a useful additional facility is an independent 6.3 V a.c. output for the heater filaments of vacuum tubes.

For safety reasons, a built-in resistor should limit the short-circuit current to 3 mA maximum, and the smoothed output current, generally limited to about 60 µA, should have a ripple content of less than one per cent.

Radioactive materials

Radioactive substances are now commonly found in schools and the purchase, storage and use of such materials are governed by the DES Memorandum, 2/76 *The Use of Ionising Radiations in Educational Establishments*. This Memorandum, together with its associated *Notes for Guidance*, is essential reading for any teacher who wishes to undertake work involving radioactive sources in an educational establishment in England or Wales. Teachers in Scotland are referred to SED Circular No. 689, *Ionising Radiations in Schools, Colleges of Education and Further Education Establishments* and to Circulars Nos 852 and 882.

The radioactive sources generally available from normal suppliers are shown in Table 5.17. Modern physics courses require pure α, β and γ sources as well as the usual radium source which provides all three types of radiation.

It should be noted that the SI unit for the activity of a radioactive source was sanctioned by the 15th General Conference of Weights and Measures (CGPM 1975) to be the becquerel (Bq) instead of the curie (Ci).

By definition:

$1\ Bq = s^{-1}$ (i.e. disintegrations per second)

and $1\ Ci = 3.7 \times 10^{10}\ Bq$

hence $1\ \mu Ci = 3.7 \times 10^{4}\ Bq$

or $1\ \mu Ci = 37\ kBq$

It follows that a source of 5 µCi should now be relabelled as having an activity of 185 kBq. The Ci and µCi will gradually become obsolete.

Source	Strength	Emission
Americium 241	185 kBq (5 µCi)	Mainly α Some low energy γ
Plutonium 239	185 kBq (5 µCi)	Mainly α Small amount of γ
Strontium 90	185 kBq (5 µCi)	β only
Cobalt 60	185 kBq (5 µCi)	γ only
Radium 226	185 kBq (5 µCi)	α, β and γ

Table 5.17 Radioactive 'closed' sources

Closed sources have the active material enclosed in a metal foil fixed in a cylindrical metal mount ending in a 4 mm diameter stem, with a wire gauze cap protecting the 'radiation window'.

Attention is drawn to the following advice about the storage of radioactive materials, given in DES Administrative Memorandum 2/76, paragraph 28.

'When not in use closed sources, open sources in tablet form and bottles containing open sources in solution should be kept in one or more metal containers, which should be retained either in a locked store or in a locked drawer or cupboard. The container and bottles should be clearly marked 'Radioactive substances'; the store, drawer or cupboard should also be clearly marked, preferably with the symbol for ionising radiations as in British Standard 3510:1968 *A basic symbol to denote the actual or potential presence of ionising radiation*. Sources should not be stored (i) in the same room as material which is highly flammable or (ii) near a position which is regularly occupied by the same person.'

Ratemeters

The inclusion of a built-in loudspeaker to give audible indication of the random nature of radiations is a particularly useful feature of a ratemeter, particularly for demonstration purposes. Generally two count-ranges are sufficient, e.g. 0 to 50 counts/second and 0 to 250 counts/second, each range having two time constants.

A ratemeter should have an internal EHT supply giving smoothed voltages in the range 300 V to 500 V for the Geiger-Müller tube, and it is preferable if the moving-coil meter is calibrated directly for the two count ranges to avoid scale multiplication factors.

Refrigerators

A refrigerator is now an important item of equipment in school science teaching and a variety of models is available from the usual laboratory suppliers. Refrigerators may be used for preserving foodstuffs and chemical and biological materials and for producing ice. If ice is likely to be consumed in substantial quantities, it may be more economic to purchase an ice-cube maker than to use the freezing compartment of a conventional refrigerator. Similarly, it is uneconomic to purchase a refrigerator which is so large that only a small part of its capacity is likely to be used regularly.

Refrigerators generally require no maintenance, although they should be located in as clean an atmosphere as possible. It is helpful if the refrigerator is fitted with a mains indicator light. Flammable solvents must not be placed in open containers inside refrigerators. Explosive concentrations of vapour/air mixture can form from very small amounts of solvent and these mixtures may be ignited by sparks from thermostats or micro-switches. New refrigerators for laboratory use should, therefore, have all spark sources placed outside the cold chamber.

If it is desired to deep freeze animal or plant tissue on a significant scale, a low temperature refrigerator ($-20\,°C$ to $-15\,°C$) should be purchased.

Resistance wire (see Wire)

Resistors

Non-inductively wound resistors, with an accuracy of the order of 0.1 per cent or 0.2 per cent, can be purchased in a Bakelite case or wound on a plastic bobbin in a transparent tube. The range of resistance of such single value resistors generally includes the following (in ohms) 0.1, 0.2, 0.5, 1, 2, 5, 10, 20, 50, 100, 200, 500 and 1000.

Resistors for school use are usually designed to work continuously at a power rating of 0.5 W, so that the maximum continuous current will depend on the value of the resistor. If resistors are to retain their stated resistance, it is essential that the maximum allowed current is not exceeded.

For certain more advanced experiments, it is convenient to have a large range of 16 or more resistors in a resistance substitution box, selection being made by means of a rotary selector switch. Again, the resistors are usually rated at 0.5 W continuous working, and a typical range of values starts at $100\,\Omega$ and increases by a constant ratio factor of about $2\times$ to $10\,M\Omega$.

Decade resistance boxes, in which four-in-line 0–9 decades give a range of, e.g. $1–9999\,\Omega$ ($\times 1\,\Omega$), are also manufactured. If 4 mm tapping points are provided between each dial, the box can be used as a potential divider or as ratio arms.

Respirometers

Several forms of respirometer are available from commercial suppliers. Most of these may be improvised from materials readily available in school laboratories (see page 162).

Retort stand bases and rods

Retort stand rods are made of cast iron or a strong, low density alloy of aluminium. Bases may be made of cast iron or an alloy of zinc. Most modern retort stands are covered with white enamel or a corrosion-resisting paint. However, retort stands are usually kept in the laboratory so that the protective coating is readily cracked or corroded. If cast iron stands show signs of rusting, they should be cleaned with emery cloth and given a coat of anti-rust paint followed by at least two coats of aluminium paint.

The storage of retort stands is often a problem in school laboratories. The problem can be eased by purchasing rods and bases as separate units. The rods can be stored horizontally (e.g. in plastic guttering fixed to a wall) and the bases kept in deep, stackable plastic trays. Similar trays can be used for the inevitable assortment of bosses, clamps, rings, burette holders, etc.

Rods and bases can be purchased in a variety of shapes and sizes. Bases may be slotted for holding boards or manometers or may be rectangular, tripod or 'A' shaped. The storage system outlined above increases the ease with which an appropriate rod and base can be selected from those available. It is essential to choose a sufficiently heavy base to minimise the risk of 'toppling' over.

For elaborate experimental apparatus, consideration should be given to the use of laboratory bench scaffolding, which may be free-standing or mounted on a bench or wall. This is relatively expensive to buy, but the range of scaffolding equipment and fittings can be increased as finance allows. The free-standing version allows a complete set of apparatus to be moved rapidly from one location to another. This may save the time of the teacher or the laboratory technician and it enables the apparatus in which a substance such as chlorine is being prepared to be moved quickly and completely to the fume cupboard. When purchasing and designing a scaffolding 'frame', the height of the fume cupboard and the width of the demonstration bench should be important considerations.

Wooden burette and funnel stands can be purchased and these are cheaper than cast iron retort stands. However, they are of more limited use as the angle at which the clamp may be held on the stand is fixed. Wooden stands are also more readily damaged by heat and water.

Rheostats

A large range of sliding contact resistances is advertised, but for most school physics experiments, it will suffice to have a class set (say one rheostat for four pupils) of the $11.6\,\Omega$, 5 A rated rheostat. The usual construction is an open 25 cm long $\times$ 3 cm diameter vitreous, enamelled tubular steel former wound with oxidised copper-nickel alloy wire, contact being made by means of a brass slider-rod with laminated metal contacts. Three 4 mm socket terminals should be fitted, so that the rheostat can be used as a series resistance or as a potentiometer.

A higher resistance rheostat, typically $330\,\Omega$, 1.2 A rating, is required to control higher voltages. It is preferable for the body of such a rheostat to be enclosed in a protective metal cover.

Safety spectacles and eye-shields

Safety spectacles and similar forms of eye-protection are becoming more widely used in school science laboratories.

Both spectacles/goggles and eye-shields are usually available in PVA, PVC or polycarbonate plastics. Some brands conform to BS 2092 *Industrial Eye-Protectors* and may be worn over normal prescription spectacles. It is helpful to be able to replace the shield in goggles as this inevitably becomes scratched with prolonged use. Adequate ventilation of goggles is essential.

If pupils are to be encouraged to wear some form of eye-protection, it is important that the teacher sets an example and that the design of spectacles/goggles be such that the pupils will willingly wear them. If safety spectacles are to be bought for general use in a laboratory, the following points are important:

  i the spectacles must offer adequate resistance to both chemical attack and impact. Spectacles usually carry the letter C and the number 1 if they offer resistance to 'Grade 1' impact and to chemical corrosion;
 ii the spectacles must be *easily* adjusted so that they will fit a wide range of head sizes;
iii at least some, if not all, of the class set of safety spectacles/goggles must be suitable to be worn over normal spectacles;
 iv a suitable storage system must be devised, e.g. a plywood box, divided into individual compartments and fitted with a perspex lid to exclude dust;
  v pupils need to be taught how to clean their safety spectacles and to care for them generally. If the storage system suggested in (iv) is adopted, each pupil may become responsible for a given pair. Pupils must be warned not to use their handkerchiefs or laboratory coats to clean them. Lens or paper handkerchief tissues should be provided for this purpose. Sterilising, cleaning cloths are also available;
 vi in the interests of hygiene, it is desirable to disinfect the spectacles from time to time. This can be done most easily by placing each pair of spectacles in turn in a polythene bag filled with gaseous sulphur dioxide and leaving it there for a few minutes.

In addition to safety spectacles/goggles for general class use, every school should have a small number of face protectors. These have clear plastic visors which may be raised, but which cover the entire face when in operation. Such protectors should always be worn where there is any danger of corrosive material splashing on to the face, e.g. when opening a new Winchester of 0.880 aqueous ammonia, and must offer adequate protection against impact. For the advice of the DES on eye protection see page 31.

Scalers

A scaler fulfils two functions.

a It is a counter of α, β and γ radiations in conjunction with a Geiger-Müller tube or solid state detector. It follows that a scaler should have an internal smoothed EHT supply of 300 V to 500 V for the G-M tube, and a PET connection for the G-M tube.

b The second important function of a scaler is as a timing device, generally to an accuracy of 1 ms, when used in conjunction with a built-in oscillator and start/stop gates operated by a photodiode assembly.

In more recent designs of scalers, the two dekatron tubes and mechanical register have been replaced by a digital display panel, generally providing a four-figure read-out with a three-position decimal point. Such instruments can also be used as frequency meters and as ratemeters, although an internal loudspeaker is not provided.

Signal generators (see Generators)

Silica ware

Silica ware is expensive to purchase and is used only occasionally in school science teaching. Silica tubing is sometimes used instead of resistance glass and is available in translucent and transparent varieties—the former being much cheaper. The principal advantages of silica are its very low coefficient of expansion and its resistance to most acids, even at high temperatures. However, it is attacked by caustic alkalis and it is more brittle than glass.

Where silica absorption cells are provided for use in spectrophotometers, the cleaning instructions issued by the manufacturer should be followed. For most other school purposes, silica ware may be adequately cleaned by immersing it either overnight in a laboratory cleaning mixture or for about 30 minutes in a mixture containing concentrated nitric(V) acid, hydrofluoric acid (40 per cent) and water in the ratio 7:2:7 by volume. The use of hydrofluoric acid requires that an eye-shield be worn and that the hands be protected by means of rubber gloves.

Sinks

Laboratory sinks are made of plastic or ceramic materials, fibreglass, or stainless steel. Stainless steel sinks are expensive but simple to instal and available in a variety of grades. Stainless steel shows excellent resistance to solvents but the resistance to attack by acids depends on the grade of steel employed in the manufacture.

Plastic sinks are made of polythene, polypropylene or, exceptionally, polyvinylchloride (PVC). They are inexpensive, easy to instal and have the advantage that glassware dropped accidentally into a plastic sink is more likely to survive the impact than with stainless steel or ceramic sinks. Plastic sinks show satisfactory resistance to acids but may be distorted by organic solvents. Such distortion may be particularly severe with sinks made of PVC. Flammability is a serious problem with some plastic sinks and the safety committee of the ASE advises that, in future, polythene or polypropylene sinks should not be fitted in laboratories and that 'PVC is unsuitable as an alternative'.

Ceramic sinks, made of earthenware or high-fired stoneware, are expensive to purchase and instal. Accidental chipping or cracking of the surface of an earthenware sink may expose the absorbent core. Fired stoneware has excellent resistance to chemical attack and, if the disadvantages of cost can be overcome, is perhaps the best material to choose.

Consideration should be given to the number, size and type of sinks needed in a laboratory complex. A large sink is needed for washing glassware and it is helpful if at least one sink is fitted with a conventional waste-disposal unit. For most routine teaching purposes, quite small sinks or 'run-aways' should be adequate.

Skeletal material

Natural skeletons are available from commercial suppliers in various forms.

a Articulated skeletons

Articulated skeletons are those in which the bones have been cleaned and bleached and then assembled and articulated on a baseboard in their natural positions relative to one another. Wire, glue, metal rods and perspex sheet may be incorporated in the assembly so as to display the bones in the best possible way. Such skeletons are normally provided with a dustproof cover made from perspex or other suitable transparent material. The larger suppliers maintain stocks covering a range of 20–30 different species representing all classes of vertebrate. Other species may also be available by making a special request to the supplier.

Human articulated skeletons are also available. These are not normally mounted in a rigid fashion on a board but are articulated in such a way as to allow free and natural movement of the joints. When ordering an articulated human skeleton it is advisable to purchase a gallows stand from which it can be suspended. This will reduce the damage caused to bones if they are placed in a box or cupboard, as well as providing an ideal means of display for the skeleton.

Articulated limb skeletons may also be obtained for a variety of animals including man.

b Disarticulated skeletons

These consist of loose bones which have been cleaned or bleached. They are normally supplied packed in cardboard boxes and are available for a similar range of species as articulated skeletons. Half human skeletons are also available in disarticulated form. These comprise the whole skull, vertebral column, sacrum and coccyx, together with half the sternum and attached ribs and the

limb bones for one side only. Separate limb, hand and foot bones are also available. Disarticulated skeletons are normally about half the price of fully articulated skeletons on a baseboard with a transparent cover.

c Display sets

Display sets of disarticulated bones of certain animals are arranged in grooves in plastic foam so that the relationship between the various bones may be seen while still permitting the removal and examination of individual bones. These sets combine some of the advantages of articulated and disarticulated skeletons.

d Skulls

Skulls are available separately for a wide range of animals (mostly mammals). These are normally supplied complete with teeth and lower jaw which is removable from the rest of the skull. Dog skulls are available with the sutures marked and the separate bones labelled. Human skulls may be obtained with the top section of the cranium removable, permitting examination of the interior.

Human skeletal material is also manufactured in plastic at about two-thirds of the cost of natural bones. Entire skeletal models are available in either life-size or miniature ($\frac{4}{10}$ life-size) versions. The life-size models are fully sculptured and are perfectly adequate for all but medical work. The miniature models are articulated only at the main joints and the hands and feet are moulded in one piece with the outlines of the bones visible. Disarticulated and separate limb, hand and foot skeletons and skulls are also available in plastic material.

In recent years techniques for clearing and embedding skeletal material in acrylic resin have been perfected. These permit a detailed examination of the material from all aspects while preventing damage due to handling. The technique is specially suitable for displaying the skeletons of embryological material (e.g. chicks and small mammals) but specimens displayed in this way are expensive. They may also be prepared in the laboratory (see page 286).

Sockets (see Leads)

Soil apparatus

Augers

Two types of auger are available for extracting soil samples from various depths. A cylindrical model withdraws unbroken samples and is thus preferable if quantitative sampling or population counts are to be made at different depths. A helical auger operates rather like a carpenter's brace and bit with the sample being removed from the soil in broken pieces. In both cases, the sample removed is 3.5–4.0 cm in diameter. Augers may be fitted with extension rods and a 'tommy bar' for rotation in the soil.

Funnels for extracting soil organisms

A number of designs of funnel have been perfected by Baermann, Berlese, Tullgren and others. These are available from suppliers, but can be improvised easily in the laboratory (see pages 156, 157, and 164). Each consists of a smooth-walled funnel into which is placed the soil sample enclosed in, or supported on, a fine mesh or sieve. An electric lamp is supported above the sample, thus providing heat and light which will drive the organisms downwards through the sample until they pass through the mesh and down the funnel to its narrow end, where they are retained in a collecting tube.

Berlese and Tullgren funnels are designed to be used with soil samples in their natural state (i.e. dry extraction). It is vitally important that the sample be dried from the upper surface downwards so that the arthropods and other organisms migrating downwards through the soil are always able to move to moister soil conditions until they finally leave the sample. If the drying out is too rapid, all the outer surfaces of the sample become dried, thus forcing the organisms towards the centre, where they eventually become incarcerated and die. In Baermann's design, the sample is enclosed in a mesh bag and suspended in water. Heat and light drive aquatic soil organisms (e.g. nematodes) down through the sample and into the funnel.

Sieves

Soil sieves are normally made in nesting sets of between three and six sieves. The largest mesh is placed at the top and the finest at the bottom fits into a metal receiver. A dried soil sample of known mass is placed in the topmost sieve and the whole nest shaken so that the finest particles pass through all the sieves and are collected in the receiver. When there is no further passage of particles, the mass of soil in each sieve can be determined. Thus the percentage of particles of various sizes in the original sample can be calculated.

Test kits

Soil testing kits are available which will indicate deficiencies in nitrogen, phosphorus and potassium as well as measure the soil pH. Soil pH may also be determined using indicator paper (see page 113) or a pH meter (see page 135).

Thermometers

A soil thermometer consists of a hollow tube made of copper or some other metal with a high conductivity, into which a mercury thermometer is placed. The temperature of the soil at a given depth may thus be measured without the risk of breaking the thermometer. A hole for the insertion of the thermometer may be prepared either with a special auger designed for the purpose or by driving a metal rod, of appropriate diameter, into the ground to the required depth.

Solid state detectors (see Geiger-Müller tubes)

Spirometers

A spirometer consists basically of a hollow chamber suspended over water. The chamber can be filled with oxygen and a subject connected to it so that the chamber rises and falls as the subject breathes in and out. The most commonly used type in physiological work is the wedge-shaped hinged-box type of which there are a number of alternative designs available from commercial suppliers. Some of these have transparent perspex float chambers and others are made from an opaque material. The former have the advantage that the relationship between the various hoses and the chamber can be easily seen.

The subject under investigation is connected to the spirometer via a mouthpiece and a flexible concertina hose fitted with valves so that inspired air is taken directly from the float chamber while expired air is passed back to it through a carbon dioxide absorber. The subject should also wear a nose clip so that there is no passage of air through the nasal aperture during the investigation. It is important that the valves operate smoothly and that the hose is of sufficient diameter to permit the subject to breathe freely.

The metal cylinders in which oxygen is supplied for industrial purposes can become contaminated and it is therefore important that only *Medical Grade oxygen cylinders* be used to fill a spirometer. The spirometer should be flushed out several times with oxygen before being filled. This removes any residue of nitrogen in the apparatus. A medical grade carbon dioxide absorber of a coarse granular form (e.g. Carbosorb or Durasorb) should be used to fill the absorbing cylinder. Soda lime or other caustic material must not be used, as dust from it may contaminate the apparatus and be taken into the subject's lungs. It is also important that the absorbing cylinder be mounted vertically. Horizontal mounting can lead to a free passage of gas through the cylinder without passing fully through the absorber.

A spirometer needs little maintenance. The chamber is normally pivoted on points or knife-edges and care should be taken not to damage these and to ensure that they are correctly located before and during use. The carbon dioxide absorber will require changing at appropriate intervals and this is made easier if a 'tell-tale' absorber is used. The apparatus should also be checked for leaks before use. This can be done by gently depressing the gas-filled chamber and ensuring that the pointed arm returns to the same level when the pressure is relaxed.

It is normal to use the spirometer in association with a kymograph or chart recorder (see pages 114 and 90). For this purpose a pointer arm to which a pen may be attached is fitted to the float chamber. In some models the pen may be a felt or nylon-tipped pen while in others the pen may use drawing ink. In either case it is important that a free flow of ink be maintained or an intermittent tracing will be obtained.

Between subjects, the mouthpiece should be soaked in a solution of an appropriate antiseptic prepared according to the manufacturer's instructions.

Stills

In the United Kingdom, the law requires that the local officer of the Board of Customs and Excise be notified when a still is to be installed in a school. Completion of the appropriate form (EX. 3) obtained from the local officer, will require details of the output and the full boiling chamber capacity of the still.

A still boils tap water and condenses the vapour as distilled water. The necessary energy may be supplied by burning gas or by an electrically heated element. The electrically heated stills are much more widely used and many schools have an automatic still of the 'Manesty' type. These are very heavy and must be bolted to a brick wall, preferably in the preparation room rather than in the laboratory itself. The still should be mounted about 2.5 metres above floor level to allow a suitable collecting vessel to stand on a bench or other support directly beneath it. A suitably located cold water tap and electrical socket are essential as is an adequate drainage system for the still overflow. It is helpful if the electric socket is set high on the wall near the level of the still itself. This ensures that the electrical leads to the still are kept as short as possible. The electric socket itself should be fitted with a red indicator lamp.

Most electrically heated stills have an automatic 'cut-out' which operates if the still overheats for any reason. Nonetheless it is sensible to turn on the water supply before switching on the heating element. Similarly, when shutting down a still, the electric power must be disconnected before the water supply is switched off.

In regions where the natural water is hard, the heating element and the boiling chamber of a still quickly acquire a coat of scale. If this scale is excessive, the automatic 'cut-out' may come into operation. The element and boiling chamber must be inspected periodically and the scale removed as necessary. In regions where the natural water is very hard, such inspections should be carried out once each term. It is possible to pre-treat the water supplied to a still by means of a suitable deionising cartridge. (See page 94).

Distillation removes not only the solids dissolved in tap water; it also expels the dissolved gases, mainly oxygen and carbon dioxide. However, distilled water is expensive

to produce and should not be used where either tap water or deionised water is adequate for the purpose in hand. It must also be remembered that distilled water which has been allowed to remain in prolonged contact with the atmosphere will redissolve some of the atmospheric gases. Such water may be unsuitable for some purposes, e.g. the preparation of the 'bicarbonate' (hydrogencarbonate) indicator (see page 174).

Stopclocks/stopwatches

The most suitable timing device for a particular experiment clearly depends on the degree of accuracy required. As a rough guide, a stopclock measures correct to 1 second and a stopwatch correct to 0.1 second, while for more specialised work, electrically operated stopclocks measuring to 0.01 second and digital timers measuring to 0.001 second are available.

A typical stopclock has a long second hand which sweeps across a 12–14 cm diameter dial marked from 0 to 60 s ($\times$ 1 s). A handwound mechanism provides a 25 h or 30 h movement on one winding. Some clocks have a large minute hand, while others have a small dial inset divided 0–60 min ($\times$ 1 min). The start/stop mechanism is activated by a lever or by buttons, and in some designs the second hand can be returned to zero by a zero reset lever. An indicator hand is a useful addition when readings have to be taken at particular times.

An average-priced stopwatch costs roughly the same as a stopclock. Typical stopwatches have a main dial marked 0–60 s ($\times$ 0.2 s) with a small subsidiary dial marked 0–30 min ($\times$ 1 min), or a main dial marked 0–30 s ($\times$ 0.1 s) with a small subsidiary dial marked 0–15 min ($\times$ 1 min). The start/stop/reset mechanism in a stopwatch is often activated by depressing the knurled button which also acts as the winder. Alternatively, separate buttons may be provided for the different functions. A stopwatch should be wound up before use and held in such a way that the start button is depressed by the part of the index finger underneath the knuckle. The thumb should not be used. The stopwatch should be left to run down in storage so that the spring is not left in tension.

Stoppers

Apart from cork, stoppers are available in rubber, polythene and neoprene. Neoprene will withstand temperatures up to about 120 °C and is resistant to most organic substances encountered in preparative chemistry at school. However, prolonged use of neoprene stoppers at temperatures above 100 °C causes the stoppers to harden.

Rubber stoppers are manufactured with or without holes. The smaller sizes are sold in packs; the larger stoppers (c. 75 mm) are available individually. As with corks, rubber stoppers may also be purchased in packs of assorted sizes. Many rubber stoppers have an identifying type number which indicates the diameter of the narrower (lower) end.

Rubber stoppers may be bored using the technique described for corks (page 93), but for routine laboratory operations, it is more convenient to purchase stoppers with one or two holes already bored. Glass tubing or thermometers may be inserted into the hole in a rubber bung using the technique described on page 109.

Strobe photography

Stroboscopic photography enables pupils to have permanent records for analysis of a ball falling under gravity or collisions between masses on an air-track or on a friction-free table.

Two methods of illumination are possible.

a By means of a Xenon stroboscope set to flash at a given rate, say 30 flashes per second.
b By means of chopped light, in which a disc with six slots rotates in front of a projector lamp at a steady rate of five revolutions per second to produce 30 illuminations per second.

The photographs can be taken with a very simple 35 mm camera set to brief exposure (B) using 20 exposure, 35 mm film—typically Kodak Plus-X Pan Film. The film is placed in a monobath developer and fixer in front of the class under normal lighting conditions. It is then cut up, and individual frames are projected onto daylight photographic paper. Wet prints can be obtained within about 10 minutes of starting the development of the exposed film. The advantage of the method is that the film can be developed without the use of a darkroom.

Details of the above method are given in a pamphlet specially provided for schools by Kodak, Ltd., entitled *Record Photography in the Classroom* (Kodak Ltd., London). A detailed description also appears in *Guide to Apparatus*, Nuffield O level Physics, (1968) Longman, pages 150–152.

Teltron tubes (see Vacuum tubes)

Test tubes

Most test tubes are manufactured in accordance with BS 3218 and are normally made either of soda glass or of borosilicate/resistance glass, the latter being more expensive. Test tubes made of more specialised materials such as silica are also available. Test tubes may be purchased with or without rims and with thin or thicker glass walls. Graduated and/or stoppered test tubes are also available.

In a school, it seems sensible to use a limited number of sizes and to maintain an adequate stock of these in resistance and/or soda glass. The following are among the sizes most commonly used in school laboratories:

125 mm × 16 mm	soda glass, for general work with aqueous solutions;
150 mm × 25 mm	soda or resistance glass; this size will serve as the 'boiling tube';
100 mm × 16 mm	resistance glass; this tube may be used as a centrifuge tube but care must be taken to choose a centrifuge tube to fit the particular demands of the given centrifuge;
50 mm × 6 mm	resistance glass; this size will serve as the 'ignition tube' and such tubes will normally be used only once.

Test tubes are still purchased in boxes of one gross, the larger sizes being sold in half or one-third gross quantities.

Students should be encouraged to clean test tubes immediately after use, using detergent and a test tube brush if necessary. However, soda glass test tubes are inexpensive items of glassware and, if they are particularly dirty, cleaning them may be an uneconomic exercise.

It is useful to be able to blow a small hole (1–1.5 mm diameter) near the lower end of a 150 × 25 mm (or 125 × 16 mm) test tube. This 'holed' test tube may be used to reduce metal oxides in a stream of hydrogen, ammonia or town gas. The hole may be blown by localised heating of the appropriate part of the test tube using a bunsen burner, followed by a steady blowing from the open end of the tube. The mouth may be placed directly over the tube or the latter may be fitted with a rubber stopper carrying a suitable long delivery tube down which air is blown. The edges of the hole may be smoothed in a bunsen flame.

'Holed' test tubes of this kind are available commercially and are generally catalogued as combustion tubes.

Test tube stands and racks

Many different varieties of test tube stands and racks are now available. Holed blocks and racks are manufactured in wood, plastic—usually poly(propene) or poly(ethene)—and rubber.

The storage of test tube racks sometimes causes difficulty. The Z-shape stackable racks, made of an aluminium alloy, overcome this problem, but are relatively expensive in comparison with the wooden or plastic varieties. Racks made of plastic-coated steel wire are useful because they can be immersed in a water bath or stood in an oven at temperatures usually up to 120 °C.

It is obviously important to ensure that the test tube racks and stands can accommodate all the sizes of test tubes that a school is likely to use.

Thermometers

A *very* large range of thermometers is available from the usual laboratory suppliers, including a number of specialised varieties, e.g. clinical, confectionery, maximum/minimum, pipe thermometers.

Most of the thermometers used in schools are general purpose liquid-in-glass instruments. These are manufactured in a wide variety of ranges and sub-divisions, e.g. $-120°$ to $+35°$ ($\times 1.0°C$) and $-10°$ to $+510°$ ($\times 5.0°C$) and may contain spirit or mercury. The most frequently used thermometers in schools are the following mercury-in-glass instruments:

$-5°$ to $+105°$ ($\times 1.0°C$)
$-5°$ to $+360°$ ($\times 1.0°C$)
$-5°$ to $+50°$ ($\times 0.1°C$)

The following modifications of general purpose thermometers are also useful in schools:

i 'stirring' thermometers. These are available in 'short' or 'long' forms and have a specially thickened bulb;
ii 'immersion' thermometers. Immersion thermometers are characterised by a scale which is confined to the upper half of the glass stem. They are therefore particularly useful for reading the temperatures of ovens, water baths, etc., where the lower part of the scale of a conventional thermometer would be obscured;
iii Beckmann thermometers. The Beckmann thermometer is probably the most expensive type of mercury-in-glass thermometer used in a school. A typical Beckmann thermometer would have a range of $6°C \times 0.01°C$ within an overall range of $-20°C$ to $+110°C$. The adjustment of a Beckmann thermometer requires considerable patience and practice and most are supplied with instructions on how to set the instrument for use within the desired temperature range.

Liquid (usually mercury)-in-glass thermometers are relatively expensive items of equipment in a school and they are only too easily broken by dropping or overheating. Thermometers should not be shaken violently or subjected to severe jolting. It sometimes happens that the mercury column in a thermometer breaks into two or more portions. The column may be reformed by cooling the thermometer in ice (or solid carbon dioxide, if necessary). If this procedure fails to cause the parts of the mercury thread to re-unite, the thermometer may be warmed in a water bath, but the cooling procedure is, in general, to be preferred.

More rarely, small amounts of water vapour, present before the thermometer was sealed, collect as a bubble and cause the mercury thread to divide. Cooling, as described above, with gentle tapping or tilting as necessary, is usually sufficient to cause the bubble to escape above the mercury meniscus and to allow the continuous mercury thread to reform.

Thermometers should be stored in the cases in which they are usually supplied. The clearly labelled cases should be stored horizontally in a drawer or in a stackable tray, with appropriate partitions to allow thermometers of a given type to be kept together.

More specialised thermometers used in school science include the following.

a Clinical thermometer

The clinical thermometer is designed to record temperatures in the range 35 °C to 42 °C, and has a special mark at 'normal body temperature', which is usually taken as 37 °C (98.6 °F) or 36.9 °C (98.4 °F). The engraved stem is short, generally no more than 10 cm long, and the walls of the bulb containing the mercury are thin so that the thermometer is quick acting. Because the volume of mercury is small, the capillary tube has to be very fine, and the front of the stem is shaped as a convex lens to magnify the thread. A constriction in the tube just beyond the bulb ensures that the mercury thread does not retreat back into the bulb until it is shaken back.

A clinical thermometer must never be washed in hot water, but should be rinsed in a suitable antiseptic solution (e.g. cetrimide solution BNF or chloroxylenol solution BPC) at a temperature below 40 °C.

b Maximum and minimum thermometer

Six's dual scale thermometer is manufactured to specification BS 2840. The scale is anodised and calibrated in the range $-20°C$ to $+55°C$ ($-10°F$ to $+130°F$), and located in a weatherproof plastic case with a sloping top. A magnet is generally provided for resetting the two indices, which move in spirit above the mercury meniscus inside the two arms of the tube.

c Resistance thermometer

The resistance thermometer consists of a length of fine platinum wire of about 2 ohms resistance, wound non-inductively on a mica frame and soldered to stout copper leads. A pair of compensating leads are provided and the whole is enclosed in a transparent glass test tube, with four terminals fitted to a panel at the top of the tube.

d Thermistor probe linked to interscale thermometer dial

The Interscale demonstration meter consists of a moving-coil galvanometer having a 16.5 cm-long pointer giving a full scale deflection with a current of 5 mA or an applied voltage of 100 mV d.c. The pointer can be centred by means of a finger-operated screw on the front of the case. A guide rail is provided across the inside of the glass fronted instrument to ensure the correct alignment of a variety of scales, incorporating shunts and multipliers for d.c. current and voltage, a.c. current and voltage, temperature and other quantities.

To measure temperature, a thermistor encapsulated in a glass probe with a protective cover is plugged into the dial unit, which is scaled $-10°C$ to $+110°C$. The accuracy of measurement is $\pm 2\,°C$. For more accurate measurements to $\pm 0.25\,°C$ in a shorter range, i.e. $15\,°C$ to $25\,°C$, a special short range thermometer dial is available.

Manufacturers' catalogues sometimes quote BSI specifications for laboratory thermometers. The relevant BSI publications are as follows:

BS 593   Laboratory thermometers
BS 791   Bomb calorimeter thermometers
BS 1365 Short-range short-term thermometers
BS 1704 General purpose thermometers
BS 2840 General purpose maximum and minimum thermometers (Six's pattern)

Three centimetre wave apparatus

Apparatus using electromagnetic waves of three centimetre (nominal) wavelength can be used to demonstrate by analogy many of the experiments for which monochromatic light (or X-rays) had to be used in the past.

The basic apparatus consists of a transmitter with resonant cavity and incorporating its own power supply, and a means of modulating the radio frequency waves with an audio signal at 100 Hz or 1000 Hz. In a more recent continental development, the three centimetre waves are modulated with 'barrel organ music'—an effective and interesting variation on the usual monotone. The receiver incorporates a silicon diode and the signal can be fed to a meter or amplifier/loudspeaker, the latter being more effective for demonstration purposes. A diode probe without resonant cavity is also available—its reduced sensitivity in a particular direction being compensated for by its ability to detect a signal from any direction in a horizontal plane.

The following are the more common accessories used with three centimetre wave apparatus.

a Turntable. A motor-driven turntable which rotates a crystal model at 30–50 revolutions per minute in a parallel beam of 3 cm waves.

b Crystal model. A crystal model made from expanded polystyrene spheres for Bragg diffraction. Polystyrene tiles may also be made up into a model of a crystal.

c Wax lenses. A plano-convex lens of wax (or a hollow plastic lens which can be filled with liquid paraffin) is used to produce a parallel beam of 3 cm waves from the transmitter, and a second lens converges the scattered waves, after Bragg diffraction, on to the receiver. 30 cm diameter lenses with a focal length of 60 cm are suitable.

d Hollow prism. Hollow perspex prisms, either 45°, 45°, 90° or 60°, 60°, 60°, to be filled with liquid paraffin.

e Set of plates. A Young's double slit may be formed using an anodised aluminium plate 21 cm × 6 cm wide placed between two 21 cm × 21 cm anodised aluminium plates to form two vertical slits each of the order of a few centimetres wide. The plates should be independently mounted on wooden supporting feet.

f Half-silvered plate. A perspex or hardboard plate, again about 21 cm × 21 cm, is analogous to a half-silvered mirror in optics.

g Polarisation grille. This consists of a mounted set of thick parallel wires which are rotated in a vertical plane between the transmitter and receiver to demonstrate polarisation effects.

Ticker tape vibrators

Ticker tape vibrators (timers) may be purchased to operate from accumulators, or more conveniently from a 6 V to 12 V a.c. supply at 50 Hz (or 60 Hz). Paper tape, gummed or plain, is drawn through guides on a small platform where a carbon paper disc is struck by a small head attached to the armature of an electromagnet, so making 50 (or 60) sharp dots on the tape every second.

More recent versions of the vibrator have accessory attachments for 4.5 mm external diameter rubber tubing for the 'pearls-in-air' experiment, and a special adjuster allows the centre of the carbon disc to be moved for multiple use of each carbon disc.

Spare packs of carbon paper discs and spare rolls of ticker tape should be kept in store. Ticker tape holders to dispense the tape from the usual 300 m length rolls are commercially available. They avoid the collapse of the roll, and can be conveniently mounted in a laboratory stand.

Vacuum pumps

If these are used properly little or no maintenance is required. Lubrication is generally necessary from time to time and should be done in accordance with the manufacturer's instructions. Vacuum pumps should be kept clean and free from dust and should not be used to extract corrosive or toxic gases.

Major difficulties in the operation of vacuum pumps should be solved in conjunction with, or preferably *by*, the manufacturer.

Vacuum tubes

It is a legal requirement (see DES 2/76) for discharge tubes to be operated at an accelerating voltage of 5 kV or less, and this is achieved in the Teltron range of vacuum tubes by having a heated cathode, so aiding thermionic emission. Cold cathode tubes, connected to a van de Graaff generator, produce X-rays when the fast moving electrons hit a target and are therefore *dangerous* and *must not be used.*

Teltron tubes are of a standard size—they are about 25 cm long with a 13 cm diameter bulb at the end of a cylindrical glass section containing the electron gun. A stand has been designed to hold all the tubes safely, and the base of the stand has a pair of sockets to hold the pair of Helmholtz coils which are required for accurate magnetic deflection of the electron beam.

The following tubes are available in the Teltron range:

i *demonstration diode.* For determination of anode characteristics and demonstration of rectifying action;

ii *demonstration triode.* For determination of anode and mutual characteristics and action of amplifier and oscillator;

iii *gas triode.* A demonstration triode filled with low pressure helium gas for investigations of ionisation potentials and production of positive ions;

iv *maltese cross tube.* For demonstration of straight line propagation. A direct light shadow from the filament is undeflected, whereas the electron beam is deflected by a magnetic field;

v *Perrin tube.* For comparison of the charge carried by electrons with the frictional charge on a polythene strip rubbed with wool;

vi *deflection tube.* Demonstration of Thomson's method for e/m, the determination of the charge/mass ratio for electrons. The narrow beam is made visible on an inclined phosphorescent screen with a grid marked on it;

vii *luminescent tube.* For illustrating the action of an electron beam on different phosphors;

viii *critical potentials tube.* To illustrate excitation potentials in helium;

ix *electron diffraction tube.* For demonstration of diffraction rings formed by an electron beam passing through a very thin layer of mounted graphite;

x *fine beam tube.* This is a tube containing two electron guns and helium gas at low pressure to make the electron beam visible. The beam from the vertical gun can be deflected into a circle by means of the Helmholtz coils for the determination of e/m, while the beam from the horizontal gun produces a spot on a luminescent screen and so the effects of electric and magnetic fields on an electron beam can be demonstrated.

Van de Graaff generators (see Generators)

Vibration generators (see Generators)

Voltmeters (see Meters)

Wash bottles

Wash bottles are available in polythene (poly(ethene)) or glass. Polythene wash bottles are operated by compress-

ing the container and are manufactured with nominal capacities of 125 cm^3, 250 cm^3 and 500 cm^3. The most obvious advantages of polythene wash bottles are their resistance to breakage and their ease of operation, which requires only one hand.

Glass wash bottles are more expensive but they are available in sizes up to 1 litre. Water is delivered from a jet by blowing down a bent glass mouthpiece. The addition of a 'pinch' valve to this mouthpiece enables a continuous flow of water to be obtained from the jet. It is, of course, a simple matter to construct a glass wash bottle using the items of glassware and other equipment normally found in a school laboratory.

Wash bottles must *always* be clearly labelled.

Waste disposal

The routine operations of school science teaching produce considerable quantities of waste material. This material is of several different types and each of the following should be kept in a separate and clearly labelled container prior to disposal.

i broken glass;

ii wet, solid waste, e.g. filter papers;

iii biological material, e.g. the remains of a dissected animal or plant, microbiological material or refuse from animal cages;

iv waste solvents;

v recoverable residues, e.g. silver, iodine;

vi radioactive material.

The disposal of waste is governed by the Control of Pollution Act, 1974, and special arrangements may need to be made with the appropriate Disposal Authority (see page 48), for radioactive, carcinogenic or highly toxic materials. Local advice on waste disposal should be sought from the Disposal Authority, from a Public Analyst or from a Public Health Inspector. Advice on the disposal of radioactive waste is given in Administrative Memorandum, 2/76, (DES, 1976) *The Use of Ionising Radiations in Educational Establishments.*

Water baths

Water baths may be purchased with or without a built-in electrical heating element. The common circular bath, fitted with a set of detachable metal rings to accommodate flasks of various sizes, is designed to be heated on a tripod or hotplate. Such baths are of considerable use in preparative chemistry but it is not difficult to improvise an adequate substitute.

Thermostatically controlled water baths are considerably more expensive but they are essential items of equipment likely to be used by more than one science department. When considering the purchase of a thermostatically controlled water bath, the following points should be considered:

  i the range over which the thermostat and heating element operate;

  ii the degree of accuracy of temperature control required;

  iii the ease with which the temperature can be set to the desired point;

  iv portability;

  v the incorporation of a 'fail-safe' device and an indicator light to show when the heater is on;

  vi the dimensions of the bath. It is important to remember that several pupils may wish to use the bath simultaneously and that the bath should be sufficiently deep to accommodate racks of test tubes, conical flasks, etc.;

  vii the availability of accessories such as adjustable trays, bath covers or a constant level device;

  viii the ease with which simple repairs may be carried out, e.g. the replacement of the heating element or the resetting of the thermostat.

It should be noted that temperature control may be enhanced by floating a layer of poly(propene) spheres on the surface of a water bath and that an 'immersion' thermometer may be of help in reading the temperature.

Weights (see Masses)

Wire

The Standard Wire Gauge (SWG) numbers are gradually being discontinued, and will be replaced by metric sizes. A range of SWG numbers with the nearest preferred metric size is shown in Table 5.18.

The wire requirements of a school physics laboratory can be various, especially if work of a project nature is being done. However, certain size wires of a particular material are used more frequently than others, and the list below shows the results of an analysis of the wire requirements specified by the Nuffield Physics Projects for work with pupils aged from eleven to eighteen years. Such a list may be taken to represent the basis of wire stock in a physics laboratory.

  i *copper, bare* (SWG) 14, 20, 22, 26, 32;

  ii *copper, tinned* (SWG) 26, 36;

  iii *copper, plastic covered* (SWG) 26. This wire is used extensively in the electromagnetic kit;

  iv *Constantan (Eureka), bare* (SWG) 24, 32, 34;

  v *Constantan (Eureka), insulated* (SWG) 28, 32;

  vi *Nichrome, bare* (SWG) 26;

  vii *iron, bare* (SWG) 16;

  viii *steel, bare* (SWG) 44 (0.08 mm diameter for the oil film experiment).

Bibliography

AMA, ASE, and AAM (1970) *The Teaching of Science in Secondary Schools*, John Murray.

Bradbury, S. (1973) *Peacocke's Elementary Microtechnique*, Arnold.

CLEAPSE, *Apparatus Notes*.

Department of Education and Science (1970) *Administrative Memorandum 7/70, The use of lasers in schools and other educational establishments*, HMSO.

—(1976) *Administrative Memorandum 2/76, The use of ionising radiations in educational establishments*, HMSO.

—(1976) *Administrative Memorandum 7/76 The use of asbestos in educational establishments*, HMSO.

Everett, K., Hughes, D. (1975) *A guide to laboratory design*, Butterworths.

Lewis, J. L. (ed) (1972) *Teaching school physics*, UNESCO source book, Longman/Penguin Books.

Nuffield O level Biology (1966–67) *Teachers' Guides I–V*, Longman/Penguin Books.

Nuffield Junior Science (1967) *Animals and Plants*, Collins.

Nuffield O level Physics (1968) *Guide to Apparatus*, Longman/Penguin Books.

Nuffield A level Physics (1971) *Teachers' Handbook*, Penguin.

Nuffield A level Biology (1971) *Laboratory Book*, Penguin.

Nuffield A level Physics (1973) *Apparatus Construction Drawings*, Penguin.

Nuffield O level Biology (Revised) (1974–75) *Teachers' Guides 1–4*, Longman.

Oelke, W. C. (1969) *Laboratory Physical Chemistry*, Van Nostrand.

Scottish Education Department (1968) *Memorandum 6/68 Inhalation of asbestos dust*, HMSO.

—(1968) Circular No. 689, *Ionising radiations in schools, Colleges of Education and Further Education Establishments*, HMSO.

SWG	Nearest preferred metric size (diameter) mm
12	2.65
14	2.00
16	1.60
18	1.25
20	0.90
22	0.71
24	0.56
26	0.45
28	0.40
30	0.31
32	0.28
34	0.25
36	0.20

Table 5.18 Nearest preferred metric size to SWG numbers

—(1970) Circular No. 776, *Use of Lasers in schools, Colleges of Education and Further Education Establishments*, HMSO.

—(1973) Circular No. 852, *The temporary use of ionising radiations . . . by visiting lecturers or student teachers in training*, HMSO.

—(1973) Circular No. 882, *Special precautions for the safe handling of Radium 226 closed sources of an approved type . . .*, HMSO.

Scottish Schools Science Equipment Research Centre, *Bulletins and Apparatus Reports*, issued at approximately six weekly intervals.

Welsh Office (1976) 2/76, see DES Administrative Memorandum 2/76.

—(1976) 5/76, see DES Administrative Memorandum 7/76.

Wray, J. D. (1974) *Small Mammals*, English Universities Press for the Schools Council.

Kodak (1972) *Record Photography in the Classroom*, Kodak Ltd.

Journals

Education in Chemistry, Chemical Society of London.

Journal of Biological Education, Institute of Biology.

Journal of Chemical Education, American Chemical Society.

Laboratory Equipment Digest, Morgan Grampian Ltd.

Physics Education, Institute of Physics.

School Science Review, Association for Science Education.

6 Improvisation of apparatus

Introduction

The improvisation of apparatus may involve the following.

a Role substitution, e.g. the use of a polystyrene drinking cup as a beaker or calorimeter. The original item generally requires little or no modification before it can be used to fulfil the new function(s) in the laboratory. Figure 6.1 illustrates how a plastic bottle may be sawn to provide a beaker, a filter funnel and a ring support for a round-bottomed flask.

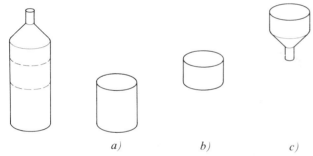

a)   *b)*   *c)*

Figure 6.1 a) A beaker, b) a 'cork ring' and c) a filter funnel improvised from a plastic bottle

b Role simulation, i.e. the construction of an item of apparatus to meet a need which, for reasons of cost and/or availability, cannot be met by commercial suppliers of laboratory equipment. Where cost is the significant factor, the intention is to produce an item to fulfil a given role at a lower cost. This saving may often be effected by using locally available materials and skills.

Apparatus may also be improvised for a purpose which cannot, at that time, be served directly by a commercially available item of equipment. This may be the case where students are involved in project work, particularly if such work has a technological bias. In these circumstances, the improvisation of apparatus has educational, rather than simply economic, objectives.

General considerations

a It is not always satisfactory to copy faithfully the design of a commercially available item of equipment. A design based upon the resources available locally may be cheaper and as satisfactory in operation. In some contexts, e.g. developing countries, the absence of a particular raw material may demand the development of a new item of equipment.

b It is possible to reduce the cash outlay on composite items of equipment by purchasing the individual components and assembling them within the school. This may be particularly advantageous if planning is sufficiently advanced to allow the components to be assembled by technical staff during school holidays.

c Improvised apparatus will be used by pupils who are accustomed to the high quality of commercially produced materials. School produced and commercially available apparatus should, therefore, differ significantly in cost but *not* in quality and reliability.

d When designing apparatus, it is important to maintain a balance between sophistication and simplicity. Increased sophistication invariably means greater cost. Excessive simplicity leads to inaccuracy or unreliability.

e Students can learn much by considering the problems of design and construction of apparatus to meet a particular need. However, the constructional skills of students and the time and resources available in a school are likely to be severely limiting factors.

f When technical or teaching staff are involved on a significant scale in improvising apparatus, the 'hidden' costs should not be ignored as it may be that the staff time could be spent more profitably in other ways.

g In a developing country, apparatus imported from overseas is expensive and often subject to unacceptable delay in delivery. The production of apparatus locally may thus be of sufficient importance in this context to be undertaken on a large scale. The production of improvised equipment on such a scale is a matter of import substitution. As such, quasi-commercial considerations apply and a purpose-built, economically self-supporting, workshop is desirable. For a discussion of the problems of manufacturing and improvising school science equipment in developing countries see Whittle, J. M. S., Local Production: Principles and Practice, *SSR* (1975), 197, **56,** 669. See also Warren, K., Lowe, N. K. (1975) *The*

Production of School Science Equipment, Commonwealth Secretariat.

h In purchasing raw materials such as timber or plastic, economies of scale can usually be made. Co-operation with the handicraft and related departments in a school is therefore sensible.

j Improvised apparatus, especially electrical equipment, should be clearly and unambiguously labelled where this is appropriate.

The following information is intended as a guide to the range and quality of a number of materials commonly used in improvising apparatus. For items such as electronic components, specialist catalogues should be consulted.

Abrasive cloths and papers

Many different kinds of abrasives are in use and the grading systems vary from one type to another. Some abrasive cloths and papers are available in 'open' and 'close' coats, these terms referring to the spacing of the abrasive grains on the paper. If available, open grain abrasives are generally more suitable for use on soft woods.

The commonest types of abrasive cloths and papers are as follows.

a Glasspaper. This is used for hand-smoothing wood and is available in sheets 9 in × 11 in (approximately 22.5 cm × 27.5 cm) in the following grades (close coat only):

00 or flour, 0, 1, $1\frac{1}{2}$, F2, M2, S2, $2\frac{1}{2}$, 3.

b Glass cloth. Glass cloth is available in sheets 9 in × 11 in in close coat only. Its uses are similar to those of glasspaper and it is available in the same grades.

c Aluminium oxide paper. Used mainly for machine sanding, it may be obtained in sheets 9 in × 11 in in both open and close coats. Aluminium oxide paper is available in the following grades:

9/0, 8/0, 7/0, 5/0, 4/0, 3/0, 2/0, 0, $\frac{1}{2}$, $1\frac{1}{2}$, 2, $2\frac{1}{2}$, 3.

Aluminium oxide cloth is available only as rolls or discs and is used for machine work. It is made in grades 400 to 24 in both open and close coats.

d Garnet paper. Garnet paper is available in sheets 9 in × 11 in and in rolls, discs and belts. It is made in both open and close coats in the following grades:

Sheets: 8/0, 7/0, 6/0, 5/0, 4/0, 3/0, 2/0, 0, $\frac{1}{2}$, 1, $1\frac{1}{2}$, 2.
Rolls: 5/0, 4/0, 3/0, 2/0, 0, 1, $1\frac{1}{2}$, 2, $2\frac{1}{2}$, 3, $3\frac{1}{2}$.

e Flint paper. 11 in × 9 in sheets are manufactured in a variety of grades. Flint paper is of limited use in smoothing wood.

f Silicon carbide paper. This is useful in smoothing metal surfaces. It is made in 9 in × 11 in sheets and in discs and rolls. Silicon carbide paper is also available in waterproof sheets of the same size and in the same grades:

400, 320, 280, 240, 220, 180, 150, 120, 100, 80 and 60.

Silicon carbide cloth is available in grades from 320 to 24.

Table 6.1 summarises comparable grades of a number of abrasives.

Glass paper or cloth	Garnet Aluminium oxide (woodworking)	Flint paper	Silicon carbide Aluminium oxide (metal working)
	—		400
	9/0		320
	8/0		280
	7/0		240
	6/0		220
00 or flour			—
0	5/0		180
1	3/0	120	120
$1\frac{1}{2}$	2/0	100	100
F2	—	—	—
M2	—	—	60
$2\frac{1}{2}$	$1\frac{1}{2}$	$1\frac{1}{2}$	40
3	2	2	36
—	$2\frac{1}{2}$	$2\frac{1}{2}$	30
—	3	3	24

Table 6.1   Comparable grades of a number of abrasives

The higher the grade number of glasspaper and glass cloth the greater the abrasiveness. When smoothing wood by hand, grade M2 glasspaper (or an equivalent) is used for preliminary smoothing and this may be followed by grade 1 or $1\frac{1}{2}$. Grade 0 is used only on delicate woods and 'flour' paper should be used for final finishing.

For most other abrasives, the degree of abrasiveness decreases with increasing grade number.

Glass

Glass used in windows, known as sheet glass, is available in three grades. The third (lowest) grade is sometimes referred to as horticultural glass because of its frequent use in greenhouses. Although the lower grades of sheet glass sometimes distort objects seen through them, they are quite adequate for some purposes in a school science laboratory.

Glass is usually referred to in terms of the weight of glass in a square foot, or of the thickness in millimetres. Table 6.2 relates thickness in millimetres to weight in ounces per square foot, the unit which is still commonly used in the United Kingdom.

Most window glass is '21 oz' grade.

Thickness mm	Weight ounces per sq. ft
1.58	15
2.1	18
2.5 bare	21
2.5 full	24
3.2	26
4.0	32
5.0	42

Table 6.2   Thickness and weight of glass

A number of specialist glasses are also available, e.g. reeded glass—which obscures vision to some degree—wired glass, toughened glass and tinted glass.

Note that sheet glass can also be purchased in a 'standard' size, e.g. 24 in × 18 in. A glass merchant may carry a large number of panes of this same size and for an item of equipment such as an aquarium, it may be possible to purchase the glass without the additional cost of cutting.

Glues

The bonding effectiveness of a glue is determined by both the nature of the glue and the chemical condition of the surfaces to be joined. Greasy or oily surfaces cannot be satisfactorily bonded and variations of temperature within a joint may also cause problems. The following types of glue are available.

a Animal glue

This is obtained from the skins and bones of animals and is sometimes known as Scotch or Salisbury glue. It may be purchased in lumps or in powder form for mixing with water in a glue pot. However, it is often more conveniently purchased ready for immediate use in tubes or tins. Some prepared glues must be applied hot.

Animal glue is relatively strong and durable and does not stain wood. However, the bonding is not waterproof and when the glue has to be used hot, arrangements must be made to prevent the glue from chilling when it is being applied to the joint. Like all glues of plant/animal origin, animal glue has a low resistance to attack by microorganisms. Proprietary brands include *Croid Aero, Croid Universal*, and *Fortil* and *Cox's*.

b Fish glue

This is made from the head, skin and cartilage of fish, although the highest quality fish glue is made from isinglass. It is an expensive glue and is usually purchased in tubes. The resistance of fish glue to water depends on the method of manufacture.

Proprietary brands include *Lepage's Liquid Glue* and *Seccotine*.

c Casein glue

This glue is made from casein precipitated from milk. Casein glues are water resistant and form strong joints but are liable to stain some woods. The glue is not 'sticky' when prepared and joints must therefore be clamped.

Proprietary brands of this relatively expensive type of glue include *Casco* and *Croid Insol*.

d Cellulose glues

These are sold in tube form and are suitable for fastening metal or plastic to wood. Cellulose-based glues are used cold, are highly water resistant and free from staining problems. A proprietary brand is *Durofix*.

e Polyvinyl acetate (PVA) glues

These comprise a very versatile range of glues which are cold-setting and are used directly from the container without a hardening agent. PVA glues are non-staining on wood and store well for several months. A PVA glue is liable to 'creep' and should not, therefore, be used to bond joints which are under stress, e.g. load bearing. Some PVA glues contain a higher than normal proportion of amyl acetate (pentyl ethanoate) and this shortens the time taken for the glue to set.

Proprietary brands include *Bondfast, Croid Fabrex, Casco PVA*, and *Redi-bond*.

f Resin glues

These glues are based upon phenol-methanal (-*formaldehyde*) or urea-methanal resins. Schools are more likely to use urea-formaldehyde based glues because application is easier.

Urea-formaldehyde glues are sold in various forms, but the two essential components are the glue itself and the hardener. The glue may be a viscous liquid or in powder form to be mixed with water. It does not begin to harden until it is in contact with the liquid hardener. The time taken for a given resin glue to set depends on the hardener, which is usually available in slow, medium and fast grades. Large work with many joints demands a slow hardener. For most purposes, a medium hardener is suitable. Note that the speed of setting increases with increasing temperature.

There are several methods of applying resin glues. One method involves the application of the glue to one surface and hardener to the surface to which it is to be joined. In other instances, manufacturers recommend that a hardener and glue be mixed before application.

Resin glues purchased as viscous liquids deteriorate with storage. This disadvantage is overcome by purchasing the glue as a powder, which will keep indefinitely if sealed. Powdered resin glues, incorporating a hardener, are also available and these are mixed with water before use. The advice of the DES on the setting of resins should also be noted (see page 39).

Resin glues are usually free from staining problems and are highly water resistant.

Proprietary brands include *Cascamite*, *Aerolite 300* and *Araldite*.

g Rubber-based glues

These 'contact' adhesives are convenient to use when large surfaces are to be joined together and a 'glue line' is not important. They are also useful in bonding wood to other materials, e.g. plastic.

The adhesive is applied to both of the surfaces to be joined and allowed to dry for a specified time, usually 5–20 minutes. The two surfaces are then positioned and pressed together. As the bond is immediate, correct positioning is of the utmost importance.

Contact adhesives are not as strong as many other glues but they have excellent water resistance properties and are free from staining problems. Proprietary brands include *Evostik*, *Bostic* and *Unistik*.

h Specialist glues

A number of glues have been developed to meet particular needs, e.g. oilseed residue glue, vegetable glue, blood albumin glue. These are not normally of importance in schools.

As a general rule, a rubber-based contact glue should be used to fasten rubber to wood. Plastics may be bonded to wood using a rubber-based adhesive or a resin glue.

j Cyanoacrylate glues

Extreme caution is needed in handling glues containing alphacyanoacrylates. These are instantaneous and powerful adhesives which can bond skin to skin. Gloves should be worn when using them and eye protection is advised to protect against involuntary wiping of the eyes. If, in spite of precautions, tissue surfaces do become stuck together and separation is not possible by simple peeling or rolling, medical advice should be sought as a matter of urgency.

The use of alphacyanoacrylate glues and access to them in a school must be rigidly controlled.

Metal framing

Much time can be saved when building larger items of equipment such as a cylinder support or a laboratory trolley, by using suitable metal framing. This is sold under a variety of trade names, e.g. *Dexion*, *Speedframe*, *Handyangle*.

The metal frame is manufactured in different cross-sectional patterns, with each pattern usually available in more than one size. It is also available in different finishes

and a wide range of accessories may be purchased, e.g. castors and adjustable feet. Metal frames are sold in standard packs or by the unit length. A small extra charge may be levied for cutting and packing individually ordered lengths. The metal may be cut using a hacksaw.

Details of individual types of metal frame are best obtained from the manufacturers.

Nails

There are dozens of different types of nail which have been developed for special purposes. The names differ from one trade to another and from one country to another. Nails, except for the largest, are sold by weight and in Imperial rather than metric lengths. Steel nails must be stored in a dry atmosphere so as to prevent rusting.

Perspex

Perspex sheet is purchased by the square measure but the sheets are often too large (e.g. 8 ft × 4 ft) for immediate requirements. Thicknesses from $\frac{1}{8}$ in to 1 in are available.

When only a small amount of perspex sheet is required, perspex 'off-cuts' may be a more economical method of purchase. These are usually sold by weight.

Perspex tubing is also available in assorted diameters up to 6 in.

Screws

Screws are usually made of mild steel or brass, but copper, gunmetal and aluminium screws are also available. In addition, they may be purchased in a variety of decorative finishes—e.g. 'blued', electroplated, copper, nickel-plated—but these are available in a smaller range of sizes. Screws are identified by length and gauge. The latter is the diameter of the shank of the screw, so that a screw of a given gauge is normally available in a variety of lengths. Gauges range from 0000 to 50, but the most commonly used gauges are from 4 to 12.

Screws may have countersunk, raised or round heads and 'Phillips' pozidrive versions are also marketed.

Screws are purchased by quantity rather than weight—bulk purchase is therefore more economical. When ordering screws, it is necessary to specify the number required, the gauge, the length, the material (steel, brass) and the type of screw head (countersunk, etc.).

Timber

Types of timber

A universally accepted distinction is made between softwoods, produced from coniferous trees, and hardwoods, derived from the broad-leaved trees which are deciduous in temperate climates. The more important softwoods are larch, manio, podo, pine, spruce ('white-wood'), red cedar and yew.

The range of hardwoods is very wide and the choice of wood for a particular task depends upon local availability and cost. Abura, for example, is an African hardwood which is acid resistant and may be used in making laboratory benches. If timber is to be purchased in a significant quantity, the advice of a woodwork colleague, a timber merchant or a carpenter should be sought.

In addition to 'natural' timbers, a number of manufactured boards are available, sometimes sold under proprietary names. Most of these are unsuitable for structural purposes. Asbestos-cement and asbestos-fibre boards are generally used in outdoor work but the more workable fibreboard is sometimes useful in making heat-resistant surfaces for laboratory benches. Fibre building boards are made of various materials such as wood, cane or straw and are sold in a variety of thicknesses. The commonest types are:

i insulating or 'block' board, nominal thickness not less than 12 mm;
ii wallboard, 4–10 mm thick and of higher density than insulating board;
iii hardboard, normally 3 mm or 5 mm thick and of higher density than wallboard;
iv plastic laminates, made from compressed paper sheets impregnated with resin and available in a variety of colours and simulated wood grains.

The plastic laminates are easy to clean and have a moderate resistance to heat and corrosive chemicals. They are usually glued to a wood base with a contact or resin adhesive (see page 153).

Plywoods are available in a range of thicknesses (4–12 mm) and qualities. Full specifications will be found in a timber merchant's catalogue. Both fibreboards and plywoods are available with a variety of veneers on one or both sides.

Purchasing timber

Timber is sold in a number of ways.

a Metre super. A metre super is a square metre and the wood may be of any dimension that will give this area. Thickness is not considered. Thus, pieces of wood 1 m × 1 m, 2 m × 0.5 m, 4 m × 0.25 m are all 'one metre super'.

b Metre run. Width and thickness of the wood are ignored in quoting the price of timber per metre. Softwoods are usually sold on this basis.

c Metre cube. Here the cost is related to the price of a solid metre cube of the timber or to any other shape of 1 m³ volume.

d In the round. Log timber is usually sold per cubic metre.

When buying timber from a merchant, the following considerations are important:

i it is cheaper to select suitable wood and cut the material oneself to the sizes required than to ask the timber merchant to do this;
ii a small alteration in design can sometimes make a substantial difference to cost;
iii the distinction between nominal and actual sizes should always be borne in mind. A 25-mm-thick piece of wood may measure only 22 mm after planing;
iv when measuring and calculating the wood required, adequate allowance must be made for trimming;
v ensure that dimensions are given unambiguously. In the timber trade, the order of specification is usually thickness, length and width;
vi it is cheaper to design equipment which uses the standard nominal sizes of wood because less wood is wasted as 'off-cuts';
vii when presenting an order for timber, it is helpful to group together all pieces of the same wood and thickness.

Standard sizes of softwood

It is still possible to purchase softwood in Imperial measurements, e.g. 2 in × 1 in but this is becoming increasingly rare as metrication proceeds. Softwood is available in the sizes shown in Table 6.3 on page 156.

Sources of information about the improvisation of apparatus

Before embarking on the design and/or construction of an item of laboratory equipment, it is obviously sensible to see whether it has been improvised before and, if so, by what means. The following index is intended to help in this task. Abbreviations used in the index are explained at the beginning of the index.

For information about working with wood, plastics, metals, etc. see the Bibliography on page 164.

List of abbreviations used in index

BSCS: Biological Sciences Curriculum Study, 1964. *Innovation in Equipment and Techniques for the Biology Teaching Laboratory,* Heath

ISTEP—Biology; Chemistry; Physics: Inexpensive Science Teaching Equipment Project, 1972. *Guide to Constructing Inexpensive Science Teaching Equipment,* University of Maryland. 3 vols

JCE: Journal of Chemical Education, American Chemical Society

Thickness mm	Width mm								
	75	100	125	150	175	200	225	250	300
16	✓	✓	✓	✓					
19	✓	✓	✓	✓					
22	✓	✓	✓	✓					
25	✓	✓	✓	✓	✓	✓	✓	✓	✓
32	✓	✓	✓	✓	✓	✓	✓	✓	✓
38	✓	✓	✓	✓	✓	✓	✓	✓	✓
44	✓	✓	✓	✓	✓	✓	✓	✓	✓
50	✓	✓	✓	✓	✓	✓	✓	✓	✓
63		✓	✓	✓	✓	✓	✓		
75		✓	✓	✓	✓	✓	✓	✓	✓
100		✓		✓		✓		✓	✓
150				✓		✓			✓
200						✓			
250								✓	
300									✓

Table 6.3   Standard sizes of softwood

JBE: Journal of Biological Education, Institute of Biology
NAB: Nuffield A level Biology, 1971. *Laboratory Book, a Technical Guide.* Penguin Books
NAC: Nuffield A level Chemistry, 1970. *Teachers' Guide II,* Penguin Books
NAP: Nuffield A level Physics, 1973. *Apparatus Construction Drawings.* Penguin Books
NCS: Nuffield Combined Science, 1970. *Teachers' Guide III.* Longman/Penguin
NJS: Nuffield Junior Science, 1967. *A Sourcebook of Information and Ideas.* Collins

NOB: Nuffield O level Biology 1966 and 1967. *Teachers' Guide.* Volume as indicated. Longman/Penguin
NOBR: Nuffield O level Biology (Revised) 1974 and 1975. *Teachers' Guide.* Volume as indicated, Longman
NSS: Nuffield Secondary Science, 1972. *Apparatus Guide,* Longman
SSR: School Science Review, Association for Science Education
UNESCO: 1973. *New UNESCO Sourcebook for Science Teaching.* UNESCO, Paris

Index

Materials, testing apparatus, *SSR*, 1972, 187, **54**, 346

Maze, human learning, *SSR*, 1973, 188, **54**, 513; for rodents, *SSR*, 1976, 201, **57**, 707

Measuring cylinder, *UNESCO*, 28; *ISTEP*, *Chemistry*, 64

Measuring device, tadpoles, etc., *NOBR*, 1, 76

Melting point apparatus, *SSR*, 1966, 163, **47**, 831; *SSR*, 1961, 148, **42**, 522; *UNESCO*, 40

Mesoscope, *SSR*, 1968, 170, **49**, 165

Metal: structure model, *SSR*, 1972, 185, **53**, 755; *SSR*, 1968, 168, **49**, 480; twisting apparatus, *NSS*, 290

Microbalance, *NOB*, IV, 68; *NOBR*, 2, 199; *ISTEP*, *Physics*, 22; *SSR*, 1972, 186, **54**, 125; maxi, *SSR*, 1968, 170, **49**, 184

Micro-boiling point apparatus, *JCE*, 1967, **44**, 43

Microgenerator (gas), *ISTEP*, *Chemistry*, 249

Microprojector, *UNESCO*, 25; *SSR*, 1975, 198, **57**, 77; field work, *SSR*, 1965, 160, **46**, 686; vertical, *SSR*, 1971, 181, **52**, 966

Microscope, *ISTEP*, *Biology*, 14; *SSR*, 1968, 169, **49**, 856; model, *NSS*, 241; hot stage, *JCE*, 1965, **42**, 91; polarising, *SSR*, 1966, 164, **48**, 179

Microtome, hand, *ISTEP*, *Biology*, 35; *BSCS*, 45

Millikan's apparatus, *SSR*, 1965, 159, **46**, 445; *NAP*, 28; analogue, *SSR*, 1964, 157, **45**, 638

Model: arm, *NSS*, 277; *NOB*, IV, 106; eye, *NSS*, 244; lung(s), *UNESCO*, 143; *SSR*, 1968, 171, **50**, 388; ribs, *NOBR*, 2, 67

Models, molecular, *see* Spacefilling, Skeletal, etc.

Morse, signalling circuit, *NJS*, 62

Mortar and pestle, *ISTEP*, *Chemistry*, 120

Moth cage, *NSS*, 192

Motor: electric, *UNESCO*, 104; *ISTEP*, Physics, 212; *NJS*, 46 and 78; linear induction, *SSR*, 1968, 170, **49**, 142

Mounts, *NJS*, 115

Museum jars (worms), *SSR*, 1974, 195, **56**, 303

Neon oscillator, *NSS*, 230

Nephelometer, *SSR*, 1970, 176, **51**, 637

Nesting boxes, *UNESCO*, 151

Nets, collecting, *ISTEP*, *Biology*, 53

Newton balance, *SSR*, 1970, 179, **52**, 407

Newtonmeter, *SSR*, 1971, 182, **53**, 183

Nocturnal dial, *SSR*, 1972, 185, **53**, 793

Nuclear analogue, *SSR*, 1967, 166, **48**, 895

Observation cell, insect mating, *NOB*, V, 158

Olfactometer, *SSR*, 1974, 195, **56**, 318

Open field apparatus (mouse behaviour), *NAB*, 124

Optical bench, *UNESCO*, 116; board, *ISTEP*, *Physics*, 119

Optical isomerism, model, *SSR*, 1969, 175, **51**, 367

Optical tunnel, *SSR*, 1964, 157, **45**, 652

Optics kit, *NAP*, 37

Orrery, *SSR*, 1963, 155, **45**, 229; conic section, *SSR*, 1964, 157, **45**, 634; earth/moon, *SSR*, 1966, 163, **47**, 839

Oscillator: audio, *SSR*, 1964, 158, **46**, 205; hacksaw blade, *NAP*, 14; neon, *NSS*, 230; sine wave, *SSR*, 1975, 197, **56**, 774; *SSR*, 1964, 158, **46**, 195

Oscilloscope, *SSR*, 1963, 154, **44**, 716; beam splitter, *SSR*, 1974, 193, **55**, 793; *SSR*, 1974, 192, **55**, 555; demonstration, *SSR*, 1966, 163, **47**, 776; double beam, *SSR*, 1967, 165, **48**, 537; from TV set, *SSR*, 1966, 163, **47**, 818

Oven, thermostatic, *SSR*, 1973, 188, **54**, 537

Oxidation indicator apparatus, *ISTEP*, *Chemistry*, 258

Ozoniser, *SSR*, 1954, 128, **36**, 96; Brodie, *SSR*, 1960, 145, **41**, 534

Paraffin burner, *NSS*, 291

Pendulum, simple timer, *ISTEP*, *Physics*, 50

Periscope, *NSS*, 240, *NJS*, 141; *SSR*, 1971, 181, **52**, 969

Pestle and mortar, *ISTEP*, *Chemistry*, 120

Petri dish, *ISTEP*, *Chemistry*, 113; display unit, *SSR*, 1974, 195, **56**, 303

pH: meter, *SSR*, 1973, 189, **54**, 766; scales, *JCE*, 1967, **44**, 330

Phase diagrams, 3D, *SSR*, 1969, 175, **51**, 380

Phase difference simulator, *NAC*, II, 267

Photochemical reactor, *JCE*, 1970, **47**, 122

Photographic dark box, *SSR*, 1971, 180, **52**, 658

Photometer, *SSR*, 1963, 154, **44**, 708; *NJS*, 157; density, *SSR*, 1968, 171, **50**, 335

Photopotometer, *SSR*, 1960, 145, **41**, 505

Photosynthometer, *SSR*, 1972, 185, **53**, 739

Piezoelectricity, demonstration of, *UNESCO*, 189

Pinhole camera, *SSR*, 1961, 147, **42**, 332; *NSS*, 238, *NJS*, 157

Pipette, *ISTEP*, *Chemistry*, 67; calibration of, *BSCS*, 15; plastic, *SSR*, 1966, 162, **47**, 486; rinser, *BSCS*, 19; transfer of microorganisms, *NOBR*, 4, 66; 50 drops cm^{-3}, *NOB*, II, 25

Planetarium, *SSR*, 1967, 165, **48**, 555

Plankton: concentrator, *SSR*, 1966, 162, **47**, 481; net, *SSR*, 1962, 150, **53**, 440

Plant: collecting apparatus, *ISTEP*, *Biology*, 136; cutter, standard lengths, *NAB*, 125; growth area, *BSCS*, 39; growth marker, *BSCS*, 44; press, *ISTEP*, *Biology*, 140

Plastic: laboratory ware, *SSR*, 1963, 154, **44**, 732; syringes, *ISTEP*, *Biology*, 257; utilities, *SSR*, 1966, 162, **47**, 494

Platinum electrode, *NAC*, II, 284

Point frame, *NAB*, 125

Polarimeter, *NAC*, II, 271; *SSR*, 1967, 166, **48**, 790; *SSR*, 1964, 158, **46**, 163; *SSR*, 1963, 155, **45**, 175; demonstration, *SSR*, 1967, 166, **48**, 888; for OHP, *SSR*, 1973, 191, **55**, 324

Polariscope, *NAC*, 270; *JCE*, 1970, **47**, 699

Polarising microscope, *SSR*, 1966, 164, **48**, 179

Transfer pipette, *ISTEP*, *Biology*, 224

Transformer: analogue, *SSR*, 1972, 185, **53**, 775; iron wire, *ISTEP*, *Physics*, 140; sheet iron, *ISTEP*, *Physics*, 147; variable output, *ISTEP*, *Physics*, 153

Trap: animal, *UNESCO*, 156; aquatic, *ISTEP*, *Biology*, 73; aquatic fungi, *NOB*, III, 189; pitfall, *SSR*, 1974, 192, **55**, 523; small vertebrate, *ISTEP*, *Biology*, 119

Tripod, *ISTEP*, *Chemistry*, 84; stand, *SSR*, 1966, 163, **47**, 773; storage, *SSR*, 1968, 168, **49**, 492

Trolley catapult, *NSS*, 302

Trolley (force and motion), *ISTEP*, *Physics*, 61; *NSS*, 264

Tropisms, apparatus for studying, *UNESCO*, 172

Tullgren funnel, *NSS*, 186; *NOBR*, 3, 104; *UNESCO*, 156

Tuning fork (transistor maintained), *SSR*, 1962, 151, **43**, 708

Tweezers, *see* Forceps

Tyndall Effect (OHP), *SSR*, 1972, 184, **53**, 585

Ultraviolet irradiation box, *NAB*, 147

Vacuum: apparatus (syringe), *ISTEP*, *Physics*, 99; distillation apparatus, *SSR*, 1968, 169, **49**, 809; sublimation, disposable apparatus, *SSR*, 1968, 169, **49**, 809

Vapour density apparatus, semi-micro, *SSR*, 1949, 111, **30**, 240; *JCE*, 1965, **42**, 336

Vibrator, sound, *NJS*, 205

Victor Meyer apparatus, *SSR*, 1972, 187, **54**, 318; *SSR*, 1967, 166, **48**, 802

Viscometer apparatus, *SSR*, 1961, 147, **42**, 312

Vivarium, *ISTEP*, *Biology*, 151; *NJS*, 87; woodlouse, *SSR*, 1969, 172, **50**, 566

Voltaic cell (A1/C), *SSR*, 1963, 156, 45, 451

Voltameter, *SSR*, 1968, 170, **49**, 161; *SSR*, 1964, 158, **46**, 154; *SSR*, 1962, 151, **43**, 706

Voltmeter, *NJS*, 49; attracted disc, *SSR*, 1961, 147, **42**, 306; high impedance, *SSR*, 1972, 187, **54**, 341; high resistance, *SSR*, 1968, 171, **50**, 347; one step, *SSR*, 1971, 182, **53**, 164

Volume gauge, *NSS*, 201

Volumeter, *see* Respirometer

Volumetric flask, *see* Flask, volumetric

Wallchart holder, *SSR*, 1973, 188, **54**, 565

Wash bottle, *ISTEP*, *Chemistry*, 114; hot, *SSR*, 1966, 163, **47**, 774

Watch glass, *ISTEP*, *Chemistry*, 122

Water bath, *ISTEP*, *Chemistry*, 189; constant temperature, *BSCS*, 79

Water clock, *see under* Clock

Water circuit board, *SSR*, 1971, 180, **52**, 633

Water lens, *SSR*, 1976, 200, **57**, 543

Wave machine, *SSR*, 1968, 169, **49**, 512

Wave models, *NSS*, 271

Weather vane, *NJS*, 250

Weighing devices, *UNESCO*, 23

Weightlessness, model, *UNESCO*, 220

Wheatstone bridge, *SSR*, 1976, 201, **57**, 443

Wheel and axle, *NJS*, 183

Whiteboard, *SSR*, 1970, 177, **51**, 929

Wind generator, *NSS*, 273

Wind sock model, *NJS*, 251

Wind vane, *UNESCO*, 222

Windmill, card, *NSS*, 212

Winding coil, *NJS*, 52

Wire loop (micro-organism transfer), *NOBR*, 4, 66

Workbench, portable, *NJS*, 129

Wormery, *NJS*, 88; *UNESCO*, 162; *NCS*, 220; *NOBR*, 1, 226; *ISTEP*, *Biology*, 168; *NSS*, 249; *SSR*, 1968, 170, **49**, 104

Yeast: apparatus for reaction with sugar solution, *SSR*, 1970, 178, **52**, 64; apparatus for measuring expansion of dough, *SSR*, 1974, 194, **56**, 90

Yo-yo, *NSS*, 218

Zoetrope, *SSR*, 1969, 175, **51**, 346

Zone refining apparatus (organic compounds), *JCE*, 1968, **45**, 116

Bibliography

Books

American Peace Corps (1968) *Science Teachers' Handbook.*

Archenhold, W. F. (1977) *References for Physics Teachers*, University of Leeds.

Association for Science Education (various dates) *Science Masters' Books*, Series, I to IV, John Murray.

Bainbridge, J. W., *et al.* (1970) *Junior Science Sourcebook*, Collins.

Barthelemy, R. E., *et al.* (1964) *Innovation in Equipment and Techniques for the Biology Teaching Laboratory*, Heath.

Bedford, J. R. (1971) *Metalcraft, Theory and Practice*, John Murray.

Bowker, M. K. (1968) *Making Elementary Science Apparatus*, Nelson.

Brydson, J. A., Saunders, K. J. (1970) *Experimental Plastics Technology*, Methuen

Bulman, A. D. (1966) *Experiments and Models for Young Physicists*, John Murray.

Clarke, P. J. (1973) *Plastics for schools*, Mills and Boon.

Coulson, E. H., *et al.* (1971) *Tubes and Beakers: Chemistry for Young Experimenters*, Doubleday.

Gosden, M. S. (1973) *Biological References from the School Science Review and the Journal of Biological Education*, University of Leeds.

Gunston, B. (1974) *Shaping Metals*, Macdonald.

Hayward, C. H. (1974) *The Woodworker's Pocket Book*, Evans.

Inexpensive Science Teaching Project (1972) *Guide to Constructing Inexpensive Science Teaching Equipment*, University of Maryland.

Institute of Craft Education (1974) *Buyer's Guide to Craft Equipment and Materials.*

Jenkins, E. W. (1974) *A Bibliography of Resources for Chemistry Teachers*, University of Leeds.

Joseph, A., *et al.* (1961) *A Sourcebook for the Physical Sciences*, Harcourt Brace.

Knudsen, J. W. (1966) *Biological Techniques*, Harper and Row.

Melton, R. F. (1972) *Elementary Economic Experiments in Physics*, CEDO, London.

Morholt, E., *et al.* (1966) *A Sourcebook for the Biological Sciences*, Harcourt Brace.

Nuffield Science Teaching Projects:

O level Biology (1966, 1967) *Teachers' Guides I–IV*, Longman/Penguin.

O level Biology (1974, 1975) Revised: *Teachers' Guides I–IV*, Longman.

A level Biology (1971) *Laboratory Book, a Technical Guide*, Penguin.

A level Chemistry (1970) *Teachers' Guide II*, Penguin/Longman.

Combined Science (1970) *Teachers' Guide III*, Penguin/Longman.

Junior Science (1967) *A Sourcebook of Information and Ideas*, Collins

Secondary Science (1972) *Apparatus Guide*, Longman

Oughton, F. (1975) *Wood Technology*, Macdonald.

Richardson, J. S., Cahoon, G. P. (1951) *Methods and Materials for Teaching General and Physical Science*, McGraw-Hill.

Schools Council, Integrated Science Project (1974) *Technicians' Manual*, 1, 2, 3, Longman/Penguin.

Schools Council *et al.* (1972/73) *Science 5/13*, MacDonald Educational, especially (i) *Working With Wood*, Stages I and II and *Background Information*; (ii) *Science from Toys*, Stages 1 and 2 and *Background Information*; (iii) *Metals*, Stages 1 and 2, and *Background Information*.

Strong, C. L. (1960) *The Scientific American Book of Projects for the Amateur Scientist*, Fireside Books.

UNESCO (1973) *New UNESCO Sourcebook for Science Teaching*, UNESCO, Paris.

Warren, K., Lowe, N. K. (1975) *The Production of School Science Equipment*, Commonwealth Secretariat.

Journals

Education in Chemistry: Chemical Society, London.

Journal of Biological Education: Institute of Biology, London.

Journal of Chemical Education: American Chemical Society.

Physics Education: Institute of Physics, London.

The Physics Teacher: American Association of Physics Teachers.

School Science and Mathematics: School Science and Mathematics Association, Inc., USA.

School Science Review, Association for Science Education, Hatfield, UK.

The Science Teacher: National Science Teachers' Association, USA.

Science Teacher, Junior Club Publications, London.

See also the *Bulletins* published by the Scottish Schools Science Research Centre and CLEAPSE.

7 Laboratory reagents and stains

Grades of reagent

The catalogue of a chemical supplier indicates that many laboratory reagents are available in more than one grade of quality. The following descriptions are likely to be encountered in the school context.

a Analar

This trade mark is applied to analytical reagents which conform to the tests and purity standards established by **AnalaR** Standards, Ltd., and published in successive editions of *AnalaR Standards for Laboratory Chemicals*. **AnalaR** reagents are of a high degree of purity, as the following specification for 'A.R.' glacial ethanoic (acetic) acid illustrates:

Minimum assay:	99.7 %
Wt. per ml* at 20 °C:	1.048–1.050 g
Freezing point:	Not less than 16.2 °C

Maximum limits of impurities

Water—insoluble matter	satisfies criteria
Non-volatile matter	0.001 %
Chloride	0.0001 %
Methanoate (formate)	0.01 %
Sulphate	0.0001 %
Arsenic	0.00005 %
Iron	0.00002 %
Heavy metals (lead)	0.00005 %
Substances reducing dichromate (VI)	0.003 %

b Aristar

It is possible to purchase some reagents in an ultra pure grade, representing the purest grade of reagent which is commercially available. The range of reagents available in such a grade is constantly being extended.

BDH Chemicals, Ltd., market ultra high purity reagents under the trade mark ARISTAR and the following is the specification of ARISTAR ethanoic (acetic) acid.

Assay:	99.9 ± 0.1 %
Wt. per ml at 20 °C:	1.048–1.049 g
Freezing point:	Not less than 16.3 °C
Refractive index:	1.3710–1.3725

*ml is the unit used.

Maximum limits of impurities ppm

Non-volatile matter	5
Ethanal (acetaldehyde)	2
Propanone (acetone)	1
Chlorate	1
Phosphate	0.1
Silicate	0.1
Sulphate	1
Aluminium	0.1
Cadmium	0.005
Calcium	0.5
Chromium	0.01
Cobalt	0.005
Copper	0.01
Iron	0.1
Lead	0.01
Magnesium	0.5
Manganese	0.005
Nickel	0.01
Potassium	0.1
Sodium	0.5
Strontium	0.5
Zinc	0.02

Reagents of ultra high purity are unlikely to be used in schools except in special circumstances.

c Technical

The composition of a technical grade reagent is variable and not subject to detailed specification. Technical grade reagents are adequate for many purposes in school science teaching, e.g. crystal growing, studying the action of heat on nitrates.

d Pure crystalline

These reagents are intermediate in purity between the analytical grades and the best technical grades. A specification is usually provided involving a minimum assay of at least 99 per cent. Thus 'pure crystalline' potassium nitrate(V) will have a composition close to the following:

Minimum assay:	99 %

Maximum limits of impurities

Chloride	0.02%
Sulphate	0.02%
Sodium	0.5%

In addition to the above grades, a large number of reagents are produced to detailed specifications for specialised purposes, e.g. chromatography, spectroscopy, clinical analysis, biochemical assay, tissue culture.

The purchase of reagents

The cost of a given reagent increases markedly with increasing purity. At 1976 prices, 1 kg of potassium nitrate(V), technical grade, cost approximately £1. If this is given unit value, the relative price of the same reagent in other grades is summarised in Table 7.1.

Grade of potassium nitrate(V)	Quantity purchased			
	250 g	500 g	1 kg	3 kg
Technical	—	—	1	2.7
Pure, crystalline	—	0.7	1.3	3.5
ANALAR	0.7	1.20	2.3	6.3

Table 7.1   Effect of grade upon price of a reagent

Many of the reagents used in school science teaching do not need to be of 'pure, crystalline' quality and it is important to purchase a laboratory reagent in a grade which is no more than adequate for the purpose intended. However, technical grade reagents tend to be sold in rather larger quantities (1 kg and above) than the more pure grades.

Table 7.1 confirms the financial advantage of bulk purchase—as long as a substance does not deteriorate with storage. Reagents such as sodium chloride, sulphur, calcium carbonate (marble chips), and lead foil are most economically purchased in large quantities, but due consideration must be given to consequent storage problems. Large glass sweet jars are often satisfactory for storing solid reagents.

Some reagents are available in more than one physical form, e.g. sodium hydroxide is sold as flakes, pellets or sticks, each with different degrees of purity. For most purposes involving sodium hydroxide in school chemistry, purchasing the **AnalaR** grade of stick would be an extravagance. Pellets are, in any case, the most convenient form for general use, especially for preparing solutions.

Table 7.2 summarises the relative prices of different forms and grades of purity of metallic zinc. At 1976 prices, 500 g of granulated, pure zinc (arsenic free) cost approximately 90p. This is given unit value in Table 7.2.

Volumetric reagents may be purchased in ampoules or vials which produce accurately standardised solutions of specified concentration when diluted with a stated quantity of distilled water. Such reagents are convenient to use and are prepared from **AnalaR** grade materials. Their purchase inevitably involves a greater capital outlay than the preparation and standardisation of a solution by a laboratory technician, but they are a particularly convenient method of preparing accurately standardised solutions for examination purposes.

The concentration of a reagent

Where a reagent is sold or used as a solution, its concentration may be specified in a number of ways which include the following.

a Molarity. A molar, M, solution of a reagent is one which contains 1 mole of the reagent dissolved in 1 litre of solution. Thus a molar, aqueous solution of sodium chloride, NaCl, contains 58.5 g of the salt in 1 litre of solution and is represented as M NaCl. More dilute solutions are represented as 0.5 M, 0.1 M or as M/2, M/10 etc.

Many common chemical reagents are prepared and used as M or 2 M solutions unless the reagent is too

Grade and form of zinc	Quantity purchased					
	100 g	250 g	500 g	1 kg	2 kg	3 kg
Zinc metal foil, 0.35 mm	—	—	2.1	—	—	—
Zinc granulated, technical	—	—	0.85	—	—	—
Zinc granulated, pure 'arsenic free'	—	—	1	—	—	—
Zinc pellets	—	—	2.1	—	—	—
Zinc powder	—	0.6	—	1.2	—	3.2
Zinc sticks 'arsenic free'	—	—	4.0	—	—	—
Zinc sticks complexometric standard	2.3	—	—	—	—	—
Zinc wool	—	—	2.5	—	—	—
Zinc granulated ANALAR	—	—	1.2	2.2	4.0	—
Zinc powder ANALAR	—	—	1	—	—	5.2

Table 7.2   Relative prices of different forms and grades of zinc

expensive (e.g. silver nitrate(V)) or insufficiently soluble (e.g. calcium sulphate).

b Molality. A molal solution is one which contains 1 mole of solute dissolved in 1000 g of solvent. Molal concentrations are of particular use in discussing colligative properties and are expressed as mol solute/kg solvent.

c Normality. A normal, N, solution is one which contains the gram-equivalent of a solute in 1 litre of solution. The gram equivalent of a reagent depends upon the reaction in which it is involved and, for this and other reasons, it is now more common practice to specify concentrations in molarities.

d Other weight and volume ratios. Chemical catalogues sometimes quote concentrations in w/w, w/v or v/v. These usually refer to the *percentage* weight or volume of a given substance dissolved in a given weight or volume of solvent.

Aqueous solutions of hydrogen peroxide are sold as 10, 20 or 100 'volumes'. A '20 volume' solution is one which produces 20 litres of oxygen measured at s.t.p., when 1 litre is decomposed in accordance with the following equation:

$$2H_2O_2(aq) = 2H_2O(l) + O_2(g)$$

Since $2 \times H_2O_2 \equiv 2 \times 34$ g, 68 g of hydrogen peroxide will produce 22.4 litres of oxygen at s.t.p., and a '20 volume' solution must contain $(20/22.4 \times 68)$ g of hydrogen peroxide per litre. Such a solution will be $(20/22.4 \times 68/34)$M, 1.8 M or approximately 6 per cent w/v.

e Specialised units of concentration. Some of these specialised units may be re-defined in terms of SI units but, in many instances, the specialised unit is likely to be used for the foreseeable future, e.g. vitamin concentrations are sometimes expressed in international units written as 'i.u.'.

Mixing solutions

If two solutions of a given solute in the same solvent are mixed, the composition of the resultant solution may be calculated from density tables. If it is required to mix concentrated sulphuric acid (density 1.84 g cm^{-3}) and dilute sulphuric acid of density 1.40 g cm^{-3} to produce an acid of density 1.60 g cm^{-3}, density tables provide the following data.

Density g cm^{-3}	Composition per cent by weight
1.40	50.50
1.84	97.50
1.60	69.09

$(97.50 - 69.09)$ g $= 28.41$ g of 50.50 per cent acid must

therefore be mixed with $(69.09 - 50.50) = 18.59$ g of 97.50 per cent acid to produce sulphuric acid of density 1.60 g cm^{-3}, containing 69.09 per cent acid by weight.

The preparation of laboratory reagents

The following general points are important and should be considered when preparing reagents in the manner indicated below.

a Reagent bottles must be clearly and adequately labelled. Attention is drawn (i) to the labels produced by the Association for Science Education, which also give brief details of hazards, first aid procedures, etc., as appropriate, and (ii) to the use of hazard warning symbols (see page 37).

Reagents which are used in large quantities, e.g. 2 M sodium hydroxide, 2 M aqueous ammonia, may be kept in bottles which have corrosion-resistant labels (see page 84).

Reagents which decompose in light should be kept in tinted bottles and, if possible, stored in the dark.

Caustic alkalis are best stored in bottles which are sealed by means of a plastic or rubber bung rather than with a ground-glass stopper.

b Distilled or deionised water is an adequate solvent for almost all aqueous solutions. Note that distilled water which has been prepared for some time may contain a significant amount of dissolved carbon dioxide.

c When a reagent is known to deteriorate with storage, large quantities should not be prepared and a freshly prepared reagent should not be added to a nearly empty reagent bottle to 'top it up'. The bottle should be emptied, washed and rinsed before being filled with the freshly prepared reagent.

d In the following preparations, the term alcohol implies ethanol in the form of colourless industrial spirit, 74° O.P.

Alphabetical list of reagents, stains and culture media

Absolute alcohol; *see* Alcohol

Acetate buffer; *see* Buffer, acetate

Acetic acid, M
Dilute 58 cm^3 of glacial ethanoic acid (acetic acid) to 1 litre with distilled water.

Acetic alcohol; *see* Alcohol, acetic

Acetic-alcohol-formaldehyde; *see* Formal-acetic-alcohol

Acetic-carmine; *see* Carmine-acetic

Acetic-orcein; *see* Orcein-acetic

Acid alcohol; *see* Alcohol, acid

Acid fuchsin; *see* Fuchsin, acid

Acid glycerol; *see* Glycerol, acid

Adenine; chromatogram marker solution
Dissolve 0.001 g adenine in 5 cm³ 0.1 M hydrochloric acid.

Adenosine triphosphate
Purchase as disodium salt in sealed glass ampoules. Store at 4 °C.

Adhesives
a Acid-proof. Mix together 1 part rubber solution, 2 parts linseed oil and 3 parts powdered pipeclay.
b For iron. Mix together 90 parts fine iron filings, 1 part flowers of sulphur, 1 part ammonium chloride. Add a little water and stir to a paste. Use immediately.
c For perspex. Mix small pieces of perspex (e.g. sawings, filings or chips from broken perspex Petri dishes) with trichloromethane (chloroform) 1,2-dichloroethane, or ethoxyethane (diethyl ether) till a sufficiently viscous consistency is obtained. Keep in a tightly stoppered jar.
d For cellophane. Mix together 15 g propane-1,2,3 triol (glycerol) with 85 g of either gum arabic or gelatin.
e For glass. Mix together 100 g gum arabic and 25 cm³ distilled water. Add 0.2 g aluminium sulphate(VI) dissolved in 2 cm³ distilled water. Alternatively, use 'Araldite' or 'Bostic No. 1'.
f For attaching microtome sections to slides. Heat 100 cm³ distilled water to not more than 30 °C. Dissolve in this 1 g best grade gelatin. When the gelatin is completely dissolved add 2 g phenol crystals and 15 cm³ propane-1,2,3 triol (glycerol) and stir well. Filter. Sections should be floated onto the slide in 4 per cent methanal (formaldehyde solution).
See also pages 153 and 154.

Adrenalin; 0.01 per cent aqueous solution
Dissolve 0.1 g of adrenalin in distilled water and dilute to 1 litre using distilled water.

Adsorption indicator; *see* individual indicators, e.g. dichlorofluorescein, eosin

Agar
A gelatinous material extracted from marine algae, used for the preparation of solid media, usually for the culture of micro-organisms. 10–15 g agar per litre of medium is usually sufficient, but up to 20 g may be necessary if acidic additives are being used. A variety of substances may be added to agar media to provide nutrients for the micro-organisms being cultured. In some cases these specialised media are available from suppliers in tablet form. Agar itself is also available in tablet form from suppliers.

Agar; for isolation of damping-off fungi (e.g. *Rhizoctonia*)
Use plain agar prepared by dissolving 15 g agar in 1 litre of water and then autoclaving. Once the mycelium has developed (bacteria will not normally grow on this medium) transfer a little agar with mycelium to potato agar.

Agar, blackened; for growth of *Nicotiana* (tobacco) seedlings including albino strains
Add 30 g agar and 6 teaspoonfuls of carbon black to 1 litre water. Heat while stirring. Autoclave to sterilise.

Agar, blood; for culture of bacteria, e.g. from soil
Blood agar is obtainable in tablet form from biological suppliers. Add 2 tablets to 10 cm³ distilled water and soak till dissolved—about 15 mins. Autoclave to sterilise.
N.B. *If the medium is being used to culture bacteria from a liquid inoculum, then the volume of the inoculum (usually 1 cm³) is subtracted from the volume of distilled water used in making up the culture.*

Agar, cornmeal
a Add 30 g maize meal to 1 litre distilled water and boil for 15 mins with frequent stirring. Allow to stand for a few minutes and decant clear liquid. Add 2 g agar for each 100 cm³ of liquid. Warm and stir till agar is dissolved. Autoclave to sterilise.
or b Add 17 g Difco cornmeal agar (from Oxoid, Ltd.) and 1 g yeast extract to 1 litre distilled water. Warm and stir till agar is dissolved. Autoclave to sterilise.

Agar, *Drosophila*; *see* Culture media, *Drosophila*

Agar, fern gametophyte
Make up three stock solutions as follows:
A Dissolve 6 g potassium chloride, 9 g magnesium sulphate(VI)-6-water, 3.6 g sodium nitrate(V) and 0.6 g iron(III) 2-hydroxypropane-1,2,3-tricarboxylate (citrate) in 100 cm³ distilled water.
B Dissolve 10 g calcium nitrate(V)·4-water in 100 cm³ distilled water.
C Dissolve 6 g potassium hydrogenphosphate(V) in 100 cm³ distilled water.
   To prepare the agar medium, mix 10 cm³ of each of the three stock solutions and make up to 1 litre with distilled water. Add 15 g agar and warm while stirring until dissolved. Pour into Petri dishes.

Agar, frozen pea
Thaw 160 g frozen peas, then blend them in a liquidiser. Add to 1 litre distilled water together with 20 g agar. Bring to the boil, stirring continuously. Autoclave before use.

Agar, germination of small seeds
Mix 3–4 g agar with 100 cm³ distilled water, bring to the boil while stirring, then pour into suitable containers, e.g. Petri dishes, and allow to set.

Agar, glucose-1-phosphate
Add 4 g agar and 1 g glucose-1-phosphate(V) to 200 cm³ distilled water. Bring to the boil while stirring until agar is dissolved. Boil thoroughly and pour into sterile Petri dishes to a depth of 2 mm.

Agar, housefly
Mix together 100 g dried milk, 100 g dried yeast and 20 g agar. Stir into 1 litre boiling water. Pour into jars to a depth of 3 cm. The culture jars can be kept in a refrigerator until the housefly eggs are introduced.

Agar, MacConkey's; for culture of coliform bacteria
  This medium is obtainable in tablet form from suppliers (e.g. Astell). Dissolve 2 tablets MacConkey agar in $10\,cm^3$ distilled water (this takes 10–15 mins). Autoclave to sterilise.
  N.B. *If a water sample is being tested for the presence of coliform bacteria, $9\,cm^3$ distilled water should be used in making up the agar. $1\,cm^3$ of the water sample being tested is then added just before the agar gels.*

Agar, malt; for culture of yeast and other micro-fungi
  a Add 15–20 g agar and 20 g malt extract to 1 litre distilled water. Mix thoroughly. Autoclave to sterilise.
  b Malt agar is also obtainable from suppliers, e.g. Oxoid, in tablet form. Dissolve 2 tablets in $10\,cm^3$ water. Autoclave to sterilise.
  See also Agar, yeast.

Agar, Nile blue sulphate or **neutral red;** for vital staining of the early stages of chick embryos
  Mix 4 g agar with $100\,cm^3$ of boiling distilled water. Pour into a clean glass dish to a depth of 1–2 mm. Allow to dry and then place in a 1 per cent aqueous solution of Nile blue sulphate or neutral red. Leave for two days and then rinse off surplus stain with distilled water. The stained agar can be stored in a refrigerator. Sections of the agar jelly about 8 mm square are placed over the embryo for a few minutes to allow the embryo to take up the stain.

Agar, nutrient
  Add 10 g beef extract, 10 g peptone, 5 g sodium chloride and 15 g agar to 1 litre distilled water. Heat until dissolved. Adjust pH to 7.5 with sodium hydrogencarbonate if necessary. Filter while hot. Autoclave to sterilise.

Agar, nutrient; for studying effect of nutrients on bacterial growth
  This is a series of agar media containing a variety of mineral salts and glucose. The composition of each medium is shown in Table 7.3 in which concentration is expressed in $g\,l^{-1}$.
  Dissolve the ingredients (except the agar) in about $200\,cm^3$ distilled water in each case. Add the agar and

make up to 1 litre with distilled water. Heat, with continuous stirring (in a waterbath if necessary), until all the agar is dissolved. Autoclave to sterilise but do not exceed $65\,kPa$ ($10\,lb\,in^{-2}$) above atmospheric pressure or decomposition of the glucose will result.

Agar, plant growth
  a Nitrogen free. Add the following to 1 litre distilled water using only **AnalaR** grade chemicals:

1 g dipotassium hydrogenphosphate(V)

0.5 g calcium tetrahydrogen diorthophosphate(V)

0.2 g magnesium sulphate(VI)

0.1 g sodium chloride

0.01 g iron(III) chloride

15 g agar

  Bring to the boil, stirring continuously. Autoclave to sterilise.
  b With nitrogen. Add ammonium sulphate(VI) to the above mixture (0.05 g per $100\,cm^3$ of medium).

Agar, potato dextrose; for culture of moulds, e.g. *Rhizopus, Mucor*
  a For a clear medium. Wash and dice 250 g of unpeeled potatoes. Tie in a muslin bag and steam for one hour. Suspend the bag above the steamer and allow it to drip until drainage is complete. Do not squeeze the bag. Make up the liquid extract to 1 litre with tap water and add 20 g glucose. Add 15 g agar and warm while stirring until agar is dissolved. Autoclave to sterilise.
  b If a clear medium is not required. Add 15 g agar to 1 litre water. Warm and stir till dissolved. Add 200 g boiled potatoes and 20 g dextrose. Autoclave to sterilise.

Agar, Ringer; for cultivation of chick embryos in vitrio
  Add 0.2–0.25 g agar to $30\,cm^3$ Ringer solution in a small flask. Bring to the boil over a low heat to prevent the agar from sticking. Cool to 40 °C. Add $20\,cm^3$ of Ringer albumen and mix.

Agar, rye meal
  Grind 60 g rye grain for 10 mins. Allow to stand in 1 litre warm water for one hour with occasional stirring.

	Complete	Lacking magnesium	Lacking nitrogen	Lacking glucose	Lacking all
Sodium chloride	0.5	0.5	0.5	0.5	—
Magnesium chloride	0.5	—	—	0.5	—
Ammonium sulphate(VI)	2.5	2.5	—	2.5	—
Glucose	2.5	2.5	2.5	—	—
Potassium dihydrogen phosphate(V)	1	1	1	1	—
Magnesium sulphate(VI)-6-water	—	—	1	—	—
Agar	20	20	20	20	20

Table 7.3   The composition of a variety of nutrient agars

Filter through three thicknesses of medical gauze. Warm and stir until agar is dissolved. Autoclave to sterilise.

Agar, sporulation of *Aspergillus*
Add the following to 1 litre distilled water:
1 g sodium nitrate

0.5 g magnesium sulphate(VI)-6-water

0.5 g potassium chloride

1.5 g potassium dihydrogen phosphate(V)

Trace iron(II) sulphate(VI)-7-water

Trace zinc sulphate(VI)

20 g glucose

15 g agar

Warm and stir until agar is dissolved. Autoclave to sterilise.

Agar, starch; for work with amylases
Add 2 g agar to 100 cm³ cold, 1 per cent starch suspension. Stand for 10 mins. Warm, while stirring, to boiling point. Pour into Petri dishes. Alternatively the starch-agar may be autoclaved prior to pouring into the Petri dishes.

An improved recipe uses Ionagar in place of standard agar. This is water clear when set. Only a low concentration is needed and 0.5–1.0 g Ionager should be sufficient for the above recipe.

Agar, yeast
This is best bought directly from suppliers already made up in sterile Petri dishes. (See also Nuffield Biology Teachers' Guide V (1967) *The Perpetuation of Life*, Longman/Penguin, pp. 70–71).
See also Agar, malt

Agar, yeast-mannitol; for culture of *Rhizobium*
Add the following to 900 cm³ distilled water:
15 g agar
10 g mannitol
0.5 g dipotassium hydrogenphosphate(V)
0.2 g magnesium sulphate(VI)-6-water
0.1 g sodium chloride
3 g calcium carbonate
100 cm³ yeast water (*see* page 194)

Heat the mixture to boiling point with continuous stirring. Autoclave to sterilise.

Albumen, egg; for experiments on protein digestion
Mix the white of an egg with an equal volume of distilled water. Pour the mixture into 500 cm³ of water at 60 °C with continual stirring. Warm slightly (not above 80 °C) until the mixture becomes opaque. Allow to cool and filter (if necessary) through muslin, glass wool or a Buchner funnel.

Albumen, egg; stock solution; for attaching wax embedded microtome sections to slides
Mix 50 cm³ egg white with 50 cm³ 1 per cent sodium chloride solution. Centrifuge to remove precipitate if necessary.

0.2 g sodium 4-hydroxybenzenecarboxylate (sodium *p*-hydroxybenzoate) may be added to prevent the growth of micro-organisms.

Albumen, Mayer's
Shake 50 cm³ white of egg with a few drops of dilute ethanoic (acetic) acid. Add 50 cm³ propane-1,2,3-triol (glycerol) and 1 g sodium 2-hydroxybenzenecarboxylate (sodium salicylate) and shake well. Filter using a filter pump, into a clean bottle.

Albumen, Ringer; for cultivation of chick embryos in vitrio
Add the egg white from one egg to 50 cm³ of Ringer solution in a 500 cm³ stoppered flask. Shake vigorously for about one minute.

Alcohol
The nomenclature of the various types of alcohol has, in the past, been somewhat confused. The following terminology is recommended.

Ethanol is the chemical name for the alcohol having the formula C_2H_5OH and formerly known as ethyl alcohol. Bradbury (1973) and others restrict the use of the term to ethanol which is produced by synthesis (as opposed to distillation) but there seems little justification for this.

Methanol is the chemical name for the alcohol having the formula CH_3OH and commonly known as methyl alcohol.

Proof spirit contains 49.28 per cent ethanol by weight (57.10 per cent by volume) and has a relative density of 0.91976 at 15.55 °C.

Degrees over proof (°O.P.) indicates the increase in volume obtained when 100 volumes of spirit are diluted with water to obtain proof spirit. A solution of ethanol, 100 volumes of which requires 50 volumes of water to dilute it to proof spirit, is thus described as 50 °O.P.

Industrial spirit is a term used to describe alcohol of an industrial quality which usually contains between 96 per cent and 98 per cent ethanol.

Industrial methylated spirit contains about 95 per cent ethanol and 5 per cent crude methanol (wood naphtha). It is available in two grades—60 °O.P. and 74 °O.P. The latter contains 99.4 per cent alcohol (ethanol plus methanol) and is of sufficient purity for any type of work likely to be encountered in schools. It is colourless and may be obtained only under licence issued by the Customs and Excise Department.

Absolute alcohol strictly speaking applies to 100 per cent ethanol. This is very expensive and is not likely to be required in school laboratories. Throughout this book, where the need for absolute alcohol is indicated, industrial methylated spirit (74 °O.P.) may be used as a substitute, especially if it is dried before use by standing over anhydrous copper sulphate.

Purple methylated spirit is similar to industrial methylated spirit but also contains traces of paraffin, pyridine and methyl violet. It is not generally suitable as a substitute for industrial methylated spirit.

Rectified spirit refers to ethanol which has been redistilled from crude spirit. Repeated re-distillations may increase the percentage of alcohol to as high as 98 per cent.

Alcohol, absolute; see Alcohol

Alcohol, acetic (Clarke's Fluid)
Mix 25 cm³ glacial ethanoic (acetic) acid with 75 cm³ absolute alcohol.

Alcohol, acetic-formaldehyde; see Formal-acetic-alcohol

Alcohol, acid
a Add 6 drops of concentrated hydrochloric acid to 100 cm³ of the appropriate percentage solution of alcohol. 70 per cent alcohol is the most commonly used.
b **For macerating plant tissue.** Dissolve 25 cm³ of 2 M hydrochloric acid in 75 cm³ of 95 per cent alcohol.

Alcohol, alkaline
Add a few drops of 0.1 per cent sodium hydrogencarbonate solution to the appropriate percentage solution of alcohol. 70 per cent alkaline alcohol is the most commonly used.

Alcohol, ammoniated
Add a few drops M aqueous ammonia solution to 100 cm³ alcohol (70 per cent).

Alcohol, aniline
Add 1 cm³ aniline to 1 litre of 90 per cent alcohol.

Alcohol-methanal (Alcohol-formalin); fixative
Dissolve 6 cm³ of 40 per cent aqueous methanal (formaldehyde) in 100 cm³ of 70 per cent ethanol.

Alizarin red S, Dawson's, for transparency skeletal preparations
Dissolve 0.1 g Alizarin red S in a solution of 10 g potassium hydroxide in 1 litre of distilled water.

Alkaline alcohol; see Alcohol, alkaline

Alum, saturated solution; see Aluminium(III) potassium(I) sulphate(VI)-12-water, saturated solution

Aluminium(III) potassium(I) sulphate(VI)-12-water (Alum), saturated solution; for crystal growing
Dissolve 30 g of aluminium(III) potassium(I) sulphate(VI)-12-water (alum) in 100 cm² of warm water (about 50 °C). Allow to cool and seed the saturated solution with a tiny crystal of alum. Keep at constant temperature in a covered container.

Amann's medium; clearing agent for sections and small whole mounts
Dissolve 20 g of phenol in 20 cm³ of distilled water. Add 20 g of 2-hydroxypropanoic acid (lactic acid) and stir until dissolution is complete. Finally add 33.3 cm³ of propane-1,2,3-triol (glycerol).

Amino acids, chromatographic solvents, for separation; see Chromatographic solvents

4-Aminobenzenesulphonic acid
Prepare a 1 per cent solution of the reagent in 30 per cent aqueous ethanoic (acetic) acid.

Ammoniacal copper(I) chloride
Dissolve 250 g of ammonium chloride in 750 cm³ distilled water and dissolve 200 g of copper (I) chloride in the solution. Prepare an aqueous ammonia solution by diluting 60 cm³ of 0.880 ammonia solution to 250 cm³ with water. Add the diluted ammonia solution to the copper solution and the reagent is then ready for use. It may be preserved for some time by adding a spiral of clean copper wire.

Ammoniated alcohol; see Alcohol, ammoniated

Ammonium carbonate, 2 M
Dissolve 160 g in a mixture of 140 cm³ of 0.880 aqueous ammonia and 860 cm³ distilled water.

Ammonium chloride, M
Dissolve 53.5 g of ammonium chloride in 1 litre of distilled water.

Ammonium ethanedioate, 0.5 M
Dissolve 72 g of ammonium ethanedioate-1-water (ammonium oxalate) in 1 litre of distilled water.

Ammonium hydroxide; see Aqueous ammonia

Ammonium molybdate
Dissolve 40 g of molybdenum(VI) oxide (trioxide) in a mixture of 70 cm³ 0.880 ammonia and 140 cm³ distilled water. Add, with constant stirring, a mixture of 250 cm³ of concentrated nitric(V) acid and 500 cm³ distilled water. Dilute to 1 litre. Allow to stand for 48 hours, then decant the clear solution.

Ammonium oxalate, 0.5 M; see Ammonium ethanedioate

Ammonium polytetraoxomolybdate (VI); see Ammonium molybdate

Ammonium sulphate(VI), saturated
Add 750 g of ammonium sulphate(VI) to 1 litre of distilled water. Stir until a saturated solution is obtained.

Ammonium sulphate(VI); fertiliser solution
Dissolve 20 g of ammonium sulphate(VI) in 4 litres of tap water. Apply this to 1 m² soil.

Ammonium sulphide
Saturate a volume of approximately 5 M aqueous ammonia with hydrogen sulphide gas. Add an equal volume of the aqueous ammonia solution. The reagent oxidises on exposure to air, depositing elemental sulphur.

Anaesthetics; for Drosophila
Use ethoxyethane (diethyl ether).

Anaesthetics; for locusts
Place the insects in a refrigerator (not below 8 °C) for a few hours. Alternatively use ethoxyethane (diethyl ether) or trichloromethane (chloroform).

Anaesthetics; for fish and amphibia
Use ethyl 3-aminobenzenecarboxylate (MS-222 San-doz or ethyl *m*-aminobenzoate). Dissolve 0.5 g in 100 cm^3 pond water or Holtfreter's solution. Place the animals in the solution for a few minutes until immobilised. If frogs are to be pithed they should be left 10–15 minutes.

Aniline alcohol; *see* Alcohol, aniline

Aniline blue (Cotton blue lactophenol); stain fixative for fungal mycelia
Dissolve 1 g of aniline blue (water soluble) in a mixture of the following reagents:

25 cm^3 distilled water
25 cm^3 propane-1,2,3-triol (glycerol)
25 g 2-hydroxypropanoic acid (lactic acid)
25 g phenol

Aniline blue; general stain for temporary botanical mounts
Dissolve 0.5 g aniline blue (water soluble) in 99.5 cm^3 lactophenol (Amann's medium).

Aniline blue–Orange G; for use in Heidenhain's Azan staining technique
Dissolve 0.5 g aniline blue and 2 g Orange G in 100 cm^3 distilled water. Add 8 cm^3 glacial ethanoic (acetic) acid.

Aniline chloride; specific stain for lignin
Dissolve 1 g of phenylammonium chloride (aniline chloride) in 89 cm^3 of 70 per cent alcohol. Add 10 cm^3 of 0.1 M aqueous hydrochloric acid.

Aniline sulphate; specific stain for lignin
Dissolve 1 g of aniline sulphate in 89 cm^3 of 70 per cent alcohol. Add 10 cm^3 of 0.05 M sulphuric acid. Store the stain in a dark bottle.

Aniline dyes; *see* Aniline blue, Aniline sulphate etc.

Anti-coagulant; *see* Blood, anti-coagulant

Antiformin; *see* Macerating fluids, for bones

Antiseptic solution; for sterilising thermometers
Use cetrimide solution BNF or chloroxylenol solution BPC.

Aqua regia
Mix carefully 50 cm^3 of concentrated nitric(V) acid and 150 cm^3 of concentrated hydrochloric acid.

Aqueous ammonia, M
Add 66.7 cm^3 of 0.880 aqueous ammonia to 750 cm^3 of distilled water. Dilute to 1 litre.

Arsenical soap; *see* Soap, arsenical

Ascorbic acid, Vitamin C
A standard 0.1 per cent solution may be made by dissolving two 50 mg tablets in 100 cm^3 of distilled water. The tablets should be crushed in a mortar with successive small volumes of water and filtered to remove the chalk or starch base in the tablets. The filtrate is then made up to 100 cm^3.

Azan-Heidenhain's stain; for general histological staining
Boil 0.1 g of azocarmine G or azocarmine B with 100 cm^3 of distilled water. Filter when cold, and acidify the filtrate with 1 cm^3 of glacial ethanoic (acetic) acid.

Azocarmine
Dissolve 0.5 g of azocarmine B or G in a 1 per cent aqueous solution of ethanoic (acetic) acid.

Baeyer's reagent; *see* Potassium manganate(VII), for alkenic bonds

Barfeod's reagent
Dissolve 13 g copper(II) ethanoate (acetate) in 200 cm^3 1 per cent aqueous ethanoic acid.

Barium chloride, M
Dissolve 244.3 g of barium chloride-2-water in 1 litre of distilled water.

Barium diphenylamine *p*-sulphonate; redox indicator
Dissolve 0.2 g of the indicator in 100 cm^3 of distilled water.

Barium hydroxide, 0.2 M
Dissolve 63 g of barium hydroxide-8-water in 1 litre of distilled water. Filter and protect the solution from atmospheric carbon dioxide.
See also Baryta water.

Barium nitrate, 0.5 M
Dissolve 130.5 g of barium nitrate(V) in 1 litre of distilled water.

Baryta water
Dissolve 32 g of barium hydroxide-8-water in 1 litre of distilled water and allow to settle. Decant the clear liquid and store in a bottle protected with a soda lime tube.

Basic fuchsin; *see* Fuchsin, acid

Bedacryl 122X
This organic solvent can be used to coat the surface of agar plates to inhibit the growth of bacteria. The liquid is poured onto the centre of the plate and allowed to spread over the entire surface. The Petri dish is then left, without its lid, for 12 hours. It will then maintain its original appearance almost indefinitely. Bedacryl 122X is available from suppliers (e.g. Gurr's).

Bench reagents

The following are the percentage composition and approximate molarities of a number of common stock reagents.

Ethanoic (acetic) acid, glacial	99.6 %w/w	17.5 M
Aqueous ammonia, 0.880	30 % NH_3 w/w	15.5 M
Hydrochloric acid, 1.18	36 %w/w	11.6 M
Nitric(V) acid, 1.42	70 %w/w	15.8 M
Sulphuric(VI) acid	97 %w/w	18.0 M

Bench reagents are usually prepared by diluting the above stock reagents to a concentration of 1 M.

For details of other bench reagents, *see* individual entries.

Benecke's solution; for the culture of freshwater algae
In 1 litre of distilled water dissolve 0.5 g calcium nitrate, 0.1 g magnesium sulphate, 0.2 g dipotassium hydrogenphosphate and one drop of a 1 per cent solution of iron (III) chloride.

Benedict's reagent
Dissolve 100 g of anhydrous sodium carbonate and 173 g of sodium 2-hydroxypropane-1,2,3-tricarboxylate (sodium citrate) in about 600 cm³ of distilled water. Filter and dilute the filtrate to 850 cm³.

Dissolve 17.3 g of copper sulphate(VI)-5-water in 100 cm³ distilled water and dilute to 150 cm³.

When both solutions are cool, slowly add the copper sulphate solution to the carbonate/citrate mixture. Dilute, if necessary, to 1 litre.

Benzene-1,3-diol
Dissolve 0.05 g benzene-1,3-diol (resorcinol) in a mixture of 30 cm³ concentrated hydrochloric acid and 70 cm³ distilled water.

Benzene-1,3,5-triol; *see* Phloroglucinol

Benzyl alcohol; *see* Phenylmethanol

Berlese's fluid; for mounting small insects
Dissolve 5 g of glucose in 5 cm³ distilled water. Dissolve 15 g of gum arabic in 20 cm³ of distilled water. Add the glucose to the gum arabic solution and saturate the mixture with 2,2,2-trichloroethanediol (chloral hydrate). Approximately 40–160 g may be required. Heat slowly and stir the mixture. Filter through glass wool.

Best's carmine; *see* Carmine, Best's

Bicarbonate indicator, stock solution
Dissolve 0.2 g thymol blue and 0.1 g cresol red in 20 cm³ of alcohol. Weigh out 0.84 g of sodium hydrogencarbonate (bicarbonate) of ANALAR grade and dissolve it in approximately 900 cm³ of distilled

water. Add the dye solution to the salt solution and dilute to 1 litre with distilled water.

When required for use dilute 25 cm³ of the above stock solution to 250 cm³.

Biological buffers: *see* Buffer, biological

Bismarck brown Y; a general stain for bacteria, cellulose, plant tissue and living organisms
Dissolve 0.3 g of Bismarck brown Y in 100 cm³ of isotonic saline *or* distilled water *or* 95 per cent alcohol.

Bismuth(III) chloride, 0.16 M
Dissolve 53 g of bismuth(III) chloride in 1 litre of dilute hydrochloric acid made by diluting 200 cm³ of concentrated acid to 1 litre.

Bismuth(III) nitrate(V), 0.08 M
Dissolve 40 g of bismuth(III) nitrate(V)-5-water in 1 litre of dilute nitric(V) acid made by diluting 200 cm³ of concentrated acid to 1 litre.

Biuret reagent
a Prepare the following solutions.
A Dissolve 440 g of sodium hydroxide in 1 litre of distilled water.
B Dissolve 2.5 g copper(II) sulphate(VI)-5-water in 1 litre distilled water.
1 cm³ of A is added to the solution being tested, followed by 1 drop of B. A violet colour indicates the presence of protein.
or b Biuret papers. Add 1 per cent copper(II) sulphate(VI)-5-water solution drop by drop to solution A above, while stirring, till a deep blue colour is observed. Immerse filter paper in the reagent, dry it and cut into small strips.

Blackened agar; *see* Agar, blackened

Blood agar base; *see* Agar, blood

Blood, anti-coagulant; for mammalian blood
a Mix saturated solutions of potassium ethanedioate (oxalate) and ammonium ethanedioate (oxalate) in the ratio of 2:3 by volume.
4–5 cm³ of this mixture per 100 cm³ blood will prevent coagulation.
or b Dissolve 10 g sodium 2-hydroxypropane 1,2,3-tricarboxylate (citrate) in 100 cm³ water. This is sufficient to prevent coagulation of 1 litre of fresh blood.

Blood-coagulant; for stopping blood flow during dissection
Dissolve 13.5 g of hydrated iron(III) chloride in 100 cm³ of distilled water acidified with 2 cm³ concentrated hydrochloric acid.

Blood grouping sera
These are often obtainable from hospitals, haematology departments and blood banks, as well as from biological suppliers.
See also Eldon cards

Blue dextran solution

Dissolve 0.1 g of blue dextran 2000 in approximately 75 cm³ of distilled water and dilute to 100 cm³. Add further water to dilute the stock solution as required.

Borax carmine; *see* Carmine, borax

Bouin's solution; fixative

Add 5 cm³ glacial ethanoic (acetic) acid and 25 cm³ 40 per cent methanal (formaldehyde) to 75 cm³ of a saturated aqueous solution of 2,4,6-trinitrophenol (picric acid).

Brain preservative; for hardening brain tissues prior to dissection

Boil 2.5 g chromium(III) potassium(I) sulphate(VI)-12-water (chrome alum) in 77.5 cm³ distilled water until dissolved. Add 5.0 g copper(II) sulphate(VI)-5-water and 5.0 cm³ glacial ethanoic (acetic) acid. Cool and add 10 cm³ of methanal (formaldehyde) solution (4 per cent). Allow 2–3 weeks for the brain tissue to harden.

Brandwein's solution; for the culture of amoeba

In 950 cm³ of distilled water dissolve 1.2 g sodium chloride, 0.03 g potassium chloride, 0.04 g calcium chloride and 0.02 g sodium hydrogen carbonate. Make up to 1 litre with phosphate buffer (pH 6.9–7.0). This stock solution should be diluted 1:10 with distilled water before use.

Brodie's fluid substitute

Dissolve 44 g of sodium bromide, 1 g of stergene and 0.3 g of Erans blue in 1 litre of distilled water.

Bromelin solution; a protease

This may be obtained from fresh (not tinned) pineapple. Peel the fruit and place the flesh in a liquidiser/blender for 10–20 seconds. Squeeze the pulp through several layers of muslin and then filter. This enzyme extract may be substituted in many experiments requiring pancreatin as a source of proteases.

Bromine in tetrachloromethane (Carbon tetrachloride)

Mix 5 cm³ of bromine with 100 cm³ tetrachloromethane. Shake thoroughly.

Bromine water

Add 25 cm³ of bromine to 500 cm³ distilled water in a large bottle. Shake the mixture until the bromine dissolves. (The procedure should be carried out in a fume cupboard and rubber gloves worn).

Bromocresol green; indicator, pH range 3.8–5.4

Add 14.4 cm³ of 0.1 M sodium hydroxide to 1 g of the indicator and dilute to 1 litre with distilled water.

Bromocresol purple; indicator, pH range 5.2–6.8

Add 18.6 cm³ of 0.1 M sodium hydroxide to 1 g of the indicator and dilute to 1 litre with distilled water.

Bromophenol blue; indicator, pH range 3.0–4.6

Add a 1 g of indicator to 15.0 cm² of 0.1 M sodium hydroxide and dilute the solution to 1 litre with distilled water.

Bromothymol blue; indicator, pH range 6.0–7.6

Add 16.0 cm³ of 0.1 M sodium hydroxide to 1 g of the indicator and dilute the solution to 1 litre with distilled water.

Broth, malt extract

Dissolve 17 g viscous malt extract and 3 g peptone in 1 litre warm distilled water. Adjust pH to 5.4 using dilute hydrochloric acid. Autoclave to sterilise.

Broth, nutrient

Mix together 10 g beef extract, 10 g peptone and 5 g sodium chloride in 1 litre of distilled water. Heat to 65 °C and stir to dissolve. Filter.

Broth, peptone

This is available in tablet form from suppliers, e.g. Astell.

Browne's soap; *see* Soap, Browne's

Brucine; for nitrate determination

Dissolve 5 g of brucine in a mixture of 90 cm³ glacial ethanoic (acetic) acid and 10 cm³ distilled water. Add 2 cm³ of pure concentrated sulphuric acid.

Bucholz solution

Mix 80 cm³ of 1 per cent aqueous acid fuchsin with 20 cm³ 1 per cent aqueous light green. (*See* Fuchsin, acid).

Buffer, acetate

Dissolve 5.45 g sodium ethanoate (acetate) in 10 cm³ glacial ethanoic (acetic) acid. Make up to 1 litre with distilled water.

Buffer, biological

These are usually best purchased from a commercial supplier such as BDH.

(*See also* the entries under Buffer, acetate; Buffer formalin; Buffer phosphate; Buffer clinical; and Buffer, general).

Buffer, clinical

These are buffers with specific clinical applications but they are occasionally of more general laboratory use. They are best purchased from a commercial supplier, e.g. BDH.

Buffer, formalin

Add 2 g of 40 per cent methanal (formaldehyde) solution to 100 cm³ of 0.067 M phosphate buffer (*see* Buffer, phosphate). This provides a solution of methanal buffered to pH 7.3.

Buffer, general

The most convenient and rapid method of preparing some buffer solutions is to purchase buffer 'tablets' from an appropriate supplier. One tablet dissolved in a specified volume of distilled water produces a stated pH, usually at 20 °C.

Buffer, phosphate, 0.067 M, pH 7.3

Dissolve 0.18 g potassium dihydrogenphosphate(V) and 0.75 g disodium hydrogenphosphate(V) in 100 cm³ distilled water.

Buffer solutions; for complexometric titrations

a pH 2. Add 21 cm^3 of M hydrochloric acid to 20 cm^3 of M sodium ethanoate (sodium acetate) and dilute to 100 cm^3.

b pH 10. Dissolve 6.75 g of ammonium chloride in 57 cm^3 of 0.880 aqueous ammonia and dilute to 100 cm^3.

Butanedione dioxime; reagent for nickel

Dissolve 10 g in 1 litre of alcohol.

Cadmium chloride, 0.25 M

Dissolve 46 g of cadmium chloride, or 57 g of the 2½-hydrate, in 1 litre of distilled water.

Cadmium nitrate(V), 0.25 M

Dissolve 77 g of cadmium nitrate(V)-4-water in 1 litre of distilled water.

Cadmium sulphate(VI), 0.25 M

Dissolve 70 g of cadmium sulphate(VI)-4-water, or 64 g of $3CdSO_4.8H_2O$ in 1 litre of distilled water.

Calcium chloride

Anhydrous calcium chloride forms a relatively efficient and cheap drying agent. (*See* page 95.)

Calcium chloride, 0.25 M

Dissolve 55 g of calcium chloride-6-water in 1 litre of distilled water.

Calcium hydroxide solution; *see* Lime water

Calcium nitrate(V), 0.25 M

Dissolve 41.5 g of calcium nitrate(V), or 59 g of the 4-hydrate, in 1 litre of distilled water.

Calcium sulphate(VI)

This is not sufficiently soluble to make up a solution in excess of about 0.02 M. Shake an excess of the solid with 1 litre of distilled water and allow to stand. Filter.

Canada balsam

The resin is dissolved in 1,2-dimethylbenzene (*o*-xylene) until an appropriately viscous solution is obtained. As a mounting agent Canada balsam has been largely superseded by synthetic resins such as DPX and Euparol.

Carbol fuchsin; *see* Fuchsin, acid

Carmine-acetic; stain fixative for nuclear material including chromosomes

Boil the following mixture for five minutes in a reflux condenser. Filter when cool.

45 cm^3 glacial ethanoic (acetic) acid
1 g carmine
55 cm^3 distilled water

Carmine, Best's, stock solution

Dissolve 2 g of carmine, 1 g of potassium carbonate and 5 g of potassium chloride in 60 cm^3 of distilled water. Heat if necessary. To the cool solution, add 20 cm^3 of 0.880 aqueous ammonia. Filter and store in a stoppered bottle at 4 °C.

Carmine, borax

Dissolve 3 g of carmine powder in a 4 per cent solution of disodium(I) tetraborate(III)-10-water (borax) in 100 cm^3 distilled water. Simmer for 30 minutes and allow to cool before adding 100 cm^3 of 70 per cent alcohol. Filter the resulting mixture.

Carmine dyed fibrin; to demonstrate digestion

This is available from suppliers (e.g. Gurr's) and can be used as an alternative to congo red dyed fibrin.

Carnoy's fluid

Mix the following reagents in the proportions indicated.

60 cm^3 absolute alcohol
30 cm^3 trichloromethane (chloroform)
10 cm^3 glacial ethanoic (acetic) acid

Celloidin; a form of cellulose nitrate normally stored in the fibrous state under water and used for embedding tissues prior to sectioning with a mechanical microtome. It should be dried on filter paper prior to use

To make a 20 per cent solution, dissolve 120 g dry Celloidin in a mixture of 219 cm^3 absolute alcohol and 261 cm^3 ethoxyethane (ether). This stock solution may be diluted with further volumes of the solvent mixture as required.

Cellosolve; *see* 2-ethoxyethanol

Cellulose tubing; *see* Visking tubing

Cement for perspex jars

Dissolve 5 g of perspex chips in a mixture of 10 cm^3 of glacial ethanoic (acetic) acid and 100 cm^3 of 1,2-dichloroethane.

Chalkley's solution; for the culture of amoebae

In 1 litre of distilled water dissolve 1.0 g sodium chloride, 0.04 g potassium chloride and 0.06 g calcium chloride. This stock solution should be diluted 1:10 with distilled water before use.

Chloral hydrate; *see* 2,2,2-trichloroethanediol

Chlorazol black

Dissolve 1 g chlorazol black E (biological quality) in 100 cm^3 of 70 per cent alcohol.

Chlorine in tetrachloromethane (carbon tetrachloride)

Bubble chlorine gas from a generator or cylinder into tetrachloromethane as required.

Chlorine water

Pass chlorine into distilled water until the water is saturated with gas. Store the solution in a dark bottle.

Chlorobutol; to retard movement of aquatic micro-organisms

Dissolve 1 g of chlorobutol in 125 cm^3 distilled water. Add the reagent dropwise, with five-minute intervals between the drops, to the medium containing the micro-organisms.

Chlorophyll solvent
To extract chlorophyll from leaves prior to running a chromatogram, use 80 per cent aqueous acetone or alcohol.

Chlor-zinc-iodine solution; *see* Schultze's solution

Chorionic gonadotrophin; for induction of breeding in amphibia
This is usually sold in powder form in ampoules with an appropriate volume of solvent in a second ampoule. 100 i.u. and 500 i.u. ampoules are available. Suppliers sometimes require a headteacher's authority to supply chorionic gonadotrophin to a school.

Open the ampoules and transfer the solvent to the powder using a clean, disposable syringe. The solution is ready for use.

Chrom-acetic; fixative for plant tissue
The composition varies with the tissue to be fixed. Relative proportions are shown in Table 7.4.

Chromatographic solvents
a To separate amino acids. Mix together $40 cm^3$ butan-1-ol, $10 cm^3$ glacial ethanoic (acetic) acid and $15 cm^3$ distilled water.
b To separate chlorophyll pigments. Mix together $10 cm^3$ propanone (acetone) and $90 cm^3$ of 40°–60° petroleum ether, **or** mix $12 cm^3$ propanone and $100 cm^3$ 100°–120° petroleum ether.
c To separate DNA bases. Mix $170 cm^3$ propan-2-ol, $3 g$ distilled water and $41 cm^3$ concentrated hydrochloric acid.

Chrom-osmic mixture
Add $1 cm^3$ chromic(VI) acid, $1 cm^3$ of 10 per cent osmic acid and $10 cm^3$ glacial ethanoic (acetic) acid to $100 cm^3$ distilled water.

Chromium(III) chloride, 0.16 M
Dissolve 26 g of anhydrous chromium(III) chloride, or 42.6 g of the 6-hydrate, in 1 litre of distilled water.

Chromium(III) nitrate(V), 0.16 M
Dissolve 40 g of chromium(III) nitrate(V), or 64 g of the 9-hydrate, in 1 litre of distilled water.

Chromium(III) sulphate(VI), 0.16 M
Dissolve 60 g of chromium(III) sulphate(VI), or 106 g of the 15-hydrate, in 1 litre of distilled water.

Clearing agents; *see* Amann's medium, Berlese's fluid, 2,2,2-trichloroethanediol (chloral hydrate), Clove oil, 1,2-dimethylbenzene (xylene), Phenylmethanol, Terpineol, and page 259.

Clinistix
Commercially produced test cards used in screening urine for glucose. Available from pharmaceutical chemists.

Clove oil
This commonly used clearing agent can be purchased directly from suppliers.

Clove oil/glacial ethanoic (acetic) acid
Mix equal volumes of clove oil and glacial ethanoic (acetic) acid.

Cobalt chloride paper; to detect presence of water vapour
Dissolve 1 g of cobalt(II) chloride (cobaltous chloride) crystals in $20 cm^3$ of distilled water. Soak filter paper in the solution and allow to dry. Complete the drying in an oven at 100 °C. Cut the paper into strips and store in a desiccator. It may be necessary to dry in an oven or over a bunsen flame before using the papers.

Cobalt(II) nitrate(V), M
Dissolve 291 g of crystalline cobalt(II) nitrate(V) in 1 litre of distilled water.

Cobalt thiocyanate paper; for detection of humidity changes
Dissolve 5 g of cobalt(II) thiocyanate (cobaltous thiocyanate) in $20 cm^3$ of distilled water. Soak filter paper in the solution and dry as for cobalt chloride paper.

Cobalticyanide papers
Soak strips of filter paper in a solution containing 4 g of potassium hexocyanocobaltate(III) (cobalticyanide) and 1 g of potassium chlorate(VII) in $100 cm^3$ distilled water. Dry the paper overnight in a cool oven.

Colchicine; for pre-treatment of root tips to arrest mitosis
Dissolve 0.2 g of colchicine in $100 cm^3$ of distilled water.
N.B. *Though not a scheduled poison, colchicine is very toxic and should be used only by a teacher or experienced technician with extreme care.*

Tissue	Chromium(VI) oxide (g)	Glacial ethanoic (acetic) acid (cm^3)	Distilled water (cm^3)
Marine algae	1	0.4	400
Algae	2	1	300
Filamentous algae	1	1	150
Fungi and prothalli	2.5	5	72.5
Plant ovaries, root tips	7	10	83
Leaves, woody tissue	10	10	80
General formula	1	1	98

Table 7.4   The composition of chrom-acetic fixatives for various plant tissues

	Mixed in parts by volume			Added to each m^3 of mixture		
	Loam	Peat	Sand	Limestone (kg)	J.I. Base (kg)	Superphosphate of lime (kg)
Seed compost	2	1	1	0.45	—	0.9
Potting composts						
No. 1	7	3	2	0.45	2.4	—
No. 2	7	3	2	0.9	4.7	—
No. 3	7	3	2	1.35	7.0	—

Table 7.5   The composition of John Innes composts

Cole's solution; test for maltose
Dissolve 10 cm^3 of propane-1,2,3-triol (glycerol) in 100 cm^3 of aqueous copper(II) sulphate(VI) (cupric sulphate) solution.

Composts
John Innes composts may be prepared by mixing together the following ingredients in the proportions shown in Table 7.5.
Loam. Sieved through a 1 cm sieve and preferably partially sterilised by heating to 50–100 °C for 10–15 minutes;
Peat. Lightly moistened granular or fibrous peat sieved through a 1 cm sieve;
Sand. Dry, coarse sand;
Limestone. Ground limestone or chalk;
John Innes base. Made by mixing 2 parts horn and hoof meal (13 per cent nitrogen), 2 parts superphosphate of lime (18 per cent phosphorus pentoxide) and 1 part sulphate of potash (48 per cent potassium oxide);
Superphosphate of lime. The first three ingredients should be spread out on a clean, dry surface and the last three ingredients sprinkled over the top to ensure even distribution. The compost should then be turned over several times.

Congo red; indicator, pH range 3.0–5.2
Dissolve 1 g of the indicator in 1 litre of 10 per cent aqueous alcohol.

Congo red; for indicating pH changes during digestion.
Congo red is blue in acid solution, red in alkaline solution
Dissolve 1 g of Congo red in 100 cm^3 of distilled water. (If yeast is to be stained prior to feeding to *Paramecium* in order to observe changes in the pH of the food vacuole as digestion proceeds, add 2 drops of aqueous ammonia to the preparation).

Congo red dyed fibrin; to demonstrate digestion
Place fibrin (obtainable from suppliers) in an aqueous solution of Congo red. Wash away the excess dye with water.

Copper(II) acetate; as a preservative for green colour in plants
Add sufficient copper(II) ethanoate (acetate) to 250 cm^3 of 50 per cent ethanoic (acetic) acid to produce a saturated solution. Add to this 250 cm^3 formalin, then make up to 4 litres with distilled water.

Copper(II) chloride; 0.25 M
Dissolve 43 g of copper(II) chloride-2-water in 1 litre of distilled water.

Copper(II) ethanoate; *see* Copper(II) acetate

Copper(II) nitrate(V), 0.25 M
Dissolve 60 g of copper(II) nitrate-3-water in 1 litre of distilled water.

Copper(II) sulphate(VI), M
Dissolve 249.7 g of copper(II) sulphate(VI) in 1 litre of distilled water.

Copper(II) sulphate-5-water; saturated solution for crystal growing
Dissolve 40 g of copper(II) sulphate-5-water in 100 cm^3 of warm water (about 50 °C).
Allow to cool and seed the saturated solution with a tiny crystal of copper(II) sulphate-5-water. Keep at constant temperature in a covered container.

Copper(II) sulphate(VI), saturated solution; *see* Copper(II) sulphate-5-water, saturated solution

Cornmeal agar; *see* Agar, cornmeal

Cotton blue in lactophenol; *see* Aniline blue

Cresol red; indicator, pH range 7.0–8.8
Add 26.2 cm^3 of 0.1 M sodium hydroxide to 1 g of the indicator and dilute the solution to 1 litre with distilled water.

Crystal violet
Dissolve 3 g crystal violet in 20 cm^3 alcohol. Add 0.8 g ammonium ethanedioate (oxalate) and 80 cm^3 distilled water.

Culture media—agar; *see* Agar

Culture media, *Aspergillus*; *see* Agar, sporulation of *Asperguillus*

Culture media, *Bacillus Subtilis*; *see* Agar, nutrient

Culture media, Coliform bacteria; *see* Agar, MacConkey's

Culture media, *Drosophila*
A large number of recipes exist in the literature, of which the following are recommended.

a Soak 72 g oatmeal in 120 cm³ water. Dissolve 35 g black treacle in 40 cm³ water. Boil 6 g agar in 400 cm³ water. Mix all three together and add a pinch of Nipagin to prevent the growth of moulds. Stir constantly and bring to the boil and continue boiling for 15 minutes. This makes sufficient for 60 specimen tubes or 10 small milk bottles.

b Mix 100 g maize meal, 30 g powdered agar, 26 g dried yeast and 50 g brown sugar with 1600 cm³ water. Stir and bring to the boil and keep boiling till a uniform consistency is obtained. Dissolve 0.5 g Nipagin in 80 cm³ water and stir this into the medium before pouring into the culture tubes. (10 cm³ of 0.5 per cent propanoic may be used instead of the Nipagin).

'Instant' *Drosophila* media are also available from commercial suppliers. These require only the addition of water. *Drosophila* media may be stored in a refrigerator for six weeks and in a deep freeze for six months. They should be sterilised by autoclaving before storage. Instant media should be stored in dry conditions.

Culture media, fern gametophytes; *see* Agar, fern gametophytes

Culture media, houseflies; *see* Agar, housefly

Culture media, moulds; *see* Agar, potato dextrose

Culture media, *Mucor*; *see* Agar, potato dextrose or Agar, malt

Culture media, *Phytophthora infestans*
Use agar, rye meal.
    Established cultures will keep for 4–5 weeks at 20 °C.

Culture media, *Rhizobium*; *See* Agar, yeast

Culture media, *Rhizopus*
Use Agar, potato dextrose or Agar, malt.

Culture solutions (Sachs)
    Complete solution. Dissolve the following in 1 litre of distilled water:
0.25 g calcium sulphate(VI)
0.25 g calcium phosphate(V)
0.25 g magnesium sulphate(VI)
0.08 g sodium chloride
0.70 g potassium nitrate(V)
0.005 g iron(III) chloride
Lacking calcium. Replace calcium sulphate(VI) by 0.2 g potassium sulphate(VI); replace calcium phosphate(V) by 0.71 g sodium phosphate(V).
Lacking iron. Omit iron(III) chloride.
Lacking nitrogen. Replace potassium nitrate(V) by 0.52 g potassium chloride.
Lacking phosphorus. Replace calcium phosphate(V) by 0.16 g calcium nitrate(V).
Lacking sulphur. Replace calcium sulphate(VI) by 0.16 g calcium chloride; replace magnesium sulphate(VI) by 0.21 g magnesium chloride.

Lacking magnesium. Replace magnesium sulphate(VI) by 0.17 g potassium sulphate(VI).
Lacking potassium. Replace potassium nitrate(V) by 0.59 g sodium nitrate(V).

    Many other formulae for culture and nutrient solutions have been devised. For further details see McLean and Cook (1963). *See also* Knop's Solution.

Cupferron reagent; for iron analysis
    Prepare a 6 per cent aqueous solution

Cytosine; chromatogram marker solution
    Dissolve 0.001 g cytosine in 5 cm³ of 0.1 M hydrochloric acid.

Dawson's Alizarin red S; *see* Alizarin red S, Dawson's

Decalcifying fluid; for bone and other calcified tissue prior to sectioning
    a Add 125 g of the tetrasodium salt of ethylenediamine tetraacetic acid (EDTA) to 1 litre of distilled water. Bring to pH 7 by the cautious addition of concentrated aqueous sodium hydroxide.
    or b Mix 40 cm³ of 10 per cent aqueous nitric(V) acid with 30 cm³ of alcohol and 30 cm³ of very dilute chromic(VI) acid. After the tissue has been decalcified, it should be transferred to 70 per cent alcohol.

Delafield's haematoxylin; *see* Haematoxylin, Delafield's

Detergent 'bubble raft' solution; for 'illustration' of holes, close packing and dislocations of atoms in crystals
    Mix 3 cm³ of teepol or washing-up liquid, 25 cm³ of propane-1,2,3-triol (glycerol) and 100 cm³ of distilled water. A 5 mm bore glass tube is drawn into a fine jet. Bubbles are blown by connecting the jet to the gas tap, using a clip to adjust the rate of flow of gas into the above solution.

Developer, photographic; *see* Photographic developer

Diazine green; *see* Janus green B

1,4-Dichlorobenzene; for pre-treatment of root tips to inhibit chromosome clumping prior to fixation
    Dissolve approximately 5 g 1,4-dichlorobenzene (*para*-dichlorobenzene) in 500 cm³ of distilled water. Allow to stand overnight at 60 °C and filter off undissolved material.

Dichlorofluorescein; adsorption indicator
    Dissolve 0.1 g in a mixture of 70 cm³ alcohol (95 per cent) and 30 cm³ distilled water.
    Alternatively, prepare a 0.1 per cent aqueous solution of the sodium salt.

2,6-Dichlorophenolindophenol (DCPIP); test for ascorbic acid
    Dissolve 0.25 g of 2,6-dichlorophenolindophenol in 250 cm³ of distilled water.

1,2-Dimethylbenzene (Xylene)
    This is perhaps the most widely used clearing agent for microscopical work and can be purchased directly from suppliers.

Dimethyl glyoxime; see Butanedione dioxime

2,4-Dinitrophenylhydrazine
Dissolve 2 g in 15 cm³ concentrated sulphuric acid. Add 150 cm³ ethanol, dilute to 500 cm³ with distilled water. Cool and filter.

Diphenylamine; for nitrate(V) test
Dissolve 0.5 g diphenylamine in 100 cm³ of concentrated sulphuric acid diluted with 20 cm³ of distilled water.

Disinfectants
a For sterilising clinical thermometers. Cetrimide solution BNF or chloroxylenol solution BPC.
b For sterilising cereal grains (prior to use on starch agar plates). Dissolve 2 g mercury(II) chloride (mercuric chloride) in 100 cm³ alcohol.
c For sterilisation of glassware after the growth of bacterial or fungal colonies. See page 280.

Disodium hydrogenphosphate(V), 0.1 M
Dissolve 53.8 g of disodium hydrogenphosphate(V)-12-water in 1 litre of distilled water.

DNA, chromatographic solvents for separation of bases; see Chromatographic solvents

DPX; mounting medium
This may be obtained commercially or made by mixing 100 cm³ 1,2-dimethylbenzene (xylene) with 18.75 cm³ tri-p-tolyl phosphate. Add 25 g Distrene 80 (a polymer of polystyrene).

Dried milk medium; for the culture of *Paramecium* and other ciliates
Add a pinch of dried skimmed milk to 250 cm³ of boiled, cooled and filtered pond water.

Dried yeast medium; for the culture of *Paramecium* and other ciliates
Add about 2 g dried yeast to 250 cm³ pond water. Mix well and allow to stand for several hours before inoculating.

Drosophila **medium;** see Culture media, *Drosophila*

Drying agents; see Calcium chloride (anhydrous), Lithium chloride (saturated), Silica gel. See also page 95.

EDTA; 0.1 M
Dissolve 37.21 g of the disodium salt, or 41.62 g of the tetrasodium salt, in 1000 cm³ distilled water.

Egg yolk medium; for culture of ciliates
Thoroughly mix 0.5 g hard-boiled egg yolk with a little boiled pond water. Make up to 500 cm³ with boiled pond water before inoculation.

Ehrlich's haematoxylin; see Haematoxylin, Ehrlich's

Eldon cards; for identifying blood groups
These cards, impregnated with appropriate sera for ABO and Rh systems, are obtainable from suppliers.

Embalming fluid; for maintaining frogs flexible for dissection
Mix 5 cm³ of 40 per cent aqueous methanal (formaldehyde) with 100 cm³ of alcohol (60 per cent). Inject into the body cavity of the animal and immerse the animal in the solution.

Eosin; adsorption indicator
Dissolve 0.1 g of the sodium salt in 100 cm³ of distilled water.

Eosin; for demonstration of plant conducting tissues
Dissolve eosin in water till a solution the colour of red ink is obtained. Eosin is toxic to plant tissues so material must not be left in the dye for more than a few hours prior to examination. Red ink is a suitable substitute and is less toxic, as is methylene blue.

Eosin Y (Yellowish); an acid dye suitable as a cytoplasmic stain for animal and plant histology
Alcoholic solution. Dissolve 1 g of eosin Y in 99 cm³ of alcohol (75 per cent).
Aqueous solution. Dissolve 1 g of eosin Y in 99 cm³ of distilled water.
As counterstain after basic dye. Dissolve 0.5 g of eosin Y in 25 cm³ of alcohol (95 per cent). Add 75 cm³ of distilled water.
As counterstain to Mayer's haemalum. Make up a saturated solution in 90 per cent alcohol.

Epsom salts; for narcotising animals prior to killing and fixation. (See pages 269–271)
Experience shows that crude Epsom salts are preferable to magnesium sulphate itself for this purpose.

Eriochrome black T; complexometric indicator
Grind 0.1 g of the solid with 10 g of A.R. grade sodium chloride. Store in a tightly stoppered bottle.

Ethanoic acid, M; see Acetic acid

Ethanoic-alcohol; see Alcohol, acetic

Ethanoic-carmine; see Carmine-acetic

Ethanol; see Alcohol

2-Ethoxyethanol (Cellosolve); a dehydrating agent miscible with water, ethanol, clove oil or xylene
This is obtainable from suppliers and may be used in place of ethanol for dehydrating sections or whole mounts.

Ethylene glycol monoethyl ether; see 2-ethoxyethanol

Ethyl *m*-amino benzoate; for use as anaesthetic for fish and amphibia; see Anaesthetics

Ethyl 3-aminobenzenecarboxylate; for use as anaesthetic for fish and amphibia; see Anaesthetics

Euparal
A resinous mounting medium used as an alternative to Canada balsam. Sections to be mounted in Euparal do not require complete dehydration or clearing and may be transferred to the mounting medium from 95 per cent alcohol. It also 'yellows' less than Canada balsam. Euparal is obtainable from biological suppliers.

FAA; *see* Formal-acetic-alcohol

Fabil, Fuchsin, aniline blue and iodine in lactophenol
Prepare the following solutions.
A Dissolve 0.4 g aniline blue in 80 cm³ of lactophenol.
B Dissolve 0.1 g basic fuchsin in 20 cm³ of lactophenol.
C Dissolve 0.3 g iodine and 0.6 g potassium iodide in 100 cm³ of lactophenol.
Mix all of Solutions A, B and C. Allow to stand overnight and then filter. The solution should keep indefinitely. (*See* note under Fuchsin, acid)

Fast green; counterstain to safranin
Mix 50 cm³ absolute alcohol with 50 cm³ clove oil. Add 0.5 g fast green FCF and dissolve.

Fehling's solutions
Solution A. Dissolve 69.2 g copper(II) sulphate crystals in distilled water and make up to 1 litre. If the solution is cloudy, add 1–2 drops of concentrated sulphuric(VI) acid.
Solution B. Dissolve 154 g sodium hydroxide and 350 g sodium potassium 2,3,dihydroxybutane-1,4-dioate (tartrate) in 1 litre of distilled water.
For storage, keep solutions A and B separate. Mix equal volumes of A and B before use.

Fern gametophyte agar; *see* Agar, fern gametophyte

Ferric alum; *see* Iron alum

Ferric chloride; *see* Iron(III) chloride

Ferric nitrate; *see* Iron(III) nitrate(V)

Ferric sulphate; *see* Iron(III) sulphate(VI)

Ferrous ammonium sulphate; *see* Iron(II) ammonium sulphate(VI)

Ferrous sulphate; *see* Iron(II) sulphate(VI)

Feulgen stain; for staining chromosomes
Pour 100 cm³ of boiling distilled water over 0.5 g basic fuchsin. Shake well. Cool to 50 °C. Filter. Add 15 cm³ of hydrochloric acid (M) and 1.5 g of potassium metabisulphite (anhydrous). Leave for 24 h in a stoppered bottle in the dark to bleach. Then add 0.25 g animal charcoal. Shake well and filter rapidly. The filtrate should be pale straw coloured and not red or purple. Store in a tightly stoppered dark bottle in a refrigerator. The stain will keep for approximately six months. (*See* note under Fuchsin, acid)

Fibrin dyed; for demonstration of digestion by trypsin
Fibrin may be obtained commercially and dyed in an aqueous solution of Congo red. Excess dye should be washed away with distilled water.
Fibrin-carmine is available in prepared form, but is expensive.

Fixatives, for plant tissue; *see* Chrom-acetic; Formaldehyde, buffered; Alcohol, acetic; Bouin's solution; Formal-acetic-alcohol; Formol saline; Alcohol-methanal

Fluorescein solution; for observation of rays of light in water
Dissolve 1 g of fluorescein in 1 litre of water. Add between 0.5 cm³ and 5 cm³ of this stock solution to a litre of water to make up the final solution. It may prove helpful to dissolve the 1 g of fluorescein in a little alcohol before making the stock solution.

Fluorescein solution; adsorption indicator
Dissolve 0.1 g of the sodium salt in 100 cm³ distilled water.

Formal-acetic-alcohol; fixative for botanical material where critical fixation of cell contents is *not* required. Used especially for algae.
Mix 3–5 cm³ of glacial ethanoic (acetic) acid with 90 cm³ of alcohol (70 per cent or 50 per cent for delicate tissues). Add 5–7 cm³ of methanal (formaldehyde) solution (40 per cent).

Formaldehyde, buffered; for use as a fixative for animal tissues
Dissolve 4 g sodium dihydrogenphosphate(V) and 6.5 g disodium hydrogenphosphate(V) in 900 cm³ distilled water. Add 100 cm³ methanal (formaldehyde) solution (40 per cent).

Formaldehyde solution
Available commercially as a 40 per cent solution (w/v) of methanal in water known also as formalin or formol solution. Concentrations of methanal solution should be expressed in terms of their methanal content. The 40 per cent solution diluted with nine times its volume of water should thus be called 4 per cent methanal and *not* 10 per cent formalin or 10 per cent formol solution. The use of methanal solution as a fixative is discussed on page 252 and as a preservative for animal tissues on pages 269–273.

Formalin
A commercial name for a saturated aqueous solution of methanal (formaldehyde) containing approximately 40 per cent w/v methanal.
It is normal to express the concentration of methanal solutions in terms of their methanal content. Thus 'formalin' diluted with nine times its volume of water is called 4 per cent methanal and *not* 10 per cent formalin.

Formalin, neutral
Add a little disodium tetraborate(III) (borax) to the appropriate concentration of methanal (formaldehyde). Check with litmus paper to ensure that a neutral or alkaline pH is obtained.

Formol
A commercial name for a saturated aqueous solution of methanal (formaldehyde) containing approximately 40 per cent w/v methanal.
See Formaldehyde

Formol-acetic-alcohol; *see* Formal-acetic-alcohol

Formol alcohol; *see* Alcohol-methanal

Formol-saline; a fixative for animal tissues

Add 0.9 g sodium chloride and 10 cm³ of aqueous methanal (formalin) to 90 cm³ of distilled water.

After fixation—a maximum of 3–4 days—store in alcohol (70 per cent).

Freezing mixtures; all parts by weight

−12 °C	Ammonium chloride	5 parts
	Potassium nitrate(V)	5 parts
	Water	16 parts
−15 °C	Ammonium nitrate(V)	1 part
	Water	1 part
−18 °C	Ice	2 parts
	Salt	1 part
−22 °C	Ammonium nitrate(V)	1 part
	Sodium carbonate-10-water	1 part
	Water	1 part
−40 °C	Sodium sulphate(VI)-10-water	6 parts
	Ammonium nitrate(V)	5 parts
	Molar nitric(V) acid	5 parts

Frozen pea agar; *see* Agar, frozen pea

Fuchsin, acid; general stain

Fuchsin (also known as Fuchsine, Magenta and Rosaniline) is known to be carcinogenic. Its manufacture is carefully controlled and, although its use in schools is not prohibited, it is undesirable.

Gatenby's fluid; for removal of albumen from amphibian eggs

Mix together 100 cm³ of chromic(VI) acid (1 per cent), 6 cm³ of concentrated nitric(V) acid and 100 cm³ of potassium dichromate(VI) (2 per cent).

Gelatin solution; for demonstration of xylem in plant stems

Dissolve 5 g gelatin in 100 cm³ hot water. Add sufficient safranin to dye the solution deeply. Cool to 45 °C before placing young seedlings in the solution for a few minutes. Cut the stems of the seedlings under the solution, which will then enter the xylem before setting.

Germination agar; *see* Agar, germination of small seeds

Gibberellic acid paste

Dissolve 1.0 mg of gibberellic acid in 0.1 cm³ of absolute alcohol. Carefully mix with 10 g of lanolin warmed to permit thorough mixing. This preparation contains 100 ppm gibberellic acid. Dilute with lanolin to give weaker concentrations.

Gibberellic acid solution; as a stimulant to plant growth

Dissolve 0.1 g gibberellic acid in 2 cm³ alcohol. Dilute to 1 litre with distilled water. This gives a solution of 100 ppm gibberellic acid. Dilute further as required.

Gilson's fixative

Mix together 42 cm³ 95 per cent alcohol, 18 cm³ glacial ethanoic (acetic) acid, 2 cm³ nitric(V) acid, 11 cm³ of a saturated aqueous solution of mercury(II) chloride and 60 cm³ distilled water

Glucose-1-phosphate(V)

This is obtainable from commercial suppliers in 1 g ampoules. It should be stored in solid form in a refrigerator and any solutions made up for use within 1 hour.

Glucose-1-phosphate(V) agar; *see* Agar, glucose-1-phosphate

Glycerin of pepsin, BP; *see* Pepsin

Glycerol albumen; *see* Albumen, Mayer's

Glycerol; for use as a temporary mountant

Mix 50 cm³ of propane-1,2,3-triol (glycerol) with 50 cm³ of distilled water. Add 1 cm³ of a concentrated solution of thymol.

Glycerol, acid; for use as a temporary botanical mountant

Mix 50 cm³ propane-1,2,3-triol (glycerol) with 50 cm³ distilled water. Add 5 cm³ concentrated hydrochloric acid.

Glycerol jelly; mounting medium

Soak 10 g gelatin in 60 cm³ distilled water for 2 h. Add 70 cm³ propane-1,2,3-triol (glycerol) and 0.25 g phenol and heat on a water bath stirring continuously until smooth. Store in a refrigerator. Apply melted to a warm slide.

Gonadotrophin hormone; *see* Chorionic gonadotrophin

Gram's iodine; *see* Iodine solution, aqueous

Gray and Wess' medium; a low refractive index slide mounting medium

Mix together 2 g polyethenol (polyvinyl alcohol), 7 cm³ 70 per cent propanone (acetone), 5 cm³ propane-1,2,3-triol (glycerol), 5 cm³ of 2-hydroxypropanoic (lactic) acid and 10 cm³ distilled water.

Guanine; chromatogram marker solution

Dissolve 0.001 g guanine in 5 cm³ 0.1 M hydrochloric acid.

Haemalum, Mayer's acid; a general stain for animal tissues which may be counterstained with eosin

Dissolve 1 g haematoxylin, 0.2 g sodium iodate(V) and 50 g aluminium(III) potassium sulphate(VI)-12-water (potassium alum) in 1 litre distilled water. Shake until the solution is blue-violet, then add 50 g 2,2,2-trichloro-ethanediol (chloral hydrate) and 1 g propane-1,2,3-tricarboxylic acid (citric acid) and store in a dark glass bottle.

Haemalum, Mayer's

Dissolve 1 g haematoxylin in 1 litre distilled water. Add 50 g potassium aluminium sulphate(VI)-12-water and 0.2 g sodium iodate(V). Dissolve and filter.

Haematoxylin, Delafield's; a basic dye much used as a general stain for nuclei or counterstained with safranin or eosin

Add 4 g of haematoxylin to 25 cm^3 of absolute alcohol. Add this solution to 400 cm^3 of an aqueous saturated solution of iron(III) ammonium sulphate(VI)-12-water (ammonium alum).

Allow to stand for four days exposed to light in a flask plugged with cotton wool. Filter. Add 100 cm^3 propane-1,2,3-triol (glycerol) and 100 cm^3 methanol. Place in a warm, well lit situation for six weeks.

Haematoxylin, Ehrlich's; a basic dye used for general purposes in animal histology

Dissolve 2 g of haematoxylin in 100 cm^3 absolute alcohol. Add 100 cm^3 distilled water and 100 cm^3 propane-1,2,3-triol (glycerol), 10 cm^3 glacial ethanoic (acetic) acid and excess aluminium(III) potassium sulphate(VI). Leave in a stoppered bottle in sunlight. Every few days, remove the stopper for a few minutes, replace and shake well. Continue this for several weeks.

Hay infusion; for culture of protozoans

Boil 10 g chopped hay in 1 litre distilled or rain water. Allow to stand for 24 h in a flask stoppered with cotton wool. Dilute with 5 parts boiled water before inoculating.

Hayem's solution; as a dilutant for blood prior to making red cell counts

Add 0.25 g mercury(II) chloride, 2.5 g sodium sulphate(VI) and 0.5 g sodium chloride to 100 cm^3 distilled water.

Heidenhain's iron alum haematoxylin; *see* Iron alum haematoxylin, Heidenhain's

Hessler's fluid; for preservation of the colour of fruits

Dissolve 50 g zinc chloride in 1 litre water. Add 25 cm^3 methanal solution (formalin) and 25 cm^3 propane-1,2,3-triol (glycerol). If a sediment appears, decant.

Hexacyanocobaltate(III) paper; *see* Cobalticyanide papers

Holfreter's solution; for rearing amphibians

Dissolve 3.5 g sodium chloride, 0.05 g potassium chloride, 0.10 g calcium chloride, and 0.02 g sodium hydrogencarbonate in 1 litre distilled water.

Housefly agar; *see* Agar, housefly

Humidity regulating solutions

Sulphuric acid solutions. The following weights of sulphuric acid in 100 g distilled water will give relative humidities of the value indicated at 25 °C.

Weight of sulphuric acid in g per 100 g of water	Relative humidity at 25 °C
0	100
11.0	95
17.9	90
22.9	85

Weight	Relative humidity
26.8	80
30.1	75
33.1	70
35.8	65
38.4	60
40.8	55
43.1	50
45.4	45
47.7	40
50.0	35
52.5	30
55.0	25
57.7	20
60.8	15
64.5	10
69.4	5

Saturated salt solutions. The following saturated solutions will give relative humidities of the values indicated at 25 °C in an enclosed space.

Salt solution	Relative humidity per cent
Potassium dichromate	98.0
Potassium nitrate	92.5
Barium chloride	90.2
Potassium chloride	84.2
Potassium bromide	80.7
Sodium chloride	75.3
Sodium nitrate	73.8
Strontium chloride	70.8
Sodium bromide	57.7
Magnesium nitrate	52.9
Lithium nitrate	47.1
Potassium carbonate	42.8
Magnesium chloride	33.0
Potassium acetate	22.5
Lithium chloride	11.1
Sodium hydroxide	7.0

Glycerol and water mixtures. The following percentages of propane-1,2,3-triol (glycerol) in water (w/w) will give relative humidities indicated.

Relative humidity per cent	Per cent glycerol/ water (weight/weight)
90	33
80	51
70	64
60	72
50	79
40	84
30	89
20	92
10	95

Hydrochloric acid, M

Take 89.05 cm^3 of 35 per cent hydrochloric acid and add carefully to 750 cm^3 of deionised water in a large measuring cylinder. Dilute to 1000 cm^3.

Hydrochloric acid; for use in Feulgen technique for staining chromosomes

Carefully add $10\,cm^3$ concentrated hydrochloric acid to $90\,cm^3$ distilled water.

Hydrochloric acid, alcoholic; for pre-treatment of plant tissue for maceration

Mix $25\,cm^3$ of $2\,M$ hydrochloric acid with $75\,cm^3$ of alcohol (95 per cent).

Hydrogencarbonate indicator; *see* Bicarbonate indicator

Hydrogen peroxide, aqueous

This is usually purchased as '20 volume' peroxide which contains approximately 6 per cent (w/v) of hydrogen peroxide in aqueous solution. It should be stored in a cool, dark place.

Indicator paper

Various forms of absorbent paper impregnated with appropriate indicators (pH, humidity, etc.) are available from suppliers, e.g. Whatman, M & B. They are available both in booklet and roll form.

Indicators; *see under* individual indicators; phenolphthalein, fluorescein, methyl orange, etc., *see also* page 113.

Indigo-carmine solution, reduced; for the detection of oxygen

Dissolve $0.1\,g$ indigo-carmine (5,5-indigo disulphonic acid) in $100\,cm^3$ boiled and cooled distilled water. Reduce this solution to a yellowish-green colour by the cautious addition of freshly prepared 5 per cent sodium sulphinate (sodium hyposulphite).

Indole acetic acid paste

Dissolve $10\,mg$ of crystallised indole-2-acetic acid in $2\,cm^3$ of absolute alcohol. Add drop by drop to $100\,cm^3$ of distilled water. Warm $20\,g$ of lanolin over a water bath. When it is just melted, remove from heat and stir in $10\,cm^3$ of the indole acetic acid solution. Mix thoroughly. This paste should keep for months in a refrigerator.

Indole acetic acid solution

Dissolve $0.1\,g$ of indole-2-acetic acid in $2\,cm^3$ of absolute alcohol. Add $900\,cm^3$ of distilled water and warm to $80\,°C$ for five minutes to evaporate the alcohol. Make up to 1 litre with distilled water. This solution contains $100\,ppm$. Dilute further if required. The solution should be kept in a refrigerator for not more than 14 days.

Industrial spirit; *see* Alcohol

Injection medium; as a preservative which does not overharden muscle tissue

Mix together $50\,g$ phenol, $50\,cm^3$ propane-1,2,3-triol (glycerol) and $50\,cm^3$ aqueous methanal (formalin). Make up to 1 litre with distilled water.

Iodine solution, aqueous

For an approximately $0.05\,M$ solution, dissolve $12.7\,g$ of iodine in $100\,cm^3$ of distilled water containing $20\,g$ of potassium iodide. Dilute to 1 litre.

For Gram's iodine, dilute to 3 litres with distilled water.

Iodine solution, alcoholic

Dissolve $1\,g$ of iodine and $1\,g$ of potassium iodide in $100\,cm^3$ of aqueous alcohol (70 per cent).

Ionagar

This special agar is water clear when set. It is available from Astell. (*See* page 248)

Iron alum; for use in Heidenhain's iron alum haematoxylin staining technique

Dissolve $3\,g$ iron(III) ammonium sulphate(VI)-12-water (iron alum) in $100\,cm^3$ distilled water.

Iron alum haematoxylin, Heidenhain's

Dissolve $0.5\,g$ haematoxylin in $10\,cm^3$ absolute alcohol. Add $90\,cm^3$ distilled water and leave for a few days to allow the stain to ripen.

Iron(III) alum; indicator

Dissolve $10\,g$ of crystals in $100\,cm^3$ of distilled water acidified with 4 drops of concentrated sulphuric(VI) acid.

Iron(II) ammonium sulphate(VI), $0.5\,M$

Dissolve $196\,g$ of iron(II) ammonium sulphate(VI)-6-water in water containing $10\,cm^3$ of concentrated sulphuric(VI) acid and dilute to 1 litre with water.

Iron(III) ammonium sulphate(VI)-12-water; *see* Iron(III) alum

Iron(III) chloride, $0.5\,M$

Dissolve $135\,g$ of iron(III) chloride-6-water in distilled water containing $20\,cm^3$ of concentrated hydrochloric acid. Dilute to 1 litre.

Iron(III) nitrate(V), $0.25\,M$

Dissolve $101\,g$ of iron(III) nitrate-9-water in 1 litre of distilled water.

Iron(II) sulphate(VI), $0.5\,M$

Dissolve $139\,g$ of iron(II) sulphate(VI)-7-water in water containing $10\,cm^3$ of concentrated sulphuric(VI) acid and dilute to 1 litre with distilled water.

Iron(II) sulphate(VI) solution; for estimating oxygen in water

Dissolve $2.2\,g$ of iron(II) sulphate(VI)-7-water in a few cm^3 of 1 per cent sulphuric acid. Make up to 1 litre with distilled water.

Iron(III) sulphate(VI), $0.25\,M$

Dissolve $140.5\,g$ of iron(III) sulphate(VI)-9-water in distilled water containing $100\,cm^3$ of concentrated sulphuric(VI) acid and dilute to 1 litre.

Isotonic saline; *see* Saline, isotonic

Janus green B; a basic vital stain used especially for fungi and protozoa

Dissolve 0.1 g Janus green B in 1 litre of isotonic saline.

Kerosene

Coloured kerosene (domestic paraffin) can be used as manometer fluid. It may be prepared by shaking the kerosene with Sudan III or Sudan blue to give as deep a colour as possible.

Kleb's solution (modified); for the culture of *Euglena*

To 1 litre of distilled water add 0.25 g potassium nitrate, 0.25 g magnesium sulphate, 0.25 g potassium dihydrogenphosphate and 1.0 g calcium nitrate. Dissolve and then add 0.01 g bacto-tryptophane broth powder. For use in the culture of *Euglena*, dilute with 9 parts distilled water and add 20 grains of rice—which have been boiled for 5 minutes—to every litre of the diluted solution.

Knop's solution; for algae

Solution A. Dissolve 1 g magnesium sulphate(VI), 1 g potassium nitrate(V) and 1 g dipotassium hydrogenphosphate(V) in 1 litre distilled water.

Solution B. Dissolve 3 g calcium nitrate(V) in 1 litre distilled water.

Mix by adding solution **A** to solution **B**. The resulting solution may be diluted with up to 4 parts of water.

Lactophenol

Mix in equal parts *by weight* phenol (crystals), propane-1,2,3-triol (glycerol), 2-hydroxypropanoic acid (lactic acid), and distilled water.

See also Amann's medium

Lead(II) acetate, 0.5 M

Dissolve 189 g of lead(II) ethanoate-3-water (lead acetate) in 1 litre of distilled water. Add sufficient acetic (ethanoic) acid to clear the solution.

Lead(II) acetate, saturated solution; used in electrolytic deposition of lead crystals

Dissolve 30 g of lead(II) ethanoate-3-water in 100 cm³ of distilled water. Add a few drops of ethanoic acid (acetic acid).

Place two platinum or carbon (pencil leads) electrodes in the solution and use a 12 V d.c. supply to pass a current (less than 50 mA).

Lead(II) ethanoate; *see* Lead(II) acetate

Leather jacket extractor

Mix 100 g sodium *cis*-octadec-9-enoate (oleate) with 1 litre of hot water. Mix 1 part of this solution with 1 part of Jeyes' Fluid. Slowly stir in 4 parts 1,2-dichlorobenzene until a thick cream is obtained. Add a little more Jeyes' Fluid to produce a flowing consistency. The mixture can be stored in this condition.

To use: Mix 5 parts of the above with 1 part of Jeyes' Fluid. Stir 30 cm³ of this mixture into 5 litres of water. This should be sufficient to extract most of the leather jackets from 1 m² of grassland if applied with a watering can.

Leishman's stain; for blood and blood parasites

To make the dry stain. Warm together for 12 hours, 1 g methylene blue, 100 cm³ distilled water and 200 cm³ 0.5 per cent aqueous solution of sodium carbonate. Allow to stand for 10 days. Add 300 cm³ of 0.1 per cent aqueous eosin. Stand for 12 hours. Filter and wash the precipitate with distilled water until the washings are colourless. Dry.

To make a stock solution. Dissolve 0.15 g of dry Leishman's stain (either purchased from suppliers or made as above) in 100 cm³ methanol.

Light green; an acid stain used for botanical purposes, often counterstained with haematoxylin or safranin

A **saturated solution** is prepared by standing excess light green SF yellowish in alcohol (90 per cent). Filter off solid material.

Light green in clove oil. Dissolve 0.2 g light green SF yellowish in 50 cm³ of absolute alcohol. Add 50 cm³ of clove oil.

See also Bucholz solution.

Lime water

Add 5 g of calcium hydroxide to 2 litres of distilled water in a Winchester. Shake periodically over a 24-hour period. After settling, the lime water is ready for use.

Lipman's solution; *see* Saline solutions for bacteria

Litmus

Dissolve 1 g of litmus powder in 1 litre of distilled water.

Locke's solution; *see* Saline solutions for bacteria

Lubricant; for stop-cocks

Use propane-1,2,3-triol (glycerol) to prevent sticking of ground-glass parts.

Lucas reagent

Dissolve 136 g anhydrous zinc chloride in 105 cm³ concentrated hydrochloric acid.

MacConkey's agar; *see* Agar, MacConkey's

Macerating fluids

For botanical work, *see* pages 256–7.

For bones. Dissolve 150 g sodium carbonate and 100 g calcium oxochlorate(I) (calcium hypochlorite) in 1 litre distilled water.

Magnesium chloride, 0.25 M

Dissolve 51 g of magnesium chloride-6-water in 1 litre of distilled water.

Magnesium nitrate(V), 0.25 M
Dissolve 63.4 g of magnesium nitrate(V)-6-water in 1 litre of distilled water.

Magnesium sulphate(VI), 0.5 M
Dissolve 124 g of magnesium sulphate(VI)-7-water in 1 litre of distilled water.
See also Epsom salts

Malachite green
Dissolve 2 g malachite green in 100 cm³ distilled water.

Mallory's triple stain
Make up the following three solutions and store them in separate bottles.
Solution A. Dissolve 0.1 g acid fuchsin in 100 cm³ distilled water.
Solution B. Dissolve 1 g phosphomolybdic acid in 100 cm³ distilled water.
Solution C. Dissolve 0.5 g aniline blue WS, 2 g orange G and 2 g ethanedioic (oxalic) acid in 100 cm³ distilled water.

Malt agar; *see* Agar, malt

Malt extract broth; *see* Broth, malt extract

Manganese(II) chloride, 0.25 M
Dissolve 48.5 g of manganese(II) chloride-4-water in 1 litre of distilled water.

Manganese(II) nitrate(V), 0.25 M
Dissolve 72 g of manganese(II) nitrate(V)-6-water in 1 litre of distilled water.

Manganese(II) sulphate(VI), 0.25 M
Dissolve 56 g of manganese(II) sulphate(VI)-4-water in 1 litre of distilled water.

Manganous salts; *see under* Manganese(II)

Mann's methyl blue and eosin; *see* Methyl blue and eosin, Mann's

Manometer fluid
a Dissolve a little Sudan blue or Sudan III in kerosene (domestic paraffin).
or b Add a trace of liquid detergent to red ink.

Mayer's acid haemalum; *see* Haemalum, Mayer's acid

Mayer's albumen; *see* Albumen, Mayer's

Mercuric salts; *see under* Mercury(II)

Mercurous salts; *see under* Mercury(I)

Mercury(II) chloride, 0.1 M
Dissolve 27 g of mercury(II) chloride in water and dilute to 1 litre.

Mercury(II) chloride; for surface sterilisation of seeds
Dissolve 0.1 g mercury(II) chloride in 100 cm³ distilled water. Immerse dry seeds in this solution for 1 minute, then wash well with several changes of sterile water.
    N.B. *Mercury(II) chloride is exceedingly poisonous and great care should be taken in using it. 1 per cent sodium oxochlorate(I) (hypochlorite) solution or Milton may be used instead. In this case, the seeds should be immersed for 3 minutes before rinsing with sterile water.*

Mercury(I) nitrate(V)
Prepare solutions of the required concentration using 1 part of nitric(V) acid to 1 part of mercury(I) nitrate and 20 parts of distilled water.

Mercury(II) nitrate, 0.25 M
Dissolve 81 g of mercury(II) nitrate(V) in 1 litre of distilled water.

Methanal; *see under* Formaldehyde, Formalin and Formol

Methanol; *see* Alcohol

Methylated spirit; *see* Alcohol

Methyl blue and eosin, Mann's
Add 35 cm³ of a 1 per cent aqueous solution of methyl blue and 35 cm³ of a 1 per cent aqueous solution of eosin Y to 100 cm³ distilled water.

Methyl cellulose; for retarding movements of protozoa
Dissolve 10 g methyl cellulose in 100 cm³ distilled water. A ring of this solution is placed on the microscope slide and a drop of the protozoan culture placed in the centre. Care should be taken to avoid getting the solution on microscope objectives as, once dried, it is difficult to remove. Commercial wallpaper-paste (e.g. Polycell) can be used instead of methyl cellulose. Pastes containing fungicides should be avoided.

Methylene blue; a nuclear and general stain
    For living organisms. Dissolve 1 g methylene blue and 0.6 g sodium chloride in 100 cm³ distilled water.
    For dead tissue. Dissolve 0.3 g methylene blue in 30 cm³ of alcohol (95 per cent). Add 100 cm³ distilled water.

Methyl green pyronin; stain for nucleic acids, DNA—green; RNA—red
Dissolve 1 g methyl green pyronin stain in 100 cm³ distilled water.

Methyl orange; indicator, pH range 3.1–4.4
Dissolve 0.5 g of methyl orange in 1 litre of distilled water. Filter before bottling the reagent.
    Alternatively, dissolve 0.5 g of the sodium salt in 1 litre of water and add 15.2 cm³ of 0.1 M hydrochloric acid. Filter, if necessary, before bottling the reagent.

Methyl orange (screened); indicator
Dissolve 1 g of methyl orange and 1.4 g of xylene cyanol FF in a mixture of 250 cm³ absolute alcohol and 250 cm³ distilled water.

Methyl red; indicator, pH range 4.4–6.2
Dissolve 1 g of the free acid in 1 litre of hot water or in 600 cm³ of industrial spirit and dilute to 1 litre with distilled water.

Millon's reagent; test for protein
Place 1 cm³ mercury in a small beaker or evaporating basin in a fume cupboard. Add 9 cm³ concentrated nitric(V) acid. When the reaction is complete, add 10 cm³ of distilled water.

N.B. *This is a hazardous operation and teachers may prefer to purchase Millon's reagent ready prepared. Extreme caution must also be exercised when using Millon's reagent.*

Millon's reagent (Cole's modification)
N.B. *See warning note above.*
Solution A. Pour 100 cm³ concentrated sulphuric(VI) acid into about 800 cm³ distilled water, cooling the flask under the tap. Grind 100 g mercury(II) sulphate(VI) in a mortar with successive portions of the diluted acid, pouring off the resulting solution into a 1 litre flask. Make up to 1 litre and filter at the pump.
Solution B. Dissolve 5 g sodium nitrate(III) (nitrite) in 500 cm³ distilled water. This solution is somewhat unstable.

Finally, mix 2 volumes of Solution A with 1 volume of Solution B. The mixture remains stable for a few weeks only.

Molisch's reagent; see α-Naphthol

Molisch's solution; for the culture of freshwater algae
In 800 cm³ distilled water dissolve 0.2 g potassium nitrate, 0.2 g magnesium sulphate and 0.2 g potassium hydrogen phosphate. In a separate flask dissolve 0.2 g calcium sulphate in 200 cm³ distilled water; add the calcium sulphate solution to the solution of the other three salts.

Moll's solution; clearing agent
Mix 20 cm³ propane-1,2,3-triol (glycerol) with 80 cm³ of 1 per cent aqueous potassium hydroxide solution.

Monk's medium; for mounting small crustacean mouth-parts and other appendages
Mix together 5 cm³ white Karo syrup, 5 cm³ fruit pectin and 3 cm³ water. Make up fresh each time the medium is to be used or add a little thymol to preserve it.

MS-222 Sandoz; for use as anaesthetic for fish and amphibia
See Anaesthetics

Mucor culture medium; *see* Agar, potato dextrose and Agar, malt

Muller's fluid; a fixative for nervous tissue
Dissolve 2.5 g of potassium dichromate(VI) and 1 g of sodium sulphate(VI) in 100 cm³ of distilled water.
Tissues immersed in this fluid harden very slowly—small pieces of tissue taking several weeks.
The fluid should be changed daily during the first week.

Mulligan's fluid; for staining central nervous system
Add 40 g phenol, 5 g copper(II) sulphate(VI)-5-water and 1.5 cm³ of concentrated hydrochloric acid to 1 litre distilled water.

Multodisks
Sterile filter paper discs used for determining the sensitivity of bacteria to antibiotics, sulphonamides, etc. Each disc has six or eight projecting arms each of which is impregnated with a different antibacterial agent. Multodisks are available from Astell (see page 248) with various standard combinations of anti-bacterial agents. They may also be made to order with virtually any combination of agents.

Naphthalen-1-ol; *see* α-naphthol

α-Naphthol
For a 10 per cent solution dissolve 10 g α-naphthol (naphthalen-1-ol) in 100 cm³ alcohol.

Navashin's fluid (chrom-acetic-formalin); fixative used in plant cytology
Mix together 1 cm³ of ethanoic (acetic) acid, 10 cm³ of a 1 per cent solution of chromium trioxide and 5 cm³ distilled water. Immediately prior to use add 4 cm³ of 40 per cent methanal (formaldehyde) solution.

Nessler's reagent
Dissolve 3.5 g of potassium iodide in 10 cm³ of water and stir 1.25 g of mercury(II) chloride into the solution until it has also dissolved. Carefully add a saturated solution of mercury(II) chloride until a faint red precipitate persists, then dissolve 12 g of sodium hydroxide in the solution, using heat if necessary. Add one or two drops of mercury(II) chloride to the cool solution until the faint precipitate reforms. Dilute the solution to 100 cm³ and store in a dark bottle. The reagent does not keep satisfactorily.

Neutral red; a basic, non-toxic vital stain
Dissolve 0.1 g of neutral red in 1 litre of isotonic saline.

Neutral red agar; *see* Agar, Nile blue sulphate or neutral red

Nickel(II) chloride, 0.25 M
Dissolve 59 g of nickel(II) chloride-6-water in 1 litre of distilled water.

Nickel(II) nitrate(V), 0.25 M
Dissolve 73 g of nickel(II) nitrate(V)-6-water in 1 litre of distilled water.

Nickel(II) sulphate(VI), 0.25 M
Dissolve 66 g of nickel(II) sulphate(VI)-6-water in 1 litre of distilled water.

***Nicotiana* agar;** *see* Agar, blackened

Nile blue sulphate(VI); a basic dye used for vital staining of amphibian or chick embryos
Dissolve 0.1 g of Nile blue sulphate(VI) in 100 cm³ of distilled water.
See also Agar, Nile blue sulphate

Ninhydrin; for 'development' of amino-acids in chromatography
Dissolve 0.1 g ninhydrin in 100 cm³ propanone (acetone). Store the solution at 4 °C. Alternatively, ninhydrin may be purchased in aerosol form. Caution is necessary in the use of this material as ninhydrin has been suspected to be carcinogenic.

Nipagin; a fungal inhibitor used in *Drosophila* media
Nipagin (methyl or propyl hydroxy-benzoate) is available from suppliers. Concentrations of 0.5 g nipagin per 1–2 litres of medium is sufficient to inhibit fungal growth.

Nitric(V) acid, M
Dilute 62.8 cm^3 of 70 per cent nitric(V) acid to 1 litre.

Nutrient agar; *see* Agar, nutrient

Nutrient broth; *see* Broth, nutrient

Nutrient solutions; *see* Culture solutions

Orcein-acetic; stain fixative for chromosomes
Stock solution. Add 2.2 g of orcein to 100 cm^3 of glacial ethanoic (acetic) acid and reflux for several hours. Cool, filter and bottle.
Working solution. Dilute 10 cm^3 of the stock solution with 12 cm^3 of distilled water. The diluted solution deteriorates, so that no more stock solution should be diluted than is necessary for the immediate task.

The following *alternative* preparation may also be employed. Add 1–2 g (depending upon the particular batch of dye) of orcein to a mixture of 45 cm^3 glacial ethanoic (acetic) acid and 55 cm^3 distilled water. Boil under reflux for 30 mins. Shake and filter when cool. The solution should be a deep purple colour. Store in a refrigerator and centrifuge if it becomes cloudy.

Orcein-propanoic; an alternative and in some ways superior chromosome stain to orcein-acetic
Follow the alternative recipe for orcein-acetic, substituting propanoic (propionic) acid for ethanoic (acetic) acid.

Osmic acid; *see* Osmium(VIII) oxide (tetroxide)

Osmium(VIII) oxide (tetroxide); for staining or as a test for fats
This is *very poisonous*, even in the 1 per cent aqueous solution in which it is available. It is also very expensive. It is therefore not recommended for school use.

Pampel's fluid
Mix the following together:
1 cm^3 glacial ethanoic (acetic) acid, 10 cm^3 40 per cent methanal (formaldehyde) solution, 25 cm^3 96 per cent alcohol and 50 cm^3 distilled water.

Pancreatin
As sold commercially, this is a mixture of enzymes obtained from mammalian pancreases.

Para-dichlorobenzene; *see* 1,4-dichlorobenzene

Penicillin discs
Paper discs, impregnated with penicillin, are available from suppliers, e.g. Oxoid.

Peptone broth; *see* Broth, peptone

Pepsin
This proteolytic stomach enzyme is obtainable in powdered form from suppliers. Its shelf life is limited to 1–2 years.
A more stable source is 'glycerin of pepsin B.P.' which is an aqueous extract of the enzyme with propane-1,2,3-triol (glycerol) added. This is sold by pharmacists and has a shelf life of over three years.

Phenol-disulphonic acid solution; for nitrate(V) determination
Dissolve 25 g phenol (A.R.) in 158 cm^3 concentrated sulphuric(VI) acid (nitrogen free). Add 67 cm^3 of fuming sulphuric(VI) acid containing about 20 per cent sulphuric(VI) oxide (trioxide). Heat the mixture on a water bath for 2 hours.

Phenolphthalein; indicator, pH range 8.2–9.8
Dissolve 5 g of phenolphthalein in 500 cm^3 of alcohol and dilute to 1 litre with distilled water.

Phenol red; indicator, pH range 6.4–8.2
Add 28.4 cm^3 of 0.1 M sodium hydroxide to 1 g of the indicator and dilute the solution to 1 litre with distilled water.

Phenosafranin; adsorption indicator
Dissolve 0.2 g in 100 cm^3 of distilled water.

Phenosafranin; for estimation of oxygen content of water
Dissolve 1 g phenosafranin in 100 cm^3 distilled water.

Phenylamine; *see under* Aniline for Aniline blue, etc.

Phenylmethanol
A useful clearing agent for animal tissues. Specimens may be kept in it for several months without harm.

Phenylthiocarbamide papers
Dissolve 1.3 g of phenylthiocarbamide in boiling water and make up to 1 litre. Soak strips of absorbent paper in the solution and allow to dry. Cut into pieces 1 cm × 2 cm. These papers may be kept indefinitely.
N.B. More concentrated solutions should *not* be used.

Phenylthiourea (PTC); *see* Phenylthiocarbamide papers

Phloroglucinol; a specific stain for lignin
Dissolve 5 g benzene-1,3,5-triol (phoroglucinol) in 100 cm^3 of alcohol (70 per cent). Stain for 4 minutes and then add 1 drop conc hydrochloric acid.
Alternatively: dissolve 5 g phloroglucinol in 100 cm^3 alcohol (95 per cent), then gradually add concentrated hydrochloric acid until precipitation just begins. Filter.

Phosphate buffer; *see* Buffer, phosphate

Phosphorylase extract
Starch phosphorylase is an enzyme which is involved in the catalysis of the conversion of glucose-1-phosphate(V) to starch. It may be extracted from potato tuber tissue by grinding with a little water and clean, sharp sand in a mortar. Use sufficient water to allow the extract to be poured into a pair of centrifuge tubes. Do not grind too finely. Centrifuge for 2–3

minutes or until the supernatant liquid gives no blue-black colour when tested with iodine in potassium iodide solution.

Phosphotungstic acid; for use in Heidenhain's Azan staining technique

Dissolve 5 g phosphotungstic acid in 100 cm³ distilled water. The solution will not keep and should be freshly made before use.

Photographic developer

Dissolve the following in 750 cm³ distilled water at 52 °C:

2.3 g Metol

75.0 g sodium sulphate(IV) (sulphite) (anhydrous) (or 150.0 g sodium sulphite crystals)

17.0 g benzene-1,4-diol (hydroquinone)

65.0 g sodium carbonate (anhydrous) (or 176.0 g sodium carbonate crystals)

2.8 g potassium bromide

Make up to 1 litre with distilled water.

Dilute with distilled water 1:1, 1:2 or 1:3 before use according to contrast required.

Phytophthora infestans, culture medium; *see* Culture media

Picro-acid fuchsin; *see* van Gieson's stain

Picro-carmine

Mix together 100 cm³ of 2 per cent ammonium hydroxide with 100 cm³ 2,4,6-trinitrophenol (picric acid). Add 5 g carmine. Evaporate to 40 cm³, cool and filter. Evaporate to dryness. Dissolve 1 g of the residue in 100 cm³ distilled water.

Pith, elder; for holding small or delicate tissue during section cutting

The pith cuts more easily if it is stored in ethanol (70 per cent) than it does when dry or wetted with water. The outer layers, including the vascular bundles, may also be removed for easier sectioning. Expanded polystyrene may be used as an alternative to pith.

Plant growth agar; *see* Agar, plant growth

Platinised asbestos

This should *not* be used in schools. It may be replaced by platinised Kaowool or other ceramic fibre. Immerse the clean, dry fibre in 20 cm³ of aqueous platinum(IV) chloride. Remove the fibre and place it in aqueous ammonium chloride. After a few minutes, remove the fibre and heat it strongly in a crucible.

Polystyrene, expanded; *see* Pith, elder

Potassium bicarbonate; *see* Potassium hydrogen-carbonate

Potassium bromide, 0.5 M

Dissolve 60 g of potassium bromide in 1 litre of distilled water.

Potassium carbonate, 1.5 M

Dissolve 207 g of potassium carbonate in 1 litre of distilled water.

Potassium chromate(VI); indicator

Prepare a 5 per cent aqueous solution.

Potassium cyanide, 0.5 M

Dissolve 32.5 g of solid in 1 litre of distilled water.

Potassium dichromate(VI), M

Disolve 294.2 g of potassium dichromate(VI) in 1 litre of distilled water.

Potassium ferrocyanide; *see* Potassium hexacyano-ferrate(II)

Potassium ferricyanide; *see* Potassium hexacyano-ferrate(III)

Potassium hexacyanoferrate(II), 0.5 M

Dissolve 211 g of potassium hexacyanoferrate(II)-3-water in 1 litre of distilled water.

Potassium hexacyanoferrate(III); for 'reducing' haemo-globin in blood

Dissolve 5 g potassium hexacyanoferrate(III) (fer-ricyanide) in 100 cm³ of distilled water.

Potassium hexacyanoferrate(III), 0.125 M

Dissolve 41.1 g of potassium hexacyanoferrate(III) in 1 litre of distilled water. The solution does not keep well.

Potassium hydrogencarbonate solution; to maintain carbon dioxide concentration of air. *See* Sodium hydrogencarbonate solution

Potassium hydroxide, M

The solid may be purchased as pellets, flakes or sticks. To prepare a molar solution, dissolve 56 g of potassium hydroxide in 1 litre of water. (The dissolution is accompanied by the evolution of much heat.) Store the solution in a bottle with a rubber bung. Avoid using cork or ground glass stoppers.

Saturated aqueous solution. Add approximately 100 g potassium hydroxide pellets to 50 cm³ water in a round-bottomed flask while holding the flask under running water. Swirl the contents continually. Store in bottles with rubber or polythene stoppers.

Alcoholic solution. Reflux 10 g of potassium hydroxide pellets for 30 minutes with 100 cm³ ethanol. Cool and filter through glass wool.

Potassium iodide

Dissolve 100 g potassium iodide in 1 litre of distilled water to produce an approximately 10 per cent solution.

Potassium manganate(VII), M

Dissolve 158 g of potassium manganate(VII) in 1 litre of distilled water. The solution does not keep well.

Potassium manganate(VII), Baeyer's reagent; for alkenic bonds

Prepare a 1 per cent aqueous solution.

Potassium manganate(VII) solution; for extraction of earthworms from soil

Dissolve 15–20 g potassium manganate(VII) in 5 litres of water. This is sufficient for approximately 1 m² of ground.

Potassium permanganate; *see* Potassium manganate(VII)

Potassium pyrogallate solution

Place approximately 20 g potassium hydroxide pellets and approximately 5 g benzene 1,2,3-triol (pyrogallic acid) in a clean dry 100 cm³ beaker. Add 50 cm³ of boiled and cooled water and simultaneously add a layer of liquid paraffin 1–2 cm thick. Gently stir the solution with a glass rod. With speed and care, the solution should be almost colourless or mauve to pale brown. **Alternatively,** dissolve approximately 8 g benzene 1,2,3-triol (pyrogallic acid) in 5 cm³ of boiled and cooled water. Cover this with a layer of liquid paraffin 1–2 cm thick. Add 50 cm³ saturated potassium hydroxide solution which will sink below the paraffin and mix with the pyrogallic acid. Gently stir with a glass rod.

Potassium sulphate(VI), 0.25 M

Dissolve 43 g of potassium sulphate(VI) in 1 litre of distilled water.

Potassium thiocyanate, M

Dissolve 97 g of potassium thiocyanate in 1 litre of distilled water.

Potato dextrose agar; *see* Agar, potato dextrose

Pregnyl; *see* Chorionic gonadotrophin

Proof spirit; *see* Alcohol

Propane-1,2,3-triol; *see* Glycerol

Protoactinum; for exponential decay

Dissolve 1 g of uranyl(VI) nitrate in 3 cm³ of distilled water. Add 7 cm³ of concentrated hydrochloric acid. Pour into a plastic bottle adding 10 cm³ of 4-methylpentan-2-one, (*iso*-butyl methyl ketone) or pentyl ethanoate (amyl acetate). For use, shake the bottle for 10 seconds and place the window of a thin-window G-M tube opposite the upper half of the liquid.

PTC papers; *see* Phenylthiocarbamide papers

Purple methylated spirit; *see* Alcohol

Pyrogallol solution; *see* Potassium pyrogallate solution

Red ink

This is a suitable substitute for eosin for demonstrating plant conducting tissue. It is in fact preferable to eosin as it is less toxic to plant tissue.

Rectified spirit; *see* Alcohol

Relative humidity solutions; *see* Humidity regulating solutions

Resazurin; for testing the freshness of milk

Obtainable in tablet form from the suppliers. Resazurin is very sensitive to light and milk samples being tested should be protected from bright sunlight.

Resorcinol; *see* Benzene-1,3-diol

***Rhizobium* suspension**

Crush root halves from fresh clover plants in a little distilled water. Allow any solid material to settle, then decant the clear fluid which contains the *Rhizobium*.

***Rhizopus* culture medium**; *see* Agar, potato dextrose and Agar, malt

Ringer's solutions

For insects. Dissolve the following in 1 litre of distilled water.

8.1 g sodium chloride
0.74 g potassium chloride
0.22 g calcium chloride (anhydrous)
0.20 g magnesium chloride
0.33 g sodium hydrogencarbonate (bicarbonate)
0.78 g sodium dihydrogenphosphate(V)

For amphibians. Dissolve 6.5 g sodium chloride, 0.3 g calcium chloride (anhydrous) and 0.25 g potassium chloride in 1 litre of distilled water.

For chick embryos (Howard-Ringer solution). Dissolve 7.2 g sodium chloride, 0.27 g potassium chloride and 0.17 g anhydrous calcium chloride in 1 litre distilled water.

See also Agar, Ringer and Albumen, Ringer.

For mammals. Dissolve 8.5 g sodium chloride, 0.3 g anhydrous calcium chloride and 0.25 g potassium chloride in 1 litre distilled water.

Rubber solution

Useful for ringing temporary microscopical preparations to prevent evaporation. Obtainable in tubes from cycle shops.

Rubeanic acid

Prepare a 0.5 per cent solution in ethanol.

Rye meal agar; *see* Agar, rye meal

Sachs solution; *see* Culture solutions

Safranin; a basic dye often used as a counterstain for haematoxylin, aniline blue or fast green

Dissolve 1 g safranin O in 99 cm³ alcohol (50 per cent). *See also* Safranin and aniline blue, and Safranin and fast green.

Safranin and aniline blue; a double stain for plant tissue

Solution A. Dissolve 1 g phenylamine (aniline) blue WS in 100 cm³ 2-ethoxyethanol.

Solution B. Dissolve 1 g safranin O in 100 cm³ 2-ethoxyethanol.

Mix 48 cm³ Solution **A** with 52 cm³ Solution **B**.

Safranin and fast green; a double stain for plant tissue

Solution A. Add 3 g fast green FCF to 100 cm³ 2-ethoxyethanol (cellosolve). Heat for 30 minutes over a water bath with occasional stirring. Cool and filter.

Solution B. Dissolve 1 g safranin O in 100 cm³ 2-ethoxyethanol.

Mix 48 cm³ Solution **A** with 52 cm³ Solution **B**.

Saline, isotonic

Dissolve the appropriate amount of sodium chloride in 1 litre of distilled water according to the tissue.

For mammalian tissue: 9.0 g
For avian tissue: 7.5 g

For amphibian tissue: 6.4 g
For elasmobranch tissue: 20.0 g
For invertebrate tissue: 7.5 g
See also Ringer's solutions.

Saline solutions; for bacteria
  a Locke's Formula. Dissolve 9 g sodium chloride, 0.1 g potassium chloride and 0.2 g calcium chloride in 1 litre distilled water.
  b Lipman's Formula. Dissolve 8 g sodium chloride, 0.4 g potassium chloride, 0.4 g calcium chloride and 0.8 g magnesium chloride in 1 litre distilled water.

Schiff's reagent; for aldehydes
Dissolve 0.2 g of rosaniline in 40 cm³ of a freshly prepared solution of sulphur dioxide. Allow to stand for several hours until the pink colour disappears. Shake the solution with 0.2 g of decolourising charcoal. Filter and dilute with distilled water to 200 cm³. Store in a dark stoppered bottle.

Schiff's reagent; for staining chromosomes; *see* Feulgen stain

Schultze's solution (chlor-zinc-iodine); microchemical tests for cellulose
Dissolve 30 g zinc chloride, 5 g potassium iodide and 1 g iodine in 14 cm³ distilled water. Schultze's solution should be kept in brown stoppered bottles in the dark. It has a shelf life of only 2–3 weeks.

Sea water
Various preparations are available.
  a Suspend 0.12 g of calcium carbonate in distilled water and bubble carbon dioxide gas through the suspension until a clear solution is obtained. Add the following reagents:
27 g sodium chloride
11 g magnesium chloride-6-water
13 g magnesium sulphate(VI)-7-water
0.75 g potassium chloride
0.10 g potassium bromide
2.0 g calcium sulphate(VI)-2-water
Dilute to 1 litre.

  b Dissolve the following reagents in 1 litre of water.
23.42 g sodium chloride
0.729 g potassium chloride
2.22 g calcium chloride-6-water
10.702 g magnesium chloride-6-water
9.0 g sodium sulphate(VI)-10-water
0.21 g sodium hydrogencarbonate
0.07 g sodium bromide-2-water

  c Dissolve each of the following separately in 1 litre of distilled water.
1.4 g magnesium chloride
9.9 g magnesium sulphate(VI)-7-water
114.0 g sodium chloride
5.7 g potassium sulphate(VI)
Mix the four solutions together and make up to 6 litres

with distilled water. Aerate for 48 hours and cool in a refrigerator.

  d Prepared tablets or powders may also be purchased from suppliers and aquarists.

Seed germination agar; *see* Agar, germination of small seeds

Silica gel
This is a useful drying agent. It is best purchased in the 'indicator' or 'tell-tale' form which contains cobalt salts and hence turns from blue to pink when exhausted. It can be regenerated by heating to 105 °C overnight and used indefinitely.

Silver nitrate(V), 0.1 M
Dissolve 17.0 g of silver nitrate(V) in 1 litre of distilled water. Arrangements should be made to recover silver from residues.

Skin preservatives; *see* Soap, arsenical and Soap, Browne's

Soap, arsenical; for the preservation of skins
This is *very poisonous* and is therefore considered too hazardous for school use. Browne's soap (*see below*) can be used as an alternative.

Soap, Browne's
Boil together 220 g white curd soap and 660 g calcium carbonate. While still hot, add 4.2 g bleaching powder. Cool and add 2.8 g tincture of musk.
  Although less effective as a preservative than arsenical soap, this is relatively non-toxic.

Soap bubble solution for soap bubble experiments
Add 2.5 g of sodium *cis*-octadec-9-enoate (sodium oleate) to 200 cm³ of distilled water in a clean stoppered bottle. After a day, add 66 cm³ of propane-1, 2, 3-triol (glycerol). Shake well and leave in a dark place for one week. Syphon off the clear liquid from below the scum, add a drop of conc. aqueous ammonia and keep this stock solution in a stoppered bottle in a dark place.

Soap solution
Dissolve 100 g of Castile soap in 1 litre of 80 per cent alcohol and allow to stand overnight.

Soda lime
This may be purchased as soda lime or prepared by mixing slaked lime (calcium hydroxide) and sodium hydroxide in the ratio 2:1 by weight.
  Soda lime may be used as a carbon dioxide absorber, especially the kind impregnated with a pH indicator (e.g. Carbosorb—produced by BDH).

Sodium acetate; *see* Sodium ethanoate

Sodium alginate; for retarding movement of protozoa
Dissolve 1 g sodium alginate in 100 cm³ distilled water.

Sodium bicarbonate; *see* Sodium hydrogencarbonate

Sodium carbonate, M
Dissolve 106 g of anhydrous sodium carbonate (= 286 g of the decahydrate) in 1 litre of distilled water.

Sodium cobaltinitrite; *see* Sodium hexanitrocobaltate(III)

Sodium chloride, M
Dissolve 58.5 g of sodium chloride in 1 litre of distilled water.

Sodium ethanoate, M
Dissolve 136 g of sodium ethanoate (acetate)-3-water in 1 litre of distilled water.

Sodium hexanitrocobaltate(III), 0.16 M
Dissolve 16 g of reagent in 250 cm^3 water.

Sodium hydrogencarbonate, M
Dissolve 84 g of sodium hydrogencarbonate (bicarbonate) in 1 litre of distilled water.

Sodium hydrogencarbonate solution; to maintain carbon dioxide concentration of air
A cold saturated solution of either sodium or potassium hydrogencarbonate will give off carbon dioxide to the surrounding air. It will continue to dissociate if carbon dioxide is removed from the atmosphere—so maintaining a constant partial pressure of carbon dioxide. It provides a means of ensuring that carbon dioxide is not a limiting factor in photosynthesis. A little cotton wool in the solution will increase the surface area.
For aquatic plants. Dissolve 2 g sodium (or potassium) hydrogencarbonate in 100 cm^3 of water. Immerse the plants in this solution.

Sodium hydroxide, M
The solid may be purchased as pellets, flakes or sticks. Dissolve 40.0 g of sodium hydroxide in 1 litre of distilled water (much heat is evolved). Store in a bottle with a rubber bung.

Sodium hypochlorite solution; *see* Sodium oxochlorate(I)

Sodium nitroprusside; *see* Sodium pentacyanonitrosylferrate(II)

Sodium oxochlorate(I) solution
For surface sterilisation of seeds. Dissolve 10 cm^3 of 10 per cent w/v sodium oxochlorate(I) (hypochlorite) solution in 100 cm^3 water.
For softening chorion of locust eggs. Dissolve 25 cm^3 of 10 per cent w/v sodium oxochlorate(I) (hypochlorite) solution in 100 cm^3 water.

Sodium pentacyanonitrosylferrate(II)
Prepare a 2 per cent aqueous solution when required for use.

Sodium picrate papers; as a test for hydrogen cyanide
Make a saturated solution of 2,4,6-trinitrophenol (picric acid) by adding 100 cm^3 distilled water to excess 2,4,6-trinitrophenol (picric acid) crystals and allowing to stand for several days. Decant the saturated solution and add sufficient sodium hydrogencarbonate (bicarbonate) (usually about 0.5 g) to neutralise the acid (i.e. until addition of further sodium hydrogencarbonate produces no effervescence). Filter and soak filter paper

in the filtrate. Dry at room temperature away from sunlight. Cut paper into strips and store in a tin.
N.B. Both 2,4,6-trinitrophenol and its sodium salt are *poisonous.*

Sodium sulphate(VI), 0.1 M
Dissolve 32.5 g of sodium sulphate(VI)-10-water in 1 litre of distilled water.

Sodium sulphide, 0.25 M
Dissolve 60 g of sodium sulphide-9-water in 1 litre of distilled water.

Sodium thiosulphate(VI), M
Dissolve 248.2 g of sodium thiosulphate(VI)-5-water in 1 litre of distilled water.

Soft wax; for sealing coverslips
Mix together 1 part paraffin and 1 part Vaseline. Melt and apply around the coverslip.

Soil-plus-water medium; for the culture of some protists
Place some calcareous clay soil or some good calcareous garden loam in a large jam jar or 600 cm^3 beaker to a depth of about 2 cm. Insert a small speck of a mixture of equal parts of ammonium carbonate and magnesium phosphate. Add water to a depth of about 10 cm. Cover and steam for 1 hour on each of two successive days (do not autoclave). Allow to stand for a further day so that atmospheric gases enter the water.

Stannous chloride; *see* Tin(II) chloride

Starch agar; *see* Agar, starch

Starch solution; as indicator
Mix 20 g of soluble starch into a paste with a small quantity of cold water. Pour the paste into 250 cm^3 of boiling water. Starch solution should always be freshly prepared.

Starch solution (1 per cent)
Mix 10 g **AnalaR** soluble starch with a little distilled water in the cold. Boil 800 cm^3 distilled water and pour the starch-water mixture into the boiling water. Allow to cool and make up to 1 litre with distilled water.

Strontium chloride, 0.25 M
Dissolve 67 g of strontium chloride-6-water in 1 litre of distilled water.

Sucrose solution; for germinating pollen grains
Dissolve 34 g sucrose in 100 cm^3 distilled water. Add a trace (0.001 g) of boric(III) acid crystals and a similar amount of yeast extract.
Some species of pollen may germinate in concentrations much lower than this.
Molar Solution. Dissolve 342 g of sucrose in 1 litre of distilled water.

Sucrose solution; for fermentation
Dissolve 150 g sucrose in 1 litre distilled water.
If used for reducing the rate of discharge of contractile vacuoles in freshwater protozoa, prepare by dissolving 17 g sucrose in 1 litre of distilled water.

Sudan black; stain for fats
Add 5 g Sudan black B to $100\,cm^3$ of alcohol (70 per cent). Reflux for 20 minutes, cool and filter.

Sudan III; stain for fats
Mix $50\,cm^3$ of alcohol (70 per cent) with $50\,cm^3$ of propanone (acetone) and add 5 g Sudan III. Filter and keep in a well stoppered bottle.

Sudan IV; stain for fats
Dissolve 5 g Sudan IV in $95\,cm^3$ of alcohol (70 per cent).

Sulphanilic acid; *see* 4-Aminobenzenesulphonic acid

Sulphur dioxide, aqueous
Pass sulphur dioxide slowly into water until a saturated solution is obtained.

Sulphuric(VI) acid, M
Add $55.1\,cm^3$ of 97 per cent sulphuric(VI) acid (slowly and with constant stirring) to about $750\,cm^3$ distilled water. Dilute to 1 litre when cool.

Sulphurous acid; *see* Sulphur dioxide, aqueous

Susa (Heidenhain's formula); general fixative for animal tissues
To $80\,cm^3$ distilled water add $4\,cm^3$ glacial ethanoic (acetic) acid, $20\,cm^3$ 40 per cent methanal (formaldehyde) solution, 4.5 g mercury(II) chloride, 0.5 g sodium chloride and 2 g trichloroethanoic acid.
N.B. This fixative is *very poisonous*.

Terpineol
This may be used as a clearing agent as a substitute for cedarwood oil which is far more expensive. Tissues may be transferred to it from 80 per cent alcohol without precipitation. It may also be sold as lilacene or artificial oil of lilac. It should not be confused with terpinol.

Tes-tape; for detecting reducing sugars
Available from suppliers and chemists as a roll of impregnated paper. Produced for detecting glucose in urine.

Thiourea solution; for inhibition of thyroid action in amphibia
Add 2.5 g thiocarbamide (thiourea) to warm pond water and make up to 1 litre. This solution should be used so as to achieve a final concentration of 0.04 per cent, i.e. approximately 1 part thiourea solution:5 parts pond water.

Thymine; chromatogram marker solution
Dissolve 0.001 g thymine in $5\,cm^3$ 0.1 M hydrochloric acid.

Thymol blue; indicator, pH ranges 1.2–2.8 and 8.0–9.6
Add $21.5\,cm^3$ of 0.1 M sodium hydroxide to 1 g of the indicator and dilute to 1 litre with distilled water.

Thymolphthalein; indicator, pH range 9.3–10.5
Dissolve 0.4 g of solid in $600\,cm^3$ of industrial spirit and add, with stirring, $400\,cm^3$ of distilled water.

Thyroxin solution
Dissolve 0.1 g thyroxin in $10\,cm^3$ 0.1 M sodium hydroxide solution. Add $90\,cm^3$ pond water. Add $1\,cm^3$ of this solution to 1 litre pond water. This stock solution now has a dilution of 1 ppm. Solutions of lower concentration can be made up by dilution.

Tin(II) chloride, 0.25 M
Dissolve 56 g of tin(II) chloride-2-water in $100\,cm^3$ of concentrated hydrochloric acid and dilute to 1 litre with distilled water. The solution may be preserved by adding a small amount of metallic tin to the stock bottle.

Tollen's reagent
Mix equal volumes of 10 per cent aqueous silver nitrate and 10 per cent aqueous sodium hydroxide. Add sufficient concentrated aqueous ammonia to just re-dissolve the precipitate.

2,2,2-Trichloroethanediol solution; for clearing temporary preparations and plant material
Dissolve 128 g of 2,2,2-trichloroethanediol (chloral hydrate) in $80\,cm^3$ of distilled water.

2,2,2-Trichloroethanediol solution; for narcotisation of small marine animals
Dissolve 0.1 g of 2,2,2-trichloroethanediol (chloral hydrate) in $100\,cm^3$ of sea water.

2,2,2-Trichloroethanediol/iodine solution; for testing for starch in delicate plants, e.g. *Spirogyra*, bryophytes
Dissolve 50 g of 2,2,2-trichloroethanediol (chloral hydrate) in $20\,cm^3$ of iodine in dilute potassium iodide solution.

Uracil; chromatogram marker solution
Dissolve 0.001 g in $5\,cm^3$ 0.1 M hydrochloric acid.

van Gieson's stain
Boil $5\,cm^3$ acid fuchsin and then filter. Add $100\,cm^3$ of a saturated solution of 2,4,6-trinitrophenol (picric acid). (see note under Fuchsin, acid).

Visking tubing
This tubing is made from regenerated cellulose and is available in various diameters. Once it has been wetted it can be stored in clean water in a screw-topped jar and used many times. It can be obtained from commercial suppliers.

Vitamin C; *see* Ascorbic acid

Wheat grain infusion; for culture of protozoans
Boil $100\,cm^3$ pond water. Cool, then add 3–5 grains of wheat. Allow to stand for 2 days before inoculating.

Wright's stain; for blood films
The stain is best purchased dry. Dissolve 0.1 g dry stain in $60\,cm^3$ methanol (100 per cent—neutral and propanone (acetone) free).

Xylene; *see* 1,2 Dimethylbenzene

Yeast agar; *see* Agar, yeast

Yeast-mannitol agar; *see* Agar, yeast-mannitol

Yeast water

Add 20 g fresh yeast to 200 cm³ distilled water in a flask. Plug flask and allow to stand for 2 hours at room temperature.

Heat in pressure cooker—autoclave for fifteen minutes at 100 kPa (15 lb/in⁻²) above atmospheric pressure. Leave to stand for several days. Pipette off clear supernatant liquid.

Zenker-formol; fixative for animal tissues

Dissolve 5 g mercury(II) chloride, 2.5 g potassium dichromate(VI) and 1 g sodium sulphate in 100 cm³ distilled water. Just before use add 5 cm³ of 40 per cent methanal (formaldehyde) solution. N.B. This fixative is *very poisonous.*

Zinc amalgam

Add 5 cm³ of 25 per cent aqueous sulphuric(VI) acid to 20 cm³ of mercury in a beaker. Add 12 g of zinc granules and warm to dissolve the zinc in the mercury. Allow to cool before use.

(CAUTION: *Mercury is very toxic.* Carry out the preparation in a fume cupboard).

Zinc nitrate(V), 0.25 M

Dissolve 74 g of zinc nitrate(V)-6-water in 1 litre of distilled water.

Zinc sulphate(VI), 0.25 M

Dissolve 72 g of zinc sulphate(VI)-7-water in 1 litre of distilled water.

Bibliography

AnalaR Standards for Laboratory Chemicals (various eds.), *AnalaR* Standards, Ltd.

Bradbury, S. (1973) *Peacock's Elementary Microtechnique*, Arnold.

BDH (1973) *Biological Stains and Staining Methods*, BDH Ltd.

Grimstone, A. V., Skaer, R. J. (1972) *A Guide to Microscopical Methods*, CUP.

Guy, K. (1973) *Laboratory Organisation and Administration*, Butterworths.

Handbook of Chemistry and Physics (various editions), Chemical Rubber Co., Ohio.

Joseph, A., Brandwein, P. F., Morholt, E. (1966) *A Sourcebook for the Physical Sciences*, Harcourt, Brace and World.

Knudsen J. W. (1966) *Biological Techniques*, Harper Row.

McLean, R. C., Cook, W. R. I. (1963) *Plant Science Formulae*, Macmillan.

Merck, E. A., *Tables for the Chemistry Laboratory*, E. Merck, A.G., Darmstadt.

Morholt, E., Brandwein, P. F., Joseph, A. (1966) *A Sourcebook for the Biological Sciences*, Harcourt, Brace and World.

Nuffield A level Biological Science (1970) *Laboratory Book*, Penguin Books.

Nuffield O level Biology (1966) *Teachers' Guides I–V*, Longman/Penguin.

Nuffield O level Biology (Revised) (1975) *Teachers' Guides 1–4*, Longman.

Oxoid, Ltd. (1973) *The Oxoid Manual*, Oxoid, Ltd.

Schwab, J. J. (1963) *Biology Teachers' Handbook*, Wiley.

Vogel, A. I. (various editions) A Textbook of Macro and Semi-Micro Qualitative Inorganic Analysis, Longman.

8 The culture and maintenance of living organisms

Every school science department will need to maintain stocks of certain species of living organisms for teaching purposes. The range of organisms and the duration for which they are kept will depend upon the educational uses to which they are to be put; the resources of space, time, accommodation and food which are available; the existence of suitable external conditions for particular species; the cost of purchase and subsequent maintenance; and the possible hazards involved in maintenance.

A survey by Kelly and Wray (1971) indicated that representatives of over forty species (or groups of organisms) were used in primary schools in England and Wales while the number in secondary schools exceeded one hundred. They were of the opinion that 'a collection of about 15 types of organisms would not be unduly difficult for most secondary schools to maintain'. The educational uses to which living organisms may be put have been grouped by Wray (1974) under four broad headings:

  i biological work and enquiry;
 ii centres of interest and activity (e.g. in creative work);
iii associated work in other subjects (e.g. in mathematics or geography);
iv remedial and beneficial uses (e.g. with deprived or maladjusted pupils).

This chapter is largely concerned with the use of living organisms for biological (and other scientific) work although the organisms maintained in a science laboratory, animal house or greenhouse may also be used for other purposes. Those wishing to investigate other uses of living organisms should consult Wray (1974). Each of the following sections dealing with the care and maintenance of individual species includes notes on possible uses to which the species may be put in the context of biology (and other science) teaching.

The resources available within a science department will vary greatly from one school to another. Some schools are provided with specialist accommodation in the form of animal houses and greenhouses, while others may have to maintain stocks of living organisms in crowded laboratories which are in constant use for teaching purposes. Specialist accommodation enables a more careful control of the environment than is possible in teaching laboratories. It also permits easy access for cleaning, feeding and other maintenance without undue disturbance of classes. On the other hand, the removal of all living organisms from the teaching laboratory takes away an important centre of interest and it may be that the most suitable arrangement is to use the specialist accommodation for the long-term maintenance of stock, and to move some to the teaching laboratories at an appropriate time. Considerations in the design of greenhouses and animal houses are discussed on page 5. The maintenance of an animal should not be undertaken unless there is suitable accommodation in the form of caging, etc., which can be kept under appropriate conditions and unless adequate supplies of appropriate food can be guaranteed.

Caring for living organisms is a time-consuming activity which in some circumstances can also prove expensive. Teachers should therefore consider both these factors when deciding upon the range of organisms to be kept. It is likely that a school will wish to maintain some organisms for a limited period only, until the teaching programme is ready for them or during short-term experimental work involving the organisms (e.g. breeding experiments with *Drosophila*). Other organisms will be required for a variety of purposes throughout the year and will need to be maintained continuously. For example, if *Amoeba* is to be used for one sixth-form practical class each year it will be more economical to buy specimens annually than to maintain cultures throughout the year. On the other hand, if a wide range of genetical work is being carried out with mice involving a number of classes at different times of the year, it will be necessary to maintain stocks continuously.

Another important factor is the availability of staff for routine maintenance during school holidays, especially the long summer holiday. If the school has technical staff who normally work during these periods, maintenance of livestock should not be a problem. If this is not the case, it may be necessary to arrange a rota of technical (and teaching!) staff who will come in to carry out routine cleaning, feeding and watering at appropriate intervals. During winter holidays it may also be necessary to provide auxiliary heating facilities if the school central heating system is switched off or is operating at a lower temperature than normal. The practice of 'boarding out' animals with pupils should be avoided if possible due to the risk of infecting stock. However, if no other course of action is open, the written permission of the parents should be obtained and animals should be sent only to homes where there are no pets. Clear written instructions for feeding, cleaning and other maintenance should be provided to avoid confusion.

The maintenance of some species of organism involves possible hazards to pupils and to staff. These are discussed fully in Wray (1974–5). The following categories of animal are either liable to be infected with pathogenic organisms which are transmissible to man or are dangerous species on some other grounds. They must, therefore, not be kept in schools:

i all native wild mammals, especially hedgehogs and rodents;
ii all imported wild mammals especially primates such as monkeys;
iii all species of native wild bird;
iv all members of the parrot family (e.g. parakeets, macaws, love-birds, budgerigars and parrots);
v all poisonous reptiles, crocodiles and alligators, terrapins and tortoises;
vi all species of mammalian parasite.

A number of plant species are poisonous or bear poisonous parts and teachers should be aware of this before cultivating them in schools (see page 38 for a list of common poisonous species). The cultivation of micro-organisms also involves possible hazards and should not be carried out in the absence of appropriate precautions (see page 281).

Initial stock of some organisms (e.g. small mammals) should only be obtained from accredited suppliers. In other cases, a range of commercial suppliers is available (see page 248). Care should be taken to avoid long delays in transport, which may be due to orders being dispatched at times inconvenient to the school (e.g. immediately before or during holidays) or due to delivery facilities at the local railway station or parcels office which were not designed to cater for livestock. Such details should be ascertained before placing an order and, if necessary, arrangements made to collect specimens on arrival from the supplier. Clear instructions should be given to suppliers notifying them of the date on which delivery is required. Many firms require a period of notice in placing orders for some species. In most cases they will not dispatch orders immediately prior to a weekend, so delivery on a Monday is usually not possible.

In some cases it may be possible to start cultures with specimens obtained in the field. It should be remembered, however, that the Conservation of Wild Creatures and Wild Plants Act, 1975 affords considerable protection to a wide range of species and a number of points should be considered before collecting (see page 265).

The keeping of animals in schools is governed by laws which also apply to the treatment of animals by the public at large. These laws are designed to protect animals from wilful ill-treatment and from neglect. The main legislation which concerns school teachers is embodied in the Protection of Animals Act, 1911 which makes it an offence to permit any unnecessary suffering to be caused to any animal. Any action calculated to inflict pain is considered an offence and this is usually thought to include actions likely to cause or possibly cause pain. It should be borne in mind that these include dietary deprivation. The Cruelty to Animals Act, 1876, permits the authorisation of suitable people, by licence issued by the Home Office, to carry out experiments which might otherwise contravene the Protection of Animals Act, 1911. Under certain strictly controlled conditions this may be extended to cover demonstrations which are considered essential for the acquisition of knowledge which might save or extend life or alleviate suffering. It could not be argued that such demonstrations were necessary at school level.

If a school teacher is personally licensed by the Home Office, his experimentation with living animals must be kept private from his pupils. It is worth noting that, from the legal point of view, vertebrate embryos are not included within the terms of the Acts, unless they are normally free-living (e.g. tadpoles). Pithing a frog is not considered to be an experiment and therefore is quite legal, though the decerebration of a mammal or bird is prevented under the terms of the 1876 Act for all but those suitably licensed. Those who wish to pursue this aspect further should consult Kelly and Wray (1975), the Report of the Departmental Committee on Experiments on Animals (The Littlewood Report), or the Acts themselves.

Information on the culturing, care and maintenance of living organisms is scattered in a variety of books and journals. Of these, a number contain extensive information on many species, or are concerned with the educational use of a variety of organisms. These sources are listed below. Information on other sources of reference for each species is given at the end of each of the following sections, which deal with the care and maintenance of individual species or groups of organisms. They are arranged in approximate taxonomic order, beginning with algae, followed by other lower plants, vascular plants, protozoa, invertebrate animals and finally the vertebrates. Information on the culture of bacteria and other micro-organisms will be found on page 276.

General sources of information

Nuffield O level Biology Project (Revised) (1975) *Teachers' Guides 1–4*, Longman (Information on culture and also on educational uses).

Nuffield O level Biology Project (1966) *Teachers' Guides I–V*, Longman/Penguin. (Some information in the original project material is not in the 1975 revision).

Nuffield A level Biology Project (1971) *Laboratory Book*, Penguin Books.

Morholt, E., Brandwein, P. F. and Joseph, A. (1966) *Sourcebook for the Biological Sciences*, Harcourt, Brace and World.

Needham, J. G., Editor (1959) *Culture Methods for Invertebrate Animals*, Dover.

Oxoid Limited (1973) *The Oxoid Manual*, Oxoid Limited. (Microbiological recipes and techniques).

Universities Federation for Animal Welfare (1972) *Handbook on the Care and Management of Laboratory Animals*, E. and S. Livingstone.

Universities Federation for Animal Welfare (1968) *Humane killing of animals*, UFAW.

Wray, J. D. *et. al.* (1974 and 1975) Publications of the Schools Council Educational Use of Living Organisms Project, English Universities Press. (The series includes publications on *Animal Accommodation, Small Mammals, Organisms for Genetics, Micro-organisms, Plants and Organisms in Habitats*, as well as a general *Source Book* and a code of *Recommended Practice for Schools Relating to the Use of Living Organisms and Material of Living Origin*).

References

Conservation of Wild Creatures and Wild Plants Act, 1975, HMSO.

Cruelty to Animals Act, 1876, HMSO.

Kelly, P. J. and Wray, J. D. (1971), The Educational use of living organisms, *J. Biol. Ed.*, **5**, 5, pp. 213–218.

Kelly, P. J. and Wray, J. D. (1975) *The Educational Use of Living Organisms: a source book*, English Universities Press for the Schools Council Educational Use of Living Organisms Project.

Protection of Animals Act, 1911, HMSO.

Report of the Departmental Committee on *Experiments on Animals* (The Littlewood Report) (1965), Cmnd. 2641. HMSO.

Wray, J. D. (1974) *Recommended Practice for Schools Relating to the Use of Living Organisms and Material of Living Origin*, English Universities Press for the Schools Council Educational Use of Living Organisms Project.

References relating to the culture and maintenance of individual organisms will be found at the appropriate points in the text.

Algae

FILAMENTOUS ALGAE

A wide range of filamentous algae is commonly found in freshwater situations in the field, in aquaria, on the surface of moist flowerpots and on damp soil. These include the blue–green alga *Oscillatoria* which forms a thin, greenish-black scum on the surface of wet soil or stagnant water. Among the green algae *Spirogyra, Vaucheria, Chara, Ulothrix, Cladophora, Oedogonium,* *Nitella* and the water net *Hydrodictyon* may be encountered.

Uses

Spirogyra and other filamentous algae are commonly used to exemplify the structure of plant cells. This is unfortunate as their structure is often far from typical of plant cells as a whole. They are also used for studies of conjugation and the evolution of sexuality in plants.

Supply

Algae may be collected in the field, obtained from commercial supply houses or from the Culture Centre of Algae and Protozoa (see page 248).

Culture

Containers. For the culture of individual species, crystallising dishes, jam jars and other similar sized glass vessels should be used. Much care must be taken to ensure the cleanliness of the vessels, as with the culture of protozoa (see page 212).

Media. Molisch's solution (see page 187), Benecke's solution (see page 174) or Knop's solution (see page 185) are all equally suitable. Alternatively the algae may be grown in a vigorous *Daphnia* culture using the modified Knop's solution method (see page 222); *Spirogyra* grows particularly well if cultured in this way. A third method is simply to introduce the algae to a well established freshwater aquarium.

External conditions. Temperatures should be kept in the range 18–27 °C. Excessively high light intensities should be avoided—a north-facing window is probably ideal. Artificial light can be used in the winter—preferably a fluorescent tube with an emission spectrum including substantial proportions in both the blue and the red wavelengths. Tungsten filament lamps tend to have a rather low emission at the blue end of the spectrum which stimulates rapid growth of rather spindly plants with little supporting tissue.

Other details. Sub-culturing should be carried out when the algae are in a vigorous, thriving state. To induce conjugation in *Spirogyra*, transfer filaments to a 2 per cent cane sugar solution and leave for approximately one week. Alternatively, place the culture in bright light conditions.

References

Nuffield A level Biology Project (1971) *Laboratory Book*, Penguin Books.

.Morholt, E., Brandwein, P. F. and Joseph, A. (1966) *A Sourcebook for the Biological Sciences*, Harcourt, Brace and World.

DIATOMS AND DESMIDS

Members of these two groups of plants are commonly encountered in microscopic examination of fresh water.

Uses

As well as including them in studies to show the range of form among the lower plants, the growth and reproduction of diatoms is often studied. They are frequently plants of great beauty with elaborate sculpturing of their silicaceous shells.

Supply

Diatoms and desmids may be collected in the field, obtained from biological suppliers or from the Culture Centre of Algae and Protozoa (see page 248).

Culture

They are best cultured in the same way as that recommended for filamentous algae (see above) using Molisch's, Benecke's or Knop's solution.

VOLVOX

This colonial green alga composed of a ball of cells numbering hundreds to thousands is sometimes found in freshwater lakes and ponds.

Uses

Volvox is used in studies of the evolution of differentiation of plant cells and for its means of reproduction through the production of daughter colonies within the parent. It is also a very beautiful organism.

Supply

Volvox may be obtained from commercial suppliers or from the Culture Centre of Algae and Protozoa (see page 248). It may sometimes be found in the field.

Culture

It may be cultured in the way recommended for filamentous algae (see above) using Molisch's, Benecke's or Knop's solution. It may also be introduced to well established freshwater aquaria where it will flourish provided the tank does not contain fish which are likely to feed upon it.

CHLAMYDOMONAS, EUDORINA, PANDORINA and CHLORELLA

(For euglenoid flagellates see page 213).

Uses

These unicellular and colonial green algae are commonly used in studies of the evolution of cellular and sexual differentiation in plants. *Chlorella* has been used in many classical studies on the biochemistry of photosynthesis. It also lends itself to investigations on micro-ecosystems, where it has the advantage of rapid rates of reproduction (under ideal circumstances) and hence the production of results in a relatively short time when compared with most 'higher' organisms. (See Morholt, Brandwein and Joseph (1966).

Supply

Cultures of all these organisms are available from commercial suppliers and from the Culture Centre of Algae and Protozoa (see page 248).

Culture

Containers. Small jam jars, baby-food jars or other similar glass vessels are ideal. They should be scrupulously cleaned and covered with aluminium foil to exclude dust.

Medium. Mix 500 g good soil with 1 litre of distilled water. Steam serilise for 2 hours or autoclave for 20 minutes at 100 kPa (15 lb in^{-2}). Cool, filter and decant several times till a clear stock solution is obtained. Mix 50 cm^3 stock solution with 950 cm^3 sterile distilled water. Add 0.5 g potassium nitrate and 0.3 g sodium hydrogencarbonate. Pour into containers.

External conditions. Place the cultures in moderate light conditions and maintain at 20–24 °C.

Other details. The containers should be inoculated with a pure culture of the required alga. Sub-culturing should be carried out when existing cultures are at a high rate of growth.

Reference

Morholt, E., Brandwein, P. F. and Joseph A. (1966) *A Sourcebook for the Biological Sciences*, Harcourt, Brace and World.

Bryophytes

MOSSES AND LIVERWORTS

These widespread lower plants are found in a variety of situations. Liverworts, such as *Marchantia, Pellia, Conocephalum* and *Lunularia*, are generally found in damp situations in woods and along streamsides, together with mosses such as *Mnium*. The moss *Sphagnum* is found in acid bogs while the floating liverworts *Riccia* and *Ricciocarpus* are wholly aquatic. Other mosses, such as *Funaria*, prefer a dryer situation. For the identification of British species of mosses and liverworts, see Watson (1968).

Uses

Mosses and liverworts have long been used as examples of plants with an alternation of generations between the leafy, haploid gametophyte and the diploid sporophyte borne upon it. Moss spores germinate to form a thin branching protonema while liverwort spores grow directly into a new thallus. The 'vegetative' parts of some mosses (e.g. *Bryum argenteum*) may be dispersed by various means, and this is a common method of reproduction in liverworts such as *Marchantia* and *Lunularia* which bear gemmae. These properties can lead to experimental investigations such as those in the Nuffield O level Biology Project (Revised) (1975).

The single layer of cells forming the moss leaf offers a suitable means of studying the generalised structure of plant cells.

Moss cushions and the water squeezed from *Sphagnum* plants provide good sources of many species of protozoa and other microscopic animals (see Corbet (1973) and Corbet and Lan (1974)).

Supply

Specimens of mosses and liverworts are best collected in the field with a small amount of the soil in which they are growing. They can be kept for a few days in polythene bags until set up in a more permanent culture. They can also be obtained from commercial suppliers (see page 248).

Culture

Containers. Most species are best kept in small terraria made from plastic sandwich boxes with lids or in small-sized metal-framed aquaria with a sheet of taped glass to serve as a lid (see Figure 8.1). Alternatively, they can be grown on the floor of a Wardian case.

External conditions. Normal laboratory temperatures and medium lighting conditions are recommended. Strong sunlight should be avoided.

Setting up. Place a layer of coarse gravel 2–3 cm deep over the floor of the terrarium. Cover this with a 1 cm thick layer of sand and 2–3 cm of good, loamy soil (see Figure 8.1). The thickness of these layers can be reduced somewhat if sandwich boxes are being used as terraria. Water should be added till it comes halfway up the gravel layer. The mosses and liverworts can then be bedded down in the damp soil. Experience will determine the appropriate moisture level to maintain the plants in good condition. For liverworts and many species of mosses, the level should be kept halfway up the gravel layer. For other species of moss (e.g. *Funaria*), rather dryer conditions are preferred. If moulds develop, the lid should be removed from the terrarium and the water level reduced till they disappear.

Culturing protonema. Protonema may be grown by crushing a dry sporangium of the moss so that the spores are shed onto the surface of Knop's solution (see page 185), diluted to one-third of its normal strength. The solution should be placed in covered Petri dishes in medium light. Germination should commence in about two weeks.

Alternatively, the spores may be germinated on an agar medium. Add 2 g agar to 98 cm³ of one-third strength Knop's solution. Boil to dissolve the agar and make up to 100 cm³ with distilled water. Autoclave for 20 minutes at 100 kPa (15 lb in⁻²). Pour into sterile Petri dishes to a depth of 3–4 mm and allow to set. Sterilise the outside of the sporangium in sodium oxochlorate(I) solution (see page 192) and rinse with sterile water. Crush the sporangium with sterile forceps and scatter the spores over the medium. Germination should commence in about 10 days and young erect gametophytes begin to appear in 2–3 months.

Other details. Aquatic liverworts such as *Riccia* can be grown on the surface of a freshwater aquarium. Liverworts can also be grown from gemmae by placing them on the surface of a Knop's/agar medium (see above) or on damp soil in a terrarium.

References

Corbet, S. A. (1973) An Illustrated Introduction to the Testate Rhizopods in *Sphagnum*, with special reference to the area around *Malham Tarn*, Yorkshire, *Field Studies*, **3**, 5, pp. 801–838.

Corbet, S. A. and Lan, O. B. (1974) Moss on a roof and what lives in it, *J. Biol. Ed.*, **8**, 3, pp. 153–160.

Morholt, E., Brandwein, P. F. and Joseph, A. (1966) *A Sourcebook for the Biological Sciences*, Harcourt, Brace and World.

Nuffield O level Biology Project (Revised) (1975) *Teachers' Guide 3*, Longman.

Watson, E. V. (1968) *British Mosses and Liverworts*, Cambridge University Press.

Pteridophytes

FERNS AND OTHER PTERIDOPHYTES

About fifty species of fern and twenty species of the related clubmosses, quillworts and horse-tails are found in Britain in a wide variety of habitats. For identification of the British species see Taylor (1960) and Clapham, Tutin and Warburg (1962).

Uses

Ferns and their allies may be used as examples illustrating the variety of plant life. They also provide an excellent illustration of plants with alternation of generations and form a basis for studies on the evolution of heterospory. The introduced aquatic species *Azolla* and *Salvinia* make suitable floating plants for an aquarium.

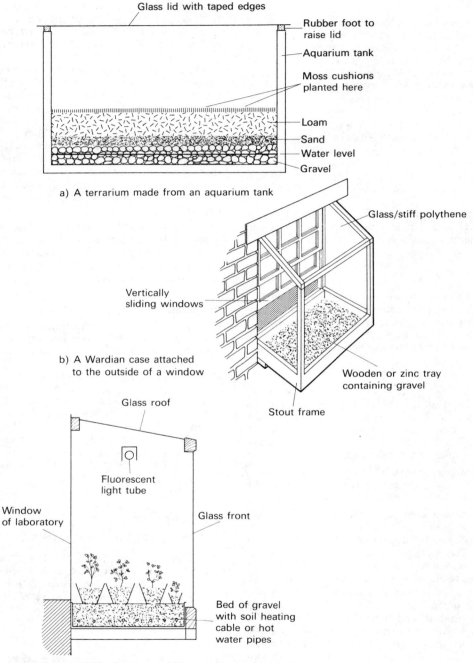

a) A terrarium made from an aquarium tank

b) A Wardian case attached to the outside of a window

c) A design of Wardian case which may be attached to the outside or the inside of a laboratory window

Figure 8.1   Artificial habitats for growing mosses, ferns and other plants normally growing in humid conditions

Supply

Ferns are best collected in the field along with a good quantity of the soil in which they are growing. Many ferns possess a sizeable rhizome and care must be taken to include this undamaged along with the rest of the plant. A limited range of species may also be obtained from commercial suppliers (see page 248).

Culture

Containers. Ferns grow particularly well in a Wardian case, which maintains a high level of humidity (see page 200). Most species will also survive in plant pots. The smaller species and those frequenting wetter habitats (e.g. *Hymenophyllum*) are best grown in terraria. The aquatic species *Azolla*, *Salvinia* and *Marsilea* should be kept in aquaria.

External conditions. Normal laboratory temperatures and medium conditions of lighting are recommended. Strong or direct sunlight should be avoided.

Setting up. If a Wardian case is being used, it should be sited out of direct sunlight (see page 200). Fill the base with a soil made by mixing 1 part coarse sand with 1 part fine peat moss and 2 parts good garden loam. Smaller terraria may be set up in the way described for mosses on page 199. If the ferns are to be grown in pots, a layer of 2–3 cm of broken pot or coarse gravel should be placed in the bottom and covered with soil of a similar mixture to that placed in Wardian cases (see above). The ferns should be planted in the soil of the terrarium or pot with plenty of the soil in which they were originally growing around the roots and rhizome. The soil in the pots should be kept moist, but not wet, by standing the pots in water at appropriate intervals.

Culturing fern prothalli. Prothalli may be grown in either a liquid or an agar medium. Fertile fronds of *Dryopteris* or *Pteridium* should be collected in late summer when the sporangia are of a black, glistening appearance. The fertile pinnules should be cut off and placed in a paper bag in a dry atmosphere and left to allow the spores to be discharged. The spores can be separated from the debris in the bag by sieving through several layers of butter muslin and then through two layers of coarse lens tissue. The spores can be stored for several years at 2 °C. To grow the prothalli, sprinkle the spores on the surface of fern gametophyte agar (see page 169) which has been previously sterilised and poured into sterile Petri dishes. Keep the Petri dishes in plastic boxes lined with wet blotting paper or with wet cotton wool to retain a high level of humidity. Keep in low light intensities. Heart- or disc-shaped prothalli should grow in about 7–8 weeks. To enable fertilisation and the subsequent growth of the sporophyte to take place, irrigate the surface of the medium with water or liquid medium (see below).

Alternatively, the spores can be sprinkled on to the surface of liquid medium. This is made up in the same way as the fern gametophyte agar (page 169) but no agar is added. The spores germinate rather more rapidly by this method, but a smaller prothallus is formed and subsequent growth of the sporophyte is not possible.

A third method involves sprinkling the spores on the surface of well-soaked blocks of peat which are then placed in sandwich boxes lined with wet blotting paper and kept in low light intensities. Young sporophytes should appear within two months.

Other details. Dehiscence of the sporangia may be observed by cutting off a small portion of fern frond 2–3 mm × 2–3 mm and placing it on a microscope slide. As the piece of leaf tissue dries out, the sporangia can be observed microscopically as they rupture.

References

Clapham, A. R., Tutin, T. G. and Warburg, E. F. (1962) *Flora of the British Isles*, Cambridge University Press.

Morholt, E., Brandwein, P. F. and Joseph, A. (1966) *A Sourcebook for the Biological Sciences*, Harcourt, Brace and World

Nuffield O level Biology Project (1966) *Teachers' Guide IV*, Longman/Penguin.

Nuffield A level Biology Project (1971) *Laboratory Book*, Penguin Books.

Taylor, P. (1960) *British Ferns and Mosses*, Eyre and Spottiswoode

Fungi

MOULDS

A range of species of mould will colonise a piece of moist bread or fruit kept in a damp atmosphere. *Aspergillus* (black-mould), *Mucor* (pin-mould), *Penicillium* (green-mould) and *Rhizopus* (bread-mould) are among those likely to appear.

Uses

Assorted cultures, such as those developing on moist bread, may be used for studies on colonisation and succession (see Nuffield O level Biology Project (Revised) (1975)). Individual species may also be used for anatomical work, and studies of saprophytism and the development of antibiotics. Some species may also be used for studies of development and the influence of the environment upon it (see Nuffield A level Biology Project (1970)).

Supply

Mixed cultures may be obtained in the way outlined above. Pure cultures may be obtained from commercial suppliers (see page 248) or by isolation from mixed populations.

Culture

Containers. Sterile Petri dishes.

Media. All the species mentioned above and many others may be cultured on any of the following agar media:

Potato dextrose agar (see page 170)
Frozen pea agar (see page 169)
Malt agar (see page 170).

Potato dextrose agar is especially suitable for *Rhizopus* and *Mucor*. Malt agar and frozen pea agar may also be used for any of the species mentioned. A specialised medium is recommended for the sporulation of *Aspergillus* (see page 171).

External conditions. Laboratory temperatures are quite suitable though development will be more rapid at 30 °C. Moderate light is recommended. Growth may be slowed considerably by placing cultures in a refrigerator. This is often a convenient method of halting growth at a particular stage of development.

Other details. Sterile culture techniques should be used throughout (see page 277). Sporulation in *Aspergillus* may be induced by culturing of a special medium deficient in organic nutrients (see page 171).

Sub-culturing should be carried out every 2–3 months though this interval may be increased by storage in a refrigerator.

COPRINUS

The ink-cap fungus grows on soil which is rich in organic matter and on the faeces of herbivores.

Uses

Apart from serving as an example of a Basidiomycete fungus, *Coprinus* may be used for work on gene segregation (see Nuffield A level Biology Project (1970)).

Supply

Cultures are best obtained from commercial suppliers (see page 248). They can be kept for up to six months in a refrigerator before sub-culturing.

Culture

Containers. A jam jar or similar sized glass vessel covered with metal foil is suitable.

Medium. Sterilised horse dung is an ideal medium on which to grow fruiting bodies from dikaryons. Dung should be spread out on newspaper to dry in a warm atmosphere (e.g. a boiler house). Once thoroughly dry it will keep indefinitely.

Technique. Add a few small pieces of dung to 50 cm³ water in the jar and cover. Autoclave for 15 minutes at 1 kg cm⁻² (15 lb in⁻²). Allow to cool and inoculate with dikaryon material. Culture at 37 °C for the first day and subsequently at 28 °C in the light. A series of inoculations should be carried out on subsequent days as the growth period is variable (7–14 days).

References

Nuffield A level Biology Project (1970) *Organisms and Populations*, Penguin Books.
Nuffield A level Biology Project (1971) *Laboratory Book*, Penguin Books.

SORDARIA

Uses

Sordaria is particularly useful for tetrad analysis in meiosis as the ascospores are arranged in linear fashion in the ascus. (See Nuffield A level Biology Project (1970)).

Supply

Cultures are best obtained from commercial suppliers (see page 248). They can be maintained for some months after arrival by storing them in a refrigerator.

Culture

Containers. Sterile Petri dishes.

Medium. Cultures should be maintained on cornmeal agar (see page 169).

Technique. Stock cultures may be maintained in the normal way using sterile technique (see page 277). To demonstrate a monohybrid cross between two strains of *Sordaria* with different coloured ascospores, inoculate the medium a week before the spores are required. Incubate at 28 °C. Perithecia containing asci will be formed along a line where the two cultures meet. A greater number of perithecia will be formed if the Petri dish is inoculated in one of the alternative ways shown in Figure 8.2. If a sufficient number of perithecia develop before the required day, the cultures should be transferred to a refrigerator where they can be kept for several months.

References

Nuffield A level Biology Project (1970) *Organisms and Populations*, Penguin Books.
Nuffield A level Biology Project (1971) *Laboratory Book*, Penguin Books.

PHYTOPHTHORA INFESTANS

This parasitic species is sometimes encountered on potato plants suffering from potato blight.

Uses

For a long time the life history of this fungus has appeared in biology and botany courses.

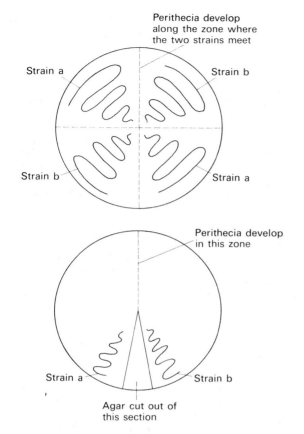

Perithecia develop along the zone where the two strains meet

Strain a

Strain b

Strain b

Strain a

Perithecia develop in this zone

Strain a

Strain b

Agar cut out of this section

Figure 8.2   Alternative methods of 'seeding' Sordaria *in a Petri dish to obtain perithecia*

Supply

Cultures may be obtained from commercial suppliers (see page 248) or from infected potato plants.

Culture

Containers. Sterile Petri dishes.

Media. Rye meal agar (page 170) should be used, though frozen pea agar (page 169) will serve as a temporary substitute. Special media are required for sporulation which is therefore best studied on the host plant.

Technique. Normal sterile techniques should be used (see page 277). Cultures should be maintained at 20 °C.

Reference

Nuffield O level Biology Project (1971) *Laboratory Book*, Penguin Books.

YEASTS

Several species of yeast may be used in school laboratories. Baker's or brewer's yeast (*Saccharomyces cerevisiae* is the species most commonly encountered, but lager yeast (*S. carlsbergensis*) and the tropical *Schizosaccharomyces pombe* are also used on occasions.

Uses

Yeasts are most commonly used for work on fermentation. *Saccharomyces cerevisiae* is a 'top' yeast which rises to the surface of the medium on which it is growing. *S. carlsbergensis* is a 'bottom' yeast which remains on the bottom of the container during fermentation.

Schizosaccharomyces pombe is a useful species on which to study population growth as new daughter cells are formed by the sudden appearance of a transverse wall dividing the elongated parent cell. The number of cells in a population count is therefore more easily determined than in species of *Saccharomyces* where budding takes place. Mutant strains of yeast (*Saccharomyces cerevisiae*) may also be used in genetic work.

Supply

Yeast in dried or cake form may be obtained from chemists, bakeries and shops specialising in the home-brew trade. Pure strains can also be obtained from the National Collection of Yeast Cultures, the Commonwealth Mycological Institute or from commercial biological suppliers (see page 248).

Culture

Containers. Sterile Petri dishes.

Medium. Malt agar (see page 170) is ideal for all the species mentioned.

Technique. Normal sterile technique should be employed and the cultures maintained at laboratory temperatures.

Lichens

A number of species of lichen are commonly found on the surfaces of stones, walls, fences and on the bark of trees. Their identification is often difficult and necessitates microscopical and chemical examination. Alvin and Kershaw (1963) may be used for identification.

Uses

Lichens may be used as indicators of airborne pollution, but this does not entail their culture in the laboratory. They also provide interesting examples of symbiosis as all lichens embody a complex relationship between a fungus and an alga.

Supply

Specimens are best collected in the field and stored temporarily in polythene bags. Excessive moisture commonly leads to the growth of moulds and should therefore be avoided.

Culture

Sections of bark or stones encrusted with lichens may be kept in terraria in much the same way as mosses and liverworts (see page 199). The lid should be kept off the terrarium for much of the time to avoid excessive moisture accumulating on the plants and the sides of the container.

Reference

Alvin, K. L. and Kershaw, K. A. (1963) *The Observer's Book of Lichens*, Warne.

Flowering plants

All school science departments will need to maintain stocks of a range of species of flowering plants for biological work. A range of possible plants, together with some of the purposes for which they may be used and references to further information, is shown in Table 8.1.

Uses

Pelargonium zonale (geranium) is commonly used for a wide variety of basic physiological work (e.g. experiments and demonstrations on photosynthesis). Other species may be used for genetic work, e.g. *Lycopersicon esculentum* (tomato), *Zea mays* (maize) etc. A wide range of species may be used for basic anatomical or morphological work. These are generally not included in Table 8.1 as suggestions can be found in any standard textbook of botany.

A further possible reason for keeping plants is that they are very attractive in the laboratory.

Supply

Local nurseries and seedsmen are convenient suppliers of many species, though once initial stocks have been obtained they may be propagated within the school in many cases. Sometimes specimens can be collected in the field, provided the guidelines outlined on page 265 are adhered to. For more specialised work with particular requirements (e.g. pure strains for genetic breeding), commercial biological suppliers are the most obvious source (see page 248 for a list of such suppliers).

Culture

Containers. Plastic or earthenware (unglazed) pots are the most suitable containers for established plants. Plastic seed trays may be used for germination and the rooting of cuttings. Good drainage is essential and, for this reason, broken crocks or small stones should be placed in the base of each pot. Some creeping plants (e.g. clovers) grow better in shallow pans.

External conditions. Many school laboratories do not provide suitable conditions for the healthy growth of plants. Atmospheric humidity tends to be too low (especially in centrally heated buildings) and the air is often loaded with dust and fumes. Heating is often irregular, lighting unilateral and often inadequate. Plants may be interfered with by pupils, which makes difficulties for long-term experimental work and watering can also prove a problem, especially at weekends. For effective maintenance of plants, steps have to be taken to overcome these problems. The greenhouse (see page 5), Wardian case (page 200), cold frames, propagating frames and polythene or other plastic or glass enclosures are all attempts to provide suitable conditions for plant growth. These conditions are discussed more fully below.

Air and soil temperatures should be kept between 10 °C and 20 °C during the daytime, though a drop of about 5 °C during the night will do no harm. The light requirements of individual species vary considerably. In most laboratories lighting is likely to be inadequate and symptoms indicating this should be borne in mind during routine examination of plants. Some air circulation is necessary, but draughts should be avoided.

Growing media. A sterilised compost reduces competition and is therefore recommended for most species (see page 178). John Innes Potting Compost No. 1 or a similar medium is suitable for starting most plants once germination has taken place. As the plants are transferred to larger pots, the medium should be replaced by John Innes Potting Compost No. 2, or in some cases, a final transfer to No. 3 may be required. Some individual species have more specialised requirements.

Germination. A number of species may be grown from seed. The seeds should be sown in seed compost (see page 178), moist sand or between layers of moist blotting paper, depending upon the species. Temperature requirements also vary as does the time taken for germination. These details are summarised in Table 8.2 on page 211.

Place the seed compost or sand in a seed box or plastic pot to within 1–2 cm of the rim. Water the compost by placing it in a shallow container of water and allowing the compost to take it up. Remove the tray or pot and allow to drain. Sow the seeds and cover with compost. The regular spacing of seeds may be facilitated by using one of the devices illustrated on page 210. The depth of the seed should be approximately equal to the maximum diameter of the seed and in general should not be greater than 1 cm. Cover the box or pot with a sheet of glass and keep in the dark until germination takes place. When the shoots appear above the level of the germination medium, the cover should be removed. It may be necessary to spray

Table 8.1   The educational uses of a variety of flowering plants

Plant species	Educational use	References
Allium cepa (onion)	Cell structure, osmotic reactions using epidermis from bulb. Example of mono-cotyledon with endospermic seed and epigeal germination	Nuffield O level Biology (Revised) (1975) *Text 2: Living Things in Action*, Longman. Bingham, C. D. (1976) *Plants*, English Universities Press for the Schools Council
Allium sativum (garlic)	Cell differentiation and chromosome behaviour during mitosis	Nuffield O level Biology (Revised) (1975) *Text 4: The Perpetuation of Life*, Longman
Allium ursinum (wild garlic)	Behaviour of chromosomes during meiosis in pollen mother cells	Comber, L. C. (1976) *Organisms for Genetics*, English Universities Press for the Schools Council
Antirrhinum majus (snapdragon)	Genetic crosses and example of continuous and discontinuous variation	Comber, L. C. (1976) *Organisms for Genetics*, English Universities Press for the Schools Council
Aquilegia spp.	Behaviour of chromosomes during meiosis in pollen mother cells	Alternative to *Allium ursinum* (see above)
Arabidopsis thaliana (Thale cress)	Genetic crosses	
Avena sativa (oat)	Tropisms (especially phototropism) exhibited by coleoptiles	Nuffield A level Biological Science (1970) *Control and Co-ordination in Organisms*, Penguin Books
Begonia rex	Artificial propagation by leaf cuttings	
Brassica nigra (mustard)	Example of dicotyledon with non-endospermous seed and epigeal germination	Bingham, C. D. (1976) *Plants*, English Universities Press for the Schools Council
Bryophyllum spp.	Asexual reproduction by plantlets. Effect of environment on variation within a clone	Bingham, C. D. (1976) *Plants*, English Universities Press for the Schools Council
Capsella bursa-pastoris (shepherd's purse)	Example of the life history of an angiosperm. Development of a plant embryo	Nuffield A level Biological Science (1970) *The Developing Organism*, Penguin Books
Cardamine hirsuta (hairy cress)	Measuring the distances to which seeds are dispersed by explosive mechanism	Nuffield O level Biology (Revised) (1975) *Teachers' Guide 2: Living Things and their Environment*, Longman
Chlorophytum spp.	Variegated leaves with no formation of starch. Example of vegetative propagation (rooting stems)	
Circaea spp. (Enchanter's nightshade)	Starch prints in photosynthesis experiments	Nuffield A level Biological Science (1970) *Maintenance of the Organism*, Penguin Books
Clematis spp.	Behaviour of chromosomes during mitosis in leaf-tip squashes. Plant climbing by means of petiole	Shambulingappa, K. G. A simple leaf-tip squash method for the study of chromosomes, *SSR*, 1966, 162, **47**, pp. 487–489
Coleus cultivars	Artificial propagation by stem cuttings. Displays continuous variation	Nuffield Secondary Science (1971) *Theme 2: Continuity of Life*, Longman
Crocus balansae	Behaviour of chromosomes during mitosis in root-tip squash	Clarke, R. A. et al. (1970) *Biology by Inquiry: Book 1* (and *Teachers' Guide*), Heinemann
Crocus spp.	Vegetative propagation by corm	
Cuscuta campestris (dodder)	Plant parasite (on *Trifolium* sp., *Lycopersicon*, *Pelargonium* etc)	Mylechreest, M., *Cuscuta campestris*, *SSR*, 1969, 172, **50**, p. 570
Drosera spp. (sundews)	Responses of insectivorous plant	

Plant species	Educational use	References
Elodea canadensis (Canadian pond-weed)	Cell structure and cyclosis in mesophyll cells.	Nuffield A level Biological Science (1970) *Maintenance of the Organism*, Penguin Books
	Gas exchange during photosynthesis and respiration	Nuffield O level Biology (Revised) (1975) *Text 2: Living Things in Action*, Longman
Endymion non-scriptus (bluebell)	Chromosome activity during meiosis in pollen mother cells	Alternative to *Allium ursinum* (see above)
Ficus elastica (India rubber plant)	Artificial propagation by air layering	
Forsythia spp.	Artificial propagation by stem cuttings	Nuffield Secondary Science (1971) *Theme 2: Continuity of Life*, Longman
Fragaria spp. (strawberry)	Example of vegetative propagation by runners	Bingham, C. D. (1976) *Plants*, English Universities Press for the Schools Council
Fuchsia spp.	Artificial propagation by stem cuttings	Bingham, C. D. (1976) *Plants*, English Universities Press for the Schools Council
Helianthus annuus (sunflower)	Example of dicotyledon with endospermous seed and epigeal germination.	Non-poisonous alternative to *Ricinus communis* (see below)
	Anatomy of dicotyledonous stem and root	
Hibiscus spp.	Behaviour of chromosomes during mitosis in root-tip squashes	Shambulingappa, K. G., A simple leaf-tip squash method for the study of chromosomes, *SSR*, 1966, 162, **47**, pp. 487–489
Hordeum spp. (barley)	Extraction of digestive enzyme.	Nuffield A level Biological Science (1970) *The Maintenance of the Organism*, Penguin Books
	Tropisms, especially phototropism in coleoptiles. Displays continuous variation	Nuffield A level Biological Science (1970) *Control and Co-ordination in Organisms*, Penguin Books
Hordeum spp. (Xantha variety of barley)	Variation resulting from environmental effects	Nuffield A level Biological Science (1970) *Organisms and Populations*, Penguin Books
Impatiens balsamina (busy Lizzie)	Autoradiography with ^{14}C. Demonstration of conducting tissue in plant stem	Nuffield A level Biological Science (1970) *The Maintenance of the Organism*, Penguin Books
Iris spp.	Formation of sugars in photosynthesis.	Nuffield O level Biology (1966) *Text III: The Maintenance of Life*, Longman/ Penguin
	Example of vegetative propagation by rhizome	Nuffield Secondary Science (1971) *Theme 2: Continuity of Life*, Longman
Lactuca sativa (lettuce var. Grand Rapids)	Effect of light on germination	Nuffield A level Biological Science (1970) *The Developing Organism*, Penguin Books
Lemna spp. (duckweeds)	Experiments on mineral requirements. Population studies on rates of growth. Exhibits photoperiodism. Example exhibiting asexual reproduction	Nuffield O level Biology (Revised) (1975) *Text 2: Living Things in Action*, Longman Rhodes, L. W. (1968) The duckweeds— their use in the high school laboratory, *American Biology Teacher*, 30, No. 7, pp. 548–551 Clarke, R. A. et al. (1970) *Biology by Inquiry: Book 2*, Heinemann

Plant species	Educational use	References
Lepidium sativum (garden cress)	Cell differentiation in root-tip squashes.	Nuffield O level Biology (Revised) (1974) *Text 1: Introducing Living Things*, Longman
	Tropisms, especially hydrotropism.	Dodd, I. A., Hydrotropic responses of roots, *SSR*, 1966, 162, **47**, pp. 476–481
	Effects of auxins on development	Nuffield O level Biology (Revised) (1975) *Text 4: The Perpetuation of Life*, Longman
Ligustrum spp. (privet)	Leaf structure including distribution of stomata	Nuffield O level Biology (Revised) (1975) *Text 2: Living Things in Action*, Longman
Lilium spp.	Example of vegetative propagation by bulb	Bingham, C. D. (1976) *Plants*, English Universities Press for the Schools Council
Lupinus spp. (lupin)	Artificial propagation by stem cuttings.	Nuffield Secondary Science (1971) *Theme 2: Continuity of Life*, Longman
	Symbiosis with *Rhizobium* in root nodules	Clarke, R. A. et al. (1970) *Biology by Inquiry: Book 2*, Heinemann
Lycopersicon esculentum (tomato)	Cell structure of fruit cells. Osmotic reactions of cells. Genetic crossing experiments.	Clarke, R. A. (et al. (1968) *Biology by Inquiry: Book 1*, Heinemann Clarke, R. A. et al. (1971) *Biology by Inquiry: Book 3*, Heinemann
	Behaviour of chromosomes during mitosis in leaf-tip squashes.	Shambulingappa, K. G., A simple leaf-tip squash method for the study of chromosomes, *SSR*, 1966, 162, **47**, pp. 487–489
	Artificial propagation by grafting.	Syrocki, J. (1971) Grafting in herbaceous stems, *American Biology Teacher*, **33**, 9, pp. 532–534
Mentha spp. (mint)	Vegetative propagation by suckers	Bingham, C. D. (1976) *Plants*, English Universities for the Schools Council
Mimosa pudica (sensitive plant)	Nastic reactions (to touch)	
Narcissus spp. (daffodils and narcissi)	Example of vegetative propagation by bulb. Osmotic reactions of tissues from stipe. Germination of pollen tube in nutrient solution	Bingham, C. D. (1976) *Plants*, English Universities Press for the Schools Council. Nuffield A level Biological Science (1970) *The Developing Organism*, Penguin Books
Nicotiana tabacum (tobacco)	Leaf prints to demonstrate the presence of starch. Genetic crosses	Nuffield O level Biology (Revised) (1975) *Text 2: Living Things in Action*, Longman Comber, L. C. (1976) *Organisms for Genetics*, English Universities Press for the Schools Council
Oenothera lamarckiana (evening primrose)	Germination of pollen tube in nutrient solution	Nuffield A level Biological Science (1970) *The Developing Organism*, Penguin Books
Ornithogalum virens	Behaviour of chromosomes during mitosis in root-tip squashes	Alternative to *Allium sativa* (see above)
Paeonia spp. (peony)	Behaviour of chromosomes during meiosis in pollen mother cells	Dyball, R. H., Ramage, H. P. and Rice, C. H. (Eds) (1962) *The Science Master's Book, Series 3, Part 3, Biology*, John Murray
Pelargonium zonale (geranium)	Leaf prints to demonstrate the presence of starch. Artificial propagation by stem cuttings	Nuffield O level Biology (Revised) (1975) *Text 2: Living Things in Action*, Longman Nuffield Secondary Science (1971) *Theme 2: Continuity of Life*, Longman
Pelargonium zonale cultivars	Exhibits continuous variation	
Pharbitis nil (Japanese morning glory)	Photoperiodism of flowering, an example of a 'short day' plant	Clarke, R. A. et al. (1971) *Biology by Inquiry: Book 3, Student Text and Teachers' Guide*, Heinemann

Plant species	Educational use	References
Phaseolus radiatus (mung bean)	Tropisms, especially geotropism of root	Nuffield A level Biological Science (1970) *Control and Co-ordination in Organisms*, Penguin Books
Phaseolus spp. (french bean, runner bean etc.)	Example of dicotyledon with non-endospermous seed and epigeal germination	
Phlox spp.	Vegetative propagation through crowns. Artificial propagation by root cuttings	Nuffield Secondary Science (1971) *Theme 2: Continuity of Life*, Longman
Pisum sativum (garden pea)	Tropism, especially hydrotropism of roots. Genetic crosses	Dodd, I. A., Hydrotropic responses of roots, *SSR*, 1966, 162, **47**, pp. 476–481 Comber, L. C. (1976) *Organisms for Genetics*, English Universities Press for the Schools Council
Polyanthus spp.	Germination of pollen tube in nutrient solution	Nuffield A level Biological Science (1970) *The Developing Organism*, Penguin Books
Primula spp.	Germination of pollen tube in nutrient solution	Nuffield A level Biological Science (1970) *The Developing Organism*, Penguin Books
Ranunculus ficaria (lesser celandine)	Vegetative propagation by root tubers	
Ranunculus spp.	Exhibits continuous variation in various parts of plant	Cameron, J. A. (1970) An introduction to the practical study of variation using *Ranunculus* species, *J. Biol. Ed.* **4**, 1, pp. 19–24
Ricinus communis (castor oil)	Example of dicotyledonous seed with endosperm and epigeal germination. *WARNING: These seeds are poisonous*	Bingham, C. D. (1976) *Plants*, English Universities Press for the Schools Council
Rosa spp. (rose)	Artificial propagation by budding	
Rubus spp. (blackberry)	Artificial propagation by layering	Nuffield Secondary Science (1971) *Theme 2: Continuity of Life*, Longman
Saintpaulia ionantha (African violet)	Artificial propagation by leaf cutting	
Saxifraga sarmentosa	Example of vegetative propagation by rooting stems	
Scilla spp.	Behaviour of chromosomes during meiosis in pollen mother cells	Alternative to *Allium ursinum* (see above)
Sempervivum spp.	Example of vegetative propagation by offsets	Nuffield Secondary Science (1971) *Theme 2: Continuity of Life*, Longman
Senecio vulgaris (groundsel)	Genetic crosses	
Solanum tuberosum (potato)	Synthesis of starch from glucose by stem tuber extract. Example of vegetative propagation by stem tubers	Nuffield O level Biology (Revised) (1975) *Text 2: Living Things in Action*. Longman. Nuffield Secondary Science (1971) *Theme 2: Continuity of Life*, Longman
Stellaria media (chickweed)	Germination of pollen on style.	Nuffield O level Biology (1966) *Text II: Life and Living Processes*, Longman/Penguin
Symphoricarpos spp. (snowberry)	Cell structure and cyclosis in cells from fruit	Clarke, R. A. et al. (1968) *Biology by Inquiry: Book 1, Students' Text and Teachers' Guide*, Heinemann

Plant species	Educational use	References
Taraxicum officinale (dandelion)	Vegetative propagation by tap-root.	Nuffield O level Biology (Revised) (1975) *Text 1: Introducing Living Things*, Longman
	Asexual reproduction by apomictic embryos	Hartmann, H. Y. and Kester, D. E. (1968) *Plant Propagation Principles and Practices*, Prentice Hall
Tradescantia spp.	Cell structure and cyclosis in cells from staminal hairs.	Clarke, R. A. et al. (1968) *Biology by Inquiry: Book 1, Student Text and Teachers' Guide*, Heinemann
	Osmotic reactions using staminal hair cells.	Clarke, R. A. et al. (1971) *Biology by Inquiry: Book 3, Student Text and Teachers' Guide*, Heinemann
	Starch prints using variegated leaves.	
	Effect of the environment on variation within a clone	Nuffield Secondary Science (1971) *Theme 2: Continuity of Life*, Longman
Tradescantia paludosa	Behaviour of chromosomes during meiosis in pollen mother cells	Savage, J. R. K., Demonstrating cell division with *Tradescantia*, *SSR*, 1967, 166, **48**, pp. 771–782
Trifolium repens (Dutch or white clover)	Example of 'one gene-one enzyme' relationship in cyanogenesis.	Darlington, C. D. and Bradshaw, A. D. (Eds) (1963) *Teaching Genetics in School and University*, Oliver & Boyd
	Multiple alleles causing variation in leaf colour and patterning.	Nuffield A level Biological Science (1970) *Organisms and Populations*, Penguin Books
	Exhibits discontinuous variation.	Darlington C. D. and Bradshaw, A. D. (Eds) (1963) *Teaching genetics in school and University*, Oliver & Boyd
	Exhibits symbiosis with *Rhizobium* in root nodules.	Clarke, R. A. et al. (1970) *Biology by Inquiry: Book 2, Student Text and Teachers' Guide*, Heinemann
	Vegetative propagation by runners	Bingham, C. D. (1976) *Plants*, English Universities Press for the Schools Council
Triticum spp. (wheat)	Tropisms, especially phototropism in coleoptiles.	Alternative to *Avena* (see above)
	Respiring germinating seeds for respirometry	
Tulipa spp. (tulip)	Behaviour of chromosomes during meiosis.	Alternative to *Allium ursinum* (see above)
	Vegetative propagation by bulb	
Utricularia spp. (bladderwort)	Example of aquatic insectivorous plant	
Vicia faba (broad bean)	Example of non-endospermous dicotyledon with hypogeal germination	
Zea mays (maize)	Genetic crosses (a wide range of prepared cobs available from suppliers).	Nuffield A Level Biological Science (1970) *Organisms and Populations*, Penguin Books.
	Example of endospermous monocotyledon with hypogeal germination.	Bingham, C. D. (1976) *Plants*, English Universities Press for the Schools Council.
	Respiring material for respirometry etc. (germinating seeds)	

210   School science laboratories

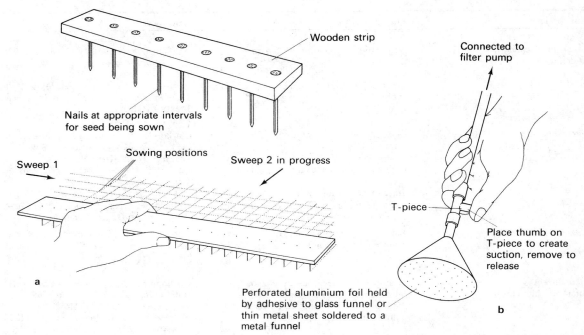

Figure 8.3   Devices for ensuring the even spacing of seeds. a) 'Sweeping' with a sowing comb. b) A sowing funnel which works by suction

with a fungicide at this stage to prevent infection. Transfer the seedlings to pots or deeper seed boxes soon after the seed leaves or first leaves open out. With large species it may be necessary to transfer to larger pots again at a later stage.

Cuttings. A number of species (e.g. *Pelargonium, Tradescantia, Impatiens, Coleus*) may be propagated from stem cuttings. Remove short pieces of shoot in early summer by cutting just below a node with a sharp knife. Then remove some of the leaves at the base of the cutting. Insert the base of the cutting into a small hole in a mixture of compost and coarse sand placed in either a pot or a seed box. Maintain at 20–25 °C. Plants should have rooted within 2–5 weeks. In some species the use of hormonal rooting compounds may speed up the formation of roots. The instructions provided with these compounds should be strictly adhered to. Further details will be found in the Nuffield Junior Science Project (1967).

Watering. It is impossible to give an exact indication of the amount of water required by plants other than stating that it must be adequate to maintain healthy growth. In most cases, water should be provided when the surface of the soil is seen to be dry. In summer, when the weather is warm and the hours of daylight long, the soil should be kept permanently moist and may require daily attention in some cases. In colder winter conditions less water is required and just enough should be provided to prevent

wilting. The frequency of watering can be reduced by any of the following:

i grouping the plants together, which will increase the humidity of the microclimate and hence reduce the demand for water;
ii sinking the plant pots to their rims in moist peat;
iii placing the plant pots in some form of enclosure (see page 200);
iv passing a wick of glass wool or similar material from the soil level through the soil till it protrudes from the bottom of the pot. Rest the pot on pebbles standing in water in a suitable container so that the lower end of the wick is permanently in water; thus the entire wick, and hence the soil, is kept moist.

However, it should be borne in mind that overwatering is as frequent a cause of poor growth as under-watering.

Most plants will recover quickly from wilting when they are watered provided they have not been in the wilted state for too long. In some cases a plant may wilt when the soil is wet; this may be due to the roots having rotted because of overwatering. A possible cure is to put the pot in a warm place with the soil almost dry in the hope that the plant will form new roots. At other times a plant may wilt when the soil is apparently moist. This may be due to the lower parts of the pot being dry although the surface of the soil is wet. In this case it is best to stand the pot in

Genus	Common name	Germination medium	Germination temperature (°C)	Germination time (days)
Allium	Onion, garlic	Paper, sand, compost	10–20	5–14
Avena	Oat	Paper*, sand, compost	20	4–10
Beta	Beet	Sand, compost	20–30	7–14
Brassica	Mustard, cabbage, etc.	Paper, compost	20–30	3–10
Cucurbita	Cucumber, marrow, etc.	Sand, compost	20–30	5–14
Helianthus	Sunflower	Paper, sand, compost	20–30	4–10
Hordeum	Barley	Paper*, sand, compost	20	3–10
Lycopersicon	Tomato	Paper, compost	20–30	3–10
Nicotiana	Tobacco	Paper, compost	20–30	5–14
Phaseolus	Runner bean	Paper*, sand, compost	20	4–10
Pisum	Pea	Paper*, sand, compost	20	3–10
Raphanus	Radish	Paper, compost	20–30	3–10
Spinacia	Spinach	Sand, compost	20–30	4–14
Trifolium	Clover	Paper, sand, compost	20	3–10
Triticum	Wheat	Paper*, sand, compost	20	3–10
Vicia	Broad bean	Paper, sand, compost	20	5–10
Zea	Maize	Paper*, sand, compost	20–30	4–10

* Paper only suitable for germination if subsequent growth is not required.

For paper germination, place seeds between several layers of moist filter paper or blotting paper. Smaller seeds may be germinated in Petri dishes but larger seeds (e.g. beans, peas), may be sandwiched between the paper and the sides of a jam jar with 2–3 cm water in the bottom of the jar.

Table 8.2 Germination details of some plants commonly grown in school laboratories

water to the level of its rim until the soil becomes thoroughly moist.

Fertilisers. The use of nutrient solutions is recommended when plants have been in the same pot for some months. Suitable fertilisers are available from gardening shops and other commercial suppliers in either liquid or powder form and the manufacturer's instructions should be followed closely. Fertilisers should not normally be used in the winter months when little growth is taking place.

Spraying. If the plants show any sign of competition from other organisms (e.g. insects or fungi), they should be sprayed with a suitable pesticide. The manufacturer's instructions should be followed closely. Spraying should not be carried out in a room in which animals are housed.

Other details. Indoor plants require repotting when they become 'pot-bound', i.e. when the roots almost fill the available space and little soil is visible when the plant is knocked out of the pot. To remove a plant from its pot, place one hand over the soil with two fingers on either side of the plant stem. Turn the plant over and tap the base or rim of the pot sharply. The soil should come out in one clean ball, held together by the roots of the plant. Repot in the next size of pot. If a good deal of soil is visible the plant probably does not need repotting.

Plants should be examined regularly for the following signs of poor growth, ill health or mechanical damage:

i abnormal colouring of the stem and leaves, especially a yellowing of the leaves;
ii abnormal growth of the stem so that it becomes thin with elongated internodes;
iii withering of leaves or parts of leaves and leaf fall at an abnormal time of year;
iv wilting of the leaves and stem;
v growth deformities of any part of the plant;
vi failure to form flowers or of flower buds to open; abnormal flower fall at either the bud or open flower stage.

If any of these symptoms develop, they are probably due to poor maintenance procedure and this should, therefore, be carefully reconsidered.

Details of particular plants

ALLIUM

If root tips are required for squashes demonstrating mitosis, garlic provides an ideal material. Onions or shallots may also be used but are generally less reliable

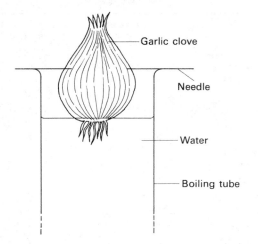

Figure 8.4   An arrangement for sprouting roots from a garlic clove

and slower to produce roots. Stick three pins into a single clove of garlic and rest them on the rim of a boiling tube full of water so that the base of the clove is below the water level (see Figure 8.4). Roots should start to appear within two days and be in a usable state within four days.

CACTI AND OTHER SUCCULENTS

These are normally sold in pots and may be transferred to a sandy soil in a large shallow container to form a desert garden. The plants should not be set deeper than they were in their pots or rotting may occur. Coarse sand or small pebbles may be placed over the surface of the soil to ensure that moisture does not collect around the bases of the plant stems as this will also cause rotting.

In summer, succulents may be generously watered at intervals of a fortnight, provided there is adequate drainage. In the winter months, watering should be kept to a minimum so that shrivelling is just prevented. Although many cacti can withstand low temperatures, it is best to keep them in a warm place. The flat-stemmed succulents known as the Christmas cactus (*Zygocactus*) and the Easter cactus (*Schlumbergera*) require similar conditions and careful watering but should be kept in a shady situation.

CUSCUTA (dodder)

This parasitic species with spindly yellow stems, vestigial leaves and no roots may be sown with clover, tomatoes or broad beans. The seeds have a long and variable dormancy which must be broken by the drastic treatment of placing them in concentrated sulphuric acid for 20 seconds and then rinsing thoroughly with distilled water. The seeds may then be sprinkled onto moist filter paper in a Petri dish and left until germination has produced a seedling some 2 cm long. The seedlings are then placed around the base of clover plants sown 6–8 weeks earlier, or tomato or broad bean plants 15–25 cm high. If the seeds are sown in May, they will produce plenty of vegetative material within two months and should flower soon afterwards. (Seeds of the American species *Cuscuta campestris* can be obtained from the Official Seed Testing Station of the NIAB (see page 249).

References

Bingham, C. D. (1976) *Plants*, English Universities Press for the Schools Council.

Nuffield Junior Science Project (1967) *Animals and Plants*, Collins.

Nuffield A level Biology Project (1971) *Laboratory Book*, Penguin Books.

Protozoans

GENERAL

Various species of protozoan are commonly used for microscopic studies of their structure, movement, feeding and osmoregulation. Some species may also be useful in behavioural studies.

Pure cultures are best obtained from commercial suppliers or from the Culture Centre of Algae and Protozoa (see page 249). Mixed cultures may be grown from pond water, soil and other sources. Examination of these cultures over a period of time can provide a useful study of succession. Furthermore, such mixed cultures can be used as a source of individual species. The culture is transferred to a Petri dish and examined with a binocular microscope. Individuals are selected, withdrawn using a clean pipette, and transferred to the appropriate culture medium. However, this is a time-consuming process and may have to be carried out several times before a pure culture is obtained.

The culture requirements of individual species are outlined below, but the following general points apply for all species of protozoan.

a Glassware must be scrupulously clean for good results. All detergent must be well rinsed off, the final rinse being with the culture medium itself. Items such as teat-pipettes used in the transfer of organisms should be reserved specifically for the purpose. This is best done by inserting the pipette through a rubber bung which is, in turn, inserted into a clean test tube.

b In general, tap water is unsuitable for culture. Although chlorine may be removed by leaving the water standing for a few days, the same is not true of copper and other toxic materials. Rainwater, spring water or pond water are all suitable and may be 'pasteurised' by heating to about 30 °C for a few minutes. Distilled water, preferably glass-distilled, is also suitable. The water should be aerated by shaking before use.

c The pH of the medium should be maintained around 7 or *very slightly* on the alkaline side of neutrality. Some cultures have a tendency to turn acid, in which case dilute sodium hydroxide solution can be added *drop by drop* until neutrality is regained. Alternatively, $10–20 cm^3$ of phosphate buffer (pH 7) can be added to each $1000 cm^3$ of culture when it is first set up.

d Temperatures should preferably be maintained in the range 18–22 °C.

e Unless photosynthetic species are involved, e.g. *Euglena, Volvox*, cultures are best kept in moderately dim light to minimise the growth of algae.

f Culture vessels should always be kept covered to avoid contamination. Narrow-necked vessels can be plugged loosely with cotton weel; larger vessels, e.g. crystallising dishes, beakers or plastic boxes can be covered with a Petri dish or other suitable lid, slightly raised to allow circulation of air. Water which has evaporated can be replaced by distilled or rain water.

g Culture vessels should be stored well away from noxious chemicals. A laboratory in which a gas such as ammonia is frequently prepared is not suitable.

h If cereals are used in the culture, care must be taken to use grain which has not been treated with pesticide.

j It is usually advisable to sub-culture at about the time a culture reaches its peak (often approximately every 4–6 weeks).

A wide range of culture procedures exists in the literature. Some of these are described below, but a very detailed consideration is outside the scope of a publication of this kind. Those wishing to proceed further are referred to Needham (1959). For the identification of protozoa, Jahn and Jahn (1949) is recommended.

References

Jahn, T. L. and Jahn, F. F. (1949) *How to Know the Protozoa*, Wm. C. Brown.

Needham, J. G. (1959) *Culture Methods for Invertebrate Animals*, Dover.

EUGLENOIDS

A number of species of *Euglena* and several other related genera can be maintained under laboratory conditions.

Uses

Euglena is commonly studied as an example of a protist with features of both the animal and plant kingdoms. Its behaviour with respect to light—including light of different frequencies—and its method of locomotion can also be examined in the school situation.

Supply

Euglenoids can often be collected in the wild from puddles in areas frequented by cattle or other livestock. The bright green colour of these puddles is often due to large populations of *Euglena*, but such a source is highly unlikely to provide a pure culture of one species. If single species cultures are required they may be obtained from biological suppliers or from the Culture Centre of Algae and Protozoa (see page 250). *Euglena spirogyra* is a particularly large species (80–180 µm) and is recommended for anatomical studies. *Astasia longa* is a saprozoic species, while *Peranema trichophorum* feeds holozoically.

Culture

Containers. Large beakers make suitable culture vessels. Alternatively, conical flasks, lightly plugged with cotton wool or foam plastic bungs may be used. All glassware must be clean (see **a** page 212).

Medium. A soil-plus-water medium (see page 192) or a modified Kleb's solution (see page 185) are recommended.

External conditions. Cultures should be kept in a well-illuminated situation, but out of direct sunlight. Temperatures should be in the range 18–22°C.

Food. If a soil-plus-water medium is being used, no further food need be added. In the case of Kleb's solution, 3–4 grains of rice per $100 cm^3$ solution should be added when the culture is set up; the rice grains should be boiled for 5–10 minutes before being added to the medium. If *Astasia* is being cultured, 10 g of peptone and 2 g of dipotassium hydrogenphosphate(V) should be added to each litre of Kleb's solution. Media of this type are not suitable for *Peranema*; if this genus is to be cultured, the method described for amoebae using Brandwein's solution should be used (see pages 175 and 214).

Other details. Sub-culturing should be carried out every 2–3 months.

References

Morholt, E., Brandwein, P. F. and Joseph A. (1966) *A Sourcebook for the Biological Sciences*, Harcourt, Brace and World.

Nuffield O level Biology Project (Revised) (1975) *Teachers' Guide 3*, Longman.

AMOEBAE

Amoeba proteus is the most commonly used species, though *A. dubia* is of a similar size and probably equally suitable for most purposes. *Pelomyxa carolinensis* (sometimes called *Chaos chaos* or *C. pelomyxa*) is a much larger species, but its more granular endoplasm, numerous contractile vacuoles and multinucleate structure make it less suitable for elementary anatomical work. It is also rather more difficult to culture. The related shelled Rhizopods *Arcella* and *Difflugia* and Heliozoans such as *Actinosphaerium* are commonly found in fresh water and may also be cultured in the ways described below.

Uses

Amoeba has long been studied in schools and in higher education. It is generally used for anatomical work and for studies of the acellular level of organisation. Its feeding, osmoregulation and mode of locomotion may also be observed.

Supply

Careful examination of the debris at the bottom of a temperate aquarium or of a pond may yield a small number of specimens. Scrapings from the underside of the leaves of the water lily *Nymphaea* and from submerged aquatics such as *Elodea* or *Myriophyllum* often yield large numbers of amoebae. If pure cultures or a more certain supply is needed, the biological suppliers or the Culture Centre of Algae and Protozoa (see page 250) maintain continuous stocks.

Culture

Containers. Crystallising basins with loose fitting lids make suitable culture vessels. The lid of a standard 10 cm glass Petri dish makes an ideal cover for a 250 cm³ crystallising basin.

Media. A wide variety of recommended media exists in the literature. The most successful are probably Brandwein's solution (see page 175), Chalkley's solution (see page 176) or a hay infusion (see page 183). If *Difflugia* is being cultured, small amounts of very fine sand should be added with which this species can construct its shell.

External conditions. Cultures should be kept at temperatures around 18–20 °C and should in no circumstances be allowed to exceed 25 °C. Medium to dim light conditions are desirable; sunlight should certainly be avoided as it will also bring about excessive rises in temperature.

Food. If Brandwein's or Chalkley's solutions are being used, the following procedure should be adopted. Pour a thin layer (1–2 mm) of warm 1 per cent non-nutrient agar into the culture vessel. Embed 4–5 uncooked rice grains in the agar and allow it to set. The amoebae are now introduced along with 20 cm³ of the water in which they have been cultured or from which they have been collected. An equal volume of Brandwein's or Chalkley's solution is also added. Over the next three days, the total culture volume is made up to 100 cm³ by daily additions of the appropriate culture solution.

Pelomyxa and *Actinosphaerium* require large ciliates as a food source. In both cases, a pipetteful of a good *Paramecium* culture should be added to the rice agar

culture before adding the amoebae. If a hay infusion is being used, it should be allowed to stand for 24 hours after boiling. Large numbers of *Chilomonas* or *Colpidium* should then be added and the culture left for 2–3 days before introducing the amoebae. If the numbers of ciliates decline, half the culture solution should be replaced by fresh hay infusion inoculated with *Colpidium* or *Chilomonas*. Four grains of rice or boiled wheat can also be added for each 100 cm³ culture.

Other details. In most cases, sub-culturing should be carried out every two months when the population of amoebae is at its maximum. In the case of *Difflugia* it may be possible to maintain a good culture simply by adding the organism to pond water rich in green filamentous algae such as *Spirogyra*.

Amoebae tend to congregate on the sides and bottom of the culture vessel. If specimens are required for examination they should be removed with a pipette. Some workers place a very thin layer of sterilised sand at the bottom of a culture; the amoebae then congregate among sand grains with which they can be removed for examination. The sand grains have the added advantage of keeping the coverslip off the microscope slide, hence allowing freedom of movement for the amoebae.

References

Morholt, E., Brandwein, P. F. and Joseph, A. (1966) *A Sourcebook for the Biological Sciences*, Harcourt, Brace and World.

Needham, J. G. (1959) *Culture Methods for Invertebrate Animals*, Dover.

PARAMECIUM and other free-living ciliates

A wide selection of ciliate species can be found in fresh water, especially if it is rich in decaying organic matter. A number of species are commonly used in schools.

Uses

Paramecium caudatum is probably the most widely studied protozoan species and is much used in schools. It provides a well documented example of ciliate organisation, while its behaviour, reproduction, feeding and osmoregulation are all worthy of investigation. *P. bursaria*, although smaller than *P. caudatum*, is of interest because of the symbiotic *zoochlorellae* found in its endoplasm. It is also commonly used for studies of conjugation. *Spirostomum* is an elongated species, some 2–3 mm in length and is especially suitable for behavioural work.

Vorticella and *Stentor* are stalked forms which are frequently found attached to submerged waterweeds or to floating duckweeds (*Lemna*). Both species can detach themselves from the substrate and become free-swimming. When they are in their attached state, they are especially suitable for work on feeding.

Podophrya is a smaller stalked form particularly suitable for observations of the contractile vacuole.

Colpidium and *Chilomonas* are small, free-swimming species frequently cultured to provide food for amoebae and other larger protozoa.

Didinium is a voracious feeder upon other ciliates including *Paramecium* and provides a good example of a protozoan which undergoes encystment.

Supply

All of the species mentioned above are frequently encountered in fresh water. A careful microscopic examination of the leaves of submerged plants, either from a natural or aquarium situation will usually yield a wide selection of species. Tropical aquaria are especially rich sources. Pure cultures of most of these species may be obtained from the Culture Centre of Algae and Protozoa or from commercial biological suppliers (see page 250).

Culture

Containers. 250 cm³ crystallising basins with 10 cm glass Petri dish lids as covers, or conical flasks lightly plugged with cotton wool are suitable as culture vessels.

Media
a Brandwein's solution (see page 175) is suitable for most species.
b A hay infusion (see page 183) is also recommended for *Paramecium* and *Spirostomum*.
c An egg yolk medium (see page 180) is especially suitable for *Vorticella* and *Stentor* as well as *Paramecium* and *Spirostomum*.
d A wheat grain infusion (see page 193) is widely recommended for a variety of species.
e Dried yeast medium (see page 180) and a dried milk medium (see page 180) are also suitable for most of the species mentioned above.

External conditions. The temperature range is rather less critical for most ciliates than it is for amoebae, although temperatures in the region of 20 °C are ideal. Relatively low levels of illumination are recommended.

Food. If Brandwein's solution is being used, the same procedure should be adopted as described for amoebae, with the difference that about *eight* uncooked rice grains should be added per 100 cm³ of culture. With all the other media, sufficient food is provided by the medium itself in the case of the smaller species. For the larger species (e.g. *Stentor*, *Spirostomum* as well as *Paramecium caudatum* and *Podophrya*), a pipetteful of *Chilomonas* culture should be added when setting up. *Didinium* must be supplied with food such as *Paramecium*.

Other details. In most of the methods described above, a

rich culture should be obtained within 2–3 weeks. Subculturing should be carried out soon afterwards. If *Podophrya* is being cultured, a few strands of undyed silk fibre provide an ideal substrate to which this species will become attached.

References

Hawkins, P. W., *Vorticella*, A Suitable Protozoan for Class Practical Work, *School Science Review*, 1973, 191, **55**, pp. 308–313.

Morholt, E., Brandwein, P. F. and Joseph, A. (1966) *A Sourcebook for the Biological Sciences*, Harcourt, Brace and World.

Needham, J. G. (1959) *Culture Methods for Invertebrate Animals*, Dover.

Coelenterates

HYDRA

Several species are suitable for school culture and use:

Chlorohydra viridissima The colour of the green hydra is due to the presence of symbiotic algae (*Zoochlorella*) in the endodermal cells. The extended body may be up to 30 mm in length with the tentacles relatively short—never longer than the body. It is hermaphrodite with the testes usually maturing first.

Hydra oligactis This grey–brown species has 4–6 very long tentacles and a body length of up to 30 mm. The sexes are separate.

Hydra vulgaris (*Hydra fusca*) The common hydra is a yellowish-brown hermaphrodite with a body length of up to 20 mm. When fully extended, the tentacles are roughly twice as long as the body.

Uses

Hydra is most frequently used for the following purposes:

  i to illustrate a particular level of organisation;
 ii to demonstrate asexual reproduction in animals;
iii to observe response to stimulus—particularly in feeding.

Chlorohydra viridissima with *Zoochlorella* may also be used as an example of symbiosis.

Supply

Hydra may be collected from ponds, canals, slow-moving rivers and streams where they are found attached to waterweed and stones. They may also be found on leaves which have fallen into ponds in autumn. Biological suppliers generally stock cultures of hydra which may be described in their catalogues as simply 'green' or 'brown' (see page 250).

Culture

Containers. Glass crystallising dishes holding 75–250 cm³ are suitable culture vessels. The medium should be maintained at a depth of 3–6 cm and the vessels should be kept covered (e.g. with an inverted Petri lid). All glassware must be scrupulously clean and should be changed every 2–3 weeks.

Medium. Suitable water is essential for successful laboratory culture.

There are a number of recommended artificial media, such as that used by Holt (1973). But, for most purposes, water from a pond known to contain hydra is probably ideal.

The suitability of the medium can be tested by placing a few specimens in it. Full expansion of the body and tentacles indicates that it is suitable. The use of tap water should be avoided.

External conditions. Culture vessels must be maintained in a cool environment—temperatures above 20 °C can prove fatal. *Chlorohydra viridissima* requires some sunlight. Cultures will also survive for several months in a domestic refrigerator where they require no feeding. This may prove particularly useful during holiday periods. In these cold conditions, sex organs may develop.

Food. Hydra is a voracious feeder and for successful culture there must be a plentiful supply of food. *Daphnia*, *Cyclops* or *Artemia* are readily taken. Cultures maintained solely on *Daphnia* frequently undergo 'depression' which begins with the contraction of the body and tentacles and ends in death. *Artemia* (see page 222) is therefore preferable if cultures are to be maintained for more than a few weeks. The food should be collected in a small sieve, thoroughly washed in the hydra culture medium and fed live to the hydra. After 1–2 hours, excess food should be removed with a teat pipette. Within 5–24 hours of feeding, half the culture medium should be poured away and replaced with fresh medium.

For maximum reproductive activity, the cultures should be fed daily and then the hydra will bud about once a week. However, the cultures will survive for several weeks without feeding, although no reproduction occurs under these conditions.

Reference

Holt, S., Continuous Culture of Hydra, *School Science Review*, 1973, 187, **54**, pp. 309–310.

SEA ANEMONES

Two species of sea anemone are widely distributed and suitable for school laboratory maintenance: *Actinia equina* and *Anemonia sulcata*. The former retracts its tentacles into the column in adverse conditions and when it is no longer covered with water; the latter is unable to do so.

Uses

Anemones may be used for anatomical purposes as examples of their phylum, for behavioural work involving feeding responses and for an examination of nematocysts and their discharge.

Supply

They are best collected fresh from a rocky shore although they may be available from marine biological suppliers (see page 250). It is important that animals are obtained in good condition if they are to be maintained in the laboratory. Transport from the shore to the school may be the biggest problem and the apparatus described by St. Aubrey (1969), which helps to maintain a suitably low temperature, is recommended.

Culture

Containers. As with all marine cultures, vessels made of glass or plastic should be used, or failing this, a porcelain sink. Metal-framed aquaria, unless of stainless steel, are to be avoided. The container should be of at least 10 litre capacity and preferably a good deal larger.

Medium. A minimum of 10 litres of fresh sea water is essential; this can be supplemented by artificial sea water (see page 191). Suitable rocks and stones from the shore should be added, but sand and mud are to be avoided. The water and stones should be left in the tank for a week before introducing the animals. At the end of this period any bacteria should be removed from the surface with blotting paper. A water filtering device should then be set up. This will provide calcium ions as well as filtering the water. Green algae, such as *Enteromorpha* and *Ulva*, may also be introduced—preferably already attached to stones.

External conditions. The main problems are aeration and temperature control. The filtering device and the plants should solve the former, while the siting of the aquarium away from heating and direct sunlight should help to keep the temperature down to the required range of 6–15°C.

Food. Small pieces of meat, fish and cheese will be taken as will pieces of filter paper spread with 'Marmite'.

Other details. It may be found most convenient to maintain anemones in association with other littoral organisms in a marine aquarium (see page 284).

Reference

St. Aubrey, Sheila N. D., The Maintenance and Transportation of Marine Intertidal and Planktonic Animals for Laboratory Use, *School Science Review*, 1969, 174, **51**, pp. 103–106.

Platyhelminths

PLANARIA

Eight species of planarian occur in fresh water in Britain and they are easily distinguished (see Nuffield O level

Biology Project (1966)). *Polycelis nigra* is probably the easiest to culture under laboratory conditions.

Uses

Planaria are especially useful for work on regeneration and for behavioural work investigating kineses. They may also be used in studying the acoelomate triploblastic level of organisation.

Supply

They may be collected from ponds, lakes or canals by examining the under surfaces of stones and leaves. Stream-living species can be obtained by baiting the edges of the habitat with pieces of raw meat (5 cm × 5 cm × 1 cm are of a suitable size). At intervals of approximately 20 minutes the planaria are removed from the underside of the meat.

They may also be obtained from most biological suppliers (see page 249). The availability of different species may depend upon the time of year.

Culture

Containers. Glass, enamel or crockery dishes are suitable. The worms should be transferred to clean dishes every week.

Medium. Pond water is the best medium in which to maintain cultures of planaria. Tap water has also been found suitable in some cases, though heavily chlorinated water is to be avoided. The water should be maintained at a depth of 3–10 cm and changed 2–3 times per week after feeding. It is important that the water should not be allowed to become foul or in any way contaminated.

External conditions. Planaria should be kept in the dark and in cool conditions. Temperatures above 24 °C should certainly be avoided and for many British species even this may well prove too high.

Food. The worms should be fed on thin strips of raw beef liver 1–3 times per week. After 1–2 hours, the excess food should be removed and the water replaced. Chopped earthworms may also prove a suitable food if beef liver is not available. Planaria which are regenerating should not be fed.

Other details. Planaria rarely reproduce sexually under laboratory conditions. Regeneration, however, can be brought about by cutting off the posterior portion of the worm with a clean, stainless steel razor blade. These tail pieces should be transferred to a separate container and allowed to stand without feeding for 4 weeks. Particular care should be taken to ensure that the water remains clean during this period. After this time they should have attained adult shape and be ready for feeding and growth.

Reference

Nuffield O level Biology Project (1966) *Teacher's Guide III*, Longman/Penguin, pp. 223–226.

FASCIOLA AND OTHER TREMATODES

The liver fluke still commonly infects sheep in some parts of the British Isles. Other flukes are frequently encountered parasitising amphibians.

Use

For studying adaptations to the parasitic mode of life.

Supply

Biological suppliers maintain stocks of preserved specimens of *Fasciola* and sometimes of other species. Living specimens of mammalian parasites must not be used in the school situation for fear of infecting the pupils.

Polystoma, *Haplometra* and other species may be obtained during dissection of the frog, toad or other amphibians, (see Cox (1971)).

Culture

Laboratory culture of parasitic flatworms of any species is extremely difficult and in many cases impossible. It is also inadvisable to maintain such cultures in schools.

Reference

Cox, F. E. G. (1971) Parasites of British Amphibians, *J. Biol. Ed.*, **5**, 1, pp. 34–51.

TAENIA AND OTHER CESTODES

Human infection with *Taenia* is uncommon in Britain today. Dogs, however, may become infected with a number of species of tapeworm, e.g. *Taenia pisiformes*, *Echinococcus granulosus*, *Dipylidium caninum* and others. As these parasites may also be transmitted to human hosts, live material must not be used in schools. The cylindrical tapeworm *Nematotaenia dispar* is commonly found in amphibian guts and, under appropriate supervision, may be used in school work.

Use

For studying adaptations to a parasitic mode of life.

Supply

Preserved specimens of a number of species, including those named above (except *N. dispar*) are obtainable from biological suppliers.

For the reasons mentioned above, living material derived from mammalian hosts should not be used.

Culture

As with trematodes, laboratory culture is both difficult and undesirable in schools.

Nematodes

ANGUILLULA ACETI

The vinegar eel-worm is a non-parasitic species adapted to living in conditions of low pH. It is viviparous and transparent so that all stages of development can be examined in utero. Gross anatomical details may be seen with a 4 mm objective. The worms are about 2 mm long when fully grown.

Uses

Apart from demonstrating general nematode structure and movement, *Anguillula* may also be used to observe stages in embryonic development from the fully formed young worm (near the vagina of the adult female) to the fertilised egg.

Supply

Anguillula is best obtained from biological suppliers (see page 249).

Culture

Containers. Glass culture jars of approximately 1 litre capacity are suitable. A thin sheet of polythene, pierced with 10–20 small holes, should be secured over the mouth of the jar to reduce evaporation.

Medium. The worms are cultured in natural bulk cider vinegar which is obtainable from biological suppliers. If it is obtained from other sources (e.g. industrial vinegar plants), care must be taken to ensure that it has not been chemically treated to inhibit the growth of bacteria and yeast.

External conditions. The temperature should be maintained in the range 20–30 °C

Food. As the worms feed on the bacteria and yeast in the vinegar, no external source of food is needed.

Other details including sub-culturing. Initial cultures should be allowed to multiply for several weeks before being divided between two fresh culture jars of cider vinegar. A little fresh vinegar should be added to the cultures from time to time to replace that lost through evaporation.

After several weeks, the population of worms should have increased enormously and sub-culturing is then necessary. A small portion of the culture is transferred to jars containing fresh cider vinegar.

Reference

Gerrard & Haig, Ltd. (1972) Nuffield Biology O level Catalogue, pp. 24–26.

ASCARIS LUMBRICOIDES

The pig roundworm is a frequent parasite all over the world. It can also infect man. A remarkable feature of *Ascaris* and a number of other parasitic nematodes is the ability of their eggs to withstand adverse conditions including immersion in formaldehyde (methanal) solution. For this reason, even preserved specimens should always be handled with care.

Uses

Ascaris is frequently used as an example of nematode structure. Its size makes it ideal for this purpose. It may also be considered from the point of view of its adaptations to the parasitic mode of life.

Supply

Preserved material may be purchased from biological suppliers. As it can infect man, living specimens must never be used in school.

Culture

Parasitic nematodes must not be maintained live in schools—whatever the species.

OTHER PARASITIC NEMATODES

A variety of species is likely to be encountered during the dissection of the dogfish and frog.

SOIL NEMATODES

These are the easiest living nematodes to obtain if cultures of *Anguillula aceti* are not being maintained on a regular basis. The number of species living in the soil is very large and their identification is not normally a matter of concern at school level.

Uses

The chief use of soil nematodes in the school situation is for the demonstration of nematode movement. Larger specimens may also be suitable for microscopic examination of general structure. It is also worth stressing to pupils that the majority of nematode species are free living. Parasitic forms such as *Ascaris* are in the minority.

Supply

Soil nematodes are best obtained by placing a small quantity of soil rich in humus on a piece of gauze or in a sieve and rinsing it in a beaker of warm water. Large numbers of nematodes will sink to the bottom of the beaker from where they can be removed with a teat pipette. They may also be extracted with a Baermann funnel (see page 143).

Culture

Containers. They are best cultured in sterile Petri dishes.

Medium. Nutrient agar (see page 170) is an ideal medium.

External conditions. Any normal laboratory conditions avoiding extremes of temperature, are suitable.

Food. This is provided by the medium, but bacteria and fungal spores may also be engulfed.

Other details. A drop or two of a mixed population of soil nematodes, extracted in the way outlined above, is placed on the nutrient agar plate. Within 5–10 days, enormous numbers will have been produced. The majority move about on the surface, but some will burrow in the medium. If cultures are to be maintained they can easily be sub-cultured every 2–3 weeks.

Annelids

POMATOCEROS TRIQUETER

This tube-living polychaete is found attached to rocks and stones on the lower shore and in the sub-littoral zone around the Atlantic coast of Europe, North America and Africa. Related species are found in the other oceans. The tube is calcareous and reaches a maximum length of 50 mm.

Use

Pomatoceros has been found particularly useful for demonstrating fertilisation of a living egg. For details of this technique see Nuffield O level Biology Project (Revised) (1975).

Supply

Specimens may either be collected in the field or purchased from suppliers who deal in marine material (see page 250).

Culture

Containers. Plastic or glass containers should be used, as metal ones will be corroded by sea water. The size will depend upon the size of the rocks to which the worms are attached. As temperature is an important factor here, the volume of water which can be maintained within a suitable range will have to be considered.

Medium. The worms are kept in sea water, which can either be collected fresh or made up in the laboratory (see page 191). Good aeration of the water is essential.

External conditions. As with many marine organisms, the main problem is to maintain the culture at a sufficiently low temperature; 4–12 °C is a suitable range and in many schools this will necessitate using the main compartment of a refrigerator during most times of the year.

Food. The simplest food to use is brine-shrimp (*Artemia*) eggs. These can be scattered on the surface of the water and allowed to hatch. Very small quantities of yeast or Horlicks suspension may also be used, but excess should be avoided as the water will become foul.

Other details. Although sexually mature *Pomatoceros* can be found at all times of the year, they occur in largest numbers during spring and autumn.

Reference

Nuffield O level Biology Project (Revised) (1975) *Teachers' Guide 1*, Longman.

NEREIS

This large polychaete is common around the shores of the British Isles. There are several species of which *N. diversicolor* is one of the more common and most suitable for school use and laboratory culture.

Uses

Nereis is a much under-used school laboratory organism. It is especially useful for work on locomotion, behaviour and osmoregulation, but simple experiments of an endocrinological nature can also be performed on this worm. For a fuller treatment, see Evans and Golding (1965).

Supply

The worms are most easily collected by digging in estuarine mud above mid-tide level. They burrow to a maximum depth of about 30 cm and so can easily be reached with a spade or fork. Living material is also available from marine biological suppliers (see page 250).

Cultures

Containers. As with all marine species, glass or plastic vessels should be used. Metal-framed aquaria are not suitable unless the frames are made of stainless steel. *Nereis* is thigmotactic and lengths of glass tubing must be provided into which they can burrow; without this they will roll themselves into a ball and die from asphyxiation. The tubing should be of slightly larger diameter than the worms.

Medium. 50–70 per cent sea water is ideal. This can either be collected from the shore and diluted with tap water or made up in the laboratory (see page 191) and diluted. It must not be allowed to become foul and should be changed at sufficiently frequent intervals to prevent this.

External conditions. The temperatures should be maintained below 18 °C.

Food. In spite of its jaws, *N. diversicolor* is not a carnivorous species but normally feeds on diatoms in the mud and perhaps also on larger green algae such as *Ulva*. In the laboratory, it readily takes Bemax placed at the end of the worm's tube, when reversal of the pharynx can be observed. Care must be taken to ensure that the Bemax is not allowed to foul the water. Changing the water after feeding is a good practice to adopt.

Reference

Evans, S. M. and Golding, D. W., *Nereis* as an Experimental Animal in Schools, *School Science Review*, 1965, 161, **47**, pp. 144–147.

EARTHWORMS

There are 25 species of earthworm found in Britain, all belonging to the family Lumbricidae. Of these, the best known, and that most commonly described in text-books, is *Lumbricus terrestris*. Two other members of the genus *Lumbricus*, three species of *Allolobophora*, *Eisenia foetida* and one or two other species are also common and widely distributed. For identification of the British earthworms see Cernosvitov and Evans (1947). In some parts of the world oligochaetes which are not members of the family Lumbricidae are also referred to as earthworms.

Uses

For many years *Lumbricus* has been used as an example of the phylum Annelida. With modern courses, not based upon a type system, its use in this context is decreasing. At an elementary level in schools it is frequently investigated from the point of view of its external features and for work on its behaviour, burrowing, feeding and locomotion. Earthworms also lend themselves to work of an ecological or physiological nature. (See Miles (1963)).

Supply

Living specimens may be obtained from suppliers or in the field. In the latter case they may be dug up and collected by hand or obtained by pouring one of a number of solutions over the soil. 5 cm^3 of 40 per cent methanal (formaldehyde) solution in 1 litre water, 2–3 g potassium manganate(VII) per litre of water and dilute solutions of detergent have all been found suitable. Approximately 5 litres should be used per m^2 of soil surface. The worms will emerge within 10–20 minutes, though successive applications of the solution will always result in more worms appearing. A copper electrode implanted in the earth can also be used for extraction (see Miles (1963)). As soon as the worms are collected they should be rinsed in water.

For identification of the worms collected, consult Cernosvitov and Evans (1947).

Preserved specimens may also be obtained from the usual suppliers.

Culture

Containers. Large biscuit tins (25 cm × 25 cm × 25 cm) or dustbins are suitable culture vessels depending on the number of worms to be maintained. The vessel should be covered with a sheet of perforated polythene. They may also be kept in observation wormeries (see page 164).

Medium. Mix together 3 parts of medium loam soil with 1 part granular peat. If necessary, water with a suspension of calcium carbonate to make the mixture just alkaline, and moisten to make a good tilth. The mixture should then be placed in the containers over a layer of activated charcoal.

External conditions. The main problem in schools may be maintaining a suitably low temperature, as 10 °C should not be exceeded.

Food. The most convenient food is oatmeal, which can be sprinkled on the surface. It is gradually removed by the worms from below. The exhaustion of the supply is easily detectable. Horse manure or leaf mould can also be used, but is less convenient.

Other details. The number of worms per container depends upon the species. Five *Lumbricus terrestris* per litre of soil is a suitable number, but *Eisenia foetida* may be kept at densities of four times this value. If cocoons are required, the worms should be allowed to settle in the containers for at least six weeks before the surface layers are picked over to search for them. Cultures of this kind can be maintained for well over a year.

References

Cernosvitov, L. and Evans, A. C. (1947) Lumbricidae (Annelida), *Linn. Soc. Lond. Synops. Brit. Fauna*, 6, 36.
Miles, H. B., Some Aspects of the Biology of the Lumbricidae, *School Science Review*, 1963, 155, **45**, pp. 83–92.

ENCHYTRAEID WORMS

These small whiteworms are commonly found in compost heaps, sometimes in enormous numbers.

Uses

Enchytraeids are particularly useful as food for leeches, certain species of small fish and for amphibians (e.g. *Xenopus* tadpoles in the early stages of metamorphosis).

Supply

They can often be bought from pet shops or stores dealing in tropical fish. Alternatively, they may be collected in the field.

Culture

Containers. Probably the most convenient vessel is a large biscuit tin with a few holes pierced in the lid.

Medium. A 5 cm layer of damp peat over the bottom of the tin is suitable.

External conditions. These are not critical but the optimum temperature is about 20 °C.

Food. Furrows are made in the peat into which bread, soaked in milk, is placed. More is added from time to time.

Other details. A few worms are placed in the bread and covered with pieces of flat glass, slate or stone. Within a few weeks, the worms will have formed large colonies.

LEECHES

These are commonly found in freshwater ponds, lakes and rivers. With the exception of the fish leeches—which in most cases are marine—they are fairly easy to maintain.

Uses

Leeches are mainly used as an example of their taxonomic group—the Hirudinea and for observations on their locomotion.

Supply

They may be available sometimes from biological suppliers (see page 249), but are easy to collect in the field where they may be found attached to, or crawling on, waterweed.

Culture

Containers. Small glass vessels such as crystallising dishes can be used. Alternatively, aquaria are suitable for larger numbers.

Medium. Pond water should be used as leeches are susceptible to chlorine and other minerals likely to be present in tap water. A few sprigs of *Elodea* or other suitable waterweed should be added to each container. The water should be changed at the slightest sign of contamination.

External conditions. The containers should be kept out of the sun in a moderately cool situation.

Food. A few living water snails, e.g. *Lymnaea* should be added to the water and replaced as they are eaten.

Other details. Faeces and the dead bodies of leeches or snails should not be allowed to remain in the water. The true blood-sucking and medicinal leeches require a blood meal about once every six months and are therefore not suitable for school laboratory culture over long periods.

Molluscs

TERRESTRIAL SNAILS AND SLUGS

A number of species of snail (e.g. *Helix aspersa, H. pomatia* and *H. (Cepaea) nemoralis*) and of slug (*Agriolimax reticulatus, Limax flavus* and *Arion circumscriptus*) are either widely distributed or easily available and are suitable for school laboratory culture.

Uses

They may be used in behavioural work, for example in simple feeding preference tests, for observation of their locomotion and as examples of their taxonomic group—the Gastropoda.

Supply

Snails and slugs are easily collected in the field—especially in damp hedgerows. Snails are only likely to be found in areas where the soil is of high pH. They may also be obtained from biological suppliers (see page 249).

Culture

Containers. A suitable vivarium can easily be made using a large aquarium (e.g. 30 cm × 20 cm × 20 cm or 45 cm × 30 cm × 30 cm) with a sheet of glass for a cover. The lid should be slightly raised, to allow some circulation of air, and its edges should be taped.

Habitat. The bottom of the container should be covered with a layer of aquarium gravel 2–3 cm deep. Over this should be placed a layer of soil 5 cm thick, kept alkaline with a sufficient amount of lime. A thin layer of leaf mould is added along with one or two stones and pieces of rotten wood. Clumps of moss may also be added to provide shelter for smaller snails and to retain moisture. Conditions should be kept moist but must not be too wet.

External conditions. These are not critical but a cool situation is preferred. Direct sunlight is to be avoided. Some slugs thrive better in darkness.

Food. Snails thrive on leaves of lettuce and bran or porridge oats along with the food they can obtain from the leaf mould. Slugs may be fed on raw potatoes, turnips, lettuce, cabbage and other vegetables.

AQUATIC SNAILS

A variety of species of snail are suitable for aquaria under either tropical or temperate conditions.

Uses

Aquatic snails are important in a balanced aquarium. They help to keep down the level of algae and the very young snails provide food for certain species of fish. Their locomotory, feeding and breathing mechanisms show interesting adaptations.

Supply

Suitable snails can be purchased from biological suppliers (see page 249) or high street aquarists. The following species are recommended for cold-water aquaria: ramshorn (*Planorbis planorbis*), flat ramshorn (*P. complanatus*, white ramshorn (*P. albus*), wandering pond snail (*Limnaea pereger*), ear pond snail (*L. auricularia*), dwarf pond snail (*L. truncatula*) and the moss bladder snail (*Physa fontinalis*).

High street aquarists generally stock a range of species suitable for tropical aquaria.

Culture

The culture of aquatic snails requires the very minimum of effort. They will breed readily in aquaria (see page 281 for details of setting up), feeding off the algae and larger plants in the tank. If the snail population becomes too great they will severely hamper the growth and development of the more delicate water plants.

Arthropods–crustaceans

DAPHNIA

This well known brachiopod is common in freshwater ponds, lakes and canals, especially where there is an abundance of decomposing plant or animal matter. The stronger swimming *Simocephalus* and the copepod *Cyclops* are frequently found in similar habitats.

Uses

Daphnia is a useful organism for a number of purposes. As the carapace is transparent, both peristalsis and the heartbeat may easily be observed microscopically. In the latter case, investigations may be carried out on the variation in the rate of heartbeat with changes in external conditions. The young develop inside the female carapace, so this too can be observed. *Daphnia* may also be used in behavioural work and forms a convenient food for hydra, tadpoles and some small fish.

Supply

Most tropical fish shops keep regular supplies of *Daphnia*, as do the normal biological supply houses (see page 249). They may also be collected in the field.

Culture

Maintaining stocks of *Daphnia* over an extended period is not easy as both food and oxygen supplies have to be maintained at a high level. However, the methods outlined below have proved successful and can also be used for the culture of *Cyclops* and *Simocephalus*.

Containers. Any glass vessel with a large surface area in relation to its capacity is suitable.

Medium. Ideally *Daphnia* should be cultured in the same water in which it was found. Fresh pond water or tap water which has stood overnight are possible alternatives.

External conditions. The temperature should be kept as low as normal laboratory conditions permit and under no circumstances should be allowed to exceed 20 °C. The water must also be kept well aerated.

Food. Daphnia normally feeds on infusoria, bacteria and microscopic algae. They can be maintained in laboratory conditions on a little yeast or Horlicks suspension giving no more than the *Daphnia* can remove in an hour or two, otherwise the water may become foul. It is better to feed 'little and often' rather than to overfeed infrequently. Another method is to place a jar of Knop's solution (see page 185) in strong light. Inoculate with unicellular algae from a green aquarium. Allow to stand for 2–3 days and then add the *Daphnia* together with a few cubic millimetres of hard-boiled egg yolk. A little yeast suspension may also be added at this stage. The yeast and egg should be mixed with a little of the culture solution and the mixture added to the main vessel. A little more food should be added at weekly intervals.

Other details. Cultures are usually supplied in polythene bags and they can be kept in these containers for several weeks by placing in the large compartment of a laboratory refrigerator.

ARTEMIA

The brine shrimp, which reaches a maximum length of about 12 mm, occurs in salt lakes, salt marshes and salt pans in which brine is concentrated. The eggs float on the surface of the brine and do not hatch until after drying. They may therefore be stored in a dry state, under which conditions the eggs may remain viable for several years.

Uses

Artemia is particularly useful as a source of food for hydra.

Supply

The eggs may be obtained from biological suppliers (see page 249) and from tropical fish food stockists.

Culture

Containers. Any glass, plastic or enamel dish which provides a large surface area is suitable.

Medium. Natural or artificial sea water make suitable media. Alternatively, place 10–15 g sodium chloride in 1 litre of water.

External conditions. These are not critical. 20 °C has been found to be a suitable temperature.

Food. A small quantity of yeast suspended in the appropriate volume of water to compensate for evaporation provides a simple and convenient food. This should be floated on the surface of the medium every 2–3 days.

Other details. If a continuous culture is not being maintained and larvae are simply required as food for hydra, then feeding is not necessary. The larvae will hatch in one or two days and can be removed from the culture with a fine sieve. They should be thoroughly washed in fresh water and in the hydra culture medium before being fed to the hydra.

ASELLUS

Several species of freshwater louse are widely distributed in the British Isles. Of these, *A. aquaticus*, which is common in ponds and slow-moving streams, is probably the most suitable for laboratory culture.

Uses

Asellus is particularly useful for observations of circulation in a small invertebrate. The movement of the blood in the gills is easily seen. It may also be used in simple experimental work on respiration and animal behaviour.

Supply

They may be collected in the field or obtained through commercial suppliers (see page 249).

Culture

Containers. Aquaria, sinks or large plastic containers are suitable.

Medium. Pond water is suitable. It should contain plenty of rotting leaves and wood.

External conditions. The temperature should not be allowed to rise excessively. Relatively constant conditions can be maintained if the containers (e.g. old sinks) are sunk in the ground. Diurnal fluctuations are thus minimised.

Food. As *Asellus* is a detritus feeder, it will obtain a plentiful supply of food from the rotting vegetation in the medium. Elm and lettuce leaves appear to be particularly suitable.

Other details. If a continuous culture is to be maintained, then gravid females may be collected from late March to May.

WOODLICE

There are over forty species of woodlouse found in Britain, of which half-a-dozen or so are very common and widely distributed. Sutton (1972) provides a useful key. *Armadillium vulgare*, *Oniscus asellus*, *Philoscia muscorum* and *Porcellio scaber* are all suitable for laboratory culture.

Uses

Their widespread distribution and ease of culture makes woodlice useful laboratory animals for a range of simple experimental work, particularly on respiration and behaviour.

Supply

Although obtainable from the normal supply houses, woodlice are very easily collected from under stones, logs and fallen leaves, where conditions are both dark and damp.

Culture

Containers. Glass or plastic containers, some 20 cm × 15 cm × 10 cm are suitable. Plastic lunch boxes are ideal; in this case several holes should be drilled in the lid to provide ventilation. If a glass lid is being used, it should be raised along one side by placing two short lengths of split plastic tubing over the edge of the glass container. The outside of the container should be painted black.

Habitat. Humid conditions are essential. These can be provided by covering the floor of the container with a 2–3 cm layer of moist sawdust or of plaster of Paris which is then moistened. In either case, a thin layer of aquarium gravel should be added. Alternatively, a 2–3 cm layer of moistened sand and peat moss with some leaf mould can be used. Several stones, or better still, pieces of broken plant pot should be placed above the moist layer. The woodlice will congregate beneath these.

External conditions. A reasonably cool temperature (certainly not above 20 °C) and darkness (provided by painting the container) are all that are required.

Food. One or two pieces of rotting wood or bark will provide a long-term supply of food. Small pieces of raw carrot or potato should also be added occasionally.

Other details. More successful cultures will be maintained if they are restricted to members of one species. Conditions should not be allowed to become too crowded or recently moulted individuals will be killed by the others.

References

Sutton, S. L. (1972) *Woodlice*, Ginn.
Williams, R. J., A Woodlouse Vivarium, *School Science Review*, 1969, 172, **50**, pp. 566–568.

CARCINUS

The shore crab is found in great numbers on British shores, particularly those which are rocky. Specimens with a carapace width of up to 5 cm are common.

Uses

Carcinus is under-used in school laboratories. It lends itself particularly to work on behaviour including simple experimental work on learning. (See Palmer (1969)).

Supply

The crabs can be gathered easily on field trips to the sea shore. They may also be obtained from marine biological suppliers (see page 250). If they are placed in a polythene bag with some seaweed they will survive for about a week.

Culture

Containers. Aquaria of plastic or glass are suitable for larger specimens. Metal-framed aquaria should not be used unless they are constructed of stainless steel. Jam jars or similar containers are suitable for smaller specimens.

Habitat. A few small rocks placed on the floor of the tank will encourage the crabs to set up territories. Sea water should then be added to a depth of 2–5 cm. Seaweed is to be avoided as it will rot and make the water foul. The water, which may either be collected fresh or made up (see page 191) should be changed whenever it becomes cloudy or begins to smell.

External conditions. Aeration of the water is desirable, but not essential. The containers should be kept in as cool a part of the laboratory as possible and certainly well away from central heating pipes or outlets.

Food. The crabs seem to prefer live food and this helps in preventing the water becoming foul. *Tubifex* worms, available from pet shops, are ideal. Alternatively, raw meat in small pieces, earthworms or similar material may be used.

Other details. The larger the crab, the longer the life in the laboratory. Populations of crabs of mixed sizes and overcrowding should be avoided because of the risk of cannibalism.

Reference

Palmer, V. A., The Shore Crab (*Carcinus maenas*) as an Experimental Animal in the School Laboratory, *School Science Review*, 1969, 173, **50**, pp. 807–815.

Insects and myriapods

ANAGASTER (EPHESTIA) KUHNIELLA

The Mediterranean flour moth and several related species are common pests of stored flour and other cereals and cereal products.

Uses

Flour moths can be used in experimental work on behaviour (e.g. on phototaxis). They are parasitised by the ichneumon *Devorgilla* and it is for studies of this example of biological control that *Anagaster* is normally used in schools.

Supply

They may be purchased from biological suppliers, especially those specialising in insects (see page 249). On occasions they may also be obtained from warehouses and other places where flour is stored.

Culture

Laboratory culture is relatively easy and large numbers may be bred in a short space of time.

Containers. A screw-top jar with a ventilated lid forms an ideal culture vessel.

External conditions. Room temperatures of around 20 °C are quite satisfactory, though the life cycle is completed rather more rapidly (in about one month) at 25 °C. All stages can be retarded with little risk of injury by chilling to around 4 °C. A relative humidity of around 60–70 per cent is desirable.

Food medium. Wholemeal flour or porridge oats with a little powdered dried yeast (5 per cent) are recommended. The food should be sterilised for 3 hours at 70–80 °C prior to use, to prevent infection of the culture by mites. It should then be poured into the culture vessel to a depth of 5–8 cm. A specimen tube half filled with water and lightly plugged with cotton wool should be inserted upright in the medium to provide the necessary humidity. Wetting of the cotton wool or of the medium itself should be avoided, otherwise moulds will develop.

Other details. *Anagaster* cultures may fail owing to competition from beetles and/or mites. Effective sterilisation is the cure for this. They may also become parasitised by a range of protozoan and fungal species. In this case, the culture should be destroyed and the equipment thoroughly washed and sterilised by autoclaving before being used again. Every few months, when the culture becomes matted with the threads produced by the larvae, sub-culturing should be carried out.

References

Nuffield O level Biology Project (1966) *Teachers' Guide IV*, Longman/Penguin.
Nuffield A level Biology Project (1971) *Laboratory Book*, Penguin Books.

ANTS

Ants may be maintained in the laboratory on a temporary basis.

Uses

Ants provide a good example of insects exhibiting polymorphism and a complex social organisation.

Supply

A small colony may be collected in the wild. Care should be taken to include a queen as well as workers.

Culture

Containers. A large battery jar or moulded glass aquarium provides a suitable observation colony container. It should be half filled with slightly moistened sandy soil. A more elaborate type of container is described in Needham (1959).

External conditions. Laboratory temperatures are quite adequate. The container should be kept covered or in the dark so that the ants make their galleries alongside the glass. The container should be stood in a basin of water to provide a moat which will prevent the ants escaping.

Food. Fresh vegetables, breadcrumbs and honey provide a good diet. The addition of a dead insect (e.g. locust) from time to time is advisable. It is important to remove uneaten food before it becomes mouldy.

Reference

Needham, J. G. *et al.* (1959) *Culture Methods for Invertebrate Animals*, Dover Publications.

APANTELES GLOMERATUS

This ichneumon commonly parasitises the larvae and pupae of the cabbage white butterfly (*Pieris brassicae*).

Uses

The parasitism of a commercial pest and the subsequent reduction in numbers of the pest itself provides a good example of parasitism and of biological control.

Supply

Apanteles is sometimes available from biological supply houses (see page 249). Alternatively, if a number of larvae and pupae of *Pieris brassicae* are collected a proportion is likely to be parasitised by *Apanteles*.

Culture

This depends upon the maintenance of a stock of *Pieris brassicae* (see page 226).

Containers. Any containers suitable for the rearing of butterflies can be used (see page 226).

External conditions. A room temperature of around 20 °C is suitable. A relative humidity of 70 per cent should be maintained.

Food. The adults will feed on honey diluted with a little water and placed in a shallow dish. As the larvae are entirely parasitic, the adults must be provided with newly hatched *Pieris brassicae* larvae in which they can oviposit.

Other details. The degree of parasitism of a batch of larvae may be altered by changing the proportion of larvae to adult *Apanteles* when oviposition is taking place. The larvae emerge from the *Pieris* larvae when the latter pupate and they themselves pupate shortly afterwards after spinning a cocoon. Adults emerge in 2–3 weeks. The life span of the adults may be extended by keeping them in a cool dark area.

Reference

Nuffield A level Biology Project (1971) *Laboratory Book*, Penguin Books.

APHIDS

Greenfly (*Aphis* spp.) and blackfly (*Myzus* spp.) are common pests of a number of garden plants and are only too well known to gardeners.

Uses

They may be kept as an example of an insect exhibiting parthenogenesis and vivipary. There is also a degree of dimorphism among the females, with winged and wingless forms. Behavioural work on food preferences can be carried out as can studies of distribution. The feeding mechanism and the rates of reproduction at different temperatures can also be investigated.

Supply

There should be no difficulty in collecting specimens from host plants during the summer months. Roses are a common source of greenfly while blackfly may be obtained from *Nasturtium* and broad bean plants. In winter, specimens may be obtainable from greenhouses.

Culture

Containers. The cylindrical perspex insect cages with a metal base and perforated metal lid are ideal. The host plants should be grown in pots and placed on the metal bases of the cages. The aphids can then be transferred to the hosts with a paint brush. For small plants a glass chimney (the type made for hurricane lamps) can be stood around the plant with muslin secured over the top with an elastic band.

External conditions. These are not critical, but temperatures in the range 20–30 °C are recommended.

Food. The host plant provides the food and hence must be grown in pots prior to the collection of the aphids.

References

Nuffield A level Biology Project (1971) *Laboratory Book*, Penguin Books.
Nuffield O level Biology Project (Revised) (1974) *Teachers' Guide I*, Longman.

AQUATIC INSECTS

A wide range of aquatic insects can be maintained in aquaria under laboratory conditions. For further details see page 284.

BEES

The honey bee (*Apis mellifera*) is cultivated in many parts of the world for its honey. This species lives in hives or colonies numbering as many as 100 000 individuals. Beekeeping should be started in a school only if there is a genuine interest on the part of both the staff and the pupils, as it is a long-term activity involving a certain amount of hazard if it is not practised correctly. Before starting it is advisable to contact the local Beekeepers' Association or an adviser on apiculture. This can lead to observation of an experienced beekeeper and some involvement with the work before a more permanent commitment to beekeeping is made.

Uses

Bees are one of the best known examples of a social insect with a complex behavioural organisation involving polymorphism and division of labour. They can also prove a source of finance through the sale of the honey.

Supply

The local Beekeepers' Association may be able to put schools in contact with a keeper who has a swarm he is prepared to give to an enthusiastic beginner. Alternatively, they can be purchased from suppliers (see page 249).

Culture

Containers. Various designs of hive are available from commercial suppliers of apiarist equipment. If more than one colony is contemplated it is best to start with two identical hives so that parts are interchangeable. Old hives should be thoroughly cleaned by scraping and then flamed with a blowlamp to ensure the destruction of any disease spores. An observation hive fitted to a laboratory window is an invaluable addition in the school situation. It should carry at least three frames which will house a colony large enough to maintain itself through the summer. A syrup of sugar and water will have to be provided during the winter for a colony of this size.

External conditions. Natural temperatures are suitable in the United Kingdom, though the foraging season may be rather shorter in more northern parts of the country. The siting of a hive requires some thought as it needs to be well away from possible interference.

Other details. The details of culture and maintenance including handling, etc., are best obtained from a specialist book on the subject (see References) or from a local beekeeper.

References

Butler, C. G. (1961) Ministry of Agriculture, Fisheries and Food Bulletin No. 9, *Beekeeping*, HMSO.

Manley, R. O. (1948) *Beekeeping in Britain*, Faber & Faber.

Young Farmers Booklet No. 2, *Beekeeping*, Evans.

Publications of the Bee Research Association, Hill House, Chalfont St. Peter, Gerrard's Cross, Buckinghamshire.

BUTTERFLIES AND MOTHS

A large range of butterflies and moths, both temperate and tropical, are suitable for laboratory culture.

Uses

Butterflies and moths are normally kept as examples of insects exhibiting complete metamorphosis in their life cycles. They may also be kept for studies on ichneumons and other animals which parasitise them.

Supply

They may be obtained as eggs, larvae (caterpillars) or pupae (chrysalids) from biological suppliers and from those specialising in insect specimens (see page 249). The more common species may be collected in the field.

Culture

Containers. For the eggs and early larval stages, transparent plastic boxes are a convenient form of container. For the larger larval stages, a cage is often found to be more satisfactory. Cylindrical plastic insect cages are obtainable from suppliers; these have a metal base and a perforated metal lid.

Alternatively, a cage may be made with a wooden frame and panels of netting. Old beam balance boxes make ideal cages if some of the glass panels are replaced with netting. Wood and netting cages have the great advantage that they can be used for imagos as well as larvae and they can also be stood in the greenhouse or in bright sunlight without fear of overheating. As an alternative method, cheesecloth sleeves may be placed over parts of the plants containing the larvae either in the wild or on potted plants in the laboratory or greenhouse.

External conditions. If the insects are being kept in plastic boxes, direct sunlight should be avoided and they are best kept in a north-facing situation. If netting-covered boxes are used, the aspect is less critical. For temperate species such as the Large White Butterfly (*Pieris brassicae*), room temperatures are quite suitable although the optimum temperature for development in this case is 24 °C.

Tropical species can tolerate higher temperatures, but levels above 32 °C should be avoided. Dampness can be a problem, especially in plastic boxes which should be lined with a sheet of absorbent paper. This should be removed daily, along with any uneaten plant material and frass, and replaced by another sheet.

Food. Most butterflies and moths are highly specific in their food requirements. Suppliers will give details of the food plants of the particular species they provide. Care should be taken to ascertain that a continuous supply of food will be available before commencing the culture of a particular species.

In general, it is easier to rear larvae on growing plants than on leaves removed from the parent plant. Potted plants can be invaluable for this purpose as the pots can be stood in the cage. Food can be stood in water so that it remains fresh a little longer. However it should be remembered that larvae which fall from the plant must be able to return to it—they cannot swim through a jam jar full of water! For this reason, some of the plant material should be allowed to touch the sides of the cage. If separate leaves or sprigs of food are being provided in a plastic box, then this must be replaced daily; if plant material stood in water is being used, then fresh food every 2–3 days should prove sufficient assuming there is always food available. When transferring the larvae from one piece of plant material to another, they may be allowed to crawl onto a paint brush and from this onto the fresh food: they should not be 'brushed' from one plant to another as this can cause damage

Other details. Larvae should not be allowed to pupate in plastic boxes, but should be transferred to a cage or allowed to pupate in sleeves. Depending on the species, pupation may take place on the food plant or on the roof, side or base of the cage. Some moth larvae burrow into the soil at this stage, and in these cases potted plants or a suitable container of soil must be provided. In general, pupae should not be moved if they are secured by silk pads.

In laboratory conditions, parasites are not normally a problem as the parasitic flies and wasps do not have access to the larvae. However, virus and bacterial infections may prove a problem and can be cured only rarely. They are best avoided by preventing overcrowding, dampness or unclean conditions. For this reason, the container should be cleaned out daily.

References

Goodden, R. (1971) *Butterflies*, Paul Hamlyn.
Nuffield Junior Science (1967) *Animals and Plants*, Collins.

CALLIPHORA

The larvae of bluebottles are commonly used as bait by fishermen, who call them 'gentles'. The following notes also apply to similar flies, such as the housefly (*Musca*).

Uses

They are commonly used in simple behavioural work involving choice chambers. The adult flies may also be useful for demonstrating their feeding mechanism. Some amphibians and reptiles will take the larvae as food. Flies also provide an example of an insect with complete metamorphosis.

Supply

'Gentles' or maggots may be purchased from shops specialising in fishing equipment and bait. In summer they may be obtained by placing some rotten meat or tinned dog or cat meat in the open air so that the adult flies lay their eggs on it. The meat should be placed out of the reach of dogs and cats.

Culture

The main problem with the culture of blowflies is the smell generated by the meat. For this reason they cannot be kept in inhabited rooms. Most schools will probably not wish to culture flies on an extensive scale. However, they may wish to raise adults in small numbers from maggots which have been purchased at an advanced stage of development (see below).

Containers. For large-scale culture, a cage consisting of a wooden frame 40–50 cm in each dimension and covered with muslin provides a suitable container. One side should contain a door large enough for inserting a hand. Escape of the flies can be kept to a minimum if the door is placed on the side furthest from the source of light and is hinged to open outwards. A loose curtain of muslin or mosquito netting is then hung inside the door. Dishes containing the food for the larvae and adults can be placed inside the cage.

External conditions. A temperature of about 25 °C is required for continuous culture.

Food. Tinned cat or dog meat is suitable for the larvae and does not smell as strongly as fresh meat. Adults can be supported indefinitely on cane sugar or on diluted milk placed in a shallow dish. If sugar is used, then water must also be provided. This can be done by means of a simple drinking fountain consisting of an upturned beaker of water stood on a filter paper inside a Petri dish.

Other details. For small-scale rearing, the meat, containing eggs and/or larvae, may be placed in a jar to a depth of 2–3 cm. The jar should then be covered with muslin to prevent further eggs being deposited on the meat. When the larvae reach a length of 15 mm, the jar should be filled to within 2–3 cm of the top with sawdust and the muslin replaced. The larvae migrate into the sawdust to pupate. If the flies are being fed to amphibians or reptiles, the sawdust containing larvae and pupae can be placed in the vivarium at this stage. The vivarium must be 'fly-proof' to prevent the escape of the adults when they hatch. Larvae or pupae may be maintained alive for several weeks in a laboratory refrigerator though development is suspended until higher temperatures are restored.

CHORTHIPPUS

The common field grasshopper (*Chorthippus bruneus*) frequents dry habitats in Britain during the summer months.

Uses

It is sometimes used as an alternative to the locust for testis squashes to show meiosis during spermatogenesis. Some people prefer it to the locust as it has a smaller number of chromosomes (in the female of *C. bruneus* $2n = 18$, in the female locust $2n = 24$; males have one chromosome less in each case).

Supply

The grasshoppers can be collected as nymphs or adults between June and August from dry fields, roadside verges and other similar situations.

Culture

Culture methods are basically the same as those for locusts—except in the treatment of the eggs—if an all-the-year-round supply is required.

Containers. Similar cages to those employed for locusts should be used (see page 233).

External conditions. A temperature of around 25 °C should be maintained by a suitable arrangement of electric lamps. It is important to keep at least one lamp in the same part of the cage as the insects, as they require a source of radiant heat. A low humidity (around 40 per cent) is essential for successful culture.

Food. Natural grass should be supplied. If this is unavailable, it can be grown in seed trays. A quick-growing variety such as *Lolium multiflorum* var Westernworth is recommended for this.

Other details. Chorthippus resembles a locust in that the eggs are laid in moist sand, which must be kept moist to prevent the egg pods drying out. *Chorthippus* naturally overwinters in the egg stage, but this can be broken by the following technique:

  i remove pods from the sand by sieving;
 ii keep them in water at 25 °C for two weeks. In this and the following stages, the egg pods are most conveniently kept in sieves so that they can be easily removed from the water;
iii chill in water to 5 °C and keep at this temperature for 1–3 months;
 iv return to water at 25 °C for 10 days; incubation, which takes a fortnight, begins at this stage;

 v tip onto moist filter paper in Petri dishes. The first nymphs should hatch in a few days.

Reference

Nuffield A level Biology Project (1971) *Laboratory Book*, Penguin Books.

COCKROACHES

Two species of cockroach are commonly found in the British Isles. *Blatta orientalis* is almost black in colour and the female has vestigial wings. *Periplaneta americana* is somewhat larger and dark golden brown in colour. The latter is probably the most suitable for school laboratory culture.

Uses

Recent changes in biological curricula have meant that cockroaches are less frequently used for anatomical work than in the past. However, they are still useful insects to maintain for a range of experimental work on behaviour, locomotion, digestion, etc. They are a good example of an insect with incomplete metamorphosis and they also frequently contain a range of parasites.

Supply

Cockroaches are available from general biological suppliers and from some firms specialising in the supply of insects (see page 249).

Culture

The main problems of laboratory culture centre on preventing escape and providing conditions which minimise cannibalism.

Containers. For long-term culture, a large cage is recommended. A suitable design has been devised by Paice (1968) and is shown in Figure 8.5. Food and water are easily provided through the funnels. Some experimentation will be required to determine the power of lamps needed to provide a suitable temperature. Frass falls through the perforated floor and can be cleaned out every few months. The vertical boards form surfaces over which the nymphs and adults can disperse and between which the egg capsules are laid. An alternative design is given by Rochford (1957).

A much simpler means of culture involves the use of a small or medium sized aquarium or vivarium—an old cracked and leaking one is quite satisfactory. The aquarium is covered with a sheet of glass and is placed on an inverted biscuit tin fitted with a lamp to provide a heat source. 1–2 cm of sawdust is placed in the bottom along with a Petri dish of water and some crumpled newspaper. The sawdust should be changed annually.

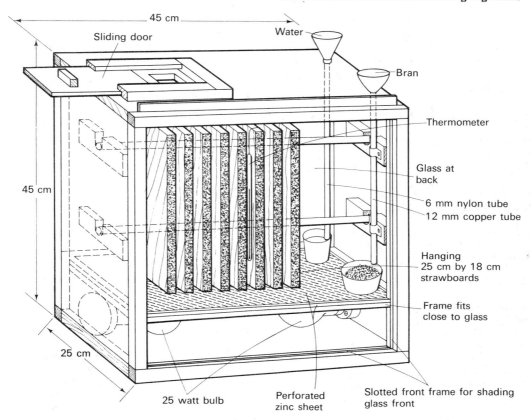

Figure 8.5   A breeding cage for cockroaches constructed in 12 mm blockboard

External conditions. The temperature should be kept around 25–30 °C for successful maintenance and breeding. The level of activity of the insects drops with falling temperature. This can be brought about deliberately (to around 10 °C) to facilitate capture and handling. Humid conditions should be maintained inside the cage at all times.

Food. Bran or porridge oats, together with a few pieces of carrot, potato and apple make a suitable diet.

Other details. Under optimum conditions, the life cycle of *Periplaneta* takes 4–6 months. With either of the techniques described, a mixed culture of nymphs and adults should be obtained within a year if it is started with 8–12 adults and a preponderance of females.

References

Paice, P. A. M., Cockroach breeding, *School Science Review*, 1968, 171, **50**, pp. 332–3.
Rochford, P. J., A Breeding Observation Chamber for the American Cockroach, *School Science Review*, 1957, 136, **38**, pp. 450–2.

Wells, R. V., A Note on the Laboratory Culture of Cockroaches, *School Science Review*, 1961, 148, **42**, pp. 520–1.

DEVORGILLA CANESCENS

This ichneumon is parasitic in its larval stages upon the larvae of the flour moth *Anagasta kuhniella*.

Uses

It is normally kept as an example of a parasite and to demonstrate biological control in the laboratory. Some behavioural work can also be carried out on the preference of the female for oviposition in *Anagasta* as opposed to other similar larvae (see Nuffield O level Biology Project (1966)).

Supply

Devorgilla may be obtained from general biological suppliers and from those specialising in insects (see page 249).

Culture

Provided *Anagasta* can be successfully cultured, there should be no great problem in maintaining stocks of *Devorgilla*. The females lay their eggs parthenogenetically, males being very rare. This greatly simplifies their laboratory culture.

Containers. The adults have a short life of less than a week and at this stage should be kept on their own in cylindrical perspex insect cages. The larvae will obviously be kept along with the *Anagasta* larvae which they are parasitising.

External conditions. A temperature of around 25 °C is suitable for both larvae and adults.

Food. The adults will take a little of the juice from soaked and split raisins. The larvae are parasitic on *Anagasta* throughout their lives.

Other details. After the emergence of the adults, they should be placed in a container with some *Anagasta* larvae in which they should oviposit. They will normally do this in larvae of any stage and, if only one larva is presented, the female will lay many eggs in it—though only one *Devorgilla* larva will feed to maturity. Infected *Anagasta* larvae should always be kept separate from uninfected stock. Manipulation of adult *Devorgilla* can be made easier by lowering their temperature to around 5 °C.

Reference

Nuffield O level Biology Project (1966) *Teachers' Guide IV*, Longman/Penguin.

DROSOPHILA

There are some twenty common species of *Drosophila* found naturally in Britain and many more in most tropical countries. Shorrocks (1972) contains a useful key for their identification. *D. melanogaster* is the species normally used for genetic work.

Uses

This well documented fly is used chiefly for genetic work. Its rapid life cycle and wide range of easily identifiable and well researched mutants make it an ideal organism for this purpose. It can also be used for behavioural work.

Supply

For genetic work pure lines are essential and should be obtained from a commercial supplier (see page 249) or other reliable source. Most commercial suppliers also provide segregated cultures of males and virgin females at an extra charge. For behavioural work, specimens may be trapped using a bottle half filled with mashed banana, pear or apple. A short length of bicycle inner tube fitted over the mouth of the bottle in the way shown in Figure 8.6 will form an effective trap. A similar device may be used to capture 'escapes' in the laboratory.

Culture

Containers. Half- or one-third-pint milk bottles are commonly used as culture vessels, but any similar sized jar may be used. Smaller scale cultures and genetic crosses may be maintained in 2.5 cm × 10 cm specimen tubes. The culture vessels should be fitted with foam plastic bungs (a type which can be autoclaved is available from suppliers). Alternatively, a bung can be improvised from a plug of cotton wool wrapped in muslin (see Figure 8.6). A piece of folded filter paper should be pushed into the food medium before the latter sets. This will absorb some excess moisture as well as providing a pupation site for the mature larvae. All culture vessels should be thoroughly washed before use; hot water containing a little disinfectant is recommended. They should then be thoroughly rinsed in cold water. Unless a ready prepared 'instant' medium is being used (see below) all containers and plugs should be sterilised prior to use. This can be done by gradually heating in an oven to 150 °C and then maintaining them at that temperature for 1–2 hours. Alternatively, they can be autoclaved at 100 kPa (15 lb in^{-2}) for 15 minutes. Steam sterilisation for 30 minutes is also satisfactory.

External conditions. At 25 °C the life cycle lasts about 10 days and this is probably the most suitable temperature for genetic work in schools. Flies will often not mate successfully for 1–2 days after etherising and this effectively increases the life cycle. Stock cultures will survive well at a room temperature of 20 °C.

Food medium. Various mixtures have been devised and most research workers have their own favourite formulae. Several are listed on page 178. An 'instant' medium is also available from suppliers, which is time-saving as it can be stored dry indefinitely, and the vessels need no sterilisation prior to the introduction of the medium and the flies. The dry food is placed in the culture vessel and a measured volume of water added. Instant medium is naturally more expensive than laboratory prepared foods, but for small-scale culture its simplicity may make it a worthwhile buy.

   Media should be 'seeded' with a little dry, powdered yeast or with a few drops of a thick yeast/water suspension before the introduction of flies.

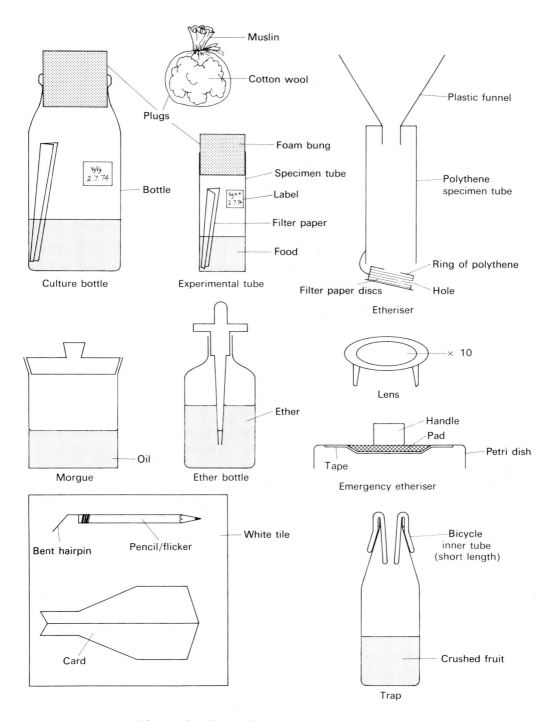

Figure 8.6   *Apparatus required for breeding* Drosophila

Other details. Strickberger (1962) has given the following details of the life cycle.

Hours	Days (approx)	Stage
0	0	Egg laid
22	1	First instar hatches from egg
47	2	First moult (second instar)
70	3	Second moult (third instar)
118	5	Formation of puparium
122	5	'Prepupal' moult (fourth instar)
130	$5\frac{1}{2}$	Pupa: eversion of head, wings and legs
167	7	Pigmentation of eyes within pupa
214–5	9	Adult emerges from puparium and gradually expands wings

Female flies will begin to mate within 10–12 hours of emergence. As most genetic work involves carefully controlled crosses between mutant strains and their progeny, *virgin* females are essential. Experimental culture tubes must, therefore, be cleared of adult flies three times each day to ensure the virginity of the females. The flies must be immediately sexed and then the males and females placed in separate tubes prior to crosses being set up. Alternatively, the virgins can be used immediately. Teachers intending to carry out genetic investigations are recommended to consult one or more of the references listed below.

Figure 8.6 shows some of the equipment necessary for genetic work with *Drosophila*.

References

Haskell, G. (1961) *Practical Heredity with Drosophila*, Oliver and Boyd.

Nuffield O level Biology Project (Revised) (1975) *Text and Teachers' Guide 4*, Longman

Shorrocks, B. (1972) *Drosophila*, Ginn.

Strickberger, W. M. (1962) *Experiments in Genetics with Drosophila*, Wiley.

FORFICULA AURICULARIA

The common earwig is a familiar insect to most people.

Uses

Earwigs are of interest in school science because the female exhibits some degree of 'parental care' both for the eggs and the nymphs for a short time after they hatch.

Supply

They are best collected from suitable sites such as rotten tree stumps and from beneath large stones.

Culture

They may be maintained in exactly the same way as woodlice (see page 223).

LITHOBIUS

This and other genera of centipedes are commonly found under large stones and beneath the bark of old rotten trees. They are voraciously carnivorous.

Uses

As well as providing an example of the Chilopoda, centipedes have an interesting and elaborate courtship behaviour.

Culture

Containers. A plastic sandwich box about 25 cm × 10 cm × 10 cm with plaster of Paris to a depth of 2 cm forms an ideal cage. A few pieces of bark and other debris should be added.

External conditions. Room temperatures of around 20 °C are quite suitable. A high level of humidity should be maintained by ensuring that the plaster of Paris is kept moist.

Food. Any small animals, such as woodlice, are suitable. Pieces of earthworm and small pieces of meat may also be taken. Excess food should not be left to decay in the container.

Other details. A certain amount of cannibalism will probably take place, but the centipedes should still breed.

LOCUSTS

Two species of locust are commonly kept in school laboratories—the desert locust (*Schistocerca gregaria*) and the migratory locust (*Locusta migratoria*). These are also the two species which cause the greatest financial loss through vast numbers devouring crops. *Schistocerca* has an invasion area covering the northern half of Africa, South Spain and Portugal, the Middle East and most of the Indian sub-continent. *Locusta* has an even wider distribution over most of the tropical and sub-tropical regions of the Old World.

Uses

Either species may be used for a wide variety of work of an experimental, anatomical and cytological nature. They are, perhaps, the most useful invertebrate to maintain in the school situation. For further details see Barrass (1964).

Supply

Locusts may be obtained from the usual biological suppliers as well as those specialising in insects (see page 249).

Culture

Locusts are relatively easy to maintain in the laboratory and, provided they are fed regularly throughout the year, they breed well.

Containers. The ideal cage for school use measures about 50 cm × 35 cm × 35 cm. Such cages, made from sheet aluminium, are available commercially. They can also be constructed from hardboard or plywood. A typical design is shown in Figure 8.7. Stoneman, Freeland and Whitmey (1973) have suggested an alternative design.

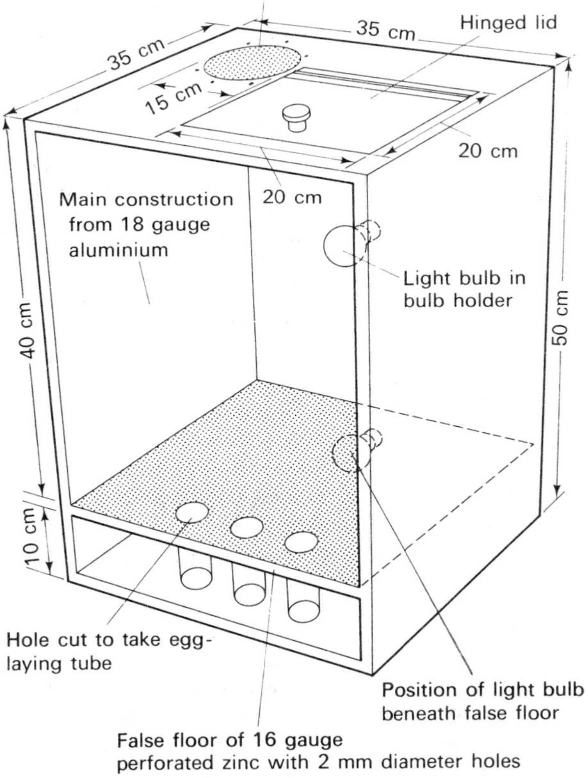

Figure 8.7   A locust cage

The entire lid should be removable for ease of cleaning, but a small panel (some 15 cm × 15 cm) should also be included to act as a door for daily maintenance and feeding. The false floor should be made of 16 gauge perforated metal and should be tight fitting. This will allow the bulk of the frass to fall through, but will prevent the escape of newly hatched first instars. The lids should be fitted with a perforated metal panel to allow free circulation of air. Suitable material should be provided from which the instars can hang as they moult. Plastic

garden netting (the very stiff variety) is ideal as it can be cleaned easily. Stiff twigs are also suitable. A source of artificial heat is essential as is the provision of egg tubes, if breeding is to take place (see below).

Smaller numbers of locusts may be maintained in perspex cylinders about 20 cm in diameter and 40 cm high, with a metal gauze panel in the roof. Once again a suitable source of heat is essential. Very small numbers can be kept in glass storage jars grouped around an electric lamp. The jars should be fitted with lids made of nylon or other similar gauze.

External conditions. For satisfactory breeding, the temperature in the cages must be maintained around 28–32 °C. Heat is best supplied by electric lamps fitted above and below the false floor. If desired the lamp above the floor can be switched off during the day, but this is not essential. The equivalent of 40–60 W below the floor and a similar power above the floor should provide sufficient heat, but this will clearly depend upon the external temperature and must be determined through trial and error. Conditions inside the cage must also be dry.

Food. The locusts should be provided with fresh grass and wheat bran or 'Bemax'. The amount of grass will depend upon the numbers being maintained. If too little is supplied, then it will be eaten almost at once and cannibalism will probably follow. If too much is given, then extra work is involved in removing the excess during cleaning. The grass should be bunched together to prevent it drying out too rapidly. If desired it can be stood in a little water. Care must be taken that it has not been contaminated with insecticides. Daily feeding with grass is ideal, but locusts will survive well over weekends if they are given an ample supply on Fridays.

During winter, or at other times when fresh grass is in short supply, the following mixture can be given:

   1 part powdered dried yeast
   10 parts dried milk
   10 parts wheat bran
   10 parts *dried* grass

Dried grass may be prepared by allowing the fresh blades to dry at room temperature for 7–10 days. It can then be finely chopped or ground and kept almost indefinitely. An artificial mixture, such as that suggested above, should always be supplemented by a certain amount of fresh plant material. If grass is totally unobtainable, then freshly grated root vegetable (e.g. carrot or turnip), shredded greens (e.g. cabbage, lettuce or celery tops) or cereal seedlings germinated for 10–14 days should be used.

Other details. The cages should be cleaned 2–3 times a week to remove excess grass, frass and any dead insects. If breeding is to take place, adults should be provided with tubes of moist sand in which the females can lay their eggs. The tubes should be at least 10 cm deep and 3 cm in diameter. Glass specimen tubes of this size can be placed

	Time for completion of one generation	Number of eggs per pod	Number of pods laid per female	Incubation period		Growth time from fledgling to adult at 32 °C
				28 °C	32 °C	
Locusta migratoria	9–10 weeks	10–30	6	16 days	11 days	4 weeks
Schistocerca gregaria	12–14 weeks	40–70	7	17 days	12 days	4–6 weeks

Table 8.3   Details of the development of the African migratory locust (*Locusta migratoria*) and the desert locust (*Schistocerca gregaria*)

inside the aluminium tubes usually supplied with commercially produced cages. The glass tubes can then be removed for incubation (see below). A fairly coarse sand is ideal and for the maximum production of live hoppers it should be heat sterilised before the addition of water in the proportions 15 parts water to 100 parts sand. More water should be added every few days to replace that lost by evaporation, until egg-laying has taken place.

The approximate amount of water to be added can be determined by weighing a moist egg tube before and after a few days evaporation in the cage. The appropriate amount of water can then be added with a disposable syringe. To ascertain whether egg-laying has taken place, pour away the topmost 1 cm of sand. This will reveal the top of the froth covering the eggs, if any have been laid.

Once egg-laying has taken place, the tubes should be covered with loose fitting lids to minimise further evaporation. (Potted meat jar lids or covers made with aluminium foil are suitable). The tubes may be kept in an incubator if the space in the cage is required for more tubes. After 10–14 days the covers should be removed and the tubes placed in a suitable cage ready for the hatching of the first instars. Further details of the development of the two species are shown in Table 8.3.

Allergy to locusts

Some people develop an allergy to locusts, the initial symptoms of which are similar to those of a common head cold. In schools, pupils are unlikely to have sufficient contact with locusts to develop the allergy unless the insects are kept permanently in a teaching laboratory. It may, therefore, be necessary to maintain them in an animal house or preparation room.

References

Barrass, R. (1964) *The Locust—A Guide for Laboratory Practical Work*, Butterworths.
Hunter-Jones, P. (1966) *Rearing and Breeding Locusts in the Laboratory*, Anti-Locust Research Centre.
Nuffield A level Biology Project (1971) *Laboratory Book*, Penguin Books.
Nuffield O level Biology Project (Revised) (1975) *Teachers' Guide 1*, Longman.
Stoneman, C. F., Freeland, P. W. and Whitmey, A., A Locust Cage and Hatchery from Plastic Aquarium Tanks, *School Science Review*, 1973, 190, **55**, pp. 86–88.

Thomas, J. G. (1963) *Dissection of the Locust*, Witherby.

SITOPHILUS

The granary weevil is a common pest of stored grain.

Uses

Sitophilus can be used for simple behavioural work, for example with choice chambers.

Supply

It is best collected from infested grain stocks.

Culture

The maintenance of this beetle is easy as it requires little attention.

Containers. Any glass or metal container with a perforated metal or netting lid is suitable. The granary weevil has been known to bore through thin plastic, so plastic containers should not be used.

External conditions. 25 °C with a relative humidity of 60–70 per cent is recommended.

Food medium. Whole wheat grains, free of any insecticide, should be used.

Reference

Nuffield A level Biology Project (1971) *Laboratory Book*, Penguin Books.

STICK INSECTS

The East Indian stick insect (*Carausius morosus*) has long been a favourite animal for classroom and laboratory culture at all school levels. More recently, several larger species have become available in Britain. Clark (1973) has shown that *Extatosoma tiaratum* is suitable for school culture. *Achrophylla wulfingi* can also be kept, the female of which grows to a length of over 30 cm.

Uses

As well as serving as an example of an insect with incomplete metamorphosis, the stick insect can be used for a wide range of behavioural and other experimental work (see Kalmus (1958)). Its colouring, shape and behaviour make it an excellent example of camouflage. *Carausius* has an unusual life history in that males are virtually unknown in captivity, the females laying their eggs parthenogenetically.

Supply

Eggs, as well as nymphs and adults, are available from commercial suppliers, especially those dealing in insects alone (see page 249).

Culture

Carausius is easily cultured and requires little attention. *Extatosoma* and the other more recent introductions require a little more care.

Containers. Almost any type of insect cage is suitable. The cylindrical perspex type with a metal base and perforated metal lid provide good all-round visibility. An old beam-balance box provides visibility and ease of cleaning through the balance box door. In this case, the glass should be removed from the top and replaced by perforated metal or mosquito netting. A larger cage is clearly needed for the very big species.

External conditions. Room temperatures of around 20 °C are quite suitable, though the more exotic species survive better at 25–30 °C. They also require a high level of humidity and regular spraying of the food plants with water using an 'atomiser' is recommended.

Food. Carausius feeds on privet, which should be supplied once a week with the twigs standing in a jar of water. To prevent eggs and nymphs dropping into the water, the jar should be covered with aluminium foil and the twigs inserted through this into the water. In winter, privet will come into leaf indoors if it is placed in water and stood in a warm place. *Carausis* is also known to take ivy. *Extatosoma* and *Acrophylla* both feed on bramble in captivity.

Other details. The cage should be cleaned every 2–3 weeks when any eggs (which are smooth and oval in shape with a well marked emergence hole at one end) should be removed and placed in a small dish in the cage. With the larger species the eggs should be placed on a little damp sand in a plastic pill box and examined at regular intervals. Eggs take anything from a few weeks to several months to hatch. The nymphal instars moult six times at approximately monthly intervals before reaching maturity.

References

Clark, J. T., *Extatosoma tiaratum*—A Monster Insect for Schools, *School Science Review*, 1973, 190, **55**, pp. 56–61.

Clark, J. T., Some Further Notes on Macleay's spectre, *School Science Review*, 1974, 192, **55**, pp. 522–523.
Kalmus, H. (1958) *Simple Experiments with Insects*, Heinemann.

TENEBRIO

Mealworms are the larvae of beetles which are pests of stored grain and grain products.

Uses

Mealworms are useful as a source of food for laboratory fish, reptiles and amphibians. They may also be useful in general experimental work and as an example of an insect with complete metamorphosis.

Supply

They may be purchased from aquarium suppliers and from some biological supply houses (see page 249).

Culture

Laboratory culture is relatively easy as the insects require little maintenance.

Containers. A large biscuit tin or sweet jar with a perforated lid makes a suitable container.

External conditions. These are not critical and normal laboratory conditions are quite suitable. The life cycle is rather slow, however, and may be speeded up by raising the temperature. At 30 °C the complete cycle takes 4–6 months.

Food medium. The container should be half filled with bran or oatmeal. A few slices of apple, carrot or potato and an occasional crust of bread should also be supplied to provide the necessary moisture as well as food for the adults when they emerge.

TRIBOLIUM

The flour beetles *Tribolium confusium* and *T. castaneum* are pests of stored cereal products.

Uses

Both species are used for work in genetics and behaviour. They also provide examples of insects with complete metamorphosis and may be used in some physiological experiments.

Supply

Both species can be obtained from commercial suppliers and this source is essential for any genetic work which demands pure lines (see page 250). Specimens for other work may be obtained on occasions from packaged flour which has become infested.

Culture

Containers. Stock cultures may be maintained in screw-top jars fitted with a muslin panel in the lid. Alternatively, muslin may be secured over the tops of the jars with elastic bands. Smaller cultures (e.g. for genetic crosses) may be kept in 2.5 cm × 10.0 cm specimen tubes with muslin covers or foam bungs.

External conditions. A temperature of 25–30 °C is needed for rapid breeding—though cultures will survive at room temperatures of around 20 °C. A relative humidity of 65–70 per cent should be maintained. This can be done by including a beaker of saturated sodium nitrate(V) solution in the incubator. Water should be added to the solution from time to time to prevent it evaporating to dryness.

Food medium. A mixture of 12–20 parts wholemeal flour and 1 part dry powdered yeast provide both a food and a medium in which the insects can burrow. Containers should be approximately half filled with the medium.

Other details. At 25 °C the life cycle takes about forty days from egg to imago. Insects can be removed from the food medium by sieving with an 8-mesh-to-the-cm sieve. Chilling in a refrigerator will immobilise the adults and make them easier to handle.

References

Head, J. J. and Dennis, N. R. (1968) *Genetics for O Level, Teachers' Guide*, Oliver & Boyd.
Nuffield A level Biology Project (1971) *Laboratory Book*, Penguin Books.

Fish

A small number of species of cold-water fish, and many tropical forms may be maintained in school aquaria. Details of setting up aquaria are given on page 281 where a consideration of suitable species will also be found.

Amphibians

Three species of frog, two species of toad and three species of newt occur naturally in Britain. In most areas, the smooth newt (*Triturus vulgaris*), the common frog (*Rana temporaria*) and the common toad (*Bufo bufo*) are the most frequently encountered. All the British species, together with the European salamander (*Salamandra salamandra*) and some more exotic forms can be kept in the manner described below. For details of maintenance of the wholly aquatic African clawed toad (*Xenopus laevis*) see page 237.

Uses

Amphibians provide excellent examples of animals in which the development of the embryo can be followed from egg-laying to metamorphosis and, with some care and patience, to maturity. In some cases the rate of development can be altered by variations in temperature or by the addition of minute traces of hormone to the water (see Nuffield O level Biology Project (Revised) (1975) and Nuffield A level Biology Project (1970)).

Most amphibians have excellent protective colouration and some variation in this takes place, due to expansion and contraction of melanocytes in the skin, according to the background colouration. This can be observed easily in the tails of anaesthetised tadpoles (see page 173 for details of anaesthetisation). The tail of the tadpole and the web of the frog's hind leg are sufficiently thin to enable observation of the flow of blood in the capillaries; again the animals should be anaesthetised for this work (see page 173).

The European salamander provides an excellent example of warning colouration. It is also of interest because the eggs develop internally so that the young are born at the larval stage.

Amphibia are also heavily infected by a range of other organisms which provide excellent examples of the parasitic relationship.

Mating and feeding behaviour can also often be observed.

The common frog (*Rana temporaria*) is much sought after for dissection; modern farming practices are also fast removing its natural habitat. These two factors have led to a marked decline in numbers in the wild in recent years. This should be borne in mind when frogs are contemplated for dissection.

Supply

Eggs and larvae are best collected in the field in spring. Adult specimens may either be collected in the field or obtained from commercial biological suppliers (see page 249). Some high street aquarists also maintain stocks of more exotic species.

Culture

The adult stages may be maintained in wet vivaria (see below). Breeding and development of the larval stages takes place in water and a freshwater aquarium should be available for this purpose. (See page 281 for details of setting up an aquarium).

Containers. Aquarium tanks are probably the best form of container in which to construct the necessary wet vivaria for the adult stages; 60 cm × 30 cm × 30 cm is a suitable size (see page 75 for details of aquaria).

As much care should be taken over the cleaning of the tanks as would be taken prior to the introduction of fish. A sheet of glass on top, raised 5 mm on rubber feet, will prevent escape of the animals, retain a high level of humidity and allow sufficient ventilation. The edges of the glass should be taped in the interests of safety.

External conditions. The vivarium should be placed in a

cool and shaded situation—never in direct sunlight. Temperatures are best kept between 15 °C and 18 °C—though salamanders and most tropical species will survive at higher temperatures. The low temperature requirements of British amphibians are a problem in centrally heated laboratories and in modern buildings where the high proportion of windows often leads to excessive heat. Considerable care should be taken when siting vivaria in such buildings.

Setting up. The floor of the aquarium tank should be covered with gravel over which a layer of good loamy soil is placed. At one end of the tank, a shallow glass dish of water should be sunk into the gravel and soil to provide a miniature pond. The soil at the other end of the tank should be sloped gradually upwards. Mosses, ferns and other shade-loving plants should be planted in the soil; a sloping 'beach' near the pond can be planted with *Sphagnum*. At least one rock should be placed in the pond and a number of suitable hiding places constructed around the vivarium using stones, pieces of bark, etc. (see Figure 8.8). The soil and gravel should be watered and kept moist at all times.

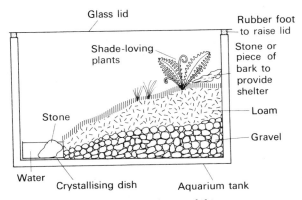

Figure 8.8   A wet vivarium for amphibians

Food. All the British amphibia are carnivorous and can be fed largely on insects varied by the addition of a few small earthworms. Mealworms and houseflies are suitable insects, the latter being best introduced at the pupal stage; they will then be eaten when they emerge as adults. Though most amphibians can survive for quite long periods without food, their condition will deteriorate and feeding on alternate days is recommended. Survival over weekends is generally no problem.

Smaller species, such as the tree frogs (*Hyla* spp.) will take *Drosophila* as will the salamanders, which will also accept *Tubifex* and *Daphnia* in the miniature pond. Chopped raw liver, dangled on a piece of thread may also be eaten on occasions. Dead, uneaten material should always be removed as soon as possible and to this end, a small sunken dish into which the food is placed may help.

Other details. All the British amphibians require water for breeding and newts will normally spend several months in the aquatic environment from spring to early summer. If breeding is to take place in the laboratory, an aquarium, well stocked with plants, must be provided as soon as the secondary sexual characteristics become more pronounced. In the case of newts, the most obvious of these characteristics is the development of a crest on the male, running from the back of the head to the tip of the tail and along the underside of the tail. Male frogs and toads develop a thickened nuptial pad on the inside of the first finger of the front legs; in the case of the toads there is a similar thickening on the second and third fingers. In all cases there is some heightening of the skin colour especially in the males.

Once egg-laying is complete, the adult frogs and toads should be returned to the wet vivarium. The early larval stages of all British amphibia have a herbivorous diet and, provided they are kept in a balanced aquarium, well-stocked with plants, feeding should not be a problem. As they get larger, the tadpoles become carnivorous, feeding on small aquatic animals. Towards the end of the larval stage, when all four legs have emerged, a stone, floating log or small sheet of cork should be provided. This will enable the animals to climb out of the water and to utilise their lungs which are developing at this time. The tail will next begin to shrink as metamorphosis takes place. The young frogs and toads can now be provided with an adult diet (see above); however, any food must be of a sufficiently small size to be taken easily into the mouth.

Breeding in newts is somewhat different in that the sperms from the male are deposited in the water enclosed in a small capsule or spermatophore. This capsule is taken up by the female after a well developed courtship display and fertilisation is internal. The eggs are generally laid singly, attached to waterweeds or the sides of the tank. The adults can be left in the aquarium along with the larvae when the latter emerge.

References

Nuffield Junior Science Project (1967) *Animals and Plants*, Collins.
Nuffield O level Biology Project (Revised) (1975) *Teachers' Guide 4*, Longman.
Nuffield A level Biology Project (1970) *The Developing Organism*, Penguin Books.

XENOPUS

The African clawed toad (*Xenopus laevis*) was formerly used extensively in human pregnancy testing. With the development of more sophisticated chemical methods of testing they are no longer used for this purpose, but are commonly maintained in the school laboratory for various aspects of biology teaching. Being both wholly aquatic and of tropical origin, it is perhaps easier to maintain a constant environment for this toad than it is for amphibians from more temperate regions, where much of the adult life is spent out of water. In centrally heated school laboratories, higher temperatures are generally easier to maintain than lower ones.

Uses

The adult toads and their larvae can be used for most of the purposes outlined for other Amphibia (see above). However, *Xenopus* does have certain advantages when compared with other species. It is relatively easy to predetermine the time of breeding as this can be induced by injections of chorionic gonadotrophin (see below). The larval stages are comparatively transparent, so that internal structures can often be viewed directly through the body wall. On the other hand, *Xenopus* is rather expensive to use for general amphibian dissections. Further information on specific uses may be found in Brown (1970).

Supply

Xenopus can be obtained from commercial biological suppliers (see page 249).

Culture

Containers. Any large tank is suitable, both for the adults and the larvae. Those made from toughened polythene or some other unbreakable plastic material are ideal. Some suppliers stock white polyethylene and polypropylene tanks measuring about 60 cm × 40 cm × 15 cm; others (e.g. Thermoplastics) stock grey plastic trays some 40 cm × 30 cm × 15 cm designed for storage purposes. Both these containers are ideal as they can be stacked, thus reducing the use of bench space. Efficient lids are essential for adult toads as they frequently jump clear of the water. Lids may be cut from plywood, hardboard or perspex, the latter permitting visibility. 'Claritex' (a plastic sheeting reinforced by wire mesh often sold in garden shops), is also suitable; this can be cut a little larger than the tank itself and then folded over the rim. 4–6 small holes drilled in the lid will increase ventilation and reduce condensation. For small numbers of tadpoles, pie dishes or other similar containers may be used. No stones, weeds, or other materials (other than the heating and aeration system) should be placed in the tanks as this will make efficient cleaning more difficult.

Medium. 8–10 cm depth of rainwater or tap water which has been allowed to stand for 24 hours to allow the escape of any chlorine is all that is required.

External conditions. The toads will survive over quite a wide range of temperature (10–30 °C) but will not feed readily at the lower end of the range. For breeding purposes, 20–23 °C is recommended with 23 °C being the optimum level. A heater and thermostat are therefore

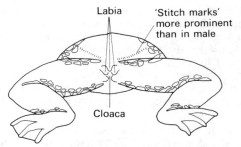

Figure 8.9   The distinguishing features of the female of Xenopus laevis

required to maintain this temperature (see page 77 for details of available heaters and thermostats). If large numbers of tadpoles are being reared, aeration is essential. A single air-tube attached to a diffuser and held in place at the bottom of the tank with a stone is a suitable

a) Picking up

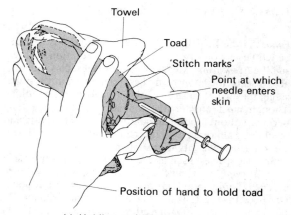

b) Holding and injecting

Figure 8.10   Picking up and holding Xenopus for injection

Stage in development	Food	Other details
TADPOLES On hatching	*Euglena* or other unicellular algal culture. Liquifry No. 1 or other similar aquarium baby fish food. Infusorians	Feed little and often throughout the larval stages—if possible several times a day although they will survive over weekends. Do not overfeed. The tadpoles should clear all the suspended food from the water. Siphon off bottom debris and change water regularly
For most of their development At metamorphosis	Liquifry No. 1 followed by No. 2. Liquidised vegetables. Tinned sieved baby food. Sieved hard-boiled egg yolk. Nettle powder. Begin to add *Tubifex* to the diet above	
ADULTS After metamorphosis (1–2 cm)	*Tubifex* and whiteworms. *Tribolium* and vestigial winged *Drosophila* can be dropped onto the surface of the water	Feed daily, though survival over weekends is usually no problem. Avoid too great a range of size of toad or some cannibalism will take place
Larger sizes	The diet mentioned above can gradually be varied with the addition of earthworms, maggots, adult blowflies, pieces of liver, heart, whitefish and fat-free meat. A full-grown adult will take about 8 g of food per feed	Feed twice a week. Remove any excess food shortly after feeding or the water will become fouled. Change the water and clean the tank 24 hours after feeding

Table 8.4   Food required at various stages of Xenopus development

arrangement. The tanks should be sited in a shady situation.

Food. Adult *Xenopus* are voracious feeders. A variety of food is suitable at all stages of development. (See Table 8.4).

Other details. In Africa, *Xenopus* is frequently found in stagnant water. However, under laboratory conditions the water must not be allowed to become putrid. The simplest method of cleaning is to prepare another tank of water at the appropriate temperature. The adults (or larvae) can then be transferred to it with the help of an aquarium net. The vacated tank can then be thoroughly cleaned. Tanks containing adult toads should be cleaned twice a week—24 hours after each feed. Fido and Robb (1971) report that tadpole tanks need only be cleaned once a fortnight if *Liquifry* is used as food. With other foods more frequent cleaning is necessary.

Breeding may be induced by injecting both sexes with chorionic gonadotrophin which is marketed under various trade-names, e.g. *Pregnyl* and *Chorulon*. Biological suppliers frequently require a Headteacher's signature before supplying these products. The toads become sexually mature when they are about two years old. The distinguishing features of the female, visible externally, are shown in Figure 8.9. The injections should be made into the dorsal lymph sac by piercing the skin of the thigh dorsally and directing the syringe needle forwards through the line of 'stitch-marks' visible on the skin surface. These marks indicate the position of the septum surrounding the dorsal lymph sac at this point. The needle should be kept just below the skin with the point directed

slightly upwards so that it does not damage internal organs. A dry towel held over the head of the toad will facilitate handling (see Figure 8.10). The details of each injection are given in Table 8.5. Immediately after the final injection, it is advisable to place the toads in a smaller tank for spawning. A suitable design is shown in Figure 8.11.

	First injection	Second injection
Time before spawning is required	48 h	2–8 h
Quantity of hormone per male	50 i.u.	100 i.u.
Quantity of hormone per female	100 i.u.	350 i.u.

Table 8.5   Details of injections of chorionic gonadotrophin to induce spawning in Xenopus

References

Brown, A. L. (1970) *The African Clawed Toad*, Butterworths.

Fido, H. S. A. and Robb, G. A., Rearing *Xenopus laevis* Tadpoles, *School Science Review*, 1971, 181, **52**, pp. 924–5.

Nuffield A level Biology Project (1971) *Laboratory Book*, Penguin Books.

Nuffield O level Biology Project (Revised) (1974) *Teachers' Guide 1*, Longman.

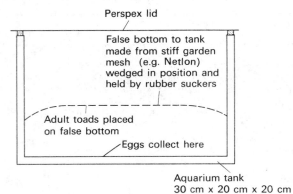

Perspex lid

False bottom to tank
made from stiff garden
mesh (e.g. Netlon)
wedged in position and
held by rubber suckers

Adult toads placed
on false bottom

Eggs collect here

Aquarium tank
30 cm x 20 cm x 20 cm

Figure 8.11   A spawning tank for Xenopus

Reptiles

TERRESTRIAL REPTILES

A selection of reptiles is available for maintenance in school laboratories. Although the specific purposes for which they can be used are rather limited, they attract a good deal of interest in the laboratory. The most commonly available species are the European green lizard and the wall lizards (*Lacerta viridis* and *L. muralis*), the British viviparous common lizard (*L. vivipara*) and the slow-worm (*Anguis fragilis*). More exotic species are also sometimes available, including chameleons and geckos. Crocodiles and alligators (however small!), and poisonous and large constrictor snakes should *never* be kept in schools. The harmless British grass snake (*Natrix natrix*) and smooth snake (*Coronella austriaca*) are also best avoided by those inexperienced in handling and feeding snakes. Tortoises and terrapins may carry diseases transmissible to man (e.g. *Salmonella*) and should not be maintained in schools.

Uses

A selection of reptiles will help to illustrate the variety of form within the class. Observation of their feeding is also a worthwhile experience. Some examples may also be used for more specific purposes; for example, the chameleons provide an excellent illustration of camouflage while the lizards are suitable for studies of tetrapod locomotion.

Supply

Stock is best obtained from biological supply houses (see page 249). It may also be possible to buy specimens from high street pet shops, but care should be taken that the supplier is reputable and is not providing infected stock.

Culture

Containers. These reptiles are best kept in a dry vivarium. Metal-framed aquaria are probably the most suitable containers (see page 75). They should be fitted with lids made of perforated zinc fixed to a wooden frame which drops over the rim of the aquarium tank. The floor of the tank should be covered with a mixture of dry soil and sand along with a little loose peat. A shallow dish of fresh water should always be available. Some of the tropical forms (e.g. chameleons) can only take water off damp vegetation so in these cases plants should be placed in the tank and regularly sprayed with a little water. However, with the European lizards mentioned above, excessive humidity in the vivarium interferes with the skin-sloughing process.

External conditions. All reptiles benefit from radiant heat, either from the sun or from an electric lamp. It is advisable to keep the vivarium in a moderately sunny situation, though prolonged direct sunlight may prove harmful in some cases. Temperatures are best kept between 20 °C and 25 °C if the animals are to remain active. All the European species normally hibernate during the winter months. If the vivarium is placed in an unheated outhouse from about mid-October the animals will become torpid and remain in this state until the spring. However, they will remain active throughout the winter if warm temperatures are maintained.

Food. It is advisable to give a selection of insects and spiders as food. *Tenebrio* larvae, maggots, cockroach and locust larval instars, and blowflies provide a suitable range of insects. Earthworms and pieces of meat dangled in front of the reptile may also be accepted on occasions.

Other details. Great care should be taken in handling the lizards and under no circumstances should they be picked up or held by the tail as this may result in the tail being shed.

Birds

Members of the parrot family (budgerigars, macaws, love-birds, parrots and parakeets) as well as many species of native British birds, transmit ornithosis and *must never* be kept in schools. Two exotic species are safely maintained in schools, the zebra finch and the Japanese quail.

THE ZEBRA FINCH (*Taeniopygia castanotis*)

This small finch has recently become quite popular in schools, partly because it is not difficult to keep.

Uses
They are used for behavioural work. For further details see Evans (1968) and Nuffield Advanced Biological Science (1970 and 1971).

Supply

Stock of birds bred in this country is best obtained from local bird dealers or from schools or other institutions already keeping the birds.

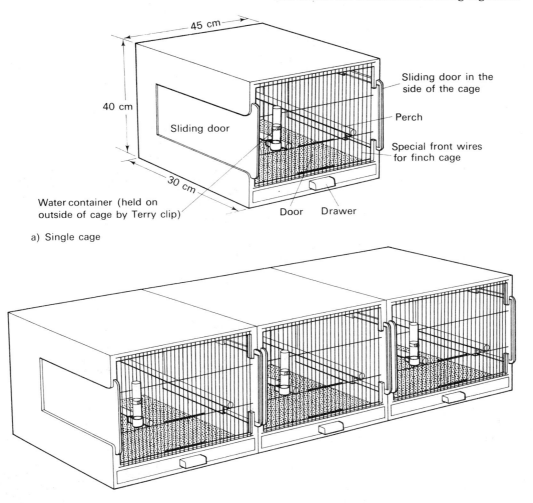

a) Single cage

b) Three cages placed side by side. They can be made continuous by removing the sliding doors between adjacent cages

Figure 8.12   A suitable design of cage for Taeniopygia castanotis *(the zebra finch)*

Culture

Containers. A suitable design of cage is shown in Figure 8.12. The sliding side panels enable the cages to be fitted together, thus making it convenient to transfer birds from one cage to another. The base of the cage should be covered with grit, sand or sawdust. A shallow dish of water will enable the birds to bathe.

External conditions. These are not critical and most laboratories are quite adequate.

Food. Yellow panicum millet is a perfectly sufficient diet, but this can be varied with other seed (e.g. white millet, small canary niger seed) and with fresh plants bearing fruit e.g. chickweed, plantains, rye grass. A cuttlefish 'bone' should also be included to provide sufficient calcium.

Other details. The cage should be cleaned out whenever it begins to get fouled with faeces. The birds can be allowed to breed when they are about one year old. In British conditions, breeding is best carried out in late spring and summer. If breeding boxes some 12 cm × 12 cm × 12 cm with a circular 4 cm diameter hole are introduced, together with some soft hay for nesting, this will often initiate breeding. Once the eggs have been laid all excess nesting material should be removed otherwise the birds may recommence building on top of the eggs.

References

Evans, S. M. (1968) The Study of Bird Behaviour in Schools, *J. Biol. Ed.* **2**, 4, pp. 373–380.

Nuffield A level Biology Project (1970) *Control and Co-ordination in Organisms*, Penguin Books.

Nuffield A level Biology Project (1971) *Teachers' Guide II*, Penguin Books.

THE JAPANESE QUAIL
(*Coturnix coturnix japonica*)

This delightful little bird has recently become available in this country and is increasingly being kept in school laboratories.

Uses

The very rapid growth to maturity of this bird makes it an ideal animal for growth studies. On hatching they weigh some 6–7 g and attain a mature weight of 120–150 g in about eight weeks. They can also be used for behavioural work.

Supply

Japanese quail are available from some biological suppliers (see page 249) or from other institutions with surplus stock.

Culture

Containers. The quail's habit of jumping 2–3 m vertically in the air when frightened necessitates containers with opaque sides and efficient lids. Tea chests, suitably modified, are acceptable. Two chests, joined together after the removal of one side from each, will provide sufficient space for two breeding trios (i.e. two males and four females). The bottom of the cage should be lined with straw. The floor space should be as clear of obstruction as possible as the birds tend to scatter the contents as they dust bathe. A suitable lid can be made by attaching fine wire mesh to a wooden frame. Handling will be made easier if the lid is in two parts, each covering half of the cage.

External conditions. Laboratory temperatures are quite adequate for the adults, though the eggs should be artificially incubated at 37.5 °C. For breeding purposes, the adults require a day/night rhythm of 18/6 hours. In winter it will therefore be necessary to extend the hours of daylight, preferably before dawn. Two 40 W tubes mounted $\frac{1}{2}$–1 m above the cage are satisfactory. Water must be provided and for adults this is most easily done with an externally fitted 'rabbit drinker'. Water for chicks should be placed in shallow dishes along with some small stones, otherwise the young birds may drown. Cameron (1969) reports that they have been known to drown in

1 mm of water. A little limestone or flint grit should be provided, preferably in a mesh covered feeder.

Food. A high protein food, such as turkey starter crumb (about 25 per cent protein) is necessary. This can be supplemented with proprietary brands of bird seed (e.g. Swoop). The food should be placed in mesh covered chicken grit hoppers otherwise the birds will scatter it, resulting in unnecessary wastage. Alternatively, the hoppers can be fitted externally.

Other details. The hens should start to lay 6–8 weeks after hatching, though the eggs may be infertile for the first two weeks. Incubation lasts $16\frac{1}{2}$ days at 37.5 °C. As the birds will not usually incubate their own eggs in the laboratory, they should be removed and placed in an incubator at 37.5 °C (see below). The eggs should be turned 3–4 times every 24 hours.

Reference

Cameron, J. A. (1969) The Japanese Quail, *J. Biol. Ed.* **3**, 2, pp. 173–179.

DOMESTIC FOWL—INCUBATING EGGS

The embryology of the chick has, in the past, been a study largely reserved for sixth-form biology specialists; much of this work was carried out using preserved material. A number of recent courses have, however, included some simple observations of the developing embryo for pupils at a much younger age (see Nuffield O level Biology). For this work to be carried out, incubation in the school laboratory is essential.

Supply

Fertile eggs should be obtained from an accredited dealer. A list of suitable firms may be obtained through local farmers' organisations. Approximately 30 per cent of eggs are likely to be infertile.

Incubators. A number of models are available and these are discussed on page 112. The Western Curfew Observation Incubator is probably the most satisfactory, but is not suitable for purposes other than egg incubation. Incubators can also be improvised cheaply (see page 160).

Conditions. The eggs should be incubated at 38 °C and the incubator should be set a day or so before the introduction of the eggs, so that conditions can stabilise. The atmosphere within the incubator must also be humid. In the case of the Curfew Incubator, a moisture tray is provided and this should be kept topped up with water. With other models, a tray of water should be placed near the bottom of the incubator.

Other details. Eggs older than seven days should not be obtained from the suppliers. They may be kept at 10–12 °C prior to the incubation period. The conditions within the incubator should be checked daily and the eggs rolled through 180° twice daily to prevent the developing embryo sticking to the inside of the shell. Each egg should be marked with the date on which incubation began and any other appropriate details. Chicks should start to hatch at around 21 days and should be kept at 28 °C for the first few days thereafter. They should be provided with water in a shallow dish and baby chick food obtainable from a corn merchant.

References

Nuffield O level Biology Project (Revised) (1974) *Text and Teachers' Guide 1*, Longman.

Mammals

A number of species of small mammal are available for school use. Of these, six are recommended, details of which are given in Table 8.6. A number of other species are commonly found in schools, e.g. Steppe lemming (*Lagurus lagurus*), Chinese hamster (*Cricetulus griseus*), spiny mice (*Acomys* spp.) and ferret (*Mustela putorius*), but their breeding is frequently unreliable, they are often difficult to handle and in most cases it is not possible to obtain stock which is guaranteed free of disease transmissible to man. Wild native mammals, especially rodents and hedgehogs should *never* be maintained in school under any circumstances, as there is a serious risk of their being infected with diseases transmissible to man and to other mammals in the school. The maintenance of any species other than those in Table 8.6 is therefore not recommended. A number of other species of mammal must never be kept in schools under any circumstances. These are discussed on page 196.

Uses

Small mammals may be put to a very wide variety of uses, both in biological and non-biological work. In the biological context they may be used for genetics, behaviour, reproduction and growth, anatomy and various aspects of physiology as well as providing examples of the mammalian class to which we ourselves also belong. Compared with many other animals commonly kept in schools, mammals are long-lived, easily handled and can be bred and maintained under school laboratory conditions with little difficulty. They are also generally attractive animals and add interest to the laboratory. Wray (1974b) examines the educational use of mammals in greater detail and his book is very fully referenced.

Supply

As far as possible, small mammals for school use should be obtained only from suppliers recognised by the Medical Research Council, whose Laboratory Animal Centre (see page 250) publishes a regular 'parade state' of stock available from accredited dealers. For all the recommended species (except Mongolian gerbils) the animals may be obtained in a number of categories. These are as follows:

* Not reared within any physical barrier but should be free from all evidence of infectious diseases especially those transmissible to man. They are quite suitable for immediate use in a school, e.g. for dissection.
** Reared under rather higher standards of management. They are likely to be more reliable for breeding purposes than * category animals and are recommended for this purpose and for behavioural work.
*** Obtained from full-term pregnant females by caesarian section and then maintained under high standards of management. They are also suitable for breeding purposes in schools and for behavioural work.
**** Also obtained by caesarian section but subsequently maintained within a barrier system which results in them being designated specific pathogen free (SPF).
***** Obtained by caesarian section and then maintained in isolation for specific research purposes.

The cost of the animals becomes progressively greater as one moves from category * to category *****. Conditions in schools are such that it will not be possible to maintain stock in a way which would justify the purchase of animals of categories **** or *****. As a general rule, therefore, category * stock should be used for dissection, while category ** or *** animals should be purchased for breeding and behavioural work.

Although Mongolian gerbils are not categorised in the way described above, they may also be purchased from suppliers which are recognised under the scheme. In this way good quality stock should be assured.

Stocks of small mammals should not be purchased from local pet shops as there is considerable risk of infection from the other animals and the people who have free access to the shops. It is also generally not advisable to obtain stock from other schools unless it is known that the animals are derived from one of the appropriate categories mentioned above and have been maintained in good conditions. Animals of unknown origin should never be accepted as gifts.

Commercial dealers may deliver stock through their own transport system. Alternatively, they may send the livestock through the British Rail Express Parcel Service to a station having facilities for storing living animals under suitable conditions. Schools must ascertain the exact details of delivery from the supplier at the time of ordering and make suitable arrangements to ensure that the transport time is kept to the minimum. Delivery over a weekend should always be avoided.

	Adult mass g		Recommended diet* (Code letters)	Daily food consumption g	Daily water intake cm³	Minimum cage size		Number per cage
	Male	Female				Ht cm	Area cm²	
Mouse *Mus musculus*	20–40	25–90	41B or FFG(M)	3–4	4–7	13	500	Trio + 1 litter
Rat *Rattus norvegicus*	200– 400	250– 300	41B or FFG(M)	15–20	24–35	25	1000	Pair
Gerbil *Meriones unguiculatus*	45– 130	50– 130	41B or FFG(M)	5–7	little in fresh vege- tables	20	1000	Pair
Syrian Hamster *Mesocricetus auratus*	90– 120	95– 140	41B or FFG(M)	10–15	8–12	20	1000	Pair or one female + litter
Guinea-pig *Cavia porcellus*	1000– 1500	850– 900	RGP or SGIV plus fresh veg.	20–30	80–150	30	2500	Pair
Rabbit *Oryctolagus cuniculus*	1500– 5000	1500– 6000	SG1 or 18 plus fresh vegetable	30–300	60–300	45	5500	One

* For all species, the diet may be supplemented by grain, carrot, apple, etc.

Table 8.6   Some details of small mammals recommended for school use

Culture

Containers. A number of designs of cage are available—these are discussed on page 86. The minimum recommended size of cage for each of the species under discussion is shown in Table 8.6. The type of cage constructed largely from painted sheet metal and frequently sold in pet shops should be avoided as the metal rapidly becomes corroded and the sharp edges and corners lead to cuts and infection which may affect both the animals and those handling the cages.

Litter and nesting materials should be provided for all the species under discussion. The litter is placed on the floor of the cage to absorb urine and the water from the faecal pellets, thus reducing the rate of decomposition. White, softwood sawdust (often obtainable free of charge from sawmills) or peat in granular form make ideal litter materials. Newspaper, 'Vermiculite', 'Sterolit', cotton wool and hardwood sawdust should be avoided. Nesting material is used by the animal for constructing a nest for itself and for its young. In this way the animal is able to maintain a suitable micro-environment within the cage. First quality meadow hay (free from thistles and pesticides), shredded white paper, wood wool and cellulose wadding make suitable nesting materials. Straw, newspaper and packing paper should not be used.

External conditions. In some schools the provision of an animal house enables the maintenance of a relatively constant environment for small mammals. Further details of animal houses can be found on page 5 and in Wray (1974a), *Animal Accommodation for Schools.* Whether the stock is being kept in special accommodation or in a teaching laboratory or preparation room, the following conditions should be maintained:

Temperature, 16–22 °C.
Relative Humidity, 40–60 per cent.
Light, normal daylight or artificial light of sufficient intensity to permit easy observation.
Noise, low general background noise; sudden noise effects should be avoided.
Ventilation, Animals should be kept in well ventilated conditions, but draughts must be avoided.

Food and Water. Food should be placed in food hoppers and not in open dishes which will become contaminated. Hoppers should be kept at least half-full otherwise they will be moved and dislodged during feeding. A number of designs of hopper are available (see page 87). Water should be provided in drinking bottles, of which there is a wide selection available (see page 87). Details of the requirements of food and water of the six recommended species can be found in Table 8.6. A daily check should be made to ensure that there is a sufficient supply of water and food for each cage of animals. Care should be taken to supply sufficient food and water for short holiday periods (e.g. weekends). During longer periods of holiday it will be necessary to make arrangements for regular feeding and cleaning.

Other details—cleaning. All cages should be cleaned and disinfected at regular intervals. Faecal pellets and litter contaminated with faeces should be removed frequently to reduce the risk of infection. Every 7–10 days, the following cleaning procedure should be carried out.

1 Remove the animals and place them in a spare cage. If possible, avoid cleaning cages containing newly born young.

2 Remove all the litter and nesting material using a paint scraper or similar instrument. The debris should be sealed in polythene bags and placed in the dustbin or, even better, incinerated.

3 Scrub the cage with hot water and liquid detergent, then rinse in clean water.

4 Immerse the cage in a suitable disinfectant for at least 15 minutes. The disinfectant should be surface active (e.g. *TASK*, Griffin *ASAB*, *TEGO*, *MHG*, Harris *BAS*, *Cetavlon*).

5 Ensure that the cage is thoroughly dry before adding fresh litter and nesting materials and reintroducing the animals.

Water bottles should be cleaned and disinfected along with the cages themselves.

Handling. Hands should be washed before and after handling small animals. Gloves need not be worn unless the animal is likely to bite, but a protective overall is advisable. Details of handling techniques for the six recommended species are given in Table 8.7. Movements should be slow and purposeful and handling careful, but firm. Animals may be reassured if the hands are rubbed in the litter before handling. An animal removed from a cage for inspection should be placed on a table with a rough surface so that it does not slip. It may be convenient to reserve a special board for this purpose which can be clamped to any table.

Sexing. Details of special handling techniques for sexing and of sexual differences in the six recommended species are shown in Table 8.8.

Breeding. The breeding of small mammals in the school situation should be carefully controlled. In general, the following code of practice should be followed.

1 Only young and healthy animals should be selected for breeding. Recommended ages are shown in Table 8.9.

2 An appropriate mating system should be adopted which will depend upon both the purpose for which it is being carried out and the species (see Table 8.10).

3 Particular care should be taken to ensure that sufficient space, food, water and nesting material are available for breeding animals.

4 Pregnant females and, in some cases, newly born young should not be handled (see Table 8.10).

5 The young should be weaned at an appropriate stage and removed from the mother (see Table 8.9).

6 Care should be taken to ensure that young animals can reach both food and water.

7 All cages should be labelled with appropriate details of parents, offspring, date of birth, etc. Separate breeding records should be kept in a file or card index.

	General handling	Handling to ascertain sex
Mouse and small rat	Grasp base of tail between thumb and forefinger. Lift and then transfer weight of the animal by placing it on the other hand or on a non-slip surface. Restrain movement by holding the base of the tail	Expose genitals by lifting the base of the tail and supporting the animal on a non-slip table or cage-top
Large rat	Grasp around shoulders with thumb under chin to control the mouth. Support weight of body with other hand held under rump	As for mouse and small rat
Gerbil and Syrian hamster	Enclose the animal with cupped hands and with the animal's head towards the handler. The animal may be further restrained by placing the thumbs over its back	Hold the animal in cupped hands and turn it over by rotating the wrists. Support the back with one hand and the head and front with the other. Gerbils can also be sexed by the method described above for mice and rats
Guinea-pig and small rabbit	Grasp around the shoulders with thumb under chin to control the mouth and head. Support weight of body with other hand held under rump	Handle in normal way then rest the animal's back on the upper right thigh. Apply gentle pressure above and below urino-genital aperture using thumb and forefinger of left hand placed above the animal's rear legs. In males the penis can be extruded by this manipulation
Large rabbit	Grasp by the scruff of the neck and the ears. Support weight of body with other hand held under rump	As for guinea pig and small rabbit

Table 8.7 Techniques for handling the six recommended species of small mammal

	General body characteristics	Urino-genital opening	Nipples and mammary glands	Penis	Behaviour
Mouse *Mus musculus*	Scrotal sacs visible in male. Enlarged size of pregnant female obvious	Distance from anus $1\frac{1}{2}$ times greater in male than in female	Obvious on female especially when lactating	Cannot be extruded by manipulation	Male examines female genital area and may nibble female
Rat *Rattus norvegicus*	Scrotal sacs visible in male. Enlarged size of pregnant female obvious	Distance from anus $1\frac{1}{2}$ times greater in male than in female	Obvious on female especially when lactating	Cannot be extruded by manipulation	Male examines female genital area and may nibble female
Gerbil *Meriones unguiculatus*	Scrotal sacs visible in male. Enlarged size of pregnant female obvious	Distance from anus $1\frac{1}{2}$ times greater in male than in female	Obvious on female especially when lactating	Cannot be extruded by manipulation	Male stands on hind feet and drums, chases female and attempts copulation frequently
Syrian hamster *Mesocricetus auratus*	Scrotal sacs visible in male. More obvious pigmented scent gland on hip of male. Enlarged size of pregnant female	Distance from anus $1\frac{1}{2}$ times greater in male than in female	Obvious on female especially when lactating	Cannot be extruded by manipulation	Female holds back flat and tail erect (lordosis) prior to copulation
Guinea-pig *Cavia porcellus*	Scrotal sacs visible in male	Distance from anus similar. Male opening circular, female opening Y-shaped	Males have prominent nipples as well as females	Can be extruded by manipulation	Female sniffs other animals and may mount other females. Male chases female who exhibits lordosis
Rabbit *Oryctolagus cuniculus*	Width of head greater on male	Distance from anus similar. Male opening circular, female opening slit-shaped	Obvious on female especially when lactating	Can be extruded by manipulation	Female flattens her body on the ground and raises her genitals to permit copulation

Table 8.8   The sexual differences in the six recommended species of small mammals

	Recommended breeding age		Oestrous cycle duration days	Oestrous duration hours	Period of gestation days	Litters per year	Number in litter	Mass at birth g	Recommended stage for weaning		Life span years
	Min weeks	Max months							Age days	Mass g	
Mouse *Mus musculus*	8–10	12	4–5	9–20	19–21	8–12	8–11	1.5	21	10–12	$1\frac{1}{2}$
Rat *Rattus norvegicus*	9–12	12–15	4–6	9–20	20–22	7–9	9–11	5–6	22	35–40	2–3
Gerbil *Meriones unguiculatus*	12	18	4–6	12–18	25–28	8–12	4–6	2.5–3.5	22	12–18	$2\frac{1}{2}$–$3\frac{1}{2}$
Syrian hamster *Mesocricetus auratus*	10–12	12	4–5	4–23	16	3–4	5–7	2	21	40	1–$1\frac{1}{2}$
Guinea-pig *Cavia porcellus*	16–24	18–24	16–19	6–15	59–72	3	3–5	90	28	180	2–3
Rabbit *Oryctolagus cuniculus*	24	24	possibly 15–16	possibly continuous	30–32	7–9	4–6	60	56	1200	4–5

Table 8.9   Breeding details of small mammals recommended for school use (from Wray, J. D. 'Small Mammals'. EUP)

	Mating system	Introduction of opposite sexes for mating	Treatment of male at birth of young	Handling of young
Mouse *Mus musculus*	Monogamous pairs. Trios (1 female with 2 males) may be used for genetic work	Introduce female to male. Observe after introduction for signs of fighting, in which case, separate	Can be left with female	May be handled shortly after birth
Rat *Rattus norvegicus*	Monogamous pairs	Introduce female to male. Observe after introduction for signs of fighting, in which case, separate	Can be left with female	Should not be handled for first week
Gerbil *Meriones unguiculatus*	Monogamous pairs	Pairs should be set up shortly after weaning and can then be left together	Can be left with female	May be handled shortly after birth
Syrian hamster *Mesocricetus auratus*	Monogamous pairs	Pair at weaning and leave together. To mate strange adults or pairs which have been separated, place male in new cage for short time, introduce female in oestrous and observe for mating. If fighting occurs, separate and try again. Separate half-hour after mating	Monogamous pairs can be left together, otherwise separate half-hour after mating	Do not handle for first 16–18 days
Guinea-pig *Cavia porcellus*	Monogamous pairs or a harem of one male with up to 12 females	Monogamous pairs can be set up at weaning. Females can be introduced to the harem	Remove male just before birth in the case of monogamous pairs	May be handled shortly after birth
Rabbit *Oryctolagus cuniculus*	Cage male and female separately	Introduce female to male's cage at oestrus (signified by swollen, red, moist vulva). Mating should occur immediately. Remove female after mating and return to her cage	Remove female after mating	Do not handle until young emerge fully furred at 28 days.

Table 8.10   Mating systems for the six recommended species

8 Uncontrolled breeding should not be allowed to take place especially if the young cannot be adequately housed or cared for.

Further details can be found in Tables 8.9 and 8.10 and in Wray (1974b), *Small Mammals*.

Health. A number of features may indicate ill health. The more obvious of these are mentioned below. However, the health of a small mammal must always be considered in relation to its overall conditions and recent history. Age, recent re-housing, normal moulting, and breeding, may all lead to changes in the appearance and behaviour of mammals which may not signify ill health. A good picture of normal appearance and behaviour can only be built up by personal experience.

i *the complete animal.* Look for obvious missing parts (e.g. claws, tail, teeth, digits), or lumps. The coat should be glossy, the eyes clear and free of abnormal exudates. The feet should be in good condition, the claws subject to normal wear but not broken. The tail should be free of scabs, blood or other evidence of bites.

ii *openings.* The nose should be clean and neither blocked nor running. The teeth should be in good condition and not overgrown. The urinogenital opening should be free of blood, unblocked and not excessively reddened. The anus should be dry and unblocked. The faeces should be firm, normal coloured and free of blood and any sign of worms.

iii *behaviour.* The animal should be alert and steady in gait. It should be breathing, feeding and drinking normally. It should not be over-aggressive or flee and hide when approached.

References

Nuffield O level Biology Project (Revised) (1975) *Teachers' Guide I*, Longman.

Nuffield A level Biology Project (1971) *Laboratory Book*, Penguin Books.

Universities Federation for Animal Welfare (1972) *Handbook on the Care and Management of Laboratory Animals*, E. and S. Livingstone.

Wray, J. (1974a) *Animal Accommodation*, English Universities Press for the Schools Council.

Wray, J. (1974b) *Small Mammals*, English Universities Press for the Schools Council.

Addresses of suppliers of living organisms

All addresses in this list have been individually checked with the suppliers concerned and were known to be correct at the time of going to press.

Micro-organisms

Astell Laboratory Service Co., 172 Brownhill Road, Catford, London SE6 2DL (Tel. 01 697 8811). Range of bacteria (School agents for Oxoid Ltd.).

Commonwealth Mycological Institute, Ferry Lane, Kew, Richmond, Surrey TW9 3AF (Tel. 01 940 4086). Micro-fungi, non-pathogenic to man and animals.

Culture Centre of Algae and Protozoa, 36 Storey's Way, Cambridge CB3 0DT (Tel. 0223 61378). Algae (except seaweeds), free-living Protozoa.

Gerrard, T. & Co., Gerrard House, Worthing Road, East Preston, West Sussex BN16 1AS (Tel. 090 62 72071/5). Wide range of micro-organisms for school use.

Griffin Biological Laboratories Ltd., Gerrard House, Worthing Road, East Preston, West Sussex BN16 1AS (Tel. 090 62 72071/5). Wide range of micro-organisms for school use.

Harris, Philip, Biological Ltd., Oldmixon, Weston-super-Mare, Avon BS24 9BJ (Tel. 0934 413063). Wide range of micro-organisms for school use.

Marine Biological Association, Supply Department, Citadel Hill, Plymouth, Devon PL1 2PB (Tel. 0752 21761). Marine plankton.

Microbiological Supplies, P.O. Box 10, Tunbridge Wells, Kent TN1 1HB (Tel. 0892 34241). Cultures of bacteria, yeasts and other fungi, viruses, algae and protozoa.

National Collection of Dairy Organisms, National Institute for Research in Dairying, Shinfield, Reading RG2 9AT (Tel. 0734 883103). Bacteria of milk and milk products.

National Collection of Industrial Bacteria, Torrey Research Station, P.O. Box 31, 135 Abbey Road, Aberdeen AB9 8DG (Tel. 0224 877071). Non-medical bacteria of industrial and scientific interest.

National Collection of Marine Bacteria, Torrey Research Station, P.O. Box 31, 135 Abbey Road, Aberdeen AB9 8DG (Tel. 0224 877071). Bacteria of marine origin.

National Collection of Plant Pathogenic Bacteria, Ministry of Agriculture, Fisheries and Food, Plant Pathology Laboratory, Hatching Green, Harpenden, Herts AL5 2BD (Tel. 058 27 5241). Indigenous plant pathogenic bacteria.

National Collection of Yeast Cultures, Brewing Research Foundation, Lyttel Hall, Nutfield, Nr. Redhill, Surrey RH1 4HY (Tel. 073 782 2272). Wide range of non-pathogenic yeast cultures.

Oxoid Ltd., See Astell Laboratory Service Co.

Philip Harris, See Harris, Philip, Biological Ltd.

Plants (other than micro-organisms)

Allwood Bros. (Hassocks) Ltd., Clayton Nursery, Hassocks, Sussex BN6 9LX (Tel. 079 18 4229). *Dianthus* species and cultivars.

Anglo Aquarium Plant Co. Ltd., Wayside, Cattlegate Road, Enfield EN2 9DP (Tel. 01 363 8548). Tropical and cold-water aquatic plants.

Bees Ltd., Sealand, Chester CH1 6BA (Tel 0244 58 501). Wide selection of herbaceous plants, trees and shrubs.

Bennett's Water Lily Farm, Putton Lane, Chickerell, Weymouth, Dorset DT3 4AF (Tel. 030 57 5150). Pond plants.

Blackmoor Nurseries, Blackmoor, Liss, Hants GU33 6BS (Tel. 042 03 3576). Fruit trees and bushes.

Blakedown Nurseries Ltd., c/o Worfield Gardens, Worfield, Bridgenorth, Shropshire (Tel 074 64 259). Cacti, alpines, heaths, bonsai trees and pot plants.

Blom, Walter & Son, Coombelands Nurseries, Leavesdon, Watford, Herts WD2 7BH (Tel. 092 73 72071). Bulbs, corms and seeds, shrubs and herbaceous plants.

Broadleigh Nurseries Ltd., Broadleigh Gardens, Barr House, Bishops Hull, Taunton, Somerset TA4 1AE (Tel. 0823 86231). Miniature bulbs etc., herbaceous plants.

Cannock Fertilisers Ltd., Cannock, Staffordshire WS11 3LW (Tel. 054 15 2727). Grass seed mixtures.

Chelsea Physic Garden, 66 Royal Hospital Road, London SW3 4HS (Tel. 01 352 5646). Wide range of unusual plant material normally only available on exchange with other botanical gardens.

Clifton Geranium Nurseries, See Earnley Gardens Ltd.

Dobie, Samuel & Son Ltd., Upper Dee Mills, Llangollen, Clwyd LL20 8SD (Tel. 0978 860119). Wide range of garden seeds and bulbs.

Drake, Messrs. Jack, Inshriach Alpine Plant Nursery, Aviemore, Inverness-shire PH22 1QS (Tel. 054 04 287). Alpine plants.

Duff, Wm. & Son (Forfar) Ltd., West Craig Nurseries, Forfar, Angus DD8 1XE (Tel. 0307 2621/2). Wide range of trees and shrubs.

Earnley Gardens Ltd., Clifton Geranium Nurseries, Cherry Orchard Road, Whyke, Chichester, Sussex (Tel. 0243 82010). Cultivars of *Pelargonium* and Geranium.

Freshwater Biological Association, **a)** Windermere Laboratory, The Ferry House, Ambleside, Cumbria LA22 0LP (Tel. 096 62 2468/9) **b)** River Laboratory, East Stoke, Wareham, Dorset BH20 6BB (Tel. 0929 462314). Range of freshwater plants, terrestrial Bryophytes and Pteridophytes.

Gamble, D. & Sons, Highfield Nurseries, Longford, Derbyshire DE6 3DT (Tel. 033 523 238). *Pelargonium* cultivars.

Gerrard, T. & Co., Gerrard House, Worthing Road, East Preston, West Sussex BN16 1AS (Tel. 090 62 72071/5). Wide range of plants for school use.

Glebe Nursery, Aboyne, Aberdeenshire AB3 5JB (Tel. 0224 322945). Seedling and transplant trees.

Gregory, C. & Sons Ltd., The Rose Gardens, Stapleford, Nottingham NG4 7JA (Tel. 0602 39 5454). Roses including miniature cultivars.

Griffin Biological Laboratories Ltd., Gerrard House, Worthing Road, East Preston, West Sussex BN16 1AS (Tel. 090 62 72071/5). Wide range of plants for school use.

Harper Tree Nursery, See Glebe Nursery.

Harris, Philip, Biological Ltd., Oldmixon, Weston-super-Mare, Avon BS24 9BJ (Tel. 0934 413063). Wide range of plants for school use.

Howell, Major V. F., Fire Thorn, 6 Oxshott Way, Cobham, Surrey KT11 2RT (Tel. 093 26 2601). Wide range of more rare seeds of botanical and horticultural interest available by subscription.

Ingwersen, W. E. Th. Ltd., Birch Farm Nursery, Gravetye, East Grinstead, Sussex RH19 4LE (Tel. 0342 810236). Alpine and hardy plants.

Jefferson-Brown, Michael, Whitbourne, Worcester WR6 5BR (Tel. 088 62 270). *Narcissus* species and cultivars, grapevines.

Mackie, A. J. (Insectivorous Botanic Garden), Arnecote Park, Bicester, Oxon OX6 0NT (Tel 086 92 42126). Insectivorous plants.

Marine Biological Association, Supply Department, Citadel Hill, Plymouth PL1 2PB (Tel. 0752 21761). Wide range of marine plants.

Muir, Ken, Honeypot Farm, Weeley Heath, Clacton on Sea, Essex CO16 9BJ (Tel. 025 583 8181/2). Strawberry cultivars, bush and cane fruit.

Murrells of Shrewsbury, Portland Nurseries, Oteley Road, Shrewsbury, Shropshire (Tel 0743 52311/2). Shrub rose cultivars including miniatures.

Official Seed Testing Station, National Institute of Agricultural Botany, Huntingdon Road, Cambridge CB3 0LE (Tel. 0223 76381). Small reference samples of seeds of weed species.

Pennell & Sons Ltd., Princess Street, Lincoln LN5 7QL (Tel. 0522 26161). Hardy nursery stock including *Clematis* cultivars.

Perry's Hardy Plant Farm, Theobalds Park Road, Enfield, Middlesex EN2 9BG (Tel. 01 363 4207). Shade-loving plants and aquatics including some rare species.

Practical Plant Genetics, 18 Harsfold Road, Rustington, Sussex BN16 2QE. Range of tomato and other seeds for genetics.

Price, Mary E., Fernhurst, Roncarbery, County Cork, Republic of Ireland. Ferns.

Reid, Ben, & Co. Ltd., Pinewood Park Nurseries, Countesswells Road, Aberdeen AB9 2QL (Tel. 0224 38744). Trees, shrubs and tree seeds.

Rochford, T. & Sons Ltd., Turnford Hall Nurseries, Turnford, Broxbourne, Herts EN10 6BH (Tel. 099 24 64512). Foliage and flowering pot plants.

Rogers, W. H. (Chandler's Ford) Ltd., Red Lodge Nursery, Chesnut Avenue, Eastleigh, Hants SO5 3HG (Tel. 042 15 2509). Conifers, including dwarf species, and ornamental shrubs.

Russell, L. R., Ltd., Richmond Nurseries, Windlesham, Surrey GU20 6LL (Tel. 0990 21411/2). Trees and shrubs.

Shirley Aquatics Ltd., Stratford Road, Monkspath, Shirley, Solihull, West Midlands B90 4EF. Tropical and coldwater plants.

Timstar Biological Suppliers, The Pinfold, Poole, Nantwich, Cheshire (Tel. 0270 626175). Range of laboratory plant material.

Tokonoma Bonsai, 14 London Road, Shenley, Radlett, Herts WD7 9EN (Tel. 092 76 7587). Bonsai trees.

Treasures of Tenbury Ltd., Burford House Gardens, Tenbury Wells, Worcs. WR15 8HQ (Tel. 0584 810777). Wide range of trees, shrubs and herbaceous plants including *Clematis* species and cultivars.

University Marine Biological Station, Millport, Isle of Cumbrae, Scotland KA28 0EG (Tel. 047 553 581/2 and 756). Marine plants.

Unwin, W. J. Ltd., Impington Lane, Histon, Cambridge CB4 4LE (Tel. 022 023 2270). Wide range of seeds.

Waterer, John, Sons & Crisp Ltd., The Nurseries, London Road, Bagshot, Surrey GU19 5DG (Tel. 0276 72288). Wide range of garden plants.

Wicks, W. C. Ltd., Lambley Nurseries, Catfoot Lane, Lambley, Nottingham NG4 4QL (Tel. 0602 268137). Foliage and flowering plants including cultivars of *Saintpaulia*.

Wills Fuchsias Ltd., The Fuchsia Nursery, Chapel Lane, West Wittering, Chichester, West Sussex PO20 8QG (Tel. 024 366 3065). *Fuchsia* species and cultivars.

Woolman, H. Ltd., Grange Road, Dorridge, Solihull, West Midlands B93 8QB (Tel. 056 45 6283). Rooted cuttings and bulbs.

Warfield Gardens Ltd., See Blakedown Nurseries Ltd.

Animals

N.B. Small mammals should only be obtained from accredited suppliers. A list of such breeders is available from the Medical Research Council's Laboratory, Animals Centre, Woodmansterne Road, Carshalton, Surrey SM5 4EF. This list is not included below.

Baxter, Ronald N. (Entomologists), 16 Bective Road, Forest Gate, London E7 0DP (Tel 01 534 6312). Lepidoptera including Saturniidae and Sphingidae.

Bleak Hall Bird Farm (Luton) Ltd., Cresta House, Alma Street, Luton, Beds LU1 2PL (Tel. 0582 23730). Birds and reptiles.

Burdett, T. A., 22 Park Hill Road, Ewell, Surrey KT17 1LS (Tel. 01 393 8181 and 01 393 7723). Amphibia including *Xenopus*, and reptiles.

Coombs, E. W. Ltd., 25 Frindsbury Road, Strood, Kent ME2 4SU (Tel. 0634 79886/7). Mealworms, crickets.

Culture Centre of Algae and Protozoa, 36 Storey's Way, Cambridge CB3 0DT (Tel. 0223 61378). Protozoa.

Freshwater Biological Association, **a)** Windermere Laboratories, The Ferry House, Ambleside, Cumbria LA22 0LP (Tel. 096 62 2468/9). **b)** River Laboratory, East Stoke, Wareham, Dorset BH2O 6BB (Tel. 0929 462314). Freshwater animals.

Gerrard, T. & Co., Gerrard House, Worthing Road, East Preston, West Sussex BN16 1AS (Tel. 090 62 72071/5). Wide range of animals for school use.

Griffin Biological Laboratories Ltd., Gerrard House, Worthing Road, East Preston, West Sussex BN16 1AS (Tel. 090 62 72071/5). Wide range of animals for schools.

Harris, Philip, Biological Ltd., Oldmixon, Weston-super-Mare, Somerset BS24 9BJ (Tel. 0934 413063). Wide range of animals for school use.

JG Animals, 19 Streatham Vale, London SW16 5SE (Tel. 01 764 4669). Amphibia and Reptiles.

Laboratory Animals Centre, *See* Medical Research Council.

Larujon Locust Suppliers, c/o Welsh Mountain Zoo, Colwyn Bay, Clwyd (Tel. 0492 2938). Locusts.

Lullingstone Silk Farm Ltd., Ayot House, Ayot St. Lawrence, Herts AL6 9BP (Tel. 0438 820221). Silkmoth eggs.

Marine Biological Association, Supply Department, Citadel Hill, Plymouth, Devon PL1 2PB (Tel. 0752 21761). Wide range of marine animals.

Medical Research Council Laboratory Animal Centre, Woodmansterne Road, Carshalton, Surrey SM5 4EF (Tel. 01 643 8000). List of accredited breeders of small mammals.

Mountain Grey Apiaries, Selby Road, Holme-on-Spalding Moor, York YO4 4EZ (Tel. 069 63 338). Hive bees.

Ponderosa Bird Aviaries, Branch Lane, The Reddings, Cheltenham, Glos. GL51 6RP (Tel. 0452 713229). Birds and mealworms.

Shirley Aquatics Ltd., Stratford Road, Monkspath, Shirley, Solihull, West Midlands B90 4E. Freshwater animals.

Southern Aviaries Ltd., Brook House Farm, Tinkers Lane, Hadlow Down, Sussex TN22 4EU (Tel. 082 585 283). Birds.

Steele, R. & Brodie, Bee Appliance Works, Warmit, Newport on Tay, Fife DD6 8PG (Tel. 082 66 728). Hive bees.

Taylor, D. J., Newmarket Cottage, Newmarket, Clay Lane, Clay Cross, Chesterfield, Derbyshire S45 9AP (Tel. 0246 863506). Reptiles, birds and hatching eggs.

Taylor, E. H. Ltd., Beehive Works, Welwyn, Herts, AL6 0AZ (Tel. 043 871 4401). Hive bees.

Thorne, E. H. (Beehives) Ltd., Beehive Works, Wragby, Lincoln LN3 5LA (Tel. 067 34 555/6). Hive bees.

Timstar Biological Suppliers, The Pinfold, Poole, Nantwich, Cheshire (Tel. 0270 626175). Range of laboratory animals including fertile hens eggs.

University Marine Biological Station, Millport, Isle of Cumbrae, Scotland KA28 0EG (Tel. 047 553 581 and 047 553 756). Marine animals.

Worldwide Butterflies Ltd., Over Compton, Sherborne, Dorset DT9 4QN (Tel. 0935 4608). Ova, larvae and pupae of Lepidoptera.

9   Biological techniques

MICROSCOPICAL TECHNIQUES

It is frequently necessary to treat material of living origin prior to microscopical examination. Such treatments, known as *microtechnique*, are designed to enable a fuller examination of the material under the microscope while still retaining it in as lifelike a condition as possible. Microtechniques usually affect the protoplasm and other cellular constituents in such a way as to give rise to 'artifacts', thus distorting the real object under examination: indeed the entire preparation may be regarded as an artifact. Picken (1955) has pointed out that 'the task of the biologist is not to avoid artifacts, but to interpret them and never to accept them for more than they are'.

A comprehensive and detailed treatment of microscopical techniques is outside the scope of this book. However, it is appropriate to outline the principles involved in the various stages of microtechnique and to mention the commoner reagents employed. Details of the preparation of these reagents will be found in Chapter 7. Those interested in furthering their knowledge in this area should consult Grimstone and Skaer (1972), Pantin (1948) or Bradbury (1973). The last of these is an invaluable reference work in the biological laboratory.

Microscopical preparations may be either *temporary* or *permanent*. The former may be examined microscopically for a relatively short time, usually a few hours but perhaps even less than one hour. Most permanent preparations, if suitably stored, should remain unchanged for many years. The majority of permanent preparations involve the following stages.

1 Fixation.
2 Sectioning, smearing or squashing. In the last case the material may require prior maceration.
3 Staining.
4 Dehydration.
5 Clearing.
6 Mounting.

These are dealt with separately below.

In addition to these six basic processes a number of other treatments may be used on occasions. These are also briefly outlined.

Fixation

The purpose of fixation is to preserve the material in such a way that decay by bacteria, moulds or autolysis is prevented and distortion of the parts caused by swelling, shrinkage, evaporation or other changes is reduced to a minimum. The material must also be fixed so as to allow subsequent treatment (e.g. sectioning, and osmotic and temperature changes), without further distortion. The cytological effects of fixation are to kill the protoplasm, to raise its refractive index from approximately that of water (1.3) to approximately that of glass (1.5) and to convert protein sols into the more viscous aqueous gels.

Fixation may be carried out by *heat treatment* (resulting in considerable distortion but particularly useful for blood smears), *low temperature techniques* (particularly suitable for histochemical work as soluble compounds are retained and proteins not denatured) or by *chemical* means.

Plant material should be fixed as soon as possible after gathering to prevent changes taking place, particularly shrinkage due to drying. Animal material should be fixed immediately after killing to prevent putrefaction or autolysis. For histological work, the material should be cut into pieces no bigger than $1\,cm \times 1\,cm \times 0.5\,cm$ and preferably smaller. The pieces should be left in about one hundred times their own volume of fixative for an appropriate time (usually 12–24 hours). For cytological work, much thinner pieces should be cut (say 1 mm thick) and left in the fixative for a correspondingly shorter time. After fixation, thorough washing of the material in an appropriate solution (see below) is essential. Always rewash stored material before subjecting it to further treatment. Table 9.1 gives details of a number of commonly used fixatives.

Sectioning

Few tissues or specimens are sufficiently thin to allow enough light to pass through them for microscopical examination. In most cases they must be sectioned, though smearing or squashing the material onto the microscopic slide or cover-slip may be a more suitable method for some tissues.

Rigid specimens, such as some plant stems, may be sectioned free-hand with reasonably satisfactory results. If the material is held firmly, it is possible, with practice, to cut sections as thin as $10\,\mu m$ (0.01 mm). A more consistent

Fixative	Uses	Time	Method of washing	Other details
Formaldehyde (buffered) —see page 181	Animal histology	16–48 h	Wash in several changes of 50 per cent, 70 per cent or 90 per cent alcohol for at least 12 h in all	Not suitable for mammalian testis due to excessive shrinkage. Tissues may be stored in the fixative. Good fixative to use prior to staining with haematoxylin
Formol-saline —see page 182	Animal histology	24–48 h	Wash in several changes of 50 per cent, 70 per cent or 90 per cent alcohol for at least 12 h in all	Opinions differ as to preference when compared with formaldehyde; it may reduce distortion brought about by osmotic changes. Almost certainly better for marine organisms
Alcohol-methanal (alcohol-formalin) —see page 172	Plant histology	15–60 min	Wash in 70 per cent alcohol, for 2–3 min	Causes some shrinkage and considerable hardening. Especially suitable for small pieces or sections. Material may be stored in fixative indefinitely
Acetic-alcohol (Clarke's Fluid) —see page 172	Cytology, especially chromosomes	30 min (Do not leave for longer)	Wash in several changes of alcohol (100 per cent). This is not necessary if carmine-acetic or orcein-acetic is being used for staining immediately	Shrinkage caused by alcohol is counter-balanced by swelling caused by ethanoic (acetic) acid. Action may be hastened by the addition of tri-chloromethane (chloroform) to form Carnoy's Fluid (see page 176)
Formal-acetic-alcohol (FAA) —see page 181	Plant material	At least 12 h	Wash in several changes of 50 per cent alcohol	The most commonly used fixative for plant material. May be left in the fixative indefinitely
Bouin's solution (aqueous) —see page 175	Mammalian histology. Delicate plant and animal material	12 h	Wash in several changes of 70 per cent alcohol for at least 24 h in all	Not suitable for organs containing large number of mucous cells or for mammalian kidney. Especially good for marine invertebrates. For embryos, Bouin's should be diluted with 25 per cent or 50 per cent of its volume with distilled water
Chrom-acetic —see page 177	Plant histology	24 h or more	Wash thoroughly in running water for 24 h. Then in distilled water for 12 h. Pass up through 50 per cent, 70 per cent and 90 per cent alcohol	A good fixative for plant tissues. Weak chrom-acetic should be used for algae, fungi, bryophytes and other delicate materials. A medium strength solution should be used for more robust tissues such as root tips, ovaries, ovules, etc. (See page 177)
Chrom-acetic formalin (Navashin's fluid) —see page 187	Plant cytology especially squashes and smears	12–24 h	Wash thoroughly in several changes of 70 per cent alcohol	Smears and root tip squashes can be well fixed by this method
Susa (Heidenhain's Fluid) —see page 193	Invertebrate histology	3–24 h	Transfer to 96 per cent alcohol and wash for 12 h	0.5 per cent iodine can be added to the alcohol in which the material is washed
Zenker-formol —see page 194	Good general animal fixative	3–18 h	Wash in running water for 12 h. Transfer to 50 per cent then 70 per cent alcohol	Helly's variation (page 194) replaces the glacial ethanoic acid with formalin which then improves its quality as a cytoplasmic fixative.

Table 9.1   Details of some commonly used fixatives

standard of section will be obtained if the specimen is held in a support formed by pieces of pith, expanded polystyrene or other suitably rigid material which can itself be easily cut. Support of this kind is essential if more flexible plant material such as thin roots or leaves are being sectioned. Animal tissues cannot normally be sectioned satisfactorily using such techniques owing to their lack of rigidity. Instead they must be permeated and embedded in paraffin wax or embedded in 'Celloidin' and then sectioned, preferably with the aid of a hand or mechanical microtome. These more elaborate techniques can also be used to produce high quality sections of plant material. The use of the mechanical microtome also enables a large number of serial sections to be cut, so providing a better appreciation of the three-dimensional structure of the specimen.

Free-hand section cutting

This is done with a section-cutting razor which must be in perfect condition and very sharp. A cheaper and perfectly satisfactory alternative is a new razor blade firmly mounted in a suitable holder. This obviates the need to sharpen the 'cut-throat' razor—a task which some people find rather difficult.

1 Place the piece of tissue to be sectioned between two pieces of pith, expanded polystyrene, carrot or some similar material which has been stored in 70 per cent alcohol.

2 Place a few drops of the liquid in which the tissue has been stored on the upper surface of the razor.

3 Keep the razor horizontal while drawing it through the supporting material and the tissue (see Figure 9.1). Develop a rhythm of long sliding movements, drawing the razor *towards* the body, cutting several sections in one series of movements.

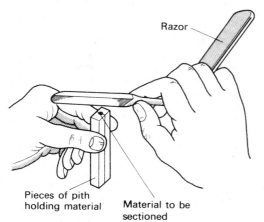

Razor

Pieces of pith
holding material

Material to be
sectioned

Figure 9.1   Cutting sections by hand for microscopical examination. The positions of the fingers and thumbs are important for safety and precision

4 Transfer the cut sections to some of the liquid in which the tissue has been stored, using a camel-hair brush.

5 Select some of the thinnest sections for subsequent staining and mounting. It is not essential to obtain complete sections of plant material. The best sections are often incomplete, but representative, sectors cut at right angles to the main axis of the piece of tissue.

Section cutting with the aid of a hand microtome

The hand microtome is a relatively cheap device in which a piece of embedded tissue can be held. On turning a knob a measured portion of tissue protrudes from the instrument and can be cut using a hand-held razor. A further turn delivers another measured quantity of tissue. The mechanism can be adjusted to give sections of $20\,\mu m$ or $30\,\mu m$ thickness and partial sections may be cut down to $10\,\mu m$ thickness.

1 Cut the material to be sectioned into lengths of 1–1.5 cm.

2 Provide support for the material by one of the following techniques:

   i hardening in 70 or 90 per cent alcohol for 24 h (suitable for some plant materials);

   ii embedding in Celloidin (see Bradbury (1973) page 66).

3 Place the material in the well of the microtome near the front edge of the platform (i.e. the side furthest from the operator).

4 Lower the microtome mechanism till the material drops just below the level of the top plate.

5 Pour molten paraffin wax (melting point 40–50 °C) into the well until the material is completely surrounded and covered. Add more wax as the first batch cools and contracts. Allow the wax to set fully.

6 Turn the microtome mechanism so that the embedded material is pushed up the well.

7 Cut away excess wax from the top of the specimen, using an old blade or scalpel, until the specimen is revealed. Cut away a further portion of wax on the side of the material furthest from the operator, so that when sections are cut, the blade will first strike the specimen itself and not the wax.

8 Flood the surface of the cutting razor and block of wax with 70 per cent alcohol.

9 Place the cutting razor on the top plate, with the sharp edge towards the operator and slide it forwards and sideways through the material (see Figure 9.2).

10 Transfer the section to alcohol, remove any loose wax from the microtome plate with a fine brush and adjust the mechanism of the microtome so that another section-thickness of material is delivered up the well ready for sectioning.

11 Repeat 8, 9 and 10 until a sufficient number of sections have been cut.

Many plant materials are best placed between two pieces of expanded polystyrene which have been trimmed

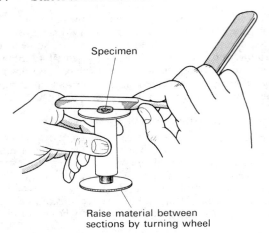

Specimen

Raise material between
sections by turning wheel

Figure 9.2   Cutting sections using a hand microtome

to fit exactly into the well of the microtome. In this case wax is not used and stages 5 and 7 above are omitted.

Section cutting using a mechanical microtome

The mechanical microtome enables a very large number of thin (8 µm–12 µm) serial sections to be cut from a piece of wax-embedded tissue. The experienced operator can cut sections down to a few micrometres thick. The material to be sectioned must first be embedded in paraffin wax in the way described below.

1 Fix and wash the piece of tissue.

2 Transfer to 70 per cent alcohol for 3–6 hours.

3 Transfer to 90 per cent alcohol for 3–6 hours.

4 Transfer to 95 per cent alcohol for 3–6 hours.

5 Transfer to three successive batches of absolute alcohol for 3–4 hours in each.

6 Transfer to xylene for 1–6 hours (delicate specimens should first be placed in a mixture of 50 per cent absolute alcohol: 50 per cent xylene for 30 minutes).

7 Melt some paraffin wax (melting point 52–56 °C) and maintain it at a temperature of not more than 5 °C above its melting point. This is best done in wide, shallow specimen tubes containing sufficient wax to cover the piece of tissue. They may be stood in a thermostatically controlled oven or in a suitable container heated by a water bath.

8 Transfer the specimen to one of the tubes of molten wax and maintain at the same temperature for one hour.

9 Transfer to a second and then to a third tube of molten wax using a warm spatula, leaving it in the wax for one hour in each case. Impregnation by the wax should now be complete.

10 Smear the inside of a watch glass or a small paper boat (see Figure 9.3) with a little propane-1,2,3-triol (glycerol) and then warm the mould. Pour some hot wax (not more than 58 °C) into the mould and introduce the impregnated

tissue into the molten wax with the aid of a pair of warm forceps. Move the specimen to the desired position and add more molten wax, if necessary, to fill the mould.

11 Hold the mould on the surface of some cold water and cool the surface of the wax by gently blowing on it.

12 As soon as a skin appears on the surface of the wax, immerse it in the cold water to cool it rapidly. This reduces crystallisation during the cooling.

13 When thoroughly cooled, remove the wax from its mould. It may be stored in this condition prior to sectioning, but if a number of blocks are to be retained they should be clearly labelled. This can be done by inserting a slip of paper bearing the appropriate details during stage 10.

The block is now ready for trimming, mounting on the microtome and sectioning. This is done in the following stages.

1 Trim away the wax from the block so that none is left along the sides of the tissue and only very little at top and bottom. It is essential that the upper and lower surfaces of the block are cut parallel to each other otherwise the sections will not come off in a continuous ribbon. Trim off the corners of the block so that the ribbon may be easily broken at the desired points. The trimmed block should now resemble that in Figure 9.4.

2 Heat the blade of an old scalpel and use this to melt the wax on the microtome chuck.

3 Press the block onto the warmed wax on the chuck and position it so that the upper and lower surfaces are parallel to the cutting edge of the microtome razor. The long axis running through the block and passing through all the sections to be cut should be at right angles to the cutting blade.

4 Allow the wax to cool and set.

5 Mount the chuck in position and set the microtome to the desired thickness of section.

6 Operate the microtome so that when the sections are cut they come off the blade in a long ribbon. This should be supported on a finger or a moist paint brush.

7 Transfer the ribbon, in convenient lengths, to a sheet of black paper well away from draughts. The ribbon may be cut between sections using a curved Swann-Morton scalpel blade (e.g. No. 22).

8 Dilute a little of a stock solution of egg albumen (see page 171) with 50 volumes of distilled water. Pipette 2–3 cm³ of this diluted solution onto a clean microscope slide. Alternatively, rub a little Mayer's albumen (page 171) onto the slide with a clean finger.

9 Place 2–3 of the wax embedded sections from the ribbon onto the pool of albumen making sure that the shiny side faces towards the glass.

10 Warm the slide slightly on a hot plate until the folds in the section flatten out, but do not allow the wax to melt. Pour off the excess albumen and wipe carefully round the sections with a clean cloth.

11 Leave the slide in an incubator at 37 °C to dry.

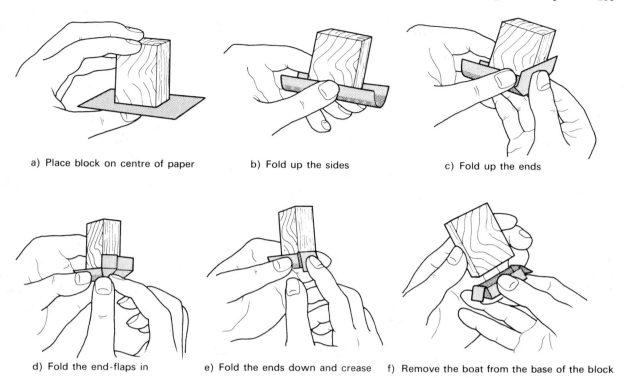

a) Place block on centre of paper   b) Fold up the sides   c) Fold up the ends

d) Fold the end-flaps in   e) Fold the ends down and crease   f) Remove the boat from the base of the block

Figure 9.3   Making a paper boat for wax embedding specimens

12 Remove the wax with xylene. This is best done in slotted vessels known as Coplin jars which hold a number of slides. A suitable alternative is a 9 cm × 3 cm specimen tube in which two slides can be placed back to back. Alternatively, the xylene may be poured gently and directly onto the slide.

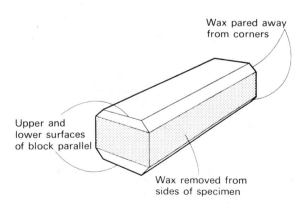

Wax pared away from corners

Upper and lower surfaces of block parallel

Wax removed from sides of specimen

Figure 9.4   The trimmed block of wax containing the embedded specimen ready for mounting on the microtome chuck

The following stages and subsequent staining, differentiation, etc., are carried out in the same way. The sections should be left in the xylene for 2 minutes to dissolve all the wax.

13 Remove the xylene with absolute alcohol (1 minute).
14 Hydrate by transferring to 90 per cent and then 70 per cent alcohol. If an alcoholic stain is being used, the sections can now be stained. If an aqueous stain is to be employed the sections should first be transferred to distilled water for 1 minute.

Possible faults

If the wax is too soft (i.e. of too low a melting point or sectioned at too high a temperature) the sections will crumble. The only remedy is to re-embed the tissue. If the wax is too hard, the ribbon will break between sections and the block should be gently warmed or coated in a wax of a lower melting point (40 °C). Other causes of the ribbon breaking are the knife or the atmosphere being too cold (an electric lamp near the blade may cure this); the knife being greasy (wipe it with xylene); the top and bottom of the block not being parallel (pare away more wax to remedy this); the xylene not being fully removed prior to the final impregnation with wax (re-impregnate with pure wax—see page 254. Ragged or torn sections may be caused by the razor being blunt or set at the wrong angle. Incorrect fixation or washing out of the fixative may also give rise to torn sections.

Smearing

Some tissues (e.g. blood) do not lend themselves to sectioning. They may be smeared in a very thick, even layer over the surface of a microscope slide or cover-slip. Other soft tissues (e.g. earthworm seminal vesicles) may, when fresh, be conveniently smeared in a similar way.

For the preparation of a blood smear the following procedure should be adopted.

1 Make sure that the microscope slides to be used are scrupulously clean.
2 Sterilise the ball of the middle finger with a cotton wool swab dipped in alcohol.
3 Use a *sterile* lancet to jab the sterilised ball of the finger. A sterile blood sampler may be used as an alternative to a lancet, but other implements, whether or not sterilised, should *never* be employed.
4 Touch a drop of the blood onto the end of a clean microscope slide, but do not allow the skin of the finger itself to touch the slide.
5 Hold the end of a second slide so that it forms an angle of about 60° with the first and touch the blood film so that it spreads along the angle between the two slides. If the second slide is drawn to the end of the first, the blood will reach the end of the slide and make subsequent handling difficult. (See Figure 9.5).
6 Push the second slide along the first so that the blood film is dragged behind the moving slide. A smear similar to that illustrated in Figure 9.5c should now be obtained.
7 Wave the slide in the air to dry the smear which is now ready for staining with Leishman's, Wright's or similar stain.

N.B. See page 43 for a discussion of safety and legal aspects of obtaining blood samples.

Other tissues such as earthworm seminal vesicles or fresh spinal cord may be smeared on the surface of a cover-slip by holding the piece of tissue in a pair of fine forceps and smearing the fluid contents onto the cover-slip.

Maceration and squashing

Root tips and other rapidly growing parts of plants are normally too firm to allow smears to be made but are not sufficiently rigid to permit sections to be cut. In these cases, maceration followed by squashing is the procedure usually followed in preparing material for staining and mounting.

The purpose of maceration is to break down the middle lamella, which lies between adjacent cells cementing them

together. The cells may then be separated by squashing, thus enabling individual examination. The middle lamella in herbaceous tissues normally consists of pectic substances. In harder, woody tissues there is an increase in

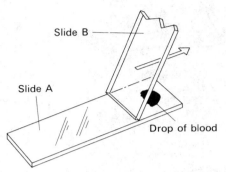

a) Place the end of slide B on the surface of slide A and pull it slowly towards the drop

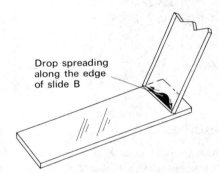

b) When the drop is contacted it will run along the whole width of slide B

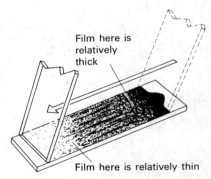

c) To make the film: push slide B quite quickly along slide A dragging the drop behind; do this once only; do not attempt to push the drop as this will result in too thin a film

Figure 9.5   Making a blood smear

lignification and this necessitates a different form of macerating treatment. There is thus a scale of severity of the macerating technique and appropriate treatment must be selected for the particular tissue being examined. The following methods are recommended.

a Root tips for mitosis

1 Soak in M hydrochloric acid at 60 °C for 6–10 minutes. This not only breaks down the middle lamellae, but also affects the chromosomes in such a way that they subsequently take up stains (such as Feulgen's) more readily.

b Fresh, sectioned delicate herbaceous tissue

1 Soak in 5 per cent aqueous chromic(VI) acid for 24 hours.
2 Wash in running water for at least 2 hours.
3 Rinse in distilled water.

c Tougher herbaceous tissue

1 Boil for a few minutes in 10 per cent aqueous potassium hydroxide solution.
2 Wash in several changes of water.
3 Carry out Method **b** above.

d Leaves

1 Remove chlorophyll by soaking the leaves in 95 per cent alcohol until decolourised.
2 Warm in 5 per cent potassium hydroxide solution (preferably on a water bath) until the epidermis separates from the mesophyll. *Do not boil.*
3 Wash in gently running water for 2 hours.
4 Soak in 10 per cent hydrochloric acid for 6 hours.
5 Wash in gently running water for 2 hours.
6 Macerated leaves may be stored in 5 per cent methanal (formaldehyde) until required.

e Woody tissue

1 Thin shavings of wood suitable for maceration may be obtained by planing a block of wood with a sharp plane.
2 Place the shavings in a mixture of equal volumes of glacial ethanoic (acetic) acid and aqueous hydrogen peroxide (20 vol) in a flask fitted with a reflux condenser. The flask should be approximately half full.
3 Reflux gently for at least 1 hour. Hard woods may need to be refluxed for as long as 8 hours. Allow to cool.
4 Neutralise with aqueous sodium carbonate.
5 Pour off the liquid, leaving the woody material behind. Wash with several changes of water.
6 The macerated woody material may be stored in FAA (see page 181) until required, or it may be stained immediately.

Once maceration is complete, it may be necessary to separate the cells either before or after staining. This may be done with the aid of fine needles (glass needles are generally preferable to steel ones as the latter frequently affect the staining reaction).

The squash technique is also designed to separate the cells. However, in this case the position of the cells after squashing should indicate their relative position in the original tissue. A good squash causes a minimum of disturbance to the relative position of the cells in the tissue—it merely flattens them out so that individual cells can be seen more clearly.

As an example, if a root-tip squash is being prepared, it should be macerated (see above) and may be stained (see page 262) before being placed on a microscope slide. The cover-slip is then added and a piece of filter paper folded around the slide and cover-slip before placing them on the bench. The centre of the cover-slip (under the filter paper) is then pressed firmly and evenly so that the tissue beneath it becomes flattened. Various workers have developed a variety of techniques for applying a suitable pressure to the cover-slip. Some press with a thumb or single finger, others tap the cover-slip with the blunt end of a pencil while others roll a cylindrical pencil over the cover-slip. The important point is that pressure should be applied vertically and all lateral sliding of the cover-slip should be avoided. Pressure applied with the thumb is probably the easiest technique for the beginner to master.

Staining

Different structures within an unstained specimen can only be distinguished if their refractive indices are different. A further factor (i.e. colour) is introduced if the specimen is stained. Thus staining is designed to make different parts of the specimen more obvious. The exact mechanisms of operation of different stains need not concern us here. It is sufficient to say that both chemical combination and physical processes (such as absorption and adsorption) are probably involved in most cases. The stain must become attached to the tissue in such a way that it is not subsequently washed out, even by the solvent in which the stain was originally dissolved.

Most stains are coloured organic dyes which are soluble in water or alcohol or both. They have been classified in various ways of which the most useful is as follows:

i *basic* or *nuclear* stains consist of a coloured organic base in combination with an acetate, sulphate or chloride radical. Such stains have an affinity for the nuclei of cells owing to the presence there of nucleic acids. Other parts of cells may also be of an acid nature and together with the nuclei are described as basiphil. Examples of such stains are Safranin and Haematoxylin;

ii *acidic* or *cytoplasmic* stains consist of a metallic base combined with a coloured organic radical. The cytoplasm of most cells tends to be basic and thus to have an affinity for acidic stains. It is therefore described as acidophil. Eosin is perhaps the most familiar example of an acidic or cytoplasmic stain;

iii *neutral* stains are normally a mixture of acidic and basic dyes.

The classification outlined above should not be viewed too rigidly. Nuclear stains will also stain the cytoplasm to some extent while cytoplasmic stains will also be picked up by some nuclear elements.

Certain dyes will not become attached to cellular material unless a mordant is used. It appears that the mordant becomes attached to the tissue to form a complex compound which can then unite with the dye to form an insoluble coloured compound. For example, iron alum is used as a mordant for haematoxylin.

Certain dyes will combine only with specific tissues or cellular components. For example phloroglucin(ol) combines only with lignin, while Schulze's solution only combines with cellulose. Such dyes are said to be *specific* for the tissue concerned.

Staining methods

A number of general staining methods are commonly used.

a Progressive staining. Dyes such as carmine and haematoxylin are taken up first by the nuclei and then by the cytoplasm. If tissue is left in the stain, it will become gradually more and more deeply overstained. Progressive staining involves leaving the tissue in the stain for just the right length of time until the nuclei and cytoplasm are stained to the desired degree. For this purpose it is often advisable to dilute the stain before use, thus reducing the risk of overstaining. If thin sections are being stained, they may be placed on the microscope stage in the staining vessel and observed till the right degree of staining is achieved.

b Retrogressive (regressive) staining. This involves deliberately overstaining the tissue and then removing the excess stain by a process of destaining or differentiation. Once again the tissue may be observed microscopically during the destaining until the desired balance is obtained. Retrogressive staining takes longer than progressive staining, but gives sharper differentiation.

c Counterstaining or differential staining. In this technique a particular part of the cell or tissue is stained with an appropriate dye. This is followed by the staining of other components with a contrasting colour. The process in fact involves the replacement of one dye with another. The resulting preparation may, for example, have all the lignin stained one colour and all the cellulose another. The terms *double staining* and *triple staining* are also used,

though these strictly apply to the use of a mixture of two or three dyes of contrasting colours.

Staining is usually carried out on sections, though smears and squashes may be more appropriate in a number of specific cases (e.g. blood smears and root-tip squashes). On occasions, bulk staining may give satisfactory results: this involves the staining of a large piece of tissue which is subsequently sectioned and mounted. Small specimens (e.g. small arthropods, medusae, colonial hydroids, etc.) may be stained and mounted entire.

A number of simple staining schedules are described on pages 261–5. These include techniques used for both temporary and permanent preparations. Those requiring a wider range of techniques should consult Bradbury (1973).

Dehydration

The media in which stained and sectioned pieces of material are generally mounted on slides are immiscible with water. The paraffin wax with which some materials are impregnated prior to sectioning is also immiscible with water. It is therefore necessary to remove all traces of water from materials before either mounting or impregnating with wax. Alcohol is generally used for this dehydration.

The naming of the various grades and types of alcohol solutions sometimes causes confusion (see page 171). For microtechnical purposes, industrial methylated spirit (74 °O.P.) is quite adequate, especially if it is dried out by standing over anhydrous copper sulphate. It may be obtained under licence issued by the Customs and Excise Department. In the staining schedules which follow (page 261) it is assumed that industrial methylated spirit will be used and the term 'absolute alcohol' is hereafter used to refer to industrial methylated spirit. Appropriate precautions should be taken to ensure that stocks of industrial methylated spirit do not take up water from the atmosphere.

If material which has previously been treated in an aqueous medium (or fresh material) is transferred directly to absolute alcohol, severe distortion and damage to tissues will result. It is therefore necessary to transfer specimens through a graded series of alcohols of increasing concentration. 70 per cent, 90 per cent, 95 per cent and 100 per cent (absolute alcohol) are those generally employed, though for more delicate tissues and bulk processing the addition of 30 per cent and 50 per cent alcohol to the series is advisable. All material should be passed through at least two changes of absolute alcohol to ensure total dehydration. Should it prove necessary to transfer material from an alcoholic to an aqueous medium (e.g. if an aqueous stain needs to be employed after one dissolved in alcohol) the graded series of alcohols should be used in reverse.

The time for which material needs to be left in each of

the alcohols depends upon the bulk of the material. Blocks of tissue prior to sectioning should be left for 3–10 hours in each concentration. Sections which have already been stained and are about to be cleared and mounted need to be left for only 1–2 minutes in each concentration, with perhaps a little longer in absolute alcohol.

Ethanol is not the only dehydrating agent employed. Propan-2-ol (*iso*-propanol), butan-1-ol (*n*-butanol) and 2-ethoxyethanol (ethylene glycol monoethyl ether— formerly known as 'Cellosolve') are also used. 2-ethoxyethanol has the advantage that its dehydration is much less violent than that of ethanol itself. Specimens may thus be transferred directly from water to 100 per cent 2-ethoxyethanol hence greatly shortening the time needed for the dehydration process.

NOTE: Diethylene dioxide (dioxan) should not be used for dehydration as it gives off a poisonous vapour which may have long-term effects.

The dehydration process for *sections* thus consists of the following.

1 Place in 30 per cent alcohol for 1 minute.
2 Transfer to 50 per cent alcohol for 1 minute.
3 Transfer to 70 per cent alcohol for 1 minute.
4 Transfer to 90 per cent alcohol for 2 minutes.
5 Transfer to 95 per cent alcohol for 2 minutes.
6 Transfer to absolute alcohol for 2 minutes.
7 Transfer to fresh absolute alcohol for 2 minutes.

Always start the dehydration process in the next higher percentage alcohol than that in which the specimen has been treated. For example, if the previous stage consisted of staining in a dye dissolved in 70 per cent alcohol, the specimens may be transferred directly to 90 per cent alcohol. The containers of absolute alcohol must always be kept covered to minimise the absorption of atmospheric moisture. For bulk tissues, the time in each concentration of ethanol should be increased to 6 hours (though 1 hour may be sufficient for concentrations up to 70 per cent). Blocks of tissue for wax embedding should not be transferred from ethanol to the wax as they are immiscible. They should be de-alcoholised (see below) or alternatively, another dehydrating agent such as butan-1-ol or propan-2-ol should be employed.

Using 2-ethoxyethanol

Transfer the specimens from water, or from whatever concentration of alcohol they are in, directly to 2-ethoxyethanol for 1–2 minutes. Bulk tissues should not be dehydrated in 2-ethoxyethanol.

Clearing

As has already been mentioned, ethanol is immiscible with the paraffin wax used for embedding specimens prior to sectioning. In a similar way, 2-ethoxyethanol is immiscible with Canada balsam and the synthetic resins

commonly used for mounting specimens on slides. It thus becomes necessary to remove the dehydrating agent from specimens (de-alcoholisation). This process is carried out with organic solvents which are miscible both with the dehydrating agents themselves and with the mounting media or embedding wax to which the specimens are subsequently to be transferred. The solvents used include xylene, clove oil and cedarwood oil. It so happens that these solvents have refractive indices which are much higher than those for the alcohols used as dehydrating agents. Tissues are thus made much more transparent by the de-alcoholisation process and thus more suitable for microscopic examination. The term 'clearing' has, therefore, come to be used for this process though it can be seen that clearing and de-alcoholisation are not quite synonymous.

Clearing agents

For temporary preparations isotonic saline is commonly used for animal tissues and 2,2,2-trichloroethanediol (chloral hydrate) for plant tissues. A number of clearing agents are available for permanent preparations, of which those most commonly used include xylene, clove oil, cedarwood oil, terpineol, Berlese's fluid and Amann's medium.

a Xylene has the disadvantage that it causes some shrinking and hardening of tissues. It also forms a white emulsion if there is the slightest trace of water remaining in a specimen which has been incompletely dehydrated. 5 g phenol dissolved in every 100 cm^3 xylene (dimethylbenzene) will reduce the occurrence of this emulsion.

b Clove oil is frequently used for plant tissues, but it does render them brittle and remove some stains such as haematoxylin and safranin. It is also necessary to remove all traces of clove oil with xylene before mounting in Canada balsam or synthetic resin.

c Cedarwood oil has in the past been considered one of the finest clearing agents for animal tissues. It is slower in its action than xylene but causes less shrinkage or hardening. Like clove oil it must be removed with xylene before mounting in Canada balsam. However its great disadvantage is that it is very expensive.

d Terpineol may be used instead of cedarwood oil and is very much cheaper.

e Berlese's Fluid (see page 174) is frequently used for clearing small whole mounts of delicate arthropods.

f Amann's Medium (see page 172) is used for clearing delicate algae, fungi and small nematodes.

Specimens to be cleared should be transferred by means of a brush directly from the final dehydrating medium to a suitable container of the chosen clearing agent. Sections already attached to slides can be cleared by dipping the whole slide in the clearing agent. The time taken for clearing will vary with the nature of the tissue involved. Sections will take only a few minutes—in the case of xylene 2 minutes is quite sufficient. Bulk tissues may take any time from 30 minutes to 24 hours.

Mounting

Specimens for microscopic examination are placed on microscope slides, surrounded by a suitable mounting medium and covered with a cover-slip. This excludes air and dust, provides the appropriate optical conditions for examination and protects the objective lens of the microscope. The ideal mounting media have refractive indices approximating to 1.54 which is the value for the contents of a cell after fixation and clearing. Temporary mounting media are normally liquid and prevent the specimen from drying up. They include isotonic saline (see page 190) for animal tissues, 30 per cent alcohol for plant tissues and glycerol (see page 182) for both animal and plant material. Mounting in glycerol should be preceded by rinsing with distilled water, but clearing is not necessary.

Permanent mounting media are normally gelatinous or resinous and include the following.

a Canada balsam in xylene. This is normally stored in wide-mouthed jars with a domed glass cover and is applied to the slide with a glass rod. It is also available in small tubes, which is perhaps a more convenient method of application. Should the balsam become too thick it can be diluted with a little xylene. Specimens mounted in balsam cannot be safely handled till the balsam is hardened; this can be speeded up by placing the newly made slides on a warm hotplate, in a low oven or other suitably warm place. In time the balsam will become a darker yellow.

b Euparal is used in the same way as Canada balsam, but has the advantage that it tolerates small amounts of water in the specimen, should dehydration be incomplete.

c DPX and various other synthetic resins are available which have the advantage that they do not yellow with time. They are used in just the same way as Canada balsam.

d Glycerol jelly or glycerol (see page 182) may be used for some delicate specimens but the cover-slip must be sealed after mounting (see below).

TECHNIQUES OF MOUNTING

Specimens are normally mounted on plain glass microscope slides measuring 7.5 cm × 2.5 cm and covered with a cover-slip (see page 130).

Normal procedure

1 Place a drop of the mounting medium in the centre of a clean slide. The drop should be of such a size as to fill the space between the cover-slip and the slide without flowing out round the edge of the cover-slip. The exact amount can only be judged with experience and will depend upon the thickness of the specimen, the size of the cover-slip and the consistency of the medium.

2 Place the specimen in the mounting medium using a very fine camel-hair brush.

3 Place the edge of a clean cover-slip on the slide to one side of the drop of medium and gently lower it onto the medium by means of a mounted needle (see Figure 9.6). Do not push the cover-slip down on to the medium but allow it to settle itself.

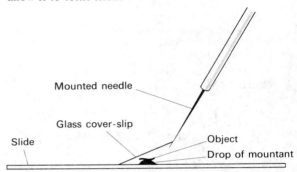

Figure 9.6   *The normal procedure for adding a cover-slip to a specimen on a slide*

Procedure for stained sections already attached to slides

It may be easier to estimate the correct quantity of mounting medium if it is applied to the cover-slip. The inverted slide, with its section attached, is then lowered on to the drop of medium.

Procedure for thick specimens

If the specimen is so thick that it is in danger of being squashed by the cover-slip or causing the cover-slip to become tilted on the slide, it must be surrounded in such a way as to support the cover-slip above the specimen itself. Supports can be made from paper, thin glass (broken cover-slips), varnish, aluminium foil and various other materials in the ways illustrated in Figure 9.7.

Procedure using glycerol jelly

1 Wash the specimen in distilled water.
2 Transfer to 40 per cent glycerol for 3 minutes.
3 Melt a little glycerol jelly on a water-bath.
4 Place the specimen on a cover-slip.
5 Add the liquid jelly using a warm pipette.
6 Invert the cover-slip on to a warm slide.
7 Allow to set, then seal with gold size (see below).

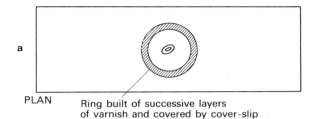

PLAN

Ring built of successive layers
of varnish and covered by cover-slip

SIDE ELEVATION

Cover-slip   Object

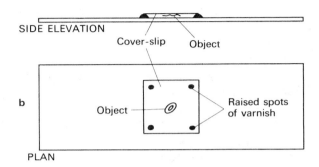

b

Object

Raised spots
of varnish

PLAN

Square frame of
millboard covered
by cover-slip

Slide

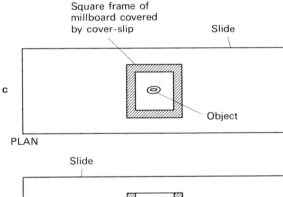

c

Object

PLAN

Slide

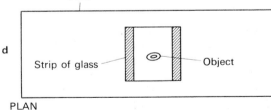

d

Strip of glass   Object

PLAN

*Figure 9.7 Methods of supporting the cover-slip when
mounting thick specimens*

Sealing

If a specimen is mounted in a medium which is liable to
evaporate (e.g. glycerol jelly), the cover-slip should be
sealed round its edge to exclude air and hence prevent
evaporation. This technique of sealing or ringing is also
sometimes used for other mounting media to produce a
fine finish to the slide. Ringing of circular cover-slips is
best carried out on a ringing turntable. A wide variety of

ringing media are mentioned in the literature. For most
purposes—other than glycerol mounts—cellulose black
enamel, cellulose varnish or shellac are suitable. Glycerol
and glycerol jelly mounts should be sealed with gold size.
The ringing medium should be applied in several thin
layers with a fine brush, allowing thorough drying
between each application. Ringing media may also be
used to build up a ring to raise a cover-slip from a
specimen.

Labelling

All prepared slides should have labels approximately
$2.5\,cm \times 2.0\,cm$ with the following details.

1 The name of the specimen or tissue and, in the case of
sections, the location and direction of the section itself.
2 The fixation and staining techniques employed.
3 The date.
4 The name of the technician.

Selected staining procedures

Aceto-carmine; *see* Carmine-acetic

Aceto-orcein; *see* Orcein-acetic

Aniline blue
A temporary preparation for plant tissues.
  1 Stain sections in aniline blue (see page 173) for 2 min.
  2 Rinse for a few seconds in distilled water.
  3 Mount in acid glycerol (see page 182).

Aniline sulphate
A temporary preparation for staining lignin in plant
tissues.
  1 Mount on microscope slide in aniline sulphate (see
page 173).
  2 If the stain evaporates from the edges of the cover-
slip, the sections may be mounted in glycerol for
continued observation.

Azan, Heidenhain's
A general histological preparation giving a vivid range of
colours. Nuclei are stained red, nucleoli yellow, collagen
blue, mucus blue, red blood corpuscles yellow, cytoplasm
pink or yellow.
  1 Stain in azocarmine (see page 173) at $60\,°C$ for 30–
60 min.
  2 Wash in distilled water.
  3 Differentiate in aniline alcohol (see page 172) while
observing under the microscope until only the
nuclei are stained.
  4 Wash in acetic alcohol (see page 172).
  5 Transfer to phosphotungstic acid (see page 189) for
1–3 h.
  6 Wash in distilled water.
  7 Stain in aniline blue-orange G (see page 173) for 1–
3 h.

8 Rinse very briefly in distilled water.
9 Dehydrate.
10 Clear in xylene.
11 Mount.

Borax carmine

A useful stain for whole mounts of animals such as medusae, hydroid colonies, small crustaceans, etc. The nuclei do not stain strongly after fixation in formaldehyde.

1 Stain the fixed specimen in borax carmine (see page 176) for 5–10 min.
2 Differentiate in acid alcohol (see page 172) until the specimen is bright and transparent when observed microscopically (usually 2–5 min).
3 Dehydrate.
4 Clear in clove oil.
5 Mount.

Carmine-acetic

Used for staining chromosomes, e.g. for mitosis in root tip or meiosis in locust testis. Temporary or permanent preparations may be made.

1 Root tips should be fixed in acetic alcohol (see page 172) for 24 h and then macerated in M hydrochloric acid for 5–6 min at 60 °C. Locust testis may be used fresh.
2 Tease the tissue apart in a drop of carmine-acetic on a microscope slide for a few minutes. A darker staining reaction will be produced if iron needles are used.
3 Cover with a cover-slip.
4 Warm the slide gently, but do not allow the stain to boil. Add more stain to the edge of the cover-slip to replace that lost by evaporation.
5 Cover the coverslip with several layers of filter paper and squash firmly but allow no sideways movement (see page 257).
6 The life of this temporary preparation may be extended somewhat by ringing the cover-slip with molten paraffin wax or rubber solution to prevent further evaporation of the stain.
7 To make a permanent preparation, place the slide and cover-slip (without wax or rubber solution seal) in acetic alcohol and leave until the cover-slip and slide separate.
8 Dehydrate slide and cover-slip separately.
9 Clear in xylene.
10 Mount.

Chlorazol black

A stain used in making permanent preparations of plant and animal material. Cell walls are stained black, nuclei may be black, yellow or green, suberin amber, chitin greenish-black, glycogen pink.

1 Bring sections to 70 per cent alcohol.
2 Stain for 5–30 min in chlorazol black (see page 176).

3 Differentiate in a dilute solution of 'Milton'.
4 Dehydrate.
5 Clear.
6 Mount.

Chlor-zinc-iodine (Schulze's solution)

A stain used in temporary preparations for the identification of cellulose. Unlignified cellulose cell walls are stained violet. Lignified, cutinised and suberised walls are stained yellow-brown, proteins brown and starch blue.

1 Mount sections directly in chlor-zinc-iodine solution (see page 191). Rather quicker reactions are obtained if the sections are placed in 5 per cent potassium hydroxide for a few minutes first.

Delafield's haematoxylin; *see* Haematoxylin, Delafield's

Ehrlich's haematoxylin; *see* Haematoxylin, Ehrlich's

Feulgen technique

This technique employs a stain which is specific for DNA and therefore makes excellent preparations of chromosome squashes.

1 Fix the material, plant or animal, in acetic alcohol for 24 h.
2 Hydrolyse in M hydrochloric acid at 60 °C for 6 min or in 5 M hydrochloric acid at room temperature for 30–60 min.
3 Rinse in distilled water.
4 Transfer to Feulgen reagent (Schiff's solution—see page 181) for 1 h.
5 Place on a slide with a drop of Feulgen reagent and macerate with the tip of a glass rod. Alternatively carry out this stage in a drop of orcein-acetic or fast green.
6 Apply a cover-slip, cover with several layers of filter paper and squash firmly but without any lateral movement (see page 257).
7 Dehydrate.
8 Clear.
9 Mount.

Double staining with orcein-acetic or fast green in stage 5 gives a greater contrast and better results than would be obtained with either Feulgen's technique or orcein-acetic used alone.

Gentian violet and Bismarck brown

Really thin sections are needed for this technique and staining with Bismarck brown must be carried out for a few seconds only for best results.

1 Stain in 1 per cent gentian violet for 15 min.
2 Transfer to Bismarck brown *for a few seconds*.
3 Wash in 70 per cent alcohol.
4 Dehydrate quickly.
5 Clear in clove oil.
6 Mount.

Haemalum, Mayer's

An aqueous nuclear stain suitable for entire small animals, bulk tissues and especially for sections of histological material.

1 Stain in Mayer's acid haemalum (see page 182)

diluted to 30 per cent with distilled water. The time necessary for this stage will depend upon the nature of the tissue, but will probably be around 5–15 min.

2 Wash in tap water until the sections are blue (about 10 min).
3 Dehydrate.
4 Clear.
5 Mount.

Alternative method
1 Stain to excess in Mayer's acid haemalum (up to 5 h for bulk tissue, but 10–20 min for sections).
2 Differentiate in acid alcohol.
3 Transfer to 50 per cent alcohol and then to 30 per cent alcohol. 30 min in each will be required for bulk tissue, several minutes only for sections.
4 Wash in tap water or in water made alkaline by the addition of a few drops of dilute, aqueous ammonia.
5 Dehydrate.
6 Clear.
7 Mount.

Haemalum, Mayer's and eosin
This technique combines the basic stain Haemalum which stains the nuclei blue, with the acidic eosin giving pink cytoplasm.
1 Stain the sections in Mayer's haemalum for 5–15 min.
3 Wash thoroughly in tap water for 10 minutes or until the sections are blue.
3 Transfer to 70 per cent and then to 90 per cent alcohol for 2 min each.
4 Stain in eosin (see page 180) for 2–5 min.
5 Wash in 90 per cent alcohol to differentiate.
6 Dehydrate.
7 Clear in xylene.
8 Mount.

Haemalum, Mayer's and van Gieson's
An alternative to the above method which shows up collagen (pink) and cytoplasm (yellow).
1 Stain and wash as in stages (1) and (2) above.
2 Stain for 2 min in modified van Gieson's stain (see page 193).
3 Rinse in water for 1 min.
4 Dehydrate.
5 Clear in xylene.
6 Mount.

Haematoxylin, Delafield's
This stain is used for plant sections and stains nuclei, unlignified tissue and chloroplasts purple, lignified and cutinised tissue yellow and cytoplasm very pale purple.
1 Stain in Delafield's haematoxylin (see page 183) for 1 min.
2 Wash in tap water.
3 Mount in glycerol.

For a permanent preparation, a retrogressive method is to be preferred.
1 Stain in Delafield's haematoxylin for 3 min.
2 Wash in tap water.
3 De-stain in acid alcohol while observing microscopically (usually less than 1 min).
4 Wash in 70 per cent alcohol for 1 min.
5 'Blue' in tap water or ammoniated alcohol (see page 172) for 1 min.
6 Dehydrate.
7 Clear in clove oil.
8 Mount.

Haematoxylin, Delafield's and safranin
This technique may be used with plant material as an alternative to the above. The safranin stains the lignified and cutinised tissues red, thus providing a contrast to the effects of the haematoxylin.
1 Stain in Delafield's haematoxylin for 3 min.
2 Wash in tap water.
3 De-stain in acid alcohol while observing microscopically (usually less than 1 min).
4 Wash in 70 per cent alcohol.
5 'Blue' in ammoniated alcohol (see page 172) or in alkaline tap water.
6 Stain in safranin (see page 190) for 5 min.
7 Differentiate in 50 per cent alcohol.
8 For a temporary preparation, transfer to distilled water and then mount in glycerol.
9 For a permanent preparation, dehydrate.
10 Clear in clove oil for 5 min.
11 Mount.

Haematoxylin, Ehrlich's
A basic stain used for general animal histology and also for smears of *Monocystis* in the seminal vesicles of earthworms.
1 If a seminal vesicle smear is being made it should be fixed in 70 per cent alcohol on the surface of the cover-slip for 5 min.
2 Stain in Ehrlich's haematoxylin (see page 183) for 15 min.
3 Differentiate in acid alcohol while observing microscopically.
4 Wash in 70 per cent alcohol.
5 'Blue' in ammoniated alcohol (see page 172).
6 Wash in 70 per cent alcohol then dehydrate.
7 Clear in clove oil for 5 min.
8 Mount.

Haematoxylin, Ehrlich's and eosin
The above technique may be modified by counterstaining with eosin.
1 Carry out stages 1–5 of the previous technique.
2 Wash in 70 per cent alcohol then transfer to 90 per cent alcohol.
3 Stain in eosin in 90 per cent alcohol (see page 180) for 1–2 min.

4 Wash in 90 per cent alcohol and dehydrate.
5 Clear in clove oil.
6 Mount.

Heidenhain's azan; *see* Azan, Heidenhain's

Heidenhain's iron haematoxylin; *see* Iron haematoxylin, Heidenhain's

Iodine solution
Used for temporary preparations of plant material to identify the presence of starch which is stained dark blue or black. All other tissues are stained yellow-to-brown except chloroplasts which may appear blue owing to presence of starch.
1 Mount in dilute iodine solution.
2 Wash in distilled water.
3 Mount in glycerol.

Iron haematoxylin, Heidenhain's
This is a particularly good stain for resolving the finest details of structure at both the cytological and histological levels. It is recommended for preparations that are to be photographed. It may be used for algae as well as for animal tissues.
1 Mordant sections in iron alum (see page 184) for 30 min to 24 h at room temperature. The process may be hastened by gentle warming.
2 Rinse in distilled water.
3 Stain in iron haematoxylin for 30 min to 24 h (i.e. for the same length of time that the sections were mordanted).
4 Differentiate in iron alum. Remove from the solution at intervals and mount in tap water prior to microscopic examination, till the nuclear and cytoplasmic details are clear.
5 Wash for several hours in running tap water.
6 Dehydrate.
7 Clear in xylene.
8 Mount.

Lactophenol cotton blue
This technique is useful for staining fungal tissue and is particularly useful for differentiating such tissues invading higher plants.
1 Mount the sections of plant tissue directly in the stain (see aniline blue on page 173).
2 Add a cover-slip.

Leishman's stain for blood smears
1 Prepare a blood smear in the way described on page 256.
2 Pipette on a few drops of Leishman's stain (see page 185), allow it to cover the smear and leave for 20 s.
3 Add an equal number of drops of distilled water, rock the slide to mix with the stain and leave for 10 min.
4 Rinse with distilled water.
5 Dry the slide by waving in the air or warming over an electric lamp.
6 Add a drop of Canada balsam and a cover-slip.

Mallory's triple stain
A good staining technique for histological work, especially for invertebrate material. It may be preceded by staining for 5 min with Delafield's haematoxylin and then washing in tap water.
1 Stain till bright red (usually 3–20 min) in Mallory's Triple Stain Solution A (see page 186).
2 Wash in distilled water.
3 Fix stain and prevent decolorisation by washing for 1 min in Mallory's Triple Stain Solution B (see page 186).
4 Wash thoroughly in distilled water.
5 Stain until blue in Mallory's Triple Stain Solution C (see page 186). This usually takes 5–20 min.
6 Wash in distilled water.
7 Dehydrate rapidly.
8 Clear in xylene.
9 Mount in DPX.

Mann's methyl blue and eosin; *see* below

Methyl blue, Mann's and eosin
A good general technique for insect sections.
1 Stain overnight in methyl blue and eosin (see page 186).
2 Rinse thoroughly but rapidly in distilled water for 30 s.
3 Differentiate in 70 per cent alcohol to which a drop of a saturated aqueous solution of orange G has been added for every cm^3 of alcohol.
4 Dehydrate.
5 Clear.
6 Mount.
Alternatively, stain for 10–30 min and differentiate in tap water.

Methylene blue
A basic dye useful for staining living material.
1 Mount the material directly in methylene blue (see page 186).

Orcein-acetic
An alternative, and in some ways preferable stain to carmine-acetic. In either case the dye may be dissolved in propanoic acid instead of ethanoic (acetic) acid.
The technique employed is identical to that for carmine-acetic (see page 262) except that orcein-acetic (see page 188) is used in stage 2.

Phloroglucin(ol)
Acidified phloroglucin is used in temporary preparations as a sensitive test for lignin which is stained red.
1 Mount sections in an alcoholic solution of phloroglucin(ol) (see page 188). Leave for 4 min.
2 Add a drop of concentrated hydrochloric acid to the edge of the cover-slip and irrigate with a piece of filter paper or blotting paper.

Safranin and fast green
This technique is highly recommended for sections of fixed plant material. Nuclei, chromosomes, cuticle and

lignin are stained red, all other tissues are stained green.

1 Stain sections in safranin (see page 190) for 1–24 h.
2 Wash in distilled water.
3 Differentiate in acetic-alcohol (1 per cent ethanoic (acetic) acid in 70 per cent alcohol).
4 Dehydrate.
5 Stain in fast green (see page 181) for $\frac{1}{2}$–4 min.
6 Differentiate in a mixture of clove oil (50 per cent), absolute alcohol (25 per cent) and xylene (25 per cent) in two stages of 5–15 min each.
7 Transfer to xylene to clear.
8 Mount.

Schulze's solution; *see* Chlor-zinc-iodine

Sudan III
A weakly acidic dye sometimes used as a stain for fats but now largely superseded by Sudan IV.

1 Mount sections of plant material in a mixture of equal volumes of Sudan III (see page 193) and glycerol.
2 Warm till the alcohol boils. Fat will be stained red.

Sudan IV
A stain used for temporary preparations to identify the presence of fats which are stained red.

1 Tease the tissue in saline on a slide.
2 Fix in aqueous formaldehyde (40 per cent) for 1 min.
3 Flood with 50 per cent alcohol.
4 Stain in Sudan IV (see page 193) for 2–3 min.
5 Wash in 50 per cent alcohol.
6 Mount in glycerol or glycerol jelly.

THE PRESERVATION OF PLANT AND ANIMAL SPECIMENS

A reference collection of plant and animal specimens is an invaluable component of a biology laboratory. In many cases, single specimens will prove sufficient but on occasions it may be necessary to preserve class sets of material. Such collections are available from commercial suppliers, but a considerable financial saving can be made if the bulk of the specimens are prepared and preserved by the school. The following notes indicate recommended procedures for a wide range of plant and animal specimens. Further details may be found in Knudsen (1966) and in the British Museum (Natural History) booklets *Instructions for Collectors* obtainable from the Museum Publications Department.

In collecting and preserving specimens a number of general points should be borne in mind.

a The Conservation of Wild Creatures and Wild Plants Act, 1975 makes it an offence to uproot any wild plants without the permission of the landowner. A number of species of plants and of animals are under additional protection and must not be collected under any circumstances. Specimens collected in the field should never include species known to be rare, either in that particular locality or nationally. Before making any collections

further information should be obtained as appropriate from one or more of the bodies listed below:
County Naturalists Trusts and local Natural History Societies;
Field Centres of the Field Studies Council;
The Biological Records Centre;
The Botanical Society of the British Isles;
The Fauna Preservation Society;
The Wildlife Youth Service;
The Institute of Biology.

If there is any doubt over a particular specimen, it should be identified *before* a decision is made as to whether it should be collected. For a list of suitable books for identification purposes see pages 290–292.

b The number of specimens collected should be only sufficient to serve the particular needs of the time and these should be processed through to the preservative without undue delay.

c Allowance should be made for the fluid in the specimen itself when calculating dilutions of formalin or alcohol to an appropriate concentration. A specimen of volume $100 \, cm^3$ preserved in $100 \, cm^3$ preservative will dilute that preservative to approximately half of its original concentration.

d Specimens which are being transferred to high concentrations of alcohol from formalin and other liquids consisting largely of water, should be passed sequentially through a graded series of alcohols (e.g. 30 per cent, 50 per cent, 70 per cent), otherwise some shrinkage and other tissue damage may take place.

e In general, specimens preserved in alcohol are more pleasant to examine closely and dissect than those preserved in formalin. If, however, formalin is the only suitable or available preservative, it may be washed out by soaking in water prior to examination or dissection.

f Each specimen should be fully labelled with the name of the species, date, nature and location of the habitat from which it was collected, the collector's name or initials and the fluid in which the specimen is preserved.

g If specimens are being preserved dry, care should be taken to ensure that they do not become infected with moulds, insects, mites or other pests.

h Specimens being preserved in liquid preservatives for display purposes should be kept in jars of a standard size or format. Museum jars may be obtained from commercial suppliers, but they are expensive. In most cases cheaper jars are quite satisfactory provided they have a wide mouth and a tight-fitting lid. Instant coffee jars are often suitable. Thin sheets of glass or perspex should be cut so that they just fit into the jar without too much freedom of movement. The specimens should then be attached to these plates with white or black thread in the manner shown in Figure 9.8. Labels may also be attached

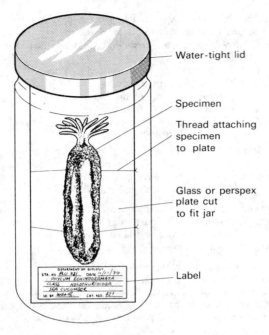

Figure 9.8   Displaying specimens in museum jars

to the plates, but they should be written in soft lead pencil or in an indian ink which is known to withstand immersion in the preservative.

The preservation of plant specimens

The majority of plant specimens are best preserved dry, pressed on herbarium sheets. There are, however, a number of exceptions and these are dealt with individually on pages 267–268.

Dry preservation on herbarium sheets

Specimens should be pressed as soon as possible after collecting to prevent wilting which will distort the appearance of most plants. The pressing of higher plants may be carried out on specimen sheets made by folding a single sheet of newspaper (60 cm × 45 cm) in two. The folded specimen sheet is opened and the plant arranged in the position in which the final specimen is to appear. It should be borne in mind that the herbarium sheets on which the specimens will eventually be mounted are of a standard size (approximately 30 cm × 42 cm) and the specimens should be so arranged to fit conveniently onto such a sheet. The roots should be arranged on the lower part of the sheet with the stem, if necessary, bent to accommodate the entire plant on the sheet. The bulk of the plant should be distributed as evenly as possible and care taken to ensure that all the floral parts as well as the

leaves and stem are visible. Appropriate labelling and other details should be placed in a notebook with a corresponding number recorded on the specimen sheet.

The folded specimen sheet is then placed between driers in a plant press. Driers may be made of felt or blotting paper measuring some 30 cm × 45 cm, though folded newspaper several layers thick makes an adequate substitute. The rate of drying can be increased by placing ventilators (made of corrugated carboard 30 cm × 45 cm) between the driers. The press can be filled with material arranged thus: ventilator, drier, specimen, drier, ventilator, drier, specimen, drier, ventilator, etc. Field presses, adequate for most school uses, can be made from laths of wood or from wood and plywood bound with string or tape. They are also available from commercial suppliers. Figure 9.9 shows three possible designs.

Once the press is filled, or all the specimens have been enclosed, a final drier and ventilator should be placed on the top and the press assembled. Gentle pressure should be applied (e.g. by kneeling on the press) while the straps are tightened. The driers should be changed every 12–24 hours which will reduce the drying time and also prevent the specimens becoming mouldy. Once the specimens are thoroughly dry, they may be arranged on sheets of herbarium paper, which should be of good quality otherwise it will yellow with age. The specimen can then be attached to the sheet with narrow strips of herbarium

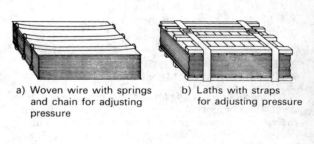

a) Woven wire with springs and chain for adjusting pressure

b) Laths with straps for adjusting pressure

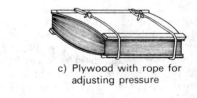

c) Plywood with rope for adjusting pressure

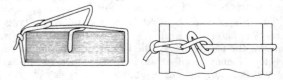

d) Method of tying a field press with a rope

Figure 9.9   Designs of plant press for use in the field

tape (clear adhesive tape is not suitable for this purpose as it shrinks, accumulates dirt and eventually becomes detached). The specimen, on its herbarium sheet, can now be stored in herbarium folders.

Aquatic plants including freshwater and marine algae

The majority of aquatic plants, including multicellular algae, can be preserved in basically the same way as that described above. The specimens should first be rinsed in clean, fresh water to remove unwanted animal material and, in the case of marine algae, salt which would subsequently hinder the drying process. The specimens are then arranged directly on the herbarium sheet and covered with a sheet of waxed paper or muslin to prevent them sticking to the drier. The material is pressed and dried in the same way as that described above.

The more delicate algal material cannot be satisfactorily arranged on the herbarium paper by the above method and the following procedure should be adopted. Float the specimen in a shallow tray of fresh water; insert the herbarium paper beneath the specimen and arrange the latter with a paint brush; carefully withdraw the paper from the tray by pulling along the bottom edge of the sheet, and allow the specimen to drop onto the sheet as the paper is withdrawn; excess water should then be drained off before the muslin is laid over the specimen and the material pressed and dried in the normal way. This technique is not suitable for the fucoid algae which may be preserved in FAA (see page 181).

Dry preservation of mosses, liverworts and lichens

These lower plants may be preserved in much the same way as that described for higher plants. All soil and grit should first be washed off the rhizoids and the plants then blotted partially dry before arranging them on several thicknesses of newspaper. Further layers of newspaper and specimens can then be placed on top with a final layer of newspaper. Liverworts should be placed directly on sheets of herbarium paper or thin card and covered with a layer of muslin before inserting between sheets of newspaper. A light pressure should be applied to the specimens, which is sufficient to press the specimens into two dimensions but not so great that it flattens the material to the point of distortion. The preserved specimens are normally kept in envelopes stored in boxes of an appropriate size (e.g. shoe boxes).

Fluid preservation of fungi, soft-bodied fruits and storage organs

Large, soft-bodied fungi are notoriously difficult to preserve satisfactorily. Low pressure freeze-drying gives good results, but suitable apparatus for this is not available to most schools, although it can be improvised

in some cases (see page 274). Specimens may be preserved in a reasonable condition for a short time by placing in FAA or in 4 per cent methanal, but the colour disappears and the tissues eventually disintegrate.

Soft-bodied fruits, fleshy stems and other organs, may be preserved by placing the specimens in 2–3 per cent methanal or FAA.

Preservation of natural colours in plants

Although some dried plants retain their natural colours to a certain extent, most will fade and will do so more quickly if they are exposed to light or to humid conditions. For display purposes it may be desirable to preserve the original colour of the plant, in which case one of the following procedures may be adopted.

a Preservation of the green colour by coppering. Make up a saturated solution of copper(II) ethanoate (copper acetate) in glacial ethanoic (acetic) acid. Mix 1 part of this saturated solution with 4 parts of water in a beaker and bring it to 100 °C in a water bath (cover the beaker during heating). Plunge the fresh green specimen into the hot solution and leave for 10 minutes. Allow the solution to cool, then wash the specimen in gently running water and preserve it in 5 per cent formalin.

b Preservation of the green colour. This method is suitable for algae as well as for higher plant material. Make up the following solution:

alcohol (50 per cent)	90 cm^3
5 per cent formalin	5 cm^3
propane-1,2,3-triol (glycerol)	2.5 cm^3
copper(II) chloride-2-water	10 g
uranyl(VI) nitrate-6-water	1.5 g
glacial ethanoic acid (acetic acid)	2.5 cm^3

Place the plant material in this solution for 3–10 days depending upon the size and thickness of the material. If the plants are of a yellowish green colour in life, then reduce the copper(II) chloride to 5 g

c Preservation of fruit colour. Preserve the material in Hessler's fluid (see page 183).

d Preservation of colour of green algae. Dissolve 0.5–1 g copper(II) ethanoate (acetate) in 100 cm^3 1 per cent methanal. Immerse the algal material in this solution for 24 hours before preserving in 2 per cent methanal.

e Preservation of flower colour prior to embedding in polyester resin. Make up a 1 per cent solution by weight of thiourea in 2-methylpropan-2-ol (*tertiary* butyl alcohol). To this solution add 2 per cent by weight sodium 2-hydroxypropane-1,2,3-tricarboxylate (citrate) for the preservation of blue or green flowers or 2 per cent by weight hydroxypropane-1,2,3-tricarboxylic (citric) acid for the preservation of red or pink flowers. Immerse the flowers in the appropriate solution for 12–24 hours. Flowers of intermediate colour should be treated with a mixture of the two solutions. The flowers will then be very fragile and should be carefully dried in an oven at 50 °C before embedding in the usual way (see page 274).

f Rapid drying with hot sand. Place sufficient dry, fine, silver sand to cover the specimen in an oven and heat to 200 °C. The sand is best put into a metal jug for this process and left in the oven for several hours to ensure thorough and even heating. Pour some of the hot sand into a deep tray and place the fresh specimen on the sand. Cover it with more hot sand, making sure that the sand penetrates the flowers. Replace the tray in the oven and turn off the heat. Leave until quite cool.

The preservation of animal material

The procedure for preserving animal specimens varies greatly from one animal group to another. In most cases, the specimen must first be killed or narcotised in such a way as to leave the tissues relaxed and suitable for display. The material is then fixed in this condition and finally preserved. This final preservation is normally made either in a solution of ethanol and methanol, usually known as 'alcohol', or in a solution of methanal (formaldehyde) known as 'formalin'. A fuller discussion of the nomenclature of the various solutions of alcohols used in biological work will be found on page 171. For preservation purposes, industrial methylated spirit, diluted to an appropriate concentration, is quite satisfactory. The term formalin also causes some confusion which is clarified on page 181. Appropriate killing, narcotising and fixing agents together with information on the preservation of a variety of animal groups, are given in Table 9.2.

Various other methods of preservation are used in more specialised circumstances and these are described on pages 274 to 276.

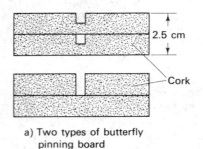

a) Two types of butterfly pinning board

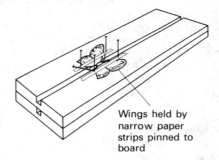

b) Butterfly pinned in position on board

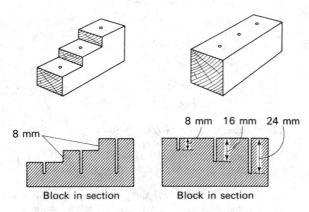

c) Two types of pinning block for beetles and other insects

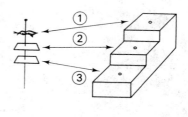

d) Using a beetle pinning block to ensure regular spacing of labels

Figure 9.10   The dry-mounting of insects for display

Table 9.2   The killing and preservation of animal specimens

Animal group	Killing or narcotising agent	Fixative	Preservative	Notes
Protozoa	—	—	—	Best preserved in smears on microscope slides which can then be stained. Precise techniques vary with the species. See Bradbury (1973)
Porifera (sponges)	70 per cent alcohol	70 per cent alcohol	70 per cent alcohol	Change the alcohol after 24 hours and again if it becomes discoloured. Sponges may also be preserved dry
Coelenterates				
Hydra	Hot Bouin's solution	Bouin's solution	70 per cent alcohol	Flood live specimens with the hot Bouin's solution starting at the base and moving up to the tentacles
Sea anemones	Clove oil in sea water	2 per cent methanal	2 per cent methanal	Place anemones individually in large jars containing 1 litre of sea water. Allow to expand, then add several large drops clove oil to the water every hour for 3 hours. Allow to stand for 10–12 hours then probe tentacles gently to ascertain that narcotisation is complete. Add formalin to form a 2 per cent methanal solution
Colonial hydroids, small medusae, scyphozoans (jellyfish), siphonophors and ctenophores	Epsom salts in sea water	2 per cent methanal	2 per cent methanal or 70 per cent alcohol	Allow specimens to extend fully in clean sea water. Add small quantities of Epsom salts to the water without disturbing the specimens. Repeat every 20–30 minutes for 3–4 hours until the tentacles fail to react to probing with a mounted needle. Add formalin to form a 2 per cent methanal solution. If preserving in 70 per cent alcohol the specimens should be transferred through 30 per cent and 50 per cent alcohol
Platyhelminths (flatworms)				
Planarians and trematodes (flukes)	2 per cent methanal or Gilson's fixative (see page 182)	2 per cent methanal or Gilson's fixative	2 per cent methanal or 70 per cent alcohol	The specimens should be placed in dry Petri dishes with just sufficient water to allow complete expansion. The fixative should then be squirted on (preferably at 50–60 °C) with a pipette
Cestodes (tape worms)	2–4 per cent methanal or FAA	2–4 per cent methanal or FAA	2 per cent methanal or 70 per cent alcohol	Specimens should be allowed to relax in cold water and then wrapped around a glass or similar support before placing in the fixative
Nematodes (roundworms)	Hot 2 per cent methanal or hot 70 per cent alcohol	Hot 2 per cent methanal or hot 70 per cent alcohol	2 per cent methanal or 70 per cent alcohol	Larger forms may be dropped directly into the hot fixative. Numerous smaller forms are best placed in a container of water before draining off as much water as possible and adding the hot fixative
Rotifers	Magnesium sulphate, menthol crystals or chloretone crystals	1 per cent osmic acid	1–2 per cent methanal or 70 per cent alcohol	Place the specimens in a watch glass under a dissecting microscope and add a few crystals of menthol or chloretone. Wait till the cilia almost stop moving, then add a few drops 1 per cent osmic acid (CAUTION!) directly on to the animals. Wash several times in water and transfer to preservative. If alcohol is being used, the specimens should be transferred through 30 per cent and 50 per cent alcohol to 70 per cent

Animal group	Killing or narcotising agent	Fixative	Preservative	Notes
Annelids (segmented worms)				
Earthworms and leeches	Alcohol, menthol crystals or magnesium sulphate	2–4 per cent methanal	70–90 per cent alcohol or 2 per cent methanal	Place worms in a shallow dish with sufficient water to cover them. Add a little alcohol, menthol crystals or magnesium sulphate every few minutes till the worms are limp and insensitive to touch. Transfer to fixative, keeping them straight, and leave overnight. Transfer to alcohol
Freshwater oligochaetes	Hot 4 per cent methanal	4 per cent methanal	70–90 per cent alcohol	Place worms in a dish with as little water as possible and allow them to extend. Pour on hot, 4 per cent methanal and allow them to become rigid before transferring to the alcohol
Polychaetes (marine bristle-worms)	Alcohol, menthol crystals or Epsom salts	FAA	70 per cent alcohol or 2 per cent methanal	Place worms in a little sea water in a shallow dish large enough to allow full expansion. When expanded, add narcotising agent a little at a time until the worm becomes insensible. Transfer to another shallow dish, add FAA and leave for 24 hours before placing in the preservative
Molluscs				
Freshwater and terrestrial snails	Epsom salts or asphyxiation	2–4 per cent methanal	70 per cent alcohol or 2 per cent methanal	To asphyxiate snails place them in a stoppered jar completely filled with water and leave for 24 hours. Alternatively, they may be placed in warm water, allowed to expand before slowly adding Epsom salts till narcotising is complete. Transfer to fixative for 24 hours and then to preservative. If alcohol is being used, the snails should be transferred via 30 per cent and 50 per cent to 70 per cent alcohol. The shells and opercula may also be preserved dry
Terrestrial slugs	Boiled and cooled water or soda-water	2–4 per cent methanal or 70 per cent alcohol	70 per cent alcohol or 3 per cent methanal	Follow the same procedure as that outlined above for terrestrial snails, but substitute soda-water or boiled and cooled water for the narcotising process
Marine snails and nudibranchs	Alcohol	4 per cent methanal or FAA	70 per cent alcohol	Allow to expand in fresh sea water, then add alcohol drop by drop over several hours to form a 10 per cent solution. When totally insensitive either add formalin to form a 4 per cent methanal solution or transfer to FAA. Finally preserve in 70 per cent alcohol after transferring through 30 per cent and 50 per cent alcohol. The shells may also be preserved dry
Bivalves	Alcohol or fresh water (for marine specimens)	Alcohol	75 per cent alcohol	Allow to expand in habitat water, then insert a wooden peg between the shells. Marine species may then be killed by transferring to fresh water and running up through 30 per cent and 50 per cent to 70 per cent alcohol. Alternatively alcohol may be added slowly till a 30 per cent solution is formed before transferring to 50 per cent and finally 70 per cent alcohol for preservation. The shells may also be preserved dry

Animal group	Killing or narcotising agent	Fixative	Preservative	Notes
Cephalopods	Very dilute formalin (less than 0.5 per cent methanal)	2 per cent methanal	2 per cent methanal or 70 per cent alcohol	Place specimen in a large container of sea water and add a very small amount of formalin to form a fraction of a 1 per cent methanal solution. Allow the specimen to swim until it dies. Transfer to a storage jar, arrange and add 2 per cent methanal. Leave overnight, then change formalin. Shelled specimens should be preserved in 70 per cent alcohol after passing up through 30 per cent and 50 per cent alcohol
Echinoderms				
Holothurians (sea cucumbers)	Epsom salts or asphyxiation in stale sea water	4 per cent methanal or 50 per cent alcohol	70 per cent alcohol	Allow to expand in sea water, then add Epsom salts a little at a time over several hours. Allow to stand overnight till they are insensible to touch. Transfer to 4 per cent methanal for several hours, wash and preserve in 70 per cent alcohol. With larger specimens some alcohol should also be injected into the body cavity. Change alcohol after 24 hours
Other echinoderms (starfish, brittle stars, sea urchins, etc.)	Epsom salts	2–4 per cent methanal	70 per cent alcohol or dry	Allow to stand in sea water in shallow pans. Add Epsom salts gradually over several hours and allow to stand overnight. Check that narcotisation is complete by adding a small drop of formalin near one of the tentacles. Transfer to a fresh pan and arrange suitably before adding 2–5 per cent neutralised methanal. Stand for 24 hours before transferring to 50 per cent alcohol for 1 hour and then 70 per cent alcohol. Alternatively specimens may be air dried in the shade, preferably in a windy position
Crustaceans				
Small freshwater and marine forms	2 per cent methanal	2 per cent methanal	70–90 per cent alcohol	Add sufficient formalin to the specimens in their habitat water to form a 2 per cent methanal solution. Transfer to alcohol for preservation
Barnacles	Chloretone or menthol crystals	2 per cent methanal	70 per cent alcohol	Place in sufficient sea water to allow full expansion. Add a large pinch of narcotising agent and allow to stand for 6 hours. Check that narcotisation is complete by adding a small drop of formalin near one of the specimens. If there is no contraction add further formalin to form a 2 per cent methanal solution. Wash and transfer to alcohol, 30 per cent, then 50 per cent and finally preserve in 70 per cent
Large crustaceans	Fresh water, asphyxiation in putrid sea water or (for tropical forms) dropping in ice cold water	Alcohol	70 per cent alcohol or neutralised 2 per cent methanal	Care is required in killing as they are apt to shed limbs as a defence mechanism. For truly marine forms fresh water or exposure to air are probably best. Hermit crabs can be removed from their shells by placing them in a mixture of fresh and sea water or by gently warming with a lighted match. Transfer up through the alcohols to 70 per cent

Animal group	Killing or narcotising agent	Fixative	Preservative	Notes
Insects				
All juvenile forms (except Orthoptera, Isoptera). Adult Protura, Collembola, Diplura, Ephemeroptera, Mallophaga, Anopleura, Thysanoptera, Siphonaptera	75–80 per cent alcohol	75–80 per cent alcohol	75–80 per cent alcohol or permanent slide with or without treatment with potassium hydroxide solution	Place the insects in the alcohol in small vials or suitable storage tubes. Change the alcohol after 24 hours. Adult Protura, Collembola, Diplura and Thysanoptera may also be made into permanent slides without using potassium hydroxide in the preparation. Adult Mallophaga, Anoplura and Siphonaptera are best made into permanent slides with prior treatment with potassium hydroxide. Adult Ephemeroptera may also be preserved dry after killing with a killing bottle
All other insects juvenile or adult	Killing bottle containing ethyl ethanoate (acetone), ethoxyethane (ether), tetrachloromethane (carbon tetrachloride) or pentyl ethanoate (amyl acetate)		Dry mount on cards or pins. Butterflies may be stored temporarily in folded envelopes prior to setting out on a pinning board. See Figure 9.10 for details of mounting technique. Boxes containing dry-mounted insects should include a little 1,4-dichlorobenzene or naphthalene to prevent attack by pests	The killing bottle should be made up by pouring 2 cm depth of plaster of Paris into a stout screw-topped or stoppered jar. When the plaster is dry it should be saturated with the liquid poison. Crumpled paper should be added next to restrict the movement of the insects before death. For butterflies and moths the paper should be such that it prevents the insects making contact with the wet plaster and hence damaging their wings. The specimens will remain limp for several days prior to mounting provided all the liquid poison does not evaporate. N.B. Cyanide killing bottles, though very effective for insects, are also very poisonous to humans and should not be used by pupils
Other arthropods including spiders, scorpions, pseudo-scorpions, ticks, mites, water mites, centipedes and millipedes	70 per cent alcohol	70 per cent alcohol	70 per cent alcohol	Specimens should be dropped directly into the alcohol, though centipedes and millipedes may first be killed in a killing bottle (see above). The smaller species of mites, etc. may be mounted on slides in Berlese's fluid or Gray and Wess' medium
Tunicates	2,2,2-trichloroethanediol (Chloral hydrate)	2 per cent methanal or FAA	2 per cent methanal or 70 per cent alcohol	Make up a 0.2 per cent solution of chloral hydrate in sea water. Place the specimens in this solution and allow to expand and become narcotised. Alternatively add a pinch of chloral hydrate each hour to the expanded specimens in sea water. Continue for 3 hours. Leave overnight then replace sea water with fixative. If preserving in alcohol transfer to 30 per cent, then 50 per cent and finally 70 per cent

Animal group	Killing or narcotising agent	Fixative	Preservative	Notes
Fish	Formalin or 70 per cent alcohol	2 per cent methanal (10 per cent for larger specimens) or 70 per cent alcohol	70 per cent alcohol or 2 per cent methanal	Place the fish in a large vessel to which a little formalin has been added. When dead, transfer to fixative. Alternatively, place directly in 70 per cent alcohol. Larger specimens should be cut along the right side of the abdominal cavity to allow entry of the fixative and preservative
Amphibians	MS-222, chloretone solution or freezing	3–4 per cent methanal	2 per cent methanal or 70 per cent alcohol	Place the specimens in a 0.05 per cent solution of MS-222 (tricain methane sulphonate) or in chloretone solution (approx. $5\,cm^3$ to 5 litres water). Allow to sink, relaxed, to the bottom, then position and transfer to fixative. Alternatively specimens may be placed in plastic bags in a freezer for 24 hours, then thoroughly thawed out, positioned and fixed. A little fixative should also be injected into the body cavity.
Reptiles	Freezing	4 per cent methanal	3 per cent methanal	Place in a plastic bag in a freezer. Leave for 24 hours, then allow to thaw out very thoroughly. Care should be taken not to damage toes while the specimen is frozen. Place in 4 per cent methanal and inject some of the fixative into the body cavity or slit the cavity to allow the fixative to enter. In the case of snakes, injections should be made every 5 cm along the length of the body. With large lizards, injections should also be made into each limb and into the base of the tail
Birds (N.B. The collection of wild birds is illegal in Britain. See also page 240)	See notes	—	Dry mounts	Injured birds may be killed by depressing the chest beneath the wings for about one minute. Other birds should not be killed. For details of dry mounting, consult Knudsen (1966)
Mammals	Carbon dioxide in a killing chamber (see page 288)	3 per cent methanal	3 per cent methanal	After placing in a humane killing chamber, specimens should be left in the chamber for at least 10 minutes after collapse to ensure that death has taken place (see page 288). The body cavity and larger muscles should be injected with fixative. For display purposes specimens are usually prepared as dry mounts. For details of this technique see Knudsen (1966)

Low pressure freeze-drying

This technique has been developed in recent years and is being used increasingly to preserve specimens for museum display. It is particularly suitable for the preservation of small, soft-bodied organisms (e.g. insect larvae and the larger fungi) where preservation in alcohol or formalin is unsatisfactory. Specimens preserved in this way have a very life-like appearance; even vertebrate eyes remain glistening and realistic.

1 Arrange the specimen in an appropriate position.
2 Place the specimen in the freezing compartment of a refrigerator, in a deep freeze, or, better still, in a Pyrex glass vessel immersed in a mixture of propanone (acetone) and solid carbon dioxide at −70 °C. This last method of rapid cooling is essential for more delicate specimens such as the larvae of many insects which may otherwise explode during the desiccation process. Leave to freeze.
3 Transfer the specimen to a desiccator fitted to a vacuum pump (see Figure 9.11). The desiccator should previously

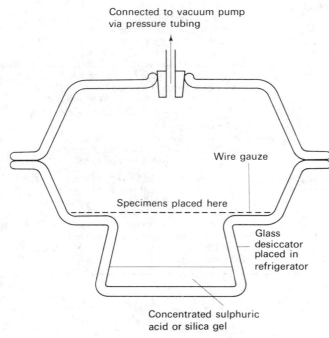

Connected to vacuum pump
via pressure tubing

Wire gauze

Specimens placed here

Glass
desiccator
placed in
refrigerator

Concentrated sulphuric
acid or silica gel

Figure 9.11 Apparatus for freeze-drying specimens at very low pressures

have been kept in the refrigerator for several hours. If the desiccation is taking place below 0 °C, then phosphorus(V) oxide or silica gel should be used. Above this temperature, concentrated sulphuric acid may be used, but the results are generally less satisfactory.

4 Reduce the pressure by switching on the vacuum pump. Ideally pressures as low as 0.05 mm mercury (c. 6 N m^{-2}) or even lower should be used, in which case a diffuser may be necessary as well as a vacuum pump.
5 Slowly restore the pressure to normal atmospheric pressure and allow the specimen to return to room temperature.

Resin embedding

In recent years, the development of resin embedding techniques has made it possible for biological specimens of a delicate nature to be made sufficiently robust to withstand frequent handling by pupils. In the past such specimens would have been preserved in alcohol or formalin thus making them much less attractive for display purposes. The specimens are embedded in a tough polyester resin resembling transparent perspex in appearance, thus enabling them to be handled and viewed from all sides. There are limits to the size of specimen which can be treated in this way (see below) and a number of precautions need to be taken to ensure satisfactory results.

Several commercial scientific suppliers maintain stocks of the resin, moulds and other materials required for the embedding. The necessary materials may also be obtained from shops specialising in handicraft supplies. The exact procedure for embedding varies slightly according to the particular type of resin being used and manufacturers generally provide detailed instructions accompanying their products. However, the following procedure is appropriate for most available resins.

1 Preparation of specimens for embedding

Dry specimens, such as shells, bones, teeth, coral and dry insects, can be embedded without prior treatment. However, some of these raise particular problems due to their sharp corners, entrapped air or very hard surfaces to which the resin does not become attached. These are dealt with individually under section 6 below. Small amounts of water, mixed with the resin, will cause pale streaks or whitening of the block which obscures the visibility of the specimen. It is therefore of the utmost importance that fresh specimens, or those that have been previously preserved in alcohol, be free of water at the time of embedding. This can be done by transferring the specimen through a graded series of alcohols ending with absolute alcohol. It can then be placed in either a clearing agent such as methyl benzenecarboxylate (benzoate) or methyl 2-hydroxybenzoate (salicylate) or in dry ether and left for several days. The specimen is then placed in a beaker of the resin to which no hardener or accelerator has been added. In this way the clearing agent or ether is replaced by the resin.

If ether is being used, the process may be aided by placing the beaker in a vacuum. The specimen is then ready for embedding in the normal way. A simpler method of removing water is to transfer the specimens through a graded series of ethane-1,2-diol (ethylene glycol) (50 per cent, then 70 per cent and finally pure anhydrous ethane-1-2,-diol) and then embed immediately. In this case the specimens should be left in each of the solutions for several hours or days depending on their size.

Plant material may need to be treated to preserve both the green colour and the colour of the flowers. This may be done by one of the methods outlined on page 267. Some plant material may also be successfully dehydrated by placing it in warm, very dry sand and leaving it in a warm place for several weeks. The sand is then carefully poured off and the specimen sprayed with hair lacquer.

2 Selection and preparation of the mould

Polythene is an ideal material for the construction of moulds and many commercially produced embedding kits employ it. Glass, sheet or enamelled metal, china or perspex moulds are also suitable. Wooden moulds are not ideal as the resin takes up the shape of the mould surface thus reproducing grain patterns which must subsequently be removed by sanding and polishing. All moulds must be scrupulously clean and free from dust. Select a mould which is large enough to allow at least 1 cm of resin all around the specimen.

Moulds should be parallel sided or with the top opening slightly larger than the base otherwise removal of the finished cast will be impossible without destroying the mould. In most cases it is advisable to wipe a little release wax (available from suppliers) over the inner surfaces of the mould. After a few minutes this should be polished. On all but the very smoothest surfaces, several layers of wax should be built up in this way before casting. During casting there is a slight linear shrinkage of the resin (about 2 per cent) and this should facilitate the removal of the finished product from the mould.

3 Casting

The resin is supplied in liquid form which, when mixed with a hardener, polymerises to form the solid glass-like embedding medium. In some cases an accelerator is also used to speed up the hardening process, but with many resins this is unnecessary. The polymerisation is an exothermic reaction and, because the resin is a poor conductor of heat, cracking may result from differential expansion of different parts of the block. For this reason it is advisable to build up the block in several layers, with no single layer being greater than 6–8 mm in thickness. Once some success has been achieved and with increasing experience, the thickness of individual layers may be increased.

Mix together the liquid resin and an appropriate amount of hardener (usually about 1 per cent of the total volume) in sufficient quantity to form a first layer in the mould some 6–8 mm thick. Cover the mould to exclude dust and place it in a warm place on a level surface and leave till the resin gels (usually 20 minutes to 1 hour depending on the temperature of the room and the proportion of hardener added).

Now place the specimen on the first layer, face downwards. If it is likely to float (the resin has a relative density of about 1.1), place a few drops of resin and hardener under the specimen and allow it to set. Pour on subsequent layers of resin and hardener until the mould is filled, allowing each to set in a warm, level place and protected from dust before adding the next. After the final layer, place a piece of glass, plastic or aluminium sheet across the top of the resin to produce a hard, flat surface. Leave for several hours (overnight, if possible) and then eject the cast from the mould. If this proves difficult, place the mould and block in a water bath at 50 °C and then in a refrigerator. The block should then become free.

4 Finishing

If adequate care has been taken over the selection and preparation of the mould and during the casting itself, the block should be free of any large protrusions. Any which do remain can be removed with a sharp knife. The sides of the block must then be ground and polished. This is done by rubbing the block on graded series of 'wet and dry' sandpapers which are kept as wet as possible. 120 grain, 240 grain and 360–500 grain paper, used in that order, are suitable in most cases, though an initial grinding with 60 grain paper may be necessary if the surfaces are very uneven. The sandpapers should be placed on a flat surface and the resin block rubbed on them to ensure a good finish. Finally polish with the special paste provided along with the resin.

5 Other details

a Coloured backgrounds. If a coloured background is required to highlight some features of the specimen, this can be obtained by mixing a little colour pigment with the final layer of resin. A range of pigments is available from suppliers.

b Labels and arrows. These may be added to specimens by typing or drawing on a thin sheet of card, cutting out the label and placing it alongside the specimen during the casting process.

c. Sufficient resin for two layers may be prepared at one time and half placed in a refrigerator while the first layer is polymerising in a warm place.

6 Problems

A number of faults and other problems are likely to occur during the preparation of resin-embedded specimens.

Some of these can be overcome by small changes in procedure and these are outlined below.

a Cracks in the block. These may be caused if too thick a layer of resin is cast at one time (see 3 p. 275). They may also occur when specimens with very sharp edges are embedded. The problem can be overcome by pouring the resin in even thinner layers or by pouring a thin layer of resin over the specimen once it has been placed on the basal layer of solid resin. Allow this thin layer to set before building up the block in the usual way.

b Air bubbles. These are usually caused by air which is within the specimen being embedded (e.g. in a shell or dry insect). To overcome this problem, place the specimen in a vacuum immediately after it has been covered by the liquid resin/hardener mixture. Alternatively, inject some of the mixture into the specimen before casting. Bubbles may also sometimes be caused by careless and over-vigorous mixing of the liquid resin and hardener.

c Silvering of the specimen. This occurs when there is poor adhesion between the resin and the surface of the specimen itself. It is most likely to happen when the specimen has a hard, shiny surface (e.g. some beetles and teeth). The resin contracts away from the surface in much the same way as it contracts away from the surface of the mould. The fault can sometimes be overcome by placing the completed block in a water bath at 70 °C and then pressing it between two very flat surfaces held in a vice.

d The resin not setting. The use of too low a proportion of hardener or casting at a low temperature will greatly increase the time that it takes for polymerisation to occur. Check that the correct mixture is being used. Room temperatures of about 20 °C are quite suitable for casting, but the polymerisation time may be reduced by gently warming. Do not exceed 40 °C. The resin will also not fully polymerise where it is in contact with the air. This leaves a slightly tacky surface to the block on its exposed surface, which may be ground away with sandpaper in the way outlined above (section 4) or better still, prevented by placing a sheet of glass, metal or polythene over the final layer or resin so that all air is excluded.

e The block not being released from the mould. The resin continues to contract for several days after casting and this may result in it becoming loose at a later time. If this does not take place, heat the block to about 90 °C in a water bath and then cool it to 0 °C or below by placing it in the freezing compartment of the laboratory refrigerator. Repeat this several times and the block will usually be released.

Agar embedding

This is a useful technique for the preservation of material which is too large for conventional microscopical treat-ment and yet too small and fragile to withstand liquid preservation in alcohol or formalin and excessive manipulation. Coelenterate medusae, teleost and amphibian eggs and embryos, and small vertebrate brains stained by Mulligan's method (see page 287) may all be satisfactorily preserved in this way.

1 Prepare a 4 per cent aqueous solution of agar in distilled water in a beaker placed in a water bath of boiling water. It is advisable to use high quality agar as this will give the greatest clarity in the finished product.
2 Cool the agar to about 45 °C and pour a thin layer into a solid watch glass and allow it to cool until it is set. The remainder of the agar should be maintained around 45 °C in its molten state.
3 Arrange the specimen on the surface of the solid agar. Specimens which have already been stained or preserved may be used, but they must be thoroughly washed in water before embedding.
4 Remove all excess moisture from both the surface of the set agar and the surface of the specimen itself. This can be done with strips of filter paper. If any surface moisture remains it is likely to prevent the adhesion of subsequent layers of agar resulting in separation of the block.
5 Cover the specimen with another layer of molten agar so that it is now embedded in the centre of a lens-shaped block of agar. Allow to set.
6 Ease the block from the solid watch glass and trim it with a sharp scalpel or razor blade into a neat rectangular shape with the specimen in the centre. At this stage the block will be opaque.
7 Transfer the block to 70 per cent alcohol (in which it may be stored if further treatment is not immediately convenient) and then to 80 per cent, 90 per cent and finally absolute alcohol leaving it for one day in each.
8 Transfer the block to a mixture of equal volumes of absolute alcohol and phenylmethanol (benzyl alcohol) and leave for a few hours before placing it in pure phenylmethanol. The block and specimen should now clear fairly quickly until they are completely transparent.
9 The blocks should be stored in phenylmethanol (benzyl alcohol) but may be removed for examination, provided they are returned to the alcohol within an hour or so to prevent drying out.

MICROBIOLOGICAL TECHNIQUES

Most micro-organisms occur in the natural environment in relatively low concentrations. To ascertain their presence and to carry out more detailed examination, it is necessary to provide ideal conditions for their multiplication so that large and visible colonies are formed. Most micro-organisms multiply very rapidly under the right conditions—in the case of some bacteria the generation time may be as short as twenty minutes—so that a single organism may multiply to form many millions within a very short time. A visible colony may thus be formed in one or two days. In some situations it may then be necessary to further sub-culture to produce a pure culture of a single species. However, this would not normally be done in schools and certainly not below sixth-form level. The techniques of microbiology involve providing the ideal conditions for the growth of the micro-organisms under study while at the same time preventing the entry of unwanted contaminants. A number of procedures are associated with these techniques and are discussed below.

Some cultures may have an unknown composition and may include micro-organisms of a pathogenic nature. It is therefore of the utmost importance that appropriate precautions are taken both to minimise the likelihood of such forms being cultured and to prevent any which are cultured being transmitted to man or to laboratory animals. A section on safety precautions and on the disposal of cultures and contaminated equipment is therefore also included in the discussion below.

Culture media

The choice of medium depends upon the organism being cultured and may be either liquid (broths) or solid (agars). They may be purchased in tablet or powder form from some general biological suppliers and also from certain specialist suppliers such as Astell Services. They may also be made up from recipes such as those given on pages 178 to 179. The use of these recipes is essential for more specialist media, but in many school situations the tablets are probably more convenient, even if a little more expensive. In general, bacteria grow more rapidly in conditions which are less acid than those promoting the fastest growth of fungi. Thus nutrient agar promotes rapid growth of many bacteria while malt agar is more suitable for most fungi.

Broths

Broths are normally kept in screw-top bottles (McCartney or Bijou bottles) but in some cases test tubes plugged with non-absorbent cotton wool and covered with aluminium foil are adequate. To prepare a broth culture from tablets, add $10\,cm^3$ distilled water to a single tablet in a McCartney bottle. Some tablets may be of a different strength and therefore require different volumes of water. Screw the lid on the bottle loosely. If the culture is being prepared in a test tube, it should be plugged with cotton wool and covered with foil. Larger volumes may be prepared in conical flasks by adding the appropriate volume of distilled water to a number of tablets, i.e. $10\,cm^3$ for each tablet. Flasks should also be plugged with cotton wool and capped with foil. The broth is now sterilised for 15 min at approximately $100\,kPa$ ($15\,lb\,in^{-2}$) above atmospheric pressure (see below). Broths should be allowed to cool before use. They may be stored in a refrigerator for many days provided the caps are retained on the bottles or other containers.

Agar cultures

These are usually kept in Petri dishes, though 'slopes' in small screw-top bottles may be required in some circumstances. Petri dishes may be made of either glass or perspex. The former may be sterilised and used repeatedly. Perspex dishes are sold pre-sterilised in packs of ten. They are intended for use once only and cannot be re-sterilised. To prepare an agar culture from tablets, add $10\,cm^3$ distilled water to 2 tablets in a McCartney bottle. Screw the lid on loosely. Test tubes (or conical flasks for larger volumes) may be used as alternatives to McCartney bottles. They should be plugged with non-absorbent cotton wool and capped with aluminium foil. The tablets should be allowed to soak for 15 min until they are dissolved. The medium is then sterilised for 15 min at approximately $100\,kPa$ ($15\,lb\,in^{-2}$) above atmospheric pressure (see below) and allowed to cool to about $45\,°C$ before pouring into a Petri dish.

Plating

The process of pouring the medium into the Petri dish (known as plating) is a critical one as it provides an opportunity for the entry of contaminating organisms. It should be carried out away from draughts and open windows and on a surface wiped with 70 per cent alcohol. Some workers favour the use of a transfer chamber (see pages 119 and 120) to reduce the chances of contamination. Figure 9.12 illustrates the main stages in plating.

If the medium has already set in the McCartney bottles it may be re-liquified by warming it on a water bath at $90\,°C$ and then transferring to a second water bath at $45\,°C$ once the medium has liquified. Plates should not be poured at temperatures above $45\,°C$ or condensation will occur inside the lid of the dish. Even at this temperature some condensation may take place, though this will be minimised if the Petri dishes are warmed slightly before plating. If condensation should occur, it may be removed by placing the dishes in the manner shown in Figure 9.13 in an incubator at $37\,°C$ for 1 h. Alternatively they may be left to dry in a similar position for 2–3 h at room temperature. Ideally the surface on which they are placed should be swabbed with 70 per cent alcohol to remove dust containing possible contaminants. Once the condensation has evaporated the lids should be replaced and the dishes stored upside down until they are used. If possible, poured plates should be left for several days before use to reveal the presence of any contaminating organisms. Contaminated plates should then be rejected.

Preventing the entry of contaminating micro-organisms

In any microbiological work, it is essential that the entry of contaminants is prevented. There are two ways in

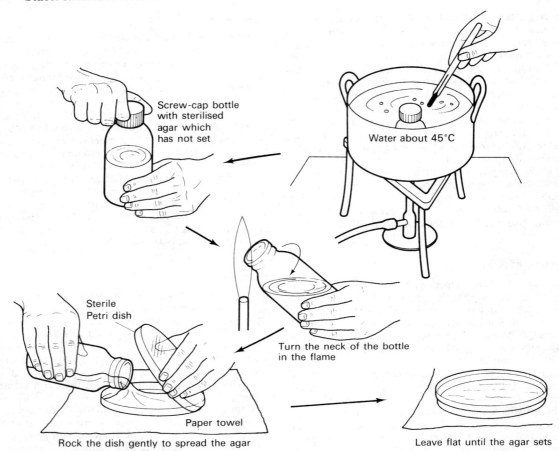

Screw-cap bottle with sterilised agar which has not set

Water about 45°C

Sterile Petri dish

Turn the neck of the bottle in the flame

Paper towel

Rock the dish gently to spread the agar

Leave flat until the agar sets

Figure 9.12   Stages in the preparation of an agar plate

which contamination often occurs. Firstly, the glassware and other apparatus with which the culture media come into contact may not be sterile. Secondly, contaminating micro-organisms may enter during plating, streaking or other stages of the culture or of the experimental process.

a Sterilisation

All glassware and other apparatus should be properly sterilised. Some disposable items (e.g. Petri dishes) are supplied in packs which have been pre-sterilised by gamma radiation. There are normally 10 Petri dishes in each polythene pack. Sufficient dishes for the purpose in

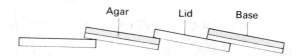

Agar   Lid   Base

Figure 9.13   Drying agar plates to remove condensation

hand should be squeezed out of the pack and the pack resealed. Syringes, spreaders and other pre-sterilised items are normally packed singly and should not be opened until immediately prior to use.

Glassware may be sterilised by dry heat in an oven or oven/incubator (see page 134). It should be placed in a cold oven and then warmed up gradually to 160°C and maintained at that temperature for 1–2 hours—if it is placed directly in a hot oven it may crack. The oven should then be switched off and allowed to cool with the glassware *in situ*. The oven door should be kept closed throughout the sterilisation and cooling process. Apparatus is more commonly sterilised in an autoclave or pressure cooker reserved for the purpose (see page 79). It is advisable to wrap glassware in aluminium foil to prevent it getting wet during the sterilisation process. Smaller items such as pipettes, spreaders, forceps, etc., should be individually wrapped and the wrapping only removed immediately prior to use.

Media, whether broths or agars, should be sterilised in

rinsed with
Sodlum thiosulphate } cleans
equipment

an autoclave or pressure cooker (see page 79). An appropriate amount of water should be placed in the autoclave or pressure cooker as directed by the manufacturer. The media should then be placed in suitable containers inside the cooker. Autoclaves are normally provided with an inner container and in the case of domestic pressure cookers, the metal vegetable containers are quite suitable. The lid is then secured and heat applied until steam issues from the escape vent. The weights (if any) are then placed in position and the escape vent closed. Heat should continue to be applied until the pressure reaches $100\,kPa$ ($15\,lb\,in^{-2}$) above atmospheric pressure. It should then be adjusted to maintain the pressure at this level for 20 min, after which the heat should be switched off and the cooker allowed to cool until the pressure has returned to atmospheric level. In the case of an autoclave this return to atmospheric pressure may be hastened by placing it over a sink and opening the escape valve to allow the release of the hot steam under pressure. The valve should then be closed, thus ensuring conditions of considerably reduced pressure inside the autoclave once it is fully cooled, and the remaining water vapour condensed. Sterile apparatus may be retained inside an autoclave or pressure cooker for several days prior to use.

b Sterile technique

The chances of entry of contaminants at later stages in the culture or experimental process can be minimised by the correct application of sterile technique. All pieces of apparatus subsequently used in the experimental process should be sterile. Forceps and loops should be flame sterilised by heating to red-heat in a Bunsen flame. Spreaders, pipettes and other small items of glassware should be retained in their sterile packs or wrappings until immediately prior to use. Petri dish lids should only be removed when this is essential for experimental procedure, and then only for as short a time as possible and in conditions unlikely to lead to the entry of unwanted micro-organisms.

Experimental procedures

Streaking

Many culture techniques require the dilution of an inoculum over the surface of an agar plate by a process known as streaking. This is carried out as follows. (See also Figure 9.14).

1 Wipe down the area of bench to be used with 70 per cent alcohol and assemble the plates to be inoculated, the inoculating culture, an inoculating loop and a lighted Bunsen burner. Place the first plate on the bench near the burner.

2 'Flame' the inoculating culture bottle by placing its neck and cap in the flame and rotating it once, keeping the

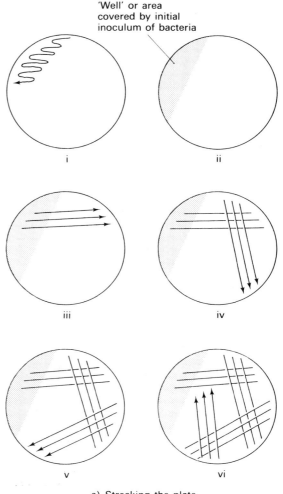

'Well' or area covered by initial inoculum of bacteria

a) Streaking the plate

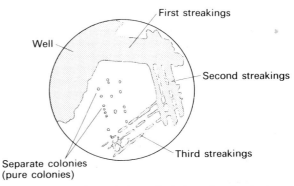

Well

First streakings

Second streakings

Third streakings

Separate colonies (pure colonies)

b) Appearance of a typical streaked plate after incubation

Figure 9.14   Stages in streaking an agar plate

bottle at an angle of about 45° so that the culture does not wet the cap. Right-handed operators should hold the bottle in the left hand. (Subsequent instructions assume a right-handed operator.) Flaming should last only a few seconds as the next stage involves unscrewing the cap.

3 Unscrew the cap with the right hand and flame the neck of the bottle in the same way, rotating it once. Replace the cap loosely.

4 Take the inoculating loop in the right hand and the culture bottle in the left. Sterilise the loop by holding it almost vertically in the flame till the loop itself glows red hot. Without pausing, use the crooked little finger of the right hand to remove the lid from the culture bottle. Cool the loop by touching the edge of the culture and then dip it into the centre of the culture. Touch the loop against the inside of the bottle to drain off excess liquid and then remove the loop. Flame and re-cap the bottle and set it down before immediately proceeding to the next stage. (In the case of agar slopes, the loop should be wiped across the surface of the slope to gather a film of bacteria.)

5 Immediately raise the lid of the Petri dish to an angle of about 45° and introduce the loop with its film of inoculum. Stroke the loop lightly over the surface of the agar as shown in Figure 9.14i. Do not dig the loop into the agar.

6 Remove and reflame the loop, closing the Petri dish meanwhile.

7 Raise the Petri dish lid once more and cool the loop by touching it on a part of the agar which is not to be used. Make three more streaks as shown in Figure 9.14iii.

8 Repeat 6 and 7 three times, flaming the loop and cooling it prior to streaking on each occasion. The plate will thus have been streaked in the manner shown in Figure 9.14vi. The intention is to spread the original loopful of inoculum in progressively greater dilutions over the surface of the plate.

9 Incubate the plate for 36–48 hours at 37 °C.

NOT RECOMMENDED AS BODY TEMP

Labelling

It is essential, both for safety reasons and for experimental procedures, that *all* cultures are fully and correctly labelled at *all* stages. As there is always a danger of lids and caps becoming accidentally exchanged, labels should be placed on the base of Petri dishes and on the glass of McCartney bottles, flasks or test tubes. A glass-writing pencil, spirit-based felt-tip pen or self-adhesive labels should be used. Gummed labels which require moistening should never be used as they are liable to become detached during sterilisation. Labels must never be licked during micro-biological work. Details on the label should include the nature of the specimen and the date and, in the case of class experiments, the pupil's name (or initials) and class.

Incubation

Prior to incubation, Petri dishes should be sealed with adhesive tape to prevent their being opened. They should be stacked upside down in the incubator, which reduces the chances of contamination as any micro-organisms which enter will fall onto the lid of the dish and not onto the culture itself; it also enables the labelled bases to be clearly seen. Bacterial cultures should be incubated at 37 °C for 36–48 hours. Fungal cultures grow rather more slowly and require lower temperatures (20–25 °C) though they will grow at 37 °C. Room temperatures are probably quite adequate.

The growth of colonies may be halted at any stage by placing the cultures in a refrigerator; growth will resume on a return to a suitable temperature. Growth may be ceased permanently by adding a few drops of 40 per cent methanal (formaldehyde) solution to a filter paper, placing this inside the lid of the Petri dish and leaving it overnight. If bacterial colonies are to be studied in the absence of fungal mycelia, it is advisable to remove the plates from the incubator after 36 hours and to inhibit further growth by one of these methods.

Inspection of plates

Plates should be left sealed with adhesive tape, for inspection by pupils. Inspection of colonies through the base of the plate is normally all that is required, but they may also be viewed through the lid. Should this be prevented by condensation, the lid should be replaced with another dry, sterile lid and the plate re-sealed with tape. This should be carried out by the teacher or by a technician and *not* by pupils.

Disposal of contaminated equipment

All cultures of micro-organisms should be destroyed by autoclaving before disposal. Cultures in plastic Petri dishes should be placed in autoclavable disposal bags (available from general biological suppliers), sealed and then autoclaved. They may then be disposed of through normal channels (refuse bins or incineration). Glass Petri dishes and other culture vessels (e.g. McCartney bottles) may be autoclaved directly but the lids of McCartney bottles should be slightly loosened before autoclaving. If they are removed from the autoclave while still hot, the now sterile cultures may be poured down the sink and rinsed away with plenty of water.

Small items of equipment (pipettes, slides, spreaders, etc.) should be immersed directly in buckets or basins of a suitable disinfectant. 'Chloros' (a 10 per cent solution of sodium oxochlorate(I)) or 'Milton' (a 1 per cent solution of sodium oxochlorate(I)) are both suitable. One volume of Chloros should be diluted with fifty volumes of tap water; one volume of 'Milton' should be diluted with five volumes of tap water. Both these disinfectants are active because of the free chlorine available in the solutions. As the chlorine escapes they diminish in effectiveness and should therefore be made up fresh daily. Their effectiveness may be measured with starch-iodide papers which turn blue in the presence of chlorine. Any colour paler than blue writing ink reflects inadequate chlorine for effective disinfection. Disinfectants such as 'Lysol' should not be used in schools because of their caustic nature. They are also inadequate for killing some spores. After disinfection, these small items of equipment should be washed in a hot solution of detergent, rinsed in running water and re-sterilised by autoclaving.

Safety precautions

Appropriate safety precautions are essential for any microbiological work. They are necessary to protect pupils, staff and laboratory animals from infection and also to protect the cultures of micro-organisms themselves from infection by unwanted contaminants. The following precautions should therefore be observed.

a All micro-organisms cultured in the school situation should be regarded as potentially pathogenic and treated as such. Pure cultures of micro-organisms should not be used unless they are listed as being suitable for use in schools—The Department of Education and Science (1977) have drawn up a list of suitable bacteria.

b Cultures should not be made from potentially dangerous sources such as faecal matter, pus and human mucus.

c Natural sources such as soil may contain pathogenic micro-organisms and it is therefore advisable to seal all cultures with adhesive tape prior to incubation (see page 280).

d Staff, technicians and pupils should wear laboratory overalls and these should be cleaned regularly.

e All 'hand-to-mouth' operations should be prohibited including moistening labels with the tongue. Smoking, eating and drinking should always be forbidden in laboratories.

f All exposed cuts should be protected with waterproof dressings.

g Refrigerators used for storing microbial cultures, etc., should never be used for storing food.

h Cultures spilled on the bench or elsewhere should be swabbed with undiluted 'Milton' or 'Chloros' diluted with 10 parts of water (i.e. a 1 per cent sodium oxochlorate(I) solution).

j Suitable containers of disinfectant should be placed in the laboratory for the receipt of items of contaminated apparatus. Pupils should be instructed carefully as to their use. (See 'Disposal of Contaminated Equipment' on page 280).

k Pupils should be trained in the correct use of microbiological techniques. They should understand their importance and with practice become habitual in their use.

l All apparatus for a particular exercise should be assembled on the bench near each working group of pupils *before* they commence the work. This will minimise pupil movement during the work. Overcrowding of pupils should be avoided.

m As far as is possible, animals should be removed from laboratories where microbiological work is to be carried out.

SETTING UP AQUARIA
Freshwater aquaria

It is most important to choose a size of aquarium suitable for the organisms it is to contain. Overcrowding is one of the most common causes of failure of aquaria. If fish are being included, then an approximate guide is as follows:

i tank without aeration or filtering—2 litres of water per cm length of fish;
ii tank with aerator, but no filter—1 litre of water per cm length of fish;
iii tank with aerator and filter—1 litre of water per 1.5 cm length of fish.
(All fish lengths exclude the tail fins.)

Hence, a 60 cm × 30 cm × 30 cm tank holding approximately 50 litres would take 25 cm of fish if neither aerator nor filter was used (i.e. 8 × 3 cm long fish or 5 × 5 cm long fish). Each cm of fish should also have at least 60 cm² surface area of water. These figures assume that the aquarium is reasonably well stocked with plants and placed in suitable illumination.

Before setting up the aquarium the tank should be washed with coarse sand and cold or warm water. Hot water should be avoided as this may soften the glazing compound in the case of metal-framed tanks. Detergents should not be used. After washing the tank, it should be rinsed several times with clean, cold water. It should then be nearly filled with water and left to stand for a day or so, by which time any leaks should be revealed. This water should then be discarded.

The tank is now ready and should be placed in the position which it is finally to occupy. This should be in conditions of moderate illumination. If the light intensity is too high, algae will grow rapidly and make the water appear green. South-facing windows should be avoided, but, if this is not possible, the sides of the aquarium facing the sun can be either covered with aluminium foil or painted green. Once in position and filled, metal-framed tanks should never be moved without emptying as this is likely to lead to slight movements between the metal and the glass sides or base, resulting in leaks.

Once in position, the bottom of the tank should be covered with a 2–4 cm layer of aquarium gravel which has been previously washed in boiling water and rinsed with cold water until it is quite clean. If an undergravel filter is being used, this should be placed in the bottom of the tank, with the pump entry tube in one of the rear corners; the gravel is then placed on top. The gravel should be sloped so that there is a greater depth (say 4 cm) at the rear of the tank than in the front (say 2 cm). Some aquarists like to add a layer of aquarium sand, say 1 cm deep, in which case rather less gravel can be used.

Next, place a large sheet of paper over the gravel (and sand if it is being used) and add water till the tank is about half full. The paper should prevent the gravel and sand from becoming stirred up by the water as it is poured in. The paper is now carefully removed. Fish pond water or 'instant pond water' in tablet form are to be preferred to tap water as the latter contains chlorine which is

Plant	Notes
Azolla spp.	A small, delicate, floating water fern
Callitriche spp.	The water starworts. Floating and submerged plants with small rounded or oval leaves in pairs
Ceratophyllum spp.	The hornworts. Rooted and submerged plants with delicate finely divided leaves in whorls
Elodea spp.	The Canadian pondweed (*E. canadensis*) and several related species. A rooted and submerged plant with small, dark green leaves in whorls of three
Fontinalis antipyretica	The willow moss. A submerged plant
Lemna spp.	The floating duckweeds which reproduce rapidly by vegetative means
Myriophyllum spp.	The water milfoils. Rooted and submerged plants with delicate, finely divided leaves in whorls of three, four or five. Require plenty of room
Potamogeton spp.	Larger rooted pondweeds with broad leaves floating on the surface. Require plenty of room
Ranunculus spp.	The crowfoots with two types of leaves, one finely divided and the other more like their close relatives the buttercups. Require plenty of room
Riccia spp.	Floating liverworts with narrow Y-shaped fronds
Utricularia spp.	The bladderworts with finely divided leaves and oval bladders in which small aquatic animals are trapped
Vallisneria spp.	Rooted grass-like plants reproducing by runners

Table 9.3   Some suitable plants for a temperate or cold-water aquarium

Plant	Notes
Azolla spp.	Small, delicate floating water ferns
Cabomba caroliniana	A rooted plant with fan-shaped, finely divided leaves
Cryptocoryne spp.	Rooted plants with broad, dark green leaf blades on long stalks
Hygrophila polysperma	A rooted plant with leaves shaped rather like privet
Lemna spp.	Floating duckweeds providing good shelter for fish fry. Reproduce vegetatively very rapidly
Limnophila sessiliflora	A rooted plant resembling *Cabomba*. Fan shaped leaves arise in whorls. Slow growing
Marsilea spp.	Rooted aquatic ferns with emergent leaves resembling a 'four-leaved' clover
Riccia spp.	Floating liverworts with narrow Y-shaped fronds providing good shelter for fish fry
Sagittaria subulata	A rooted, grass-like plant producing runners
Salvinia spp.	Floating ferns with clusters of small, almost circular leaves
Vallisneria spiralis	A rooted grass-like plant reproducing rapidly by runners
Wolffia arrhiza	A minute floating plant consisting of a globular frond some 2 mm in length

Table 9.4   Some suitable plants for a tropical aquarium

injurious to aquarium stock. However, if water must be used, then it should be left to stand for at least 24 hours before any organisms are added. This will allow the chlorine to escape. In this case it is good practice to add at least 2 litres of water from an established aquarium to the tap water.

With the tank half full, the plants can now be added. Details of suitable plants for cold-water and tropical aquaria are given in Tables 9.3 and 9.4. Larger, rooted species should be set near the back of the tank with the front left clear to provide good visibility; the roots should be carefully buried in the gravel leaving the crown of the stem visible. It may be necessary to anchor plants down until they are well rooted. In this case, a thin strip of lead, wrapped round the base of the plant can be used. Floating plants such as *Lemna* provide refuge for fish fry, but in good conditions they multiply rapidly and reduce the light intensity in the rest of the tank unless they are removed from time to time. In the early stages, it is best

Species	Food	Community tank suitability	Breeding	Temperature range	Other information
Guppy (*Lebistes reticulatus*)	Omnivorous	Adults (young may be eaten)	Viviparous (prolific)	15–32 °C	There are many types of guppy brought about by selective line breeding. If a quality such as particularly bright colouring is to be retained, care must be taken to discard poorly coloured males and to ensure early isolation of virgin females. Indiscriminate breeding will lead to a deterioration of the strain
Mollie (*Mollienisia* spp.)	Mainly vegetarian. Dry food, e.g. chopped lettuce. Browses on algae	Adults	Viviparous	21–27 °C	Gravid females have the ability to retain young for long periods, if conditions are not right
Platy (*Xiphophorus maculatus*, formerly *Platypoecilus maculatus*)	Omnivorous	Adults	Viviparous (prolific)	18–32 °C	Many variations of colour are available. Cross breeding between the colour strains usually leads to a deterioration of the colour
Swordtail (*Xiphophorus* spp.)	Omnivorous	Males tend to bully each other. One to each tank	Viviparous	22–32 °C	At first it is not possible to distinguish the sexes of this species. Later, males develop a gonopodium and the lower rays of the tail fin develop into the 'sword'
Zebra fish (*Brachydanio terio*)	Omnivorous	Large specimens may eat other fish	Bubble nester 27 °C	20–29 °C	Fairly easy to sex. The male has a longer and more pointed dorsal fin than the female. It is one of the easiest bubble nesters to breed. Once the eggs are in the nest, the male takes sole charge of the care of the fry
Siamese fighting fish (*Betta splendens*)	Mainly live food	*Only one male per tank*	Bubble nester 25 °C	20–32 °C	Two males placed side by side in glass containers will react in a most spectacular manner. A mirror held to a single male will cause a reaction. Fish will also jump to take white worms held just above the surface of the water. Their ability to take part of their air requirement at the surface enables one to confine individuals in small jars floating in the aquarium
Egyptian mouth-breeder (*Haplochromis multicolor*)	Live food	Fair. Sometimes attacks small fish	Mouth breeder 25 °C	21–27 °C	Much interest is created by the unusual method of protection of the eggs and young in the mouth of the female

Table 9.5   Some suitable fish for a tropical aquarium
Reprinted, with modifications, from Nuffield Junior Science Sourcebook *Animals and Plants*, Collins, 1967

not to introduce too many species of plant—three or four is probably quite sufficient. It is also worth remembering that the plants will grow and multiply and space should be allowed for this when setting up the aquarium

The tank can now be filled to within 2–4 cm of the top; once again a sheet of paper should be placed over the gravel and plants to prevent disturbing them. If a tropical aquarium is being set up, the heater and thermostat can now be placed in position. If these are being used as two separate units (see page 77), they should be placed at opposite ends of the tank. The aerator should be fitted next if one is being used. This should be adjusted, by means of a screwclip on the air line, to produce a continuous, fine stream of bubbles. If used in conjunction with an undergravel filter, too fast a rate of bubbling will stir up detritus and make the water cloudy.

The aquarium is now fully operational except for the introduction of the animals. It is advisable to allow at least a week for the plants to become established and, in the case of tropical tanks, for the temperature to become stabilised within acceptable limits. Tropical tanks should have a daytime temperature of about 25 °C (77 °F). Fluctuations between 21° and 27 °C (70–80 °F) are quite acceptable provided they are gradual. Night-time temperatures tend to be rather lower than those during the day when the surrounding air temperatures are higher and the lighting is on.

Fish and other animals can now be introduced. Suitable fish for a tropical tank are shown in Table 9.5 together with some details of their feeding requirements, breeding and other particulars. Recommended fish for a temperate (cold-water) tank include *Phoxinus* spp. (minnow), *Gasterosteus* spp. (stickleback), *Cyprinus* spp. (carp), *Gobio* spp. (gudgeon), *Rutilus* (roach) and *Cottus* (bullhead). The goldfish (*Carassius auratus*) has a wide tolerance to temperature and, provided changes are not too sudden, may be maintained at any level between 10 °C and 25 °C (50–77 °F). With larger specimens, the space requirements mentioned above should be kept in mind and overcrowding avoided.

Aquatic snails should also be included in the aquarium at the rate of about one snail per 2 litres of water. A wide range of species suitable for tropical tanks is available from aquarists. For temperate tanks, the ramshorn snails (*Planorbis planorbis*, *P. complanatus* and *P. albus*), the pond snails (*Limnaea auricularia*, *L. pereger* and *L. truncatula*) and the moss bladder snail (*Physa fontinalis*) may all be used. Snails will reduce the level of algae on the glass sides of the tank and will also feed upon filamentous algae. Very young snails may also serve as food for certain species of fish.

Once set up, the aquarium should require little attention other than the provision of suitable food for the animals. Various proprietary brands of fish food are available. Flake forms such as Tetra Min are recommended for tropical fish in preference to powdered food. The latter tends to sink to the bottom of the tank before it has been eaten and this can lead to fouling of the water. All fish will benefit greatly from a change of diet and although the provision of live food may not be possible on a daily basis, it should be given on occasions. Table 9.6 summarises some of the live foods which may be tried. During holidays lasting more than a week-end, long-term feeding material may be given. Available from aquarists, this slowly releases food into the water over a period of time. Alternatively, an automatic feeder may be employed (see page 76). Care should be taken not to overfeed. If flake food has not been cleared within 5 minutes of feeding, too much has been given.

If the algal population on the surface of the glass becomes too great, it can be removed using a scraper (see page 79) or a razor blade; this is probably an indication that the level of illumination is too great. Filamentous algae should be kept in check by the snails. If suspended unicellular algae cloud the water making it green, they can sometimes be removed by the addition of *Daphnia* which are then eaten by the fish. If there is a large number of fish in the tank, then some of these may have to be temporarily removed before the addition of the *Daphnia* otherwise the crustaceans will be eaten before they have a chance to remove the algae.

Dead plants and animals should be removed as soon as they are spotted otherwise they will cause the water to become polluted. Evaporation will cause a fall in the water level; this loss should be replaced from time to time. The faeces from the fish, snails and other animals should fall to the base of the tank. If an undergravel filter is being used, it should draw this waste material through the stones where bacterial action will break it down and release salts back into the water. In the case of an outside filter, the suspended waste material is removed by the polymer wool. If no filter is being used, a certain amount of partly decayed material may accumulate on the surface of the gravel. This should be removed from time to time with a dip tube or sediment remover.

Marine aquaria

Marine aquaria are generally more difficult to maintain than those containing fresh water. The combination of a sufficiently low temperature and a sufficiently high oxygen content is often difficult to achieve in the school situation. However, the effort is well worthwhile as a number of marine organisms make especially fascinating studies and enable pupils to witness their feeding and behaviour at first-hand.

The marine aquarium is set up in much the same way as the freshwater aquarium. One or two large stones should be provided for the attachment of sea anemones, etc. Either natural salt water or synthetic sea water (see page 191) may be used, though in the latter case it is always advisable to include some natural, unfiltered sea water which will provide an initial stock of small crustaceans,

Name	Maximum size	Source	Ease of culturing	Culture medium	Optimum temperature for culture	Suitability
Infusoria	Microscopic	From dealers or established tanks	Easy and quick	On lettuce in water	21 °C	For newly hatched fry of egg-laying species
Anguilluda silusia (microworms)	0.5 mm long	From dealers	Easy and fairly prolific	On finely ground oatmeal + yeast (or bread and milk)	21 °C	For fry of egg-layers aged 2 weeks +, and newly hatched live-bearers
Daphnia spp. (water fleas)	2 mm long	From dealers and ponds	Difficult to culture con-tinuously	On infusorians or 'green' water (see also page 222)	18 °C	The best general purpose food
Artemia salina (brine shrimps)	1 mm long (larvae)	From dealers as eggs	Eggs stored dry and hatched when required	Salt water or sea water (see also page 222)	21 °C	For young live-bearers rinse in fresh water before adding to the tank
Enchytraeus buchholzii (Grindal worms)	5 mm long	From dealers	Easy and prolific	On soil + baby cereal	21 °C	Good general-purpose food for adults of smaller species
Enchytraeus albidus (white worms)	15 mm long	From dealers	Easy and prolific	On soil + bread and milk (see page 220)	16 °C	Good general purpose food thought to be fattening
Tubifex spp. (mud worms)	20 mm long	From dealers and muddy ponds	Difficult to culture in quan-tity	On mud with small pieces of bread in running water	16 °C	Good general purpose food
Earthworms	15 cm long	From gardens	Difficult to culture in quan-tity	On soil with leaf mould (see also page 220)	16 °C	Chopped up as a food for adult fishes

Table 9.6 Some suitable live foods for tropical fish
Reprinted, with modifications, from Nuffield Junior Science Sourcebook *Animals and Plants*, Collins (1967)

diatoms, etc. Some of the organisms in the tank may absorb salts thus making it necessary to add a little more from time to time. A relative density of between 1.017 and 1.022 should be maintained and this can be checked with a hydrometer. Some workers like to add half a teaspoon of a mixture of 3 parts rock salt to 1 part Epsom salts to a 50 litre tank every month. Marine plants are not good oxygenators of the water and artificial aeration is therefore essential. Temperature is usually the main problem in a centrally heated school building. Levels around 10 °C are ideal and in no circumstances should they exceed 15 °C. Lighting should be provided from a fluorescent tube as incandescent lamps tend to increase the growth of micro-organisms.

Marine organisms may be collected from the coast, obtained from the Marine Biological Association or some other supplier of marine organisms (see page 249). The problems involved with the transport of marine organisms without subjecting them to excessive rises in temperature are discussed by St. Aubrey (1969). As well as seaweeds, small starfish, sea anemones, hydroids and small crabs are recommended as are a number of species of polychaete worm. Anemones may be fed on brine shrimp eggs or on small pieces of other marine organisms. All uneaten food should be removed from the tank to prevent fouling of the water. Some workers like to remove the animals being fed and to place them in a separate small container during feeding to prevent fouling.

Filter feeders, such as barnacles and many bivalve molluscs, require a continuous supply of minute organisms suspended in the water. Natural sea water is therefore essential if they are to be maintained successfully. Winkles, limpets and other molluscs which browse algae off rock surfaces should not be introduced until such algae have had time to grow and multiply on the inner surface of the glass and on the stones. The shore crab (*Carcinus*) may be fed on small pieces of meat or fish, ideally in a separate container to prevent fouling of the water. Details of the culture and maintenance of specific organisms are given in Chapter 8.

Thompson (1968) has devised a means of simulating tidal changes. Readers interested in pursuing this idea should consult his paper.

MISCELLANEOUS BIOLOGICAL TECHNIQUES

Vertebrate skeletal preparations

Standard techniques

These methods are suitable for the preparation of skulls, articulated skeletons for mounting and disarticulated bones. They are best not used on the smaller vertebrates if entire skeletons are required. These specimens (body lengths of less than 10–15 cm) are probably best prepared for skeletal display by the 'clear and stain' technique (see below).

1 Remove the bulk of the flesh from the specimen by first skinning it and then cutting through the body wall and removing the viscera. Cut away as much muscle as possible from those parts of the body where no damage to skeletal structures may result. With larger reptiles, birds and mammals, the limbs may be carefully detached from the body and treated separately, though care should be taken not to damage or lose the collar bones if these are present (in members of the cat family they are very small and not articulated with the rest of the skeleton). Carefully detach the skull from the atlas vertebra and clear away tissue from the foramen magnum to expose the base of the brain. Insert a long hypodermic needle into the foramen and push it forwards to the front of the brain. Inject water into the front of the cranial cavity which should push the brain backwards, forcing it out of the foramen magnum. If this fails, the brain should be picked out with dissecting needles and forceps. Remove the eyes and cheek muscles.

2 Remove the remainder of the flesh by one of the following techniques.

a Boil the specimen for several hours in a large pan of water until the flesh loosens from the bones. Pick the bones clean, brushing away the final traces of flesh, and wash thoroughly. Leave to dry. This is perhaps the easiest method of removing the flesh, but it does most damage to the skeleton in that the ligaments connecting the bones are softened, as are the sutures. The bone surfaces also become porous and the teeth may fall out. In younger specimens, the epiphyses at the ends of the long bones may become detached from the shafts.

b Leave the specimen in a large vessel of clean water for 2–3 weeks until the flesh is thoroughly rotted and loosened from the bones. Pour away the water along with the flesh, taking care not to lose any of the smaller bones. Wash the bones clean and allow them to dry. This method has the disadvantage of producing a very unpleasant smell lasting for many days.

c Place the specimen in clean tap water in a large container and leave for several days, changing the water daily. When the water becomes clear (usually after about three days), replace with a solution of sodium phosphate(V) (approximately 6 g per litre) and leave for 24 hours. Remove the bones and brush them with a toothbrush dipped in hot water and then in bleaching powder. Rinse in cold water. Continue to brush with bleaching powder and hot water and rinse until all the flesh is removed, but not the ligaments holding the bones in position. Finally rinse thoroughly and allow to dry. The use of rubber gloves is advisable to protect the hands while using the bleaching powder. This method is recommended if an articulated skeleton is required.

3 Degrease the specimen by leaving it to soak for 24 hours in concentrated aqueous ammonia. The long bones of larger animals should be drilled through the shaft at several points to allow the penetration of the degreasing solution. Smaller specimens may not need to be degreased.

4 Bleach the bones further if necessary by placing them in a solution of sodium oxochlorate(I) (hypochlorite) or hydrogen peroxide (6 per cent or '20 volume').

5 Assemble the skeleton, if an articulated specimen is required. The bones should be arranged in appropriate positions while the joints are still moist and pliable. If the limbs have been removed they may be attached to the axial skeleton by means of thin wire. Glue and a thin rod passing through the cervical vertebrae into the foramen magnum should be sufficient to hold the skull in position. The skeleton may be attached to card or thin plywood in the case of smaller and flatter specimens (e.g. bats, lizards, frogs, etc.).

6 Dry the bones thoroughly in a warm place.

7 Varnish the skeleton with a thin layer of clear polyurethane varnish. This will seal the pores and prevent the bones becoming dirty. It will also bring additional strength to the individual bones.

8 Store the bones in a dry place protected from dust. Articulated skeletons for display may be enclosed in perspex or glass and wood cases.

Clearing and staining techniques

These methods are ideal for skeletal preparations of small vertebrates as the bones are left intact and in the correct spatial relationship to the other parts of the skeleton. It is also a particularly good technique for embryological material.

1 Fix the specimen in 4 per cent methanal for about one

week. Material fixed in 95 per cent alcohol may also be used, but specimens preserved for longer than a week give less satisfactory results.

2 Remove the viscera and skin from larger specimens. This should not be attempted with embryos, newly born and furless small mammals and other delicate specimens.

3 Place in distilled water for 24 hours to remove formalin or alcohol. More than one change of distilled water may be used.

4 Place in a 1–2 per cent aqueous solution of potassium hydroxide for 12–24 hours, depending upon the size and delicacy of the specimen. Smaller and more delicate specimens should be left for shorter times in the more dilute solution. If the specimen is heavily pigmented, a little 3 per cent hydrogen peroxide solution may be added to the potassium hydroxide at the end of this time, and the specimens then washed for 24 hours in distilled water before returning them to fresh potassium hydroxide solution for 12 hours.

5 Add 1 cm^3 Alizarin red stock solution. Leave and observe daily until the vertebrae can be seen to have taken up the stain (usually within 24–48 hours). (See page 172.)

6 Place the specimen in fresh 2 per cent potassium hydroxide solution to which a few drops of propane-1,2,3-triol (glycerol) have been added. Leave till the specimen becomes clear with the skeletal material adequately stained (usually 3–14 days depending upon the size and nature of the material).

7 Transfer through each of the following solutions for 24 h each:

  i 20 cm^3 propane-1,2,3-triol (glycerol), 3 cm^3 of 2 per cent potassium hydroxide solution, 77 cm^3 distilled water;
  ii 50 cm^3 propane-1,2,3-triol, 3 cm^3 of 2 per cent potassium hydroxide solution, 47 cm^3 distilled water;
  iii 75 cm^3 propane-1,2,3-triol, 25 cm^3 distilled water.

8 Transfer to pure propane-1,2,3-triol (glycerol) for storage.

Mulligan's techniques for differentiating white and grey matter in vertebrate central nervous systems

This is an easy and dramatic way of differentiating the white and grey matter in vertebrate brains and spinal cords.
1 Decalcify the skull by soaking in 2–4 per cent nitric acid in 4 per cent methanal or in the acid alone after fixation in formalin.

2 Dissect out the brain and anterior part of the spinal cord by carefully removing the skull.
3 Cut the brain into slices of convenient thickness (not less than 5 mm) in the desired planes. Use a very sharp scalpel or razor blade so that a cleanly cut surface is obtained.
4 Rinse in gently running water and remove any debris with a fine brush.
5 Place in Mulligan's fluid (see page 187) for 2 minutes.
6 Wash in gently running water for 1 minute.
7 Place in 2 per cent tannic acid solution for 1 minute.
8 Wash in gently running water for a few minutes.
9 Place in 1–2 per cent ammonium iron(III) sulphate-12-water (iron alum) solution and watch until the grey matter turns black. This usually happens within 1 minute.
10 Wash in gently running water for several minutes.
11 Preserve in 70 per cent alcohol or 4 per cent methanal. The finished sections may also be embedded in polyester resin (see page 274).

Latex injection

Museum dissections in which the blood vessels have been injected with coloured latex are available from commercial suppliers. They can also be prepared in the school laboratory and the injection technique can be extended to class dissections. This not only enables the vessels to be seen more clearly, but also makes them less subject to damage than they would otherwise be. Latex injection techniques can also be used at the organ level for work with lungs and kidneys (see Hillier (1973)).

The latex normally used is a prevulcanised product containing about 60 per cent suspended solids stabilised by ammonia, which coagulates when acidified. It has naturally a pale translucent yellow appearance, but can be varied by the addition of appropriate coloured pigments. A concentration of 2–5 per cent by volume of pigment in the latex should give a satisfactory intensity of pigmentation. The unpigmented latex, pigments and a range of ready pigmented latexes are available from suppliers. Before use the latex should be strained through fine muslin or nylon to remove larger particles which might otherwise block the injection apparatus.

Injection is best carried out with a disposable plastic syringe and a needle of size appropriate for the animal concerned and the blood vessel being injected (see below). The latex is not toxic, but is difficult to remove from fabrics; for this reason appropriate protective clothing should be worn. Splashes of latex on the skin should be removed with running water. Latex which escapes from blood vessels, or otherwise obscures a specimen for dissection, can be removed with water. It is advisable to keep a wash bottle of water to hand for this purpose. It is also recommended that a pair of forceps with a swab of cotton wool dipped in 25 per cent ethanoic (acetic) acid be kept available; this can be used to coagulate any leaks which may occur.

Latex injection can be used for a variety of animals. Specimens should be fresh and the injection carried out as soon after death as possible. In the case of rats or larger animals it is advisable to remove as much blood as possible before injection (see below). If freshly killed specimens are not available it is possible to use those which have been frozen or preserved in some other way. If formalin has been used as a preservative it is advisable to inject 5 per cent aqueous ammonia into the blood vessels prior to the injection of latex. This will neutralise any acidity due to the formalin and prevent premature coagulation of the latex. Further details for the injection of a range of organisms are outlined below.

Mammals

1 Open the abdominal and thoracic cavities in the normal way and expose the heart.
2 If the animal has been freshly killed, insert a needle into the left ventricle, fit a syringe and withdraw as much blood as possible.
3 Leave the needle in position, but replace the syringe with one filled with red latex. Inject the latex while examining a loop of the intestine until it can be seen that the finer vessels are filled with latex.
4 Withdraw the syringe and needle and wipe the punctured heart with cotton wool dipped in 25 per cent ethanoic (acetic) acid.
5 Now repeat the procedure with the right ventricle, withdrawing as much blood as possible before injecting blue latex. Again monitor the injection by examining the intestine until the finer vessels are filled. It may be advisable to massage the neck so that the latex penetrates fully into the veins of the head.
6 Withdraw the syringe and needle and again wipe the punctured heart with 25 per cent ethanoic acid.
7 The hepatic portal system may be separately injected with a different colour of latex. In this case, insert the needle into the hepatic portal vein near the liver, but direct the needle away from the liver itself. Inject the latex while gently moving the intestine and examining it until the finer vessels are filled. Remove the needle and wipe with 25 per cent ethanoic acid.
8 Fix overnight by immersing in 10 per cent formalin to which 1 per cent ethanoic acid has been added.

The size of syringe used will depend upon the size of the mammal. As a rough guide, the arterial system of the mouse will require $1-2\,cm^3$ of latex, that of the rat $7-8\,cm^3$, while the rabbit will need $20-50\,cm^3$. The venous system will require slightly more latex in each case.

Amphibians

1 Open the animal in the normal way and expose the heart.
2 Insert a needle into the truncus arteriosus. Pass a piece of thread round between the heart and the conii and tie to prevent backflow and leakage of latex.

3 Fit a syringe filled with red latex and inject until the smaller arteries in the intestine are well filled. It will be necessary to direct the needle first into one and then into the other side of the arterial system to ensure full penetration by the latex.
4 Withdraw the syringe and wipe the truncus with cotton wool dipped in 25 per cent ethanoic (acetic) acid.
5 Inject the venous system with blue latex, first into the anterior abdominal vein and then into the musculocutaneous vein. In each case wipe the punctured vessel with ethanoic acid as before.
6 Remove the thread tying the heart and fix in 4 per cent methanal to which 1 per cent ethanoic acid has been added.
As a rough guide, $1-2\,cm^3$ latex will be required for the arterial system of a common frog (*Rana temporaria*) and a similar volume for the venous system.

Dogfish

1 Open the pericardial cavity to expose the heart and also expose the dorsal aorta just posterior to the point where the coeliac artery leaves it. Tie or clamp the coeliac artery.
2 Insert a syringe needle into the conus arteriosus and direct it forwards into the ventral aorta.
3 Fit a syringe containing blue latex and inject into the afferent system until the latex is clearly visible in the gills. This will normally take about $7\,cm^3$ latex. Remove the needle and wipe with 25 per cent ethanoic acid.
4 Now insert a needle into the dorsal aorta and tie the cut end of the vessel to prevent leakage. Fit a syringe containing red latex and inject till the latex is clearly visible in the gills. This will normally take about $5\,cm^3$ latex.
5 Fix by immersion overnight in 4 per cent methanal containing 1 per cent ethanoic acid.

Injection of the entire dogfish is difficult as internal leakage almost always occurs and the problem is aggravated by the large venous sinuses. Some degree of success may be achieved by very gentle injection into the caudal vein.

Earthworms

1 Obtain relaxed specimens by placing the worms in water and slowly adding alcohol until a concentration of about 15 per cent is obtained.
2 Inject up to $1\,cm^3$ latex into the dorsal blood vessel very carefully.
3 Fix overnight in 4 per cent methanal to which ethanoic acid has been added to a concentration of 1 per cent.

Humane killing of small mammals

The killing of any animal, for whatever reason, should be carried out by a technician or teacher and *never* by a pupil.

The method used should be humane and avoid any possible hazards. For this reason carbon dioxide is recommended for school use. The mammal should be placed in a killing chamber such as that illustrated in Figure 9.15 and the carbon dioxide cylinder should then be turned on slowly. The gas will fill the chamber, forcing the air out from the hole near the top. The animal should be left in the chamber for at least 10 min after it is apparently dead to ensure that death is certain. A heavy gauge polythene bag may be used as an alternative to the killing chamber, but is less satisfactory.

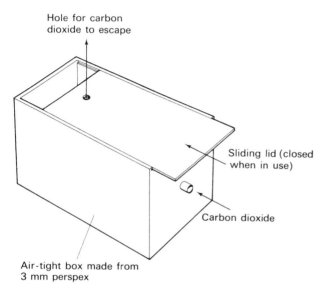

Hole for carbon
dioxide to escape

Sliding lid (closed
when in use)

Carbon dioxide

Air-tight box made from
3 mm perspex

Figure 9.15   A chamber for the humane killing of small mammals

Physical methods such as neck-breaking should be avoided as they are more liable to cause distress to pupils through their apparent brutality. Ethoxyethane and trichloromethane are both effective killing agents, but are highly flammable and are therefore best avoided. North sea or natural gas does not contain any poisonous components and hence only kills by asphyxiation and should therefore not be used. Anaesthetic drugs may not be used unless the user is licensed. For further information, the publications of the Universities Federation for Animal Welfare listed in the Bibliography are recommended.

Bibliography

Bradbury, S. (1973) *Peacocke's Elementary Microtechnique*, Arnold.

British Museum (Natural History) Various publications in the series *Instructions for Collectors*.

Conservation of Wild Creatures and Wild Plants Act, 1975, HMSO.

Department of Education and Science (1977) *The Use of Micro-organisms in Schools*, DES Education pamphlet 61, HMSO.

Fry, P. J. (1976) *Micro-organisms*, English Universities Press for the Schools Council Educational Use of Living Organisms Project.

Gliddon, R. and Whiting, H. P., Mulligan's method for permanent differentiation of grey and white matter in the central nervous system of vertebrates, *School Science Review*, 1969, 172, **50**, 572–574.

Grimstone, A. V. and Skaer, R. J. (1972) *A Guidebook to Microscopical Methods*, Cambridge University Press.

Hillier, D. V., Some simple uses of rubber latex, *School Science Review*, 1973, 191, **55**, pp. 302–305.

Knudsen, J. W. (1968) *Biological Techniques*, Harper and Row.

Morholt, E., Brandwein, P. F. and Joseph, A. (1966) *Sourcebook for the Biological Sciences*, Harcourt, Brace and World.

Nuffield O level Biology (1966–67) *Teachers' Guides I–V*, Longman/Penguin.

Nuffield Junior Science (1967) *Animal and Plants*, Collins.

Nuffield A level Biology (1971) *Laboratory Book*, Penguin Books.

Nuffield A level Biology (1971) *Teachers' Guides I and II*, Penguin Books.

Nuffield O level Biology (Revised) (1975–76) *Teachers' Guides 1–4*, Longman.

Pantin, C. F. A. (1948) *Notes on Microscopical Technique for Zoologists*, Cambridge University Press.

Picken, L. E. R., The study of minute biological structures and their significance in the organisation of cells, *School Science Review*, 1955, 129, **36**, pp. 262–268.

St. Aubrey, S. N. D., The maintenance and transport or marine intertidal and planktonic animals for laboratory use, *School Science Review*, 1969, 174, **51**, pp. 103–106.

Thompson, T. E., Experiments with molluscs on the shore and in a laboratory tidal model, *School Science Review*, 1968, 170, **49**, pp. 97–102.

Universities Federation for Animal Welfare (1968) *Humane Killing of Animals*, UFAW.

Universities Federation for Animal Welfare (1972) *Handbook on the Care and Management of Laboratory Animals*, E. & S. Livingstone.

Appendix

Books for the identification of the British fauna and flora

A wide range of books and keys is available. In many cases the identification of individual species is the work of a specialist and requires the use of detailed keys which are scattered throughout the biological literature. A comprehensive list of such keys will be found in:

Kerrich, G. J., Merkle, R. D. and Tebble, N. (Eds) (1967) *Bibliography of Key Works for the Identification of the British Fauna and Flora*, 3rd edn, Academic Press for the Systematics Association.

Another, less complete list is included in:

Bottle, R. T. and Wyatt, H. V. (Eds) (1971) *The use of Biological Literature*, 2nd edn, Butterworths.

Selected keys are also published from time to time in *Field Studies*, the journal of the Field Studies Council and by The Linnaean Society, Burlington House, Piccadilly, London W1.

Local museums and natural history societies may also be able to help with identification while, as a last resort specimens may be sent to the British Museum (Natural History) or the Royal Botanic Gardens Herbarium at Kew. However, it should be borne in mind that the officers at both these establishments are primarily engaged in research into their specialist taxonomic groups; their assistance should only be enlisted if all other attempts at identification have proved fruitless.

A number of publishers produce series of books on identification intended for the enthusiastic amateur, children and other non-specialists. These include:

Clue Books (Oxford University Press)
Field Pocket Guides (Collins)
Group Keys (Hulton)
Junior Field Guides (Ward Lock)
Kew Series (Eyre and Spottiswoode)
Natural History Series in Colour (Blandford)
Nature Books for the Pocket (Oxford University Press)
Observer's Books (Warne)
Oxford Books—of Insects, Vertebrates, etc. (Oxford University Press)
Picture Information Books (A. and C. Black)
Wayside and Woodland Series (Warne)
Young Specialist Books (Burke)

The following books are recommended for identification within the groups indicated.

Flowerless plants

Alvin, K. L. and Kershaw, K. A. (1963) *The Observer's Book of Lichens*, Warne.

Barrett, J. and Yonge, C. M. (1958) *Collins Pocket Guide to the Seashore*, Collins.

Belcher, H. and Swale, S. (1977) *A Beginner's Guide to Freshwater Algae*, HMSO.

Brightman, F. H. and Nicholson, B. E. (1966) *The Oxford Book of Flowerless Plants*, Oxford University Press.

Dickinson, C. I. (1963) *British Seaweeds*, Eyre and Spottiswoode.

Duncan, U. K. (1970) *Introduction to British Lichens*, T. Bunde and Co., Arbroath.

Hvass, E. and Hvass, H. (1966) *Mushrooms and Toadstools in Colour*, Blandford.

Hyde, H. A., Wade, A. E. and Harrison, S. G. (1969) *Welsh Ferns*, National Museum of Wales.

Jackson, A. B. (1955) *The Identification of Conifers*, Edward Arnold.

Lange, M. and Hora, F. B. (1963) *Collins Guide to Mushrooms*, Collins.

Newton, L. (1931) *A Handbook of British Seaweeds*, British Museum (Natural History).

Prescott, G. W. (1954) *How to know the Freshwater Algae*, W. C. Brown, Dubuque, Iowa, USA.

Ramsbottom, J. (1965) *A Handbook of the Larger British Fungi*, British Museum (Natural History).

Showell, J. P. (1969) *Lichens*, School Natural History Society Publication No. 38.

Step, E. (1971) *Wayside and Woodland Ferns*, Warne.

Taylor, P. G. (1960) *British Ferns and Mosses*, Eyre and Spottiswoode.

Watling, R. (1973) *Identification of the Larger Fungi*, Hulton.

Watson, E. V. (1968) *British Liverworts and Mosses*, Cambridge University Press.

Flowering plants

Chancellor, R. J. (1961) *The Identification of Seedlings of Common Weeds*, Ministry of Agriculture, Fisheries and Food, Bulletin No. 179, HMSO.

Chancellor, R. J. (1962) *The Identification of Common Water Weeds*, Ministry of Agriculture, Fisheries and Food, Bulletin No. 183, HMSO.

Clapham, A. R., Tutin, T. G. and Warburg, E. F. (1962) *Flora of the British Isles*, Cambridge University Press. (Also includes ferns and conifers).

Clapham, A. R., Tutin, T. G. and Warburg, E. F. (1968) *Excursion Flora of the British Isles*, Cambridge University Press. (Also includes ferns and conifers).

Davis, P. H. and Cullen, J. (1965) *The Identification of Flowering Plant Families*, Oliver and Boyd.

Fitter, R., Fitter, A. and Blaney, M. (1974) *The Wild Flowers of Britain and Northern Europe*, Collins.

Hubbard, C. E. (1968) *Grasses*, Penguin Books.

Jermy, A. C. and Tutin, T. G. (1968) *British Sedges*, Botanical Society of the British Isles.

Keble Martin, W. (1965) *The Concise British Flora in Colour*, Ebury Press and Michael Joseph.

McClintock, D. and Fitter, R. S. (1965) *The Pocket Guide to Wild Flowers*, Collins.

Mitchell, A. (1974) *Trees of Britain and Northern Europe*, Collins.

Philips, R. (1977) *Wild Flowers of Britain*, Pan Books.

Prime, C. T. and Deacock, R. J. (1970) *Trees and Shrubs—their Identification in Summer and Winter*, W. Heffer, Cambridge.

Vedel, H. and Lange, J. C. (1960) *Trees and Bushes in Wood and Hedgerow*, Methuen.

Invertebrates other than insects

Beedham, C. E. (1972) *Identification of the British Molluscs*, Hulton.

Cloudsley-Thompson, J. L. and Sankey, J. (1968) *Land Invertebrates*, Methuen.

Donner, J. (1966) *Rotifers*, Warne.

Eason, E. H. (1964) *Centipedes of the British Isles*, Warne.

Ellis, A. E. (1969) *British Snails: the Non-marine Gastropoda of Great Britain and Ireland*, Oxford University Press.

Jahn, T. L. and Jahn, F. (1949) *How to know the Protozoa*, W. C. Brown, Dubuque, Iowa, USA.

McMillan, N. F. (1968) *British Shells*, Warne.

Nichols, D. and Cooke, J. A. L. (1971) *The Oxford Book of Invertebrates*, Oxford University Press.

Paviour-Smith, K. and Whittaker, J. B. (1968) *A Key To the Major Groups of British Free-living Terrestrial Invertebrates*, Blackwell.

Quick, H. E. (1960) *British Slugs*, British Museum (Natural History).

Savory, T. H. (1960) *The Spiders and Allied Orders of the British Isles*, Warne.

Sutton, S. (1972) *Woodlice*, Ginn.

Insects

Beirne, B. P. (1954) *British Pyralid and Plume Moths*, Warne.

Burton, J. et al. (1968) *The Oxford Book of Insects*, Oxford University Press.

Chinery, M. (1973) *A Field Guide to the Insects of Britain and Northern Europe*, Collins.

Colyer, C. N. and Hammond, C. O. (1951) *Flies of the British Isles*, Warne.

Hicken, N. E. (1952) *Caddis*, Methuen.

Higgins, L. G. and Riley, N. D. (1970) *A Field Guide to the Butterflies of Great Britain and Europe*, Collins.

Howarth, T. G. (1972) *South's British Butterflies*, Warne.

Howarth, T. G. (1972) *Colour Identification Guide to British Butterflies*, Warne.

Linssen, E. F. (1959) *Beetles of the British Isles*, 2 vols. Warne.

Longfield, C. (1949) *Dragonflies of the British Isles*, Warne.

Lyneborg, K. (1977) *Beetles in Colour*, Blandford Press.

Marson, J. E. (1958) *Identification sheets—Insects and other Land Arthropods*, School Natural History Society.

Newman, L. H. and Mansell, E. (1968) *The Complete British Butterflies in Colour*, Ebury Press and Michael Joseph.

Ragge, D. R. (1965) *Grasshoppers, crickets and cockroaches of the British Isles*, Warne.

Shorrocks, B. (1972) *Drosophila*, Ginn.

South, R. (revised by Edelston, H. M.) (1963) *The Moths of the British Isles*, Warne.

Southwood, T. R. E. and Leston, D. (1959) *Land and Water Bugs of the British Isles*, Warne.

Step, E. (1932) *Bees, Wasps, Ants and Allied Insects*, Warne.

Stokoe, W. J. and Stovin, G. H. T. (1944) *The Caterpillars of the British Butterflies*, Warne.

Stokoe, W. J. and Stovin, G. H. T. (1949) *The Caterpillars of the British Moths*, 2 vols, Warne.

Zim, H. S. and Krantz, L. (1977) *Snails*, World's Work.

Vertebrates

Campsen, B. and Watson, D. (1964) *The Oxford Book of Birds*, Oxford University Press.

Corbet, G. B. (1976) *Finding and Identifying Mammals in Britain*, British Museum (Natural History).

Coward, T. A. (1969) *Birds of the British Isles and their Eggs*, Warne.

Lawrence, M. J. and Brown, R. W. (1973) *Mammals of the British Isles, their Tracks, Trails and Signs*, Blandford.

Muus, B. J. and Dahlstrom, P. (1971) *Freshwater Fish of Britain and Europe*, Collins.

Nixon, M. and Whiteley, D. (1972) *The Oxford Book of Vertebrates*, Oxford University Press.

Peterson, R. T., Mountford, G. and Hollom, P. A. D. (1974) *A Field Guide to the Birds of Britain and Europe*, Collins.

Smith, M. (1969) *The British Amphibians and Reptiles*, Collins.

Southern, H. N., ed. (1964) *The Handbook of British Mammals*, Blackwell for the Mammal Society of the British Isles.

Van der Brink, F. H. (1967) *A Field Guide to the Mammals of Britain and Europe*, Collins.

Some books on identification are centred on a particular ecological habit and cut across the boundaries between one taxonomic group and another. Among these, the following may be found useful:

Marine habitats

Barrett, J. and Yonge, C. M. (1958) *Collins Pocket Guide to the Seashore*, Collins.

Clayton, J. M. (1972) *The Living Seashore*, Warne.

Freshwater habitats

Clegg, J. (1965) *The Freshwater Life of the British Isles*, Warne.

Macan, T. T. (1959) *Guide to Freshwater Invertebrate Animals*, Longman.

Marson, J. E. (1968) *Water Animal Identification Sheets*, School Natural History Society.

Nuffield A level Biological Sciences (1970) *Key to Pond Organisms*, Penguin Books.

Nuffield O level Biology (1966) *Key to Small Organisms in Soil, Litter and Water Troughs*, Longman/Penguin.

The Freshwater Biological Association also publish a number of identification keys to freshwater organisms.

The following books may also be found useful for identification in various circumstances:

Bond, P. M. (1970) *Animal Tracks and Clues*, School Natural History Society.

Darlington, A. (1968) *The Pocket Encyclopaedia of Plant Galls*, Blandford.

Kirkaldy, J. F. (1963) *Minerals and Rocks in Colour*, Blandford.

Kirkaldy, J. F. (1970) *Fossils in Colour*, Blandford.

Leutscher, A. (1960) *Tracks and Signs of British Animals*, Cleaver Hume Press.

North, P. (1967) *Poisonous Plants and Fungi*, Blandford.

Index

For an alphabetical list of reagents, stains and culture media see pages 168–194

For an alphabetical index of improvised apparatus see pages 156–164

For an alphabetical list of reagents, stains and culture media see pages 168–194

For an alphabetical index of improvised apparatus see pages 156–164

For an alphabetical list of reagents, stains and culture media see pages 168–194

For an alphabetical index of improvised apparatus see pages 156–164

For an alphabetical list of reagents, stains and culture media see pages 168–194

For an alphabetical index of improvised apparatus see pages 156–164

For an alphabetical list of reagents, stains and culture media see pages 168–194

For an alphabetical index of improvised apparatus see pages 156–164

For an alphabetical list of reagents, stains and culture media see pages 168–194